Developmental Oncology

Principles and Therapy of Cancers
of Children and Young Adults

A subject collection from *Cold Spring Harbor Perspectives in Medicine*

OTHER SUBJECT COLLECTIONS FROM *COLD SPRING HARBOR*
PERSPECTIVES IN MEDICINE

*Type 1 Diabetes: Advances in Understanding and Treatment 100 Years after the Discovery
of Insulin, Second Edition*

Breast Cancer: From Fundamental Biology to Therapeutic Strategies

Aging: Geroscience as the New Public Health Frontier, Second Edition

Retinal Disorder: Approaches to Diagnosis and Treatment, Second Edition

Combining Human Genetics and Causal Inference to Understand Human Disease and Development

Lung Cancer: Disease Biology and Its Potential for Clinical Translation

Influenza: The Cutting Edge

Leukemia and Lymphoma: Molecular and Therapeutic Revolution

Addiction, Second Edition

Hepatitis C Virus: The Story of a Scientific and Therapeutic Revolution

The PTEN Family

Metastasis: Mechanism to Therapy

Genetic Counseling: Clinical Practice and Ethical Considerations

Bioelectronic Medicine

Function and Dysfunction of the Cochlea: From Mechanisms to Potential Therapies

Next-Generation Sequencing in Medicine

Prostate Cancer

RAS and Cancer in the 21st Century

SUBJECT COLLECTIONS FROM *COLD SPRING HARBOR*
PERSPECTIVES IN BIOLOGY

The Biology of Lipids: Trafficking, Regulation, and Function, Second Edition

Speciation

Synthetic Biology and Greenhouse Gases

Wound Healing: From Bench to Bedside

The Endoplasmic Reticulum, Second Edition

Sex Differences in Brain and Behavior

Regeneration

The Nucleus, Second Edition

Auxin Signaling: From Synthesis to Systems Biology, Second Edition

Stem Cells: From Biological Principles to Regenerative Medicine

Heart Development and Disease

Cell Survival and Cell Death, Second Edition

Calcium Signaling, Second Edition

Engineering Plants for Agriculture

Protein Homeostasis, Second Edition

Translation Mechanisms and Control

Cytokines

Circadian Rhythms

Immune Memory and Vaccines: Great Debates

Developmental Oncology

Principles and Therapy of Cancers of Children and Young Adults

A subject collection from *Cold Spring Harbor Perspectives in Medicine*

EDITED BY

Alejandro Gutierrez
St. Jude Children's Research Hospital

Alex Kentsis
Memorial Sloan Kettering Cancer Center
Weill Medical College of Cornell University

COLD SPRING HARBOR LABORATORY PRESS
Cold Spring Harbor, New York • www.cshlpress.org

Developmental Oncology: Principles and Therapy of Cancers of Children and Young Adults

A subject collection from *Cold Spring Harbor Perspectives in Medicine*
Articles online at www.cshperspectives.org

Executive Editor	Richard Sever
Project Supervisor	Barbara Acosta
Editorial Assistant	Danett Gil
Permissions Administrator	Carol Brown
Production Editor	Diane Schubach
Production Manager/Cover Designer	Denise Weiss
Publisher	John Inglis

Front cover artwork: Fluorescent micrograph of coronal section of 14.5-day-old developing mouse brain showing how differentiating neurons (Tuj1, red) normally express Pgbd5 (green), which when dysregulated contributes to somatic mutagenesis of childhood and young-onset solid tumors, including medulloblastomas, neuroblastomas, rhabdoid tumors, and Ewing and other chromosomally rearranged sarcomas. Image courtesy: Makiko Yamada, MD, PhD.

Library of Congress Cataloging-in-Publication Data

Names: Gutierrez, Alejandro, 1974- editor. | Kentsis, Alex, 1975- editor.
Title: Developmental oncology : principles and therapy of cancers of children and young adults / edited by Alejandro Gutierrez and Alex Kentsis.
Description: Cold Spring Harbor, New York : Cold Spring Harbor Laboratory Press, [2025] | Series: Cold Spring Harbor perspectives in medicine | "A subject collection from Cold Spring Harbor perspectives in medicine." | Includes bibliographical references and index. | Summary: "Hundreds of thousands of children develop leukemia and other cancers each year. This volume examines the immense progress made in understanding and development of therapies for pediatric cancers, which are often fatal if untreated"-Provided by publisher.
Identifiers: LCCN 2024045090 (print) | LCCN 2024045091 (ebook) | ISBN 9781621825050 (hardcover) | ISBN 9781621825067 (epub)
Subjects: MESH: Neoplasms--therapy | Child | Adolescent | Young Adult | Collected Work
Classification: LCC RC263 (print) | LCC RC263 (ebook) | NLM QZ 5 | DDC 616.99/4--dc23/eng/20241205
LC record available at https://lccn.loc.gov/2024045090
LC ebook record available at https://lccn.loc.gov/2024045091

All World Wide Web addresses are accurate to the best of our knowledge at the time of printing.

For a complete catalog of all Cold Spring Harbor Laboratory Press publications, visit our website at www.cshlpress.org.

Contents

Contents

Preface

THE SUBJECT OF THIS BOOK DEALS WITH CANCERS that affect children and young adults. Despite some improvements, many cancers remain dire diseases, and, as practicing oncologists, we feel a serious responsibility for seeking better treatments for our patients. This book aims to crystallize a central idea in our evolving field and, with that, renewed hope for patients seeking cures and scientists in search of fundamental understanding.

Developmental oncology concerns the causes of cancers that affect children and young adults, when they have not yet experienced the degree of cellular damage that inevitably accompanies our tissue regeneration and repair and is responsible for cancers as we age. Instead, cancers affecting children and young adults affect distinct populations of developing cells and tissues and occur in response to distinct biological processes that generate susceptible developmental cell states and induce somatic mutations in early life. In turn, these young-onset cancers require the development of specific therapeutic approaches to target their distinct origins and biology.

While several important books have been published on this subject over the years, we believe that recent discoveries have now recast cancers that affect children and young adults as fundamentally developmental in origin, with distinct therapeutic needs and opportunities. Thus, chapters in this book outline the central concept of developmental oncology, including mechanisms of genetic predisposition to young-onset cancers, their endogenous and exogenous somatic mutational processes, developmental and epigenetic regulation, and immune and tissue microenvironmental control. These principles explain the origins and mechanisms of pathogenesis of the most common cancers that affect children and young adults, such as leukemias and lymphomas, sarcomas, and nervous system tumors. Subsequently, the book turns its attention to the development of rational and precise therapies targeting the distinct biology of developmental cancers, including the development of combination therapies, immunologic, epigenetic, metabolic, and cellular therapies. Lastly, the book focuses on the unique demands of pediatric cancer drug development, and how this process may be enabled and accelerated.

We express our sincere gratitude to all our co-authors, and to Barbara Acosta, Richard Sever, and their colleagues at Cold Spring Harbor Laboratory Press for their commitment to scientific excellence and accessible publishing. While we could not include all contributors due to the finite constraints of a single book, we hope that *Developmental Oncology* is properly referenced to direct readers to additional knowledge in this dynamic field. This will be of interest to biologists, oncologists, and scholars in diverse scientific fields seeking to understand the fundamental causes of cancer in young people and to develop definitive therapies for their prevention and control.

ALEJANDRO GUTIERREZ
ALEX KENTSIS

Toward a Unified Theory of Why Young People Develop Cancer

Alex Kentsis

Tow Center for Developmental Oncology, Sloan Kettering Institute and Department of Pediatrics, Weill Medical College of Cornell University and Memorial Sloan Kettering Cancer Center, New York, New York 10065, USA

Correspondence: kentsisresearchgroup@gmail.com

Epidemiologic and genetic studies have now defined specific patterns of incidence and distinct molecular features of cancers in young versus aging people. Here, I review a general framework for the causes of cancer in children and young adults by relating somatic genetic mosaicism and developmental tissue mutagenesis. This framework suggests how aging-associated cancers such as carcinomas, glioblastomas, and myelodysplastic leukemias are causally distinct from cancers that predominantly affect children and young adults, including lymphoblastic and myeloid leukemias, sarcomas, neuroblastomas, medulloblastomas, and other developmental cancers. I discuss the oncogenic activities of known developmental mutators RAG1/2, AID, and PGBD5, and describe strategies needed to define missing developmental causes of young-onset cancers. Thus, a precise understanding of the mechanisms of tissue-specific somatic mosaicism, developmental mutators, and their control by human genetic variation and environmental exposures is needed for improved strategies for cancer screening, prevention, and treatment.

History of science is rich with diverse theories about the origins of human cancer. Around 400 BCE, Hippocrates favored the humoral cause of cancer due to the excess of black bile (Diamandopoulos 1996). Once human anatomy was initially defined in the 1700s with its notable lack of evidence for black bile, Stahl and Hoffman proposed that cancer is caused by the degeneration of lymph nodes. With the advent of microscopy in the nineteenth century, Müller and later Virchow showed that cancer is instead a disease of cells (DeVita and Rosenberg 2012). This is indeed the case, but what causes cancer?

By the early twentieth century, oncogenic viruses were among the first proven causes of cancer in vertebrates (Krump and You 2018). Many also recognized abnormal genomes of cancer cells, beginning with the seminal studies of Hansemann and Boveri in the early twentieth century, who documented the abnormal numbers and appearance of chromosomes in cancer cells (Bignold et al. 2006). It was not until the 1960s that chromosomal abnormalities were linked to mutational oncogene activation as the cause of cancer (Fröhling and Döhner 2008).

SOMATIC MOSAICISM IN CANCER DEVELOPMENT

Current understanding of how cancer develops is fundamentally based on a multistage mutation model. Originally introduced by Armitage and

Doll, the multistage model explains that as people age, the risk of cancer increases due to the accumulation of genetic mutations in cells (Frank 2007). These mutations can at some frequency involve genes that control cell growth and development, originally termed oncogenes and tumor suppressor genes. When specific mutations in these genes occur together, they can transform healthy cells into fully malignant cancerous ones, as documented by numerous engineered cellular and animal model systems.

According to this model, certain factors can accelerate the development of cancer. These include exposure to mutagens that can damage DNA and lead to somatic mutations, or inheriting genetic traits (alleles) that make cells less capable of repairing DNA damage, maintaining genomic integrity, or regulating cell growth and division. In older adults, the majority of cancers are believed to arise from random and stochastic mutations that accumulate over time in tissues as byproducts of continuous cell division and regeneration (Tomasetti and Vogelstein 2015; Tomasetti et al. 2017). This gradual buildup of mutations is thought to contribute to the higher cancer risk seen with increasing age.

Indeed, numerous recent studies have documented somatic genetic mosaicism in various healthy tissues, often involving somatic mutations of known tumor suppressor genes, without overt cancer development (Martincorena et al. 2015; Blokzijl et al. 2016; Lee-Six et al. 2019; Solís-Moruno et al. 2023). These precancerous cells are thought to provide a reservoir of cells somatically predisposed to malignant transformation and clinical cancer progression (Kakiuchi and Ogawa 2021).

This framework offers a cogent mechanism for the development of cancers in children and young adults. In fact, several recent studies of inferred timing of somatic mutations in adult-onset cancers have concluded that the founding genetic lesions often occurred during embryogenesis or early childhood (Gerstung et al. 2020; Pareja et al. 2022; Williams et al. 2022). During development, in a particularly susceptible cell type or state, as few as one mutation affecting a key tumor suppressor or oncogene, blocking cell differentiation, could ultimately lead to malignant transformation. Thus, premature somatic mosaicism, accelerated by inherited cancer-predisposing alleles or environmental exposure to mutagens, is the de facto cause of young-onset cancer in children and young adults.

DEVELOPMENTAL MUTAGENESIS

However, there are several pieces of evidence that suggest that premature somatic mosaicism cannot be the only cause. First, numerous genomic surveys of the somatically mutated genes in aging-associated cancers in older adults (carcinomas, glioblastomas, myelodysplastic leukemias) versus cancers that predominantly affect children and young adults (lymphoblastic leukemias, sarcomas, neuroblastomas, medulloblastomas) have documented extensive divergence of the specific oncogenes and tumor suppressor genes mutated in the aging- and young-onset cancers (Chatsirisupachai et al. 2021; Li et al. 2022; Wang et al. 2022). For example, aging-associated cancers in older adults almost always involve oncogenic somatic mutations of TP53, whereas somatic tumor TP53 mutations are vanishingly rare at diagnosis in children (Robles et al. 2016; Gröbner et al. 2018; Ma et al. 2018). In addition, cancer genomes in older adults have very high rates of single or dinucleotide substitutions, often termed mutational burden, whereas cancers from children and young adults have somatic nucleotide mutation rates that are similar to their corresponding healthy tissues (Thatikonda et al. 2023). Instead, young-onset cancers have increased rates of genome rearrangements, including deletions, translocations, and other complex DNA structural variants (Gröbner et al. 2018; Ma et al. 2018).

In addition to these fundamentally distinct molecular features of aging-associated versus young-onset cancers, epidemiologic studies of the incidence of different types of human cancer have also documented pronounced differences in the age-dependence of these cancers (for review, see Kentsis 2020; Kentsis and Frank 2020). For example, most adenocarcinomas, such as those affecting the colon and lung, exhibit monotonically increasing incidence with increasing age. In contrast, cancers that tend to affect children and

Cite this article as *Cold Spring Harb Perspect Med* doi: 10.1101/cshperspect.a041658

young adults, including lymphoblastic leuke-mias, Hodgkin lymphomas, sarcomas, neuro-blastomas, and medulloblastomas, exhibit dis-tinct peaks of incidence, as evident from the analysis of the surveillance, epidemiology, and end results (SEER) program data (Fig. 1). Thus, specific molecular features and age-dependent incidence of young-onset cancers must result from distinct primary causes from those leading to aging-associated cancers in older adults.

The occurrence of specific cancer types in childhood and young adulthood can be attribut-ed to two nonmutually exclusive primary mech-anisms. First, the presence of stem or progenitor cells with impaired DNA damage repair capabil-ities during normal tissue development, coupled with inherent DNA damage from cell division, can cause somatic mutations of essential onco-genes and tumor suppressor genes, leading to cancer initiation. Alternatively, somatic onco-genic mutations can be induced by developmen-tal mutators including nucleases and other cellu-lar processes that induce DNA mutations that are divergent from their healthy physiologic substrates. These mechanisms could lead to dif-ferent types of somatic mutations and distinct genomic signatures in the resulting young-onset tumors. This developmental mutator mecha-nism is distinct from premature somatic mosai-cism, a process primarily linked to the onset of

aging-related cancers in older individuals (see above).

The distinction between developmental mu-tators and somatic mosaicism as causes of young-onset cancers may also have a significant evolu-tionary consequence. Until the advent of modern medicine, virtually all cases of human cancer were lethal. This implies that the biological mechanisms leading to cancer in young individ-uals during their reproductive years are subject to intense negative evolutionary pressure. This pressure results in both affected individuals and the specific genetic variants they possess being eliminated from the population. Thus, prema-ture somatic mosaicism may be thought of as an accidental cause of a lethal disease, since most of its biological causes affect older adults post their reproductive age. In contrast, develop-mental mutators, by virtue of affecting children and causing young-onset cancer incidence, must also have essential developmental functions, which thereby can offset the evolutionary cost of their potential to cause cancer.

For example, the most common childhood cancer is acute lymphoblastic leukemia (ALL). ALL is frequently caused by specific chromosom-al translocations and deletions that dysregulate developmental transcription factors and other regulators of cell growth and development (Iacobucci and Mullighan 2017). In many cases, DNA breakpoints of these chromosomal trans-locations and deletions involve specific se-quences that are the substrates of the RAG1/2 DNA recombinase (Mullighan et al. 2008; Pa-paemmanuil et al. 2014). Indeed, deficiency of Rag1/2 in engineered mouse models prevents leukemia development (Swaminathan et al. 2015). Normal lymphocyte development also involves activation of the APOBEC-family de-aminase AID, which also induces somatic muta-tions that are characteristically found in many cases of lymphomas, another common blood cancer (Gu et al. 2012; Burns et al. 2013; Petter-sen et al. 2015). Remarkably, the expression of RAG1/2 and its stereotypical somatic DNA re-arrangements have also been detected in subsets of childhood acute myeloid leukemia (AML) (Boehm et al. 1987; Dyer et al. 1991; van Dongen et al. 1992; McNeer et al. 2019).

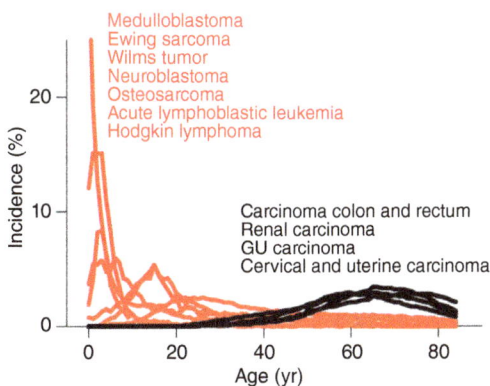

Figure 1. Age-dependent incidence of (red) young-onset versus (black) aging cancers. (GU) Genitouri-nary. (Figure based on data acquired from the SEER database of 8,662,369 malignant cases span-ning 1973–2014.)

Thus, RAG1/2 and AID are the primary developmental mutators for many blood cancers affecting children and young adults. This evolutionary cost of causing young-onset cancers is offset by the fundamental functions of RAG1/2 and AID in the healthy development of adaptive immunity, an essential evolutionary adaptation that prevents infectious disease in all vertebrate animals and humans. Similar evolutionary trade-offs have been extensively documented for many physiologic functions in diverse living organisms (Haldane and Dronamraju 1990).

While RAG1/2 and AID likely induce cancer-initiating mutations in many young-onset blood cancers, they are not expressed beyond the hematopoietic lineage, and specifically not in other cancers that characteristically affect children and young adults. Thus, additional developmental mutators must exist. Recently, PGBD5, another evolutionarily conserved nuclease with mutagenic activity in human cells, was found to be expressed in the majority of solid tumors that affect children and young adults, including various sarcomas, medulloblastomas, and neuroblastomas (Henssen and Kentsis 2018). In rhabdoid tumors, PGBD5 has been found to mediate sequence-specific somatic deletions and other DNA rearrangements, including those of the essential rhabdoid tumor suppressor gene *SMARCB1* (Henssen and Kentsis 2018). In recent studies, similar mutagenic activity of PGBD5 was also found in medulloblastomas, which is the most common childhood brain tumor (Yamada et al. 2024). In fact, most Pgbd5-deficient mice are protected from medulloblastoma-induced mutations of Ptch1 and Smo, which are penetrant tumor supressor and oncogene mutations in human medulloblastomas. Remarkably, PGBD5 is deeply evolutionarily conserved among vertebrates and is required for normal brain development, as its inherited deficiency has recently been found to cause a human intellectual disability syndrome (Jubierre Zapater et al. 2023).

Thus, as RAG1/2 is the developmental mutator nuclease for childhood blood cancers, PGBD5 appears to be a mutator nuclease for a subset of childhood solid tumors. While inflammatory signaling appears to regulate the activity of RAG1/2 in developing blood progenitor cells, consistent with the epidemiologic link between the incidence of ALL and infectious exposures in childhood, the molecular mechanisms and biological processes that dysregulate RAG1/2 and PGBD5 to cause oncogenic mutations and cell transformation are currently unknown.

TOWARD INTEGRATIVE MODELS OF CANCER DEVELOPMENT

Many tumors that affect children and young adults do not express RAG1/2, AID, or PGBD5. These distinct childhood cancers, such as astrocytomas, for example, exhibit distinct somatic mutational signatures in agreement with the developmental mutator causal model. To the extent that the expression of RAG1/2, AID, and PGBD5 is ultimately related to the specific cell lineages from which developmental tumors originate, additional developmental mutators in specific young-onset cancers may be found through the study of developmental nucleases and other developmental molecular processes that act on DNA (Kentsis and Frank 2020).

The developmental nuclease-enzyme hypothesis is especially attractive for developmental cancers with distinct mutational features. For example, mixed lineage leukemias have distinct chromosomal translocations, with topoisomerase 2 implicated in the induction of these oncogenic DNA rearrangements (Sung et al. 2006; Cowell et al. 2012). However, the mechanisms by which topoisomerase 2 or another enzyme that generates these developmental oncogenic mutations are currently not well defined. Similarly, Wilms tumors, the most common kidney tumors in children, have specific patterns of somatic deletions and duplications (Gadd et al. 2017). The causes of this distinct and extensive somatic mutagenesis in children without known genetic predisposition or genomic instability are currently unknown.

It is also possible that developmental tumorigenesis may involve heritable epigenetic defects that lead to malignant transformation, without primary somatic mutations. One can envision that developmental induction of specific metabolites or other forms of epigenetic dysregulation

Cite this article as *Cold Spring Harb Perspect Med* doi: 10.1101/cshperspect.a041658

can lead to heritable changes in gene expression and cell transformation. Remarkably, recent studies have found extensive somatic DNA imprinting mosaicism in children with sporadic Wilms tumors and hepatoblastomas (Coorens et al. 2019; Fiala et al. 2020). Similarly, distinct subsets of ependymomas currently lack known pathogenic DNA mutations (Yamaguchi et al. 2023). The causal developmental processes for such epigenetic tumorigenesis require further study.

Finally, the premature somatic mosaicism and developmental mutator causes of young-onset cancers are not necessarily mutually exclusive. First, the two processes can co-occur within the same individual, tissue, and cell of origin. Second, to the extent that most human cancers are caused by multiple somatic mutations, distinct genetic causes can cooperate in their induction. Lastly, the epidemiologic variability of the incidence of distinct cancers among different groups of individuals means that differences in environmental exposure and/or germline genetic variation can contribute to cancer risk. A precise understanding of the mechanisms of tissue-specific somatic mosaicism, developmental mutators, and human genetic variation and exposures is required for improved strategies for cancer screening, prevention, and treatment.

ACKNOWLEDGMENTS

I thank Alejandro Gutierrez for critical reading of the early draft of this manuscript, Helen Mueller, Makiko Yamada, and Damon Reed for suggestions, and Gabriella Casalena for help with figure design.

REFERENCES

Bignold LP, Coghlan BL, Jersmann HP. 2006. Hansemann, Boveri, chromosomes and the gametogenesis-related theories of tumours. Cell Biol Int 30: 640–644. doi:10.1016/j.cellbi.2006.04.002

Blokzijl F, de Ligt J, Jager M, Sasselli V, Roerink S, Sasaki N, Huch M, Boymans S, Kuijk E, Prins P, et al. 2016. Tissue-specific mutation accumulation in human adult stem cells during life. Nature 538: 260–264. doi:10.1038/nature19768

Boehm TL, Werle A, Drahovsky D. 1987. Immunoglobulin heavy chain and T-cell receptor γ and β chain gene rear-rangements in acute myeloid leukemias. Mol Biol Med 4: 51–62.

Burns MB, Temiz NA, Harris RS. 2013. Evidence for APO-BEC3B mutagenesis in multiple human cancers. Nat Genet 45: 977–983. doi:10.1038/ng.2701

Chatsirisupachai K, Lesluyes T, Paraoan L, Van Loo P, de Magalhães JP. 2021. An integrative analysis of the age-associated multi-omic landscape across cancers. Nat Commun 12: 2345. doi:10.1038/s41467-021-22560-y

Coorens THH, Treger TD, Al-Saadi R, Moore L, Tran MGB, Mitchell TJ, Tugnait S, Thevanesan C, Young MD, Oliver TRW, et al. 2019. Embryonal precursors of Wilms tumor. Science 366: 1247–1251. doi:10.1126/science.aax1323

Cowell IG, Sondka Z, Smith K, Lee KC, Manville CM, Sidorczuk-Lesthuruge M, Rance HA, Padget K, Jackson GH, Adachi N, et al. 2012. Model for MLL translocations in therapy-related leukemia involving topoisomerase IIβ-mediated DNA strand breaks and gene proximity. Proc Natl Acad Sci 109: 8989–8994. doi:10.1073/pnas.1204406109

DeVita VT Jr, Rosenberg SA. 2012. Two hundred years of cancer research. N Engl J Med 366: 2207–2214. doi:10.1056/NEJMra1204479

Diamandopoulos GT. 1996. Cancer: an historical perspective. Anticancer Res 16: 1595–1602.

Dyer MJ, Hoyle CF, Rees JK, Marcus RE. 1991. T-cell receptor and immunoglobulin gene rearrangements in acute myeloid and undifferentiated leukemias of adults: correlation with weak surface expression of CD45 and CDw52 antigens. Leuk Lymphoma 3: 257–265. doi:10.3109/10428199109107913

Fiala EM, Ortiz MV, Kennedy JA, Glodzik D, Fleischut MH, Duffy KA, Hathaway ER, Heaton T, Gerstle JT, Steinherz P, et al. 2020. 11p15.5 epimutations in children with Wilms tumor and hepatoblastoma detected in peripheral blood. Cancer 126: 3114–3121. doi:10.1002/cncr.32907

Frank SA. 2007. Dynamics of cancer: incidence, inheritance, and evolution. Princeton University Press, Princeton, NJ.

Fröhling S, Döhner H. 2008. Chromosomal abnormalities in cancer. N Engl J Med 359: 722–734. doi:10.1056/NEJMra0803109

Gadd S, Huff V, Walz AL, Ooms AHAG, Armstrong AE, Gerhard DS, Smith MA, Auvil JMG, Meerzaman D, Chen QR, et al. 2017. A children's oncology group and TARGET initiative exploring the genetic landscape of Wilms tumor. Nat Genet 49: 1487–1494. doi:10.1038/ng.3940

Gerstung M, Jolly C, Leshchiner I, Dentro SC, Gonzalez S, Rosebrock D, Mitchell TJ, Rubanova Y, Anur P, Yu K, et al. 2020. The evolutionary history of 2,658 cancers. Nature 578: 122–128. doi:10.1038/s41586-019-1907-7

Gröbner SN, Worst BC, Weischenfeldt J, Buchhalter I, Kleinheinz K, Rudneva VA, Johann PD, Balasubramanian GP, Segura-Wang M, Brabetz S, et al. 2018. The landscape of genomic alterations across childhood cancers. Nature 555: 321–327. doi:10.1038/nature25480

Gu X, Shivarov V, Strout MP. 2012. The role of activation-induced cytidine deaminase in lymphomagenesis. Curr Opin Hematol 19: 292–298. doi:10.1097/MOH.0b013e328353da3a

Haldane JBS, Dronamraju KR. 1990. *Selected genetic papers of J.B.S. Haldane*. Garland, New York.

Henssen AG, Kentsis A. 2018. Emerging functions of DNA transposases and oncogenic mutators in childhood cancer development. *JCI Insight* 3: e123172. doi:10.1172/jci.insight.123172

Iacobucci I, Mullighan CG. 2017. Genetic basis of acute lymphoblastic leukemia. *J Clin Oncol* 35: 975–983. doi:10.1200/JCO.2016.70.7836

Jubierre Zapater LJ, Rodriguez-Fos E, Planas-Felix M, Lewis S, Cameron D, Demarest P, Nabila A, Zhao J, Bergin P, Reed C, et al. 2023. A transposase-derived gene required for human brain development. bioRxiv doi:10.1101/2023.04.28.538770

Kakiuchi N, Ogawa S. 2021. Clonal expansion in non-cancer tissues. *Nat Rev Cancer* 21: 239–256. doi:10.1038/s41568-021-00335-3

Kentsis A. 2020. Why do young people get cancer? *Pediatr Blood Cancer* 67: e28335. doi:10.1002/pbc.28335

Kentsis A, Frank SA. 2020. Developmental mutators and early onset cancer. *Front Pediatr* 8: 189. doi:10.3389/fped.2020.00189

Krump NA, You J. 2018. Molecular mechanisms of viral oncogenesis in humans. *Nat Rev Microbiol* 16: 684–698. doi:10.1038/s41579-018-0064-6

Lee-Six H, Olafsson S, Ellis P, Osborne RJ, Sanders MA, Moore L, Georgakopoulos N, Torrente F, Noorani A, Goddard M, et al. 2019. The landscape of somatic mutation in normal colorectal epithelial cells. *Nature* 574: 532–537. doi:10.1038/s41586-019-1672-7

Li CH, Haider S, Boutros PC. 2022. Age influences on the molecular presentation of tumours. *Nat Commun* 13: 208. doi:10.1038/s41467-021-27889-y

Ma X, Liu Y, Liu Y, Alexandrov LB, Edmonson MN, Gawad C, Zhou X, Li Y, Rusch MC, Easton J, et al. 2018. Pan-cancer genome and transcriptome analyses of 1,699 paediatric leukaemias and solid tumours. *Nature* 555: 371–376. doi:10.1038/nature25795

Martincorena I, Roshan A, Gerstung M, Ellis P, Van Loo P, McLaren S, Wedge DC, Fullam A, Alexandrov LB, Tubio JM, et al. 2015. Tumor evolution. High burden and pervasive positive selection of somatic mutations in normal human skin. *Science* 348: 880–886. doi:10.1126/science.aaa6806

McNeer NA, Philip J, Geiger H, Ries RE, Lavallée VP, Walsh M, Shah M, Arora K, Emde AK, Robine N, et al. 2019. Genetic mechanisms of primary chemotherapy resistance in pediatric acute myeloid leukemia. *Leukemia* 33: 1934–1943. doi:10.1038/s41375-019-0402-3

Mulligan CG, Phillips LA, Su X, Ma J, Miller CB, Shurtleff SA, Downing JR. 2008. Genomic analysis of the clonal origins of relapsed acute lymphoblastic leukemia. *Science* 322: 1377–1380. doi:10.1126/science.1164266

Papaemmanuil E, Rapado I, Li Y, Potter NE, Wedge DC, Tubio J, Alexandrov LB, Van Loo P, Cooke SL, Marshall J, et al. 2014. RAG-mediated recombination is the predominant driver of oncogenic rearrangement in ETV6-RUNX1 acute lymphoblastic leukemia. *Nat Genet* 46: 116–125. doi:10.1038/ng.2874

Pareja F, Ptashkin RN, Brown DN, Derakhshan F, Selenica P, da Silva EM, Gazzo AM, Da Cruz Paula A, Breen K, Shen R,

et al. 2022. Cancer-causative mutations occurring in early embryogenesis. *Cancer Discov* 12: 949–957. doi:10.1158/2159-8290.CD-21-1110

Pettersen HS, Galashevskaya A, Doseth B, Sousa MM, Sarno A, Visnes T, Aas PA, Liabakk NB, Slupphaug G, Sætrom P, et al. 2015. AID expression in B-cell lymphomas causes accumulation of genomic uracil and a distinct AID mutational signature. *DNA Repair (Amst)* 25: 60–71. doi:10.1016/j.dnarep.2014.11.006

Robles AI, Jen J, Harris CC. 2016. Clinical outcomes of *TP53* mutations in cancers. *Cold Spring Harb Perspect Med* 6: a026294. doi:10.1101/cshperspect.a026294

Solís-Moruno M, Batlle-Masó L, Bonet N, Aróstegui JI, Casals F. 2023. Somatic genetic variation in healthy tissue and non-cancer diseases. *Eur J Hum Genet* 31: 48–54. doi:10.1038/s41431-022-01213-8

Sung PA, Libura J, Richardson C. 2006. Etoposide and illegitimate DNA double-strand break repair in the generation of MLL translocations: new insights and new questions. *DNA Repair (Amst)* 5: 1109–1118. doi:10.1016/j.dnarep.2006.05.018

Swaminathan S, Klemm L, Park E, Papaemmanuil E, Ford A, Kweon SM, Trageser D, Hasselfeld B, Henke N, Mooster J, et al. 2015. Mechanisms of clonal evolution in childhood acute lymphoblastic leukemia. *Nat Immunol* 16: 766–774. doi:10.1038/ni.3160

Thatikonda V, Islam SMA, Autry RJ, Jones BC, Gröbner SN, Warsow G, Hutter B, Huebschmann D, Fröhling S, Kool M, et al. 2023. Comprehensive analysis of mutational signatures reveals distinct patterns and molecular processes across 27 pediatric cancers. *Nat Cancer* 4: 276–289. doi:10.1038/s43018-022-00509-4

Tomasetti C, Vogelstein B. 2015. Cancer etiology. Variation in cancer risk among tissues can be explained by the number of stem cell divisions. *Science* 347: 78–81. doi:10.1126/science.1260825

Tomasetti C, Li L, Vogelstein B. 2017. Stem cell divisions, somatic mutations, cancer etiology, and cancer prevention. *Science* 355: 1330–1334. doi:10.1126/science.aaf9011

van Dongen JJ, Breit TM, Adriaansen HJ, Beishuizen A, Hooijkaas H. 1992. Detection of minimal residual disease in acute leukemia by immunological marker analysis and polymerase chain reaction. *Leukemia* 6: 47–59.

Wang X, Langevin AM, Houghton PJ, Zheng S. 2022. Genomic disparities between cancers in adolescent and young adults and in older adults. *Nat Commun* 13: 7223. doi:10.1038/s41467-022-34959-2

Williams N, Lee J, Mitchell E, Moore L, Baxter EJ, Hewinson J, Dawson KJ, Menzies A, Godfrey AL, Green AR, et al. 2022. Life histories of myeloproliferative neoplasms inferred from phylogenies. *Nature* 602: 162–168. doi:10.1038/s41586-021-04312-6

Yamada M, Keller RR, Gutierrez RL, Cameron D, Suzuki H, Sanghrajka R, Vaynshteyn J, Gerwin J, Maura F, Hooper W, et al. 2024. Childhood cancer mutagenesis caused by transposase-derived PGBD5. *Sci Adv* 10: 1–16. doi:10.1126/sciadv.adn4649

Yamaguchi J, Ohka F, Motomura K, Saito R. 2023. Latest classification of ependymoma in the molecular era and advances in its treatment: a review. *Jpn J Clin Oncol* 53: 653–663. doi:10.1093/jjco/hyad056

Cite this article as *Cold Spring Harb Perspect Med* doi: 10.1101/cshperspect.a041658

Genetic Predisposition to Hematologic Malignancies

Kayla V. Hamilton,[1] Akiko Shimamura,[1,2] and Jessica A. Pollard[1,2]

[1]Division of Pediatric Oncology, Dana-Farber Cancer Institute, Boston, Massachusetts 02115, USA

[2]Department of Pediatrics, Harvard Medical School, Boston, Massachusetts 02115, USA

Correspondence: jessica_pollard@dfci.harvard.edu

Hematologic malignancies (HMs) have been increasingly recognized in association with an underlying genetic predisposition syndrome (GPS) in individuals of all ages. It is critical for hematology and oncology providers to be aware of the diagnostic findings, physical examination findings, and aspects of family history that raise suspicion for an underlying GPS. Moreover, recognition of how somatic gene panel testing, frequently done at the time of HM diagnosis, may raise suspicion for an underlying germline condition based on the mutation profile reported, is prudent. With knowledge of an underlying germline condition, the chemotherapy used for a given HM may be impacted and the role of hematopoietic stem cell transplant more critically considered. Off-therapy monitoring after HM treatment is completed will also likely be impacted. In this work, we review key features of several GPSs associated with increased risks for HM while also outlining the diagnostic workup to identify GPSs and treatment considerations for affected patients. Armed with this knowledge, treating providers may evaluate the possibility of a GPS in patients with leukemia/lymphoma and modify their treatment plan accordingly.

Until recently, hematologic malignancies (HMs), which represent ∼25% of all pediatric cancers (Linabery and Ross 2008), were largely thought to be sporadic in nature and little was known about germline predisposition syndromes that increase the risk for leukemia or lymphoma. We now know that genetic predisposition syndromes (GPSs) play a larger role in both pediatric and adult HMs than previously understood. Although studies are limited, ∼4% of children with acute lymphoblastic leukemia (ALL) have been reported to have a GPS (Klco and Mulligan 2021), and 14%–39% of adults with a myeloid disorder had a co-occurring GPS diagnosis (DiNardo et al. 2016; Keel et al. 2016). This range reported in adults is similar to the rate of genetic predisposition in breast, colorectal, and other solid tumors.

We also know, based on natural history data, that some germline mutations, such as those in *SAMD9* and *SAMD9L*, confer HM risk early in life, whereas others, like germline *DDX41* abnormalities and short telomere disorders, may take six or seven decades to contribute to leukemic transformation (Gurnari et al. 2023). This variable timeline suggests that some hematopoietic developmental stages may be more susceptible than others to secondary somatic events that result in malignancy or that the presence of a specific germline mutation may, in itself, directly impact the risk for leukemic transformation at particular developmental stages. Moreover, the penetrance for myeloid or lymphoid disease varies based on the germline condition present (Klco and Mulligan 2021). In this work, we will review clinical symptoms and signs that are associated with GPSs that predispose to HM and highlight some of the clinical features that may be seen in affected patients. We will also discuss how the interplay between clinical history and presentation, family history, and results of somatic testing may help inform GPS risk and we will review the optimal strategies to confirm germline versus somatic origin of identified mutations. Moreover, we will describe clinical scenarios in which certain HM diagnoses warrant germline testing regardless of clinical presentation or family history. For those identified as having a GPS and associated HM, we will discuss how germline results can impact decisions regarding treatment, the role of hematopoietic stem cell transplant (HSCT) as well as potential HSCT donor considerations. Finally, we know that surveillance for HMs in a patient with a known GPS is a topic of much uncertainty and provider variability, and we will highlight some of the considerations at play when considering frequency and type of monitoring. In summary, increased awareness of the interplay between GPSs and HMs will ensure optimal medical management of patients and may provide opportunities for prevention or early intervention at the time of malignancy diagnosis.

DIAGNOSTIC CONSIDERATIONS

Clinical History and Symptoms/Signs Suggestive of GPS in Patients with HM

In keeping with increased awareness for an underlying GPS manifesting with an HM, the latest World Health Organization (WHO) leukemia classification system includes "germline predisposition to myeloid malignancies" as a provisional category (Obrochta and Godley 2018; Khoury et al. 2022). The International Consensus Classification of Myeloid Neoplasms and Acute Leukemia, a group of individuals with expertise in both clinical care and pathologic/genetic aspects of these disorders, further enhanced awareness by describing the clinical characteristics and natural history of a number of HMs that present in childhood or adulthood and are associated with underlying GPSs (Arber et al. 2022; Rudelius et al. 2023). Moreover, the National Comprehensive Care Network (NCCN) guidelines now recommend germline testing for all patients with myelodysplastic syndrome (MDS) under the age of 40. Aligned with these changes, it is imperative that treating physicians consider GPS in any child or young adult newly diagnosed with an HM and pay close attention to blood abnormalities and/or constitutional symptoms/signs that might be indicators of an underlying GPS.

When evaluating patients with HMs for an underlying GPS, many factors play a role in the risk assessment and differential diagnosis, including physical examination findings, antecedent laboratory findings, features identified on bone marrow (BM) aspirate/biopsy, and family history. In this section, we highlight several of the key features that should raise suspicion for an underlying GPS and warrant referral for genetic counseling/testing. However, it is important to note that not all GPSs associated with HM have secondary clinical features. In addition, incomplete penetrance and variable expressivity may mask awareness of a GPS in a family, even if the index case has many features of the germline disease. Conversely, an individual found to have a GPS may themselves lack the constitutional findings of that syndrome due to variable expression, making their presentation less obvious. Therefore, it is important to remember that the absence of the features described below does not rule out an underlying GPS and germline genetic testing may still be warranted.

Due to variability in phenotype and presentation for many of the GPS, we recommend performing a comprehensive medical history and physical examination paying particular attention to the features described in Table 1. Addi-

Table 1. Personal/family history and/or physical examination findings that raise suspicion of an underlying genetic predisposition syndrome (GPS)

Clinical history or examination findings	Genes and/or syndrome associated with clinical finding
Café-au-lait spots	Neurofibromatosis type 1 (*NF1*), Fanconi anemia (FA), constitutional mismatch repair deficiency (CMMRD), RASopathies
Thrombocytopenia (isolated) or bleeding disproportionate to platelet count	*RUNX1*-familial platelet disorder with associated myeloid malignancies, *ANKRD26*-related thrombocytopenia, *ETV6* thrombocytopenia and predisposition to leukemia
Pancreatic insufficiency	Shwachman–Diamond syndrome (SDS)
Early graying	Telomere biology disorders (TBDs)
Dysplastic nails	TBDs
Ataxia	Ataxia–telangiectasia (*ATM*), ataxia–pancytopenia (*SAMD9L*)
Hearing loss	*GATA2* deficiency, *SRP72*
Skeletal anomalies	FA, MIRAGE syndrome (*SAMD9*), Diamond–Blackfan anemia (DBA), SDS
Neutropenia	SDS, *ELANE*-related neutropenia
Short stature	SDS, FA, MIRAGE, ataxia–pancytopenia, DBA, RASopathies, TBDs
Pulmonary fibrosis	TBDs
Warts (many, challenging to treat)	*GATA2* deficiency
Recurrent infections, especially viral, opportunistic infection, or atypical bacteria	*GATA2* deficiency
Lymphatic abnormalities/hydrocele	*GATA2* deficiency
Squamous cell carcinomas of the head and neck, gastrointestinal (GI) tract, and other solid organ cancers	FA, TBDs
Early-onset solid tumors	Li–Fraumeni syndrome (*TP53*), CMMRD

tional intake questions may be necessary depending on the differential diagnosis for a given patient.

Nonmolecular Tests Used in the Diagnostic Workup of Patients with Suspected GPS

Clinical testing performed at the time of HM diagnosis may provide additional clues as to the risk for a GPS and raise suspicion of an underlying predisposition. A preceding history of unexplained cytopenias or elevated mean corpuscular volume (MCV) may suggest marrow dysfunction associated with a GPS. In conjunction with an abnormal complete blood count (CBC), liver function tests may aid diagnosis in some GPSs like a telomere biology disorder (TBD) or Fanconi anemia (FA). Elevation of

hemoglobin F may also be suggestive, although can be seen in the context of myeloid malignancy, and thus should be interpreted with caution (Reinhardt et al. 1998; Kudo et al. 2000; Choi et al. 2002; Alter et al. 2013; Avagyan and Shimamura 2022). Qualitative and quantitative platelet defects resulting in menorrhagia and/or easy bruising/bleeding may be a result of germline abnormalities in *RUNX1*, *ANKRD26*, and *ETV6* (Marquez et al. 2014; Hock and Shimamura 2017; Sood et al. 2017; Avagyan and Shimamura 2022). Screening for immunologic abnormalities, by testing lymphocyte subsets for T, B, and natural killer (NK) populations as well as levels of IgA, IgG, and IgM, may increase suspicion for germline *GATA2* deficiency (Spinner et al. 2014; Rossini et al. 2022). This transcription factor, which is crucial for the maintenance function of CD4$^+$ T cells, B cells, monocytes, and NK cells may impair the function of all subgroups, although, interestingly, patients with *GATA2* deficiency tend to have HM restricted to the myeloid lineage (Spinner et al. 2014; Collin et al. 2015; Bruzzese et al. 2020; Rossini et al. 2022).

More specific diagnostic studies may be prudent and include those described below. As both chromosomal breakage and telomere length can be impacted by chemotherapy, it is important to obtain these tests before treatment. If done following treatment start, a negative test may still be reassuring, although a positive result may be falsely concerning and merely reflect the impact of antecedent treatment, or, in the context of short telomeres, malignant cells that inherently have shorter length.

Chromosome Breakage

Chromosomal instability syndromes are a group of inherited disorders associated with genomic instability and breakage either spontaneously or in response to DNA-damaging agents. Two of these disorders, FA and Nijmegen breakage syndrome, may have, when exposed to the DNA cross-linking agents mitomycin C (MMC) or diepoxybutane (DEB), increased rates of breakage and form chromosomal abnormalities like radials. It is important, however, to know the limitations of this testing. Peripheral blood (PB) testing can be falsely negative as somatic reversion in hematopoietic stem cells (HSCs) can occur, mitigating the impact of DNA-damaging agents on cells derived from the BM. Skin fibroblast testing is, therefore, a gold standard for this evaluation and should be performed if the results on blood are normal but high suspicion of an underlying syndrome remains (Fargo et al. 2014). Moreover, as exposure to recent chemotherapy may also increase breakage, the risk for false-positive testing in that context must be recognized (Gutierrez-Rodrigues et al. 2023). The presence of increased chromosomal breakage with MMC or DEB, without definitive identification of a genetic abnormality, is sufficient to make the diagnosis of FA or Nijmegen breakage.

Telomere Length

Flow-FISH telomere length measurements are used to evaluate for dyskeratosis congenita and other TBDs. Fluorescent signals from telomeres of PB cell subsets are assessed on a single-cell basis by flow cytometry. Depending on the laboratory used and test selection, results can be reported for lymphocytes, granulocytes, B cells, naive and memory T cells, and NK cells; alternatively, only lymphocytes and granulocytes may be tested although the former 6-cell subset testing is the gold standard. Telomere lengths must be analyzed in relation to age-matched controls. Short telomeres in granulocytes are a nonspecific finding and are not diagnostic of a TBD. For patients with TBDs, the results often show low or very low lengths in multiple lymphocyte subsets. Length less than the first percentile in at least three different lymphocyte subsets is very sensitive and specific for TBD; results between the first and tenth percentile may also be suggestive in certain clinical contexts. Lengths >10% are unlikely to reflect a TBD; however, telomere length may be less sensitive for diagnosing a TBD in adults (Gutierrez-Rodrigues et al. 2023). Telomeres may also be shorter in patients with immunologic conditions. Telomere length testing for TBD has not been validated in patients with peripheral blasts and should be interpreted with caution. Moreover, as referenced above, chemotherapy also shortens telomere length, and/or malignant marrow cells often inherently have shorter telomeres. Thus, the interpretation of telomere length must be considered in the context of the patient's clinical status and prior treatment.

Pancreatic Trypsinogen and Pancreatic Isoamylase

Serum pancreatic trypsinogen and isoamylase are often low in patients with Shwachman–Diamond syndrome (SDS); however, the age of the patient is important in the interpretation of the laboratory results. Trypsinogen is generally low in patients with SDS before the age of 3 and then increases to the normal range in older patients. Serum isoamylase levels are low in all patients with SDS; however, the utility of this test in pa-

Cite this article as *Cold Spring Harb Perspect Med* doi: 10.1101/cshperspect.a041585

tients younger than 3 years old is limited since isoamylase levels are also low in healthy children early in life. A very low fecal elastase may indicate exocrine pancreatic dysfunction; however, normal fecal elastase levels do not rule out SDS (Burroughs et al. 2009).

Erythrocyte Adenosine Deaminase

Elevated levels of erythrocyte adenosine deaminase (eADA) raise suspicion for Diamond–Blackfan anemia (DBA). However, it is important to be aware that up to 16% of patients with DBA may have normal eADA levels (Fargo et al. 2013). Elevation of eADA may also be seen with reticulocytosis or during the recovery phase from anemia.

Bone Marrow Analysis

Analysis of the BM may provide additional clues. While dysplasia in the myeloid lineage on morphologic review may be seen with idiopathic de novo myeloid disease, some GPSs demonstrate characteristic dysmorphology at baseline that does not indicate MDS but may be a clue to the underlying germline condition. For example, characteristic dysmyelopoiesis is present at baseline in SDS, dyserythropoiesis is seen in FA, and dysmegakaryopoiesis in germline *RUNX1*, *ANKRD26*, *ETV6*, and *GATA2* (Li and Bledsoe 2023). In the context of MDS, pediatric disease may often present as a hypocellular marrow, which is in stark contrast to the hypercellularity typically seen in adult MDS. Additional morphologic features, including but not limited to the identification of micromegakaryocytes, increased reticulin fibrosis, and evidence of ring sideroblasts by iron stain, may also reflect marrow dysfunction or MDS that preceded an HM (Li and Bledsoe 2023).

Fluorescent in situ hybridization (FISH) as well as conventional cytogenetic analysis (CCA) of the BM also provide additional insights and should be performed in all new diagnoses of HM. CCA, which requires considerable manpower for analysis, is limited by the number of evaluable cells analyzed and what can be seen under a microscope. FISH analysis further facilitates the identification of cytogenetic fusions given the use of specific DNA probes. It can also identify copy number variants (i.e., submicroscopic imbalances like deletions and duplications) of a specific chromosome (Levy and Burnside 2019). In contrast to adult MDS, which is typically hypercellular, pediatric MDS is often hypocellular, so FISH is recommended even when cytogenetics are normal. Interpretation of FISH, however, must also be considered in clinical context. For example, in HM, the population of abnormal clonal cells may have a growth disadvantage compared to healthy cells. Interphase FISH testing, which does not require cell culture, is, therefore, prudent in HMs even if karyotypic testing is abnormal as it allows for the study of nondividing cells thus facilitating the identification of minor abnormal clones and/or small chromosomal deletions not detected by CCA at the DNA level (Flactif et al. 1994; Andreasson et al. 1997; Westbrook et al. 2000; Rigolin et al. 2001). Metaphase FISH, which does require culture to ensure cells are synchronized in metaphase, aids in positional information by allowing direct visualization of the chromosomes and the exact position of the signal probes on a given chromosome (Rigolin et al. 2001).

Complete or partial loss of chromosome 7 is the most common abnormality associated with germline conditions in pediatric and young adult HM and should prompt consideration of additional germline work up. Monosomy 7/deletion 7q is seen in approximately one-third of MDS or acute myeloid leukemia (AML) cases associated with FA and SDS and is particularly common in MDS/AML occurring in patients with germline mutations in *GATA2*, *SAMD9*, or *SAMD9L*. It has also been reported in HMs developing in patients with severe congenital neutropenia (SCN), *RUNX1*-familial platelet disorder (FPD), TBDs, and germline *ERCC6L2*, among others (Wlodarski et al. 2018). Trisomy 8 can also be seen with *GATA2* deficiency and gains of chromosomes 1q and 3q have been reported in FA (Cioc et al. 2010; Wlodarski et al. 2016; Avagyan and Shimamura 2022; Li and Bledsoe 2023). A recent review by Li and Bledsoe (2023) summarizes additional cytogenetic abnormalities seen in pediatric MDS/AML suggestive of an underlying germline disorder.

Importantly, FISH is restricted in its ability to detect microduplications as well as uniparental disomy and atypical rearrangements. While CCA may aid in the identification of some structural issues not identified by FISH, the use of next-generation sequencing (NGS) provides an opportunity for more comprehensive screening. This testing refers to a high-throughput technology in which billions of sequencing reactions occur, each of which is analyzed simultaneously in a given sample. In the context of DNA NGS, identification of clinically relevant point mutations, insertions, or deletion of a given gene is facilitated. NGS RNA panels (i.e., RNA-seq) are useful to identify chromosomal translocations/ fusions and gene expression profiles.

When mutations are reported by NGS, a variant allele frequency (VAF), which reflects the ratio of mutant to wild-type gene, is reported. In the context of the identification of a germline genetic variant, a high VAF (i.e., >30%) is typically seen, although exceptions occur. Awareness regarding the sensitivity of a given assay to detect a rare germline mutation, and the risk of that gene to undergo genetic reversion to a normal state is imperative as we enhance our understanding of GPSs and their association with HM. Specifically, spontaneous genetic reversion or somatic rescue for a given germline abnormality may be seen (Revy et al. 2019). This results in cells that either lack the germline mutation due to somatic reversion or develop a compensatory somatic change elsewhere in the germline mutated gene or in a related gene, resulting in functional correction of the germline defect that may then possess a growth advantage that can aid normal hematopoiesis and thereby mask detection of the inciting genetic variant (Waisfisz et al. 1999; Gross et al. 2002; Peffault de Latour and Soulier 2016). Of note, such somatic rescue of germline variants may either be adaptive (i.e., not associated with a risk for development of neoplasia) or maladaptive (i.e., increasing risk for HM). Given this potential for somatic reversion or rescue in BM-derived cells, germline testing of hematopoietic samples may not identify the germline variant and skin fibroblast testing is essential.

Similar to characteristic cytogenetic abnormalities like monosomy 7, specific NGS abnormalities in myeloid neoplasia, particularly in pediatric patients, may provide a clue to an underlying germline condition. SDS is associated with frequent somatic mutations in *EIF6*, and in MDS *TP53* mutations. MDS in FA frequently harbors *RUNX1* or RAS pathway mutations. MDS occurring in TBDs may have *TP53*, *TET2*, or *ASXL1* mutations. MDS occurring in SCN frequently harbors *CSF3R* with or without *RUNX1* mutation. *GATA2*-related MDS often harbors *STAG2* or *ASXL1* mutations (Li and Bledsoe 2023).

Germline Genetic Testing Modalities, Interpretation, and Limitations

Options for clinical germline genetic testing have evolved significantly in the last several years. Previously, testing was largely single gene or small panels, which limited the identification of genetic abnormalities associated with rare genetic syndromes. More recently larger panels that screen for several genetic conditions concurrently provide greater potential for identification of a GPS, many of which are rare and may lack associated clinical features. With the expanding options for clinical genetic testing, it is important for providers to be aware of the benefits and drawbacks of each testing approach to ensure that the optimal test is selected for a given patient in the context of their personal medical history, family history, and familial preferences regarding germline testing. Awareness of testing limitations is also prudent. For example, germline *GATA2* mutations include several recurrent pathogenic variants that are deep in an intron and would be missed on any test that does not specifically target the intronic regions. *ANKRD26* mutations are primarily within the promoter and can be missed with some testing modalities, and *SBDS* has a pseudogene that can make sequencing challenging and can lead to inaccurate results. When mutations in one of these genes are high on the differential diagnosis, it is important to evaluate the test being selected and verify with the laboratory that they have adequate coverage of all critical regions of the gene.

Awareness of atypical cell division events, which can result in aberrant gene expression is also important. For example, uniparental disomy occurs when an individual receives two copies of a chromosome, or part of a chromosome, from one parent and no copy from the other parent. This can manifest as heterodisomy, because of an error in meiosis 1, in which a pair of nonidentical chromosomes are inherited from one parent. Alternatively, isodisomy, in which a single chromosome from one parent is duplicated, due to an error in later-stage meiosis II, may also occur. Uniparental disomy is significant as it may result in large blocks of homozygosity, which can result in the expression of recessive genes with significant clinical implications. For example, isodisomy can lead to the expression of a pathogenic recessive phenotype given this chromosomal duplication. This aberration should be considered when a child has evidence of a rare recessive disorder and yet only one parent is identified as being a carrier for this genetic abnormality.

As referenced earlier, somatic genetic rescue (SGR) refers to compensatory mechanisms that may explain varied penetrance of a GPS in a kindred and/or impact the frequency of leukemic progression in these syndromes (Li and Bledsoe 2023). SGR manifests in two ways, either via genetic reversion or by the development of compensatory mechanisms that offset the deleterious phenotype (Schratz 2023). Copy neutral loss of heterozygosity (CN-LOH) is the most common cause of direct SGR in autosomal dominant disorders, leading to restoration of two wild-type gene copies by replacement of the chromosomal region harboring the germline mutation with a copy of the wild-type chromosome from the other allele. When reversion occurs, affected cells are functionally corrected and often possess a growth advantage that can aid normal hematopoiesis and potentially mask the detection of the inciting gene abnormality (Revy et al. 2019). This adaptive approach is seen in germline GATA2, FA, and TBDs (Schratz 2023).

Additional secondary adaptive mechanisms can also be protective. For example, in the context of SAMD9/SAMD9L GPS, more than 50% of individuals are found to have an SGR due to one of several mechanisms. The first, termed a "second site mutation" is the acquisition of a second SAMD9 or SAMD9L mutation in cis

that offsets the effect of the germline mutation. Removal of the mutant allele through mitotic recombination that results in uniparental isodisomy of chromosome 7 (UPD7q) may also be adaptive as can focal gene deletions of chromosome 7 that impact the mutant allele. Conversely, aneuploidy resulting in monosomy 7 due to haploinsufficiency of additional genes on chromosome 7, can facilitate driver mutations that increase the risk of leukemogenesis, in individuals with this condition (Tesi et al. 2017; Wong et al. 2018; Sahoo et al. 2021; Gutierrez-Rodrigues et al. 2023; Schratz 2023). When suspicion of syndromes with a known risk for SGR is high (e.g., germline GATA2, SAMD9, SAMD9L, RUNX1, etc.) nonhematopoietic fibroblasts are the optimal sample type for genetic testing, for both the proband and their relatives. Lastly, patients with an underlying GPS may have both a germline and somatic mutation of the clinically relevant gene, and it is imperative to discern which one of the two mutations is germline (Avagyan and Shimamura 2022).

In the context of PB or BM DNA sequencing, germline variants typically have a VAF of 30%–50% or up to 100% depending on heterozygosity versus homozygosity status. When the VAF is <30%, it is more likely that the variant is somatic, although mosaicism, SGR, and technical limitations of the testing, such as allelic dropout, can lead to germline variants being reported at a lower-than-expected VAF. When a somatic gene panel identifies a variant that can be seen in germline disease at a mutation frequency of 30% or greater, consideration of follow-up germline genetic testing is prudent (Kraft and Godley 2020; Feurstein et al. 2022a,b; Gutierrez-Rodrigues et al. 2023). Secondary somatic mutations may provide additional clues to germline risk. For example, somatic mutations in PPM1D, POT1, and the TERT promoter can be associated with TBD; EIF6 mutations are seen in SDS, transient monosomy 7 is observed in germline mutations of SAMD9/SAMD9L, and secondary somatic DDX41 mutations may be seen in conjunction with the germline abnormality.

When germline testing is indicated, providers must carefully consider not only what genetic test to order for a patient with an HM but also

Table 2. Molecular germline genetic testing panels

Type of test	When to order	Benefits	Drawbacks	Additional considerations
Targeted variant testing	Variant previously identified in family member(s), confirming germline status of variant identified on somatic testing	Low cost, easy-to-interpret results, no possibility of incidental findings	When confirming germline status of variant identified on somatic testing, the true germline mutation in the gene could be missed (e.g., the mutation identified could be the somatic second hit in a given gene that results in malignant transformation)	For recessive conditions, consider full gene analysis to rule out a second mutation that would impact risk assessment
Single gene	Phenotype highly consistent with a specific gene (e.g., neutropenia and abnormal pancreatic enzymes consistent with Shwachman–Diamond syndrome [SDS])	Potentially lower cost compared to larger panels or whole-exome sequencing (WES), good option for families who are concerned about "incidental" findings, typically covers relevant noncoding variants	If negative, may need to order a second more comprehensive germline screening test that can lead to delay in diagnosis and additional cost to family	Verify that copy number variation (CNV) analysis is done and, when relevant, known promoter/intronic variants included
Single syndrome	Phenotype highly consistent with a specific syndrome (e.g., patient with classic features of Fanconi anemia [FA] and abnormal chromosome breakage)	Potentially lower cost than larger panels or WES, good option for families who are concerned about "incidental" findings, typically covers relevant noncoding variants	If negative, may need to order a second more comprehensive germline screening test that can lead to delay in diagnosis and additional cost to family	Verify that CNV analysis is done and, when relevant, known promoter/intronic variants included
Multigene panel testing	Broad differential diagnosis	Evaluates multiple syndromes at once; has deeper sequencing of specific genes of interest compared to WES and typically covers relevant noncoding variants	Increased chance of incidental findings, more variants of uncertain significance (VUSs) identified	Verify that CNV analysis is done and, when relevant, known promoter/intronic variants included

Continued

Cite this article as *Cold Spring Harb Perspect Med* doi: 10.1101/cshperspect.a041585

Table 2. *Continued*

Type of test	When to order	Benefits	Drawbacks	Additional considerations
Whole exome sequencing	Broad differential diagnosis, prior negative panel test(s) and suspicion for underlying syndrome remains	Submitting parental samples helps with analysis and provides insight into de novo versus inherited status; rapid WES is an option for patients who need results urgently for treatment (can be as fast as 7-day turnaround)	Does not cover noncoding regions, can miss deletions/duplications, analysis is phenotype driven/ filtered (include on test requisition form if you have specific genes you want to make sure are analyzed and provide clear phenotype/clinical data), more expensive than panels	Once ordered, it can be harder to obtain insurance approval for a targeted panel test
Whole genome sequencing	Broad differential diagnosis, used when prior negative panel test(s) and suspicion for underlying syndrome remains	Covers noncoding regions, better deletion/duplication coverage than WES, can identify some structural variants, submitting parental samples helps with analysis and provides insight into de novo versus inherited status	Generally, has lower read depth/ coverage than exome (might miss mosaic results), analysis is phenotype-driven/filtered (include on test requisition form if you have specific genes you want to make sure are analyzed and provide clear phenotype/clinical data), more expensive than panels	Many insurance companies do not cover WGS
Microarray (array comparative genomic hybridization [aCGH] or single-nucleotide polymorphism [SNP] array)	Patient with developmental delays, dysmorphic features, and multiple congenital anomalies; patient with whole gene deletion identified on other testing (microarray can help determine the extent of the deletion and if other genes are involved)	Good first line test for patients with multiple congenital anomalies with broad differential diagnosis	Can miss small deletions (e.g., single gene deletions or intragenic deletions); does not identify sequence variants	Can detect consanguinity and will report expected degree of relatedness of parents— recommend discussing this during consent conversation; unlike CGH arrays, SNP arrays can detect long contiguous stretches of DNA and uniparental disomies and copy-neutral loss of heterozygosity (LOH); SNP arrays can have higher resolution depending on the number and location of SNPs utilized

the most appropriate sample type. For patients with acute leukemia who have negative MRD and no morphologic evidence of disease, germline genetic testing on blood or saliva/buccal samples can be considered. However, the blood and saliva of patients with a clonal HM that involves the BM may contain many lymphocytes, increasing the risk for classification of a somatic mutation as germline. In addition, in some instances, buccal swabs are associated with low DNA recovery, which precludes definitive results (Padron et al. 2018; Stubbins et al. 2022). Moreover, particularly in patients with a myeloid malignancy, there is the possibility of a reversion event in the blood and/or BM that can lead to a false-negative result in the blood/BM/saliva. Taken together, the ideal sample type for patients with any active acute or chronic leukemia or lymphoma with a history of BM involvement is DNA from cultured fibroblasts. Reliance on cultured fibroblasts for germline testing obviates the risk for contamination with somatic mutations derived from clonal hematopoietic cells. However, in patients with lymphoma that has not impacted the BM, blood or saliva can be considered. For patients with preceding allogeneic HSCT, cultured fibroblasts from an organ source outside of the BM is mandated. Of note, several centers are actively exploring the use of other sources of germline DNA, including DNA from fingernails and hair follicles.

With expanding options for genetic testing and interpretation, informing one's patient regarding options for genetic testing has become more nuanced. Table 2 describes the most common molecular genetic tests available, and highlights scenarios when ordering of a given test is most appropriate. The pros and cons of each modality are reviewed and additional considerations that are warranted in their context are provided.

It is important for all providers ordering genetic testing to be familiar with interpreting the genetic results and understanding the difference between pathogenic/likely pathogenic variants and variants of uncertain significance (VUSs). In the HM space, many of the GPSs described are relatively newly discovered with limited cases reported in the literature, thus functionally significant mutations may be reported as VUSs on clinical reports. However, it is important to note that the majority of VUSs are later determined to be benign findings, having no impact on the health of the patient or their family members. Typically, VUSs are not used in clinical or treatment decision making, and cascade testing is not offered to family members unless it is likely to aid in the reclassification of the variant. Because of this, it is critical for providers to review the literature and available genetic databases that comprehensively annotate previously identified VUS to discern whether a given VUS has functional data or clinical attributes in affected individuals to support its pathogenicity. Where feasible, additional functional tests (e.g., chromosome breakage for patients with VUSs in FA genes, telomere length for patients with VUSs in TBD genes) or other evaluations that might aid the diagnosis of a GPS with a syndromic phenotype (e.g., echocardiogram for a patient with suspicious VUS in a RASopathy gene) can provide further clarity regarding the clinical relevance of a VUS.

OVERVIEW OF GPSs ASSOCIATED WITH PEDIATRIC HEMATOLOGIC MALIGNANCY

With the increased use of NGS testing at the time of HM diagnosis or in studies of families with high rates of HMs, the identification of genetic abnormalities that predispose to malignancy is increasing. It is important to remember that many of these GPSs may lack a preceding abnormal phenotype. Several studies have been conducted to discern the frequency of causative germline variants in HM (Trottier and Godley 2021). The Fred Hutch published a retrospective series of all patients who underwent HSCT at their center between 1990 and 2012 and found that 13.6% of patients <45 years old who were treated for MDS were found to have suspicious variants in known germline predisposition genes (Keel et al. 2016). Targeted panel testing identified GPS in 18% of adult patients referred to MD Anderson's Hematologic Malignancy Program (DiNardo et al. 2016). A St. Jude cohort of 1120 pediatric cancer patients screened with whole genome, whole exome, and targeted sequencing found that germline mutations were seen in 8.5% of patients, 4.4% of whom were treated for leukemia. Importantly, only 23% of patients with an identified mutation had a family history of cancer predisposition (Klco and Mulligan 2021). We know that clinical penetrance of some GPS can be variable in kindreds and that some germline conditions result from de novo germline variants, particularly mutations seen in Diamond–Blackfan-associated genes, *GATA2*, *SAMD9*, the *TINF2* subset of TBDs and *TP53*, given these mutational events arise early in embryogenesis (Gonzalez et al. 2009; Davidsson et al. 2018; McReynolds et al. 2018; Trottier and Godley 2021). Because of this, the absence of a suspicious family history does not obviate the need for germline testing. Knowledge of an underlying germline condition in an HM patient may impact one's treatment approach to their malignancy and also provide awareness that allows families to make informed and personal reproductive and family planning decisions (Trottier and Godley 2021).

In Table 3, we have highlighted several GPSs that confer risk of both myeloid and lymphoid malignancy in pediatric and adult patients and have described the most common features associated with each condition. The table is separated into three distinct categories based on phenotypic presentation: (1) germline predisposition syndromes primarily associated with HMs not affecting multiple organ systems, (2) cancer predisposition syndromes with increased risk for HMs, and (3) BM failure syndromes. Category 1 encompasses syndromes where the primary phenotype is leukemia and other cancer risk or multiorgan involvement is rare. Syndromes in this category should be carefully considered for any patient presenting with a personal and family history of leukemia without a strong history of other cancers or additional medical history. Category 2 encompasses syndromes that are associated with a wide variety of cancers, including solid, central nervous system (CNS), and HMs. Syndromes in this category should be considered for patients with a personal history of a solid tumor and an HM as well as patients with a strong family history of cancer, particularly cancers diagnosed at younger than average age. Category 3 encompasses classically defined BM failure syndromes and syndromes that present with additional organ dysfunction.

While Table 3 primarily focuses on genes associated with MDS or leukemia, it is important to recognize that there are several conditions that should be considered in individuals with a strong personal and family history of lymphoma. DNA repair disorders and primary immunodeficiencies are the most common syndromes associated with lymphoma risks. In addition, EBV⁺ lymphomas during childhood should raise the possibility of genetic predisposition either due to an immunologic condition or a GPS with immunologic dysfunction. Additionally, it is important to be aware that although only a small percentage of lymphomas occur in the setting of an underlying genetic predisposition, familial aggregation of lymphoma is well established, and risks are increased for individuals with a close relative with either Hodgkin lymphoma or non-Hodgkin lymphoma. First-degree relatives of NHL, HL, and CLL patients are estimated to have a ~1.7-fold, 3.1-fold, and 8.5-fold elevated risk of developing NHL, HL, and CLL, respectively, and families should be counseled accordingly (Cerhan and Slager 2015).

Lastly, there are several genes, including *TET2*, *CSF3R SRP73*, and *MECOM*, that have recently been proposed as playing a potential role in germline genetic leukemia predisposition that is not included in the table. As the scientific discovery of GPSs associated with HM seems to be growing exponentially, we anticipate this list will continue to expand significantly.

TREATMENT CONSIDERATIONS IN CHILDREN WITH AN HM DIAGNOSIS AND UNDERLYING GPS

With the 5th edition of the WHO Classification of Hematolymphoid Tumors and with development of the 2022 International Consensus Classification of Myeloid Neoplasms and Acute Leukemias, HMs derived from germline predisposition were, for the first time, incorporated into HM classification systems (Arber et al. 2022; Khoury et al. 2022). With this heightened awareness, it is imperative that treating physicians apply their knowledge of the underlying

Table 3. Genes associated with genetic predisposition syndrome (GPS): clinical characteristics and hematologic malignancy (HM) presentation

Gene(s)	Syndrome	Inheritance pattern	Most common hematologic malignancies	Lifetime risk for hematologic malignancy	Other malignancy risks	Additional clinical features	Additional findings that support diagnosis	Additional comments	References
Germline predispositions primarily associated with hematologic malignancies not affecting multiple organ systems									
DDX41	DDX41-associated familial myelodysplastic syndrome and acute myeloid leukemia	AD	Myelodysplastic syndrome (MDS)/AML, chronic myeloid leukemia (CML), lymphoid neoplasms	Reduced penetrance	N/A	Cytopenia	Somatic DDX41 mutations affecting the other allele are common, some have deletion on 5q35.3 leading to haploinsufficient DDX41 expression	Favorable prognosis; male predominance of HM (3:1); HM typically occurs in older adults	Polprasert et al. (2015), Li et al. (2022), and Tiong et al. (2023)
CEBPA	CEBPA-associated familial acute myeloid leukemia	AD	AML	Near complete penetrance with germline amino-terminal variant, reduced penetrance with germline carboxy-terminal variant	N/A	N/A	Germline amino terminus often accompanied by somatic carboxy-terminal mutation	Favorable prognosis; pros and cons of transplant should be discussed with the family given favorable prognosis but increased risk of new leukemia	Tawana et al. (2015, 2017)
ETV6	ETV6 thrombocytopenia and predisposition to leukemia	AD	B-cell ALL, MDS/AML	30%	N/A	Thrombocytopenia	Often hyperdiploid leukemia with loss of ETV6 wild-type (WT) allele; small hyperchromatic megakaryocytes, disseminated toxic granulations, and dysplastic eosinophils in the absence of frank myelodysplasia		Feurstein and Godley et al. (2017), Di Paola and Porter (2019), and Homan et al. (2023)
RUNX1	RUNX1-familial platelet disorder with associated myeloid malignancies	AD	MDS/AML, ALL, CML, lymphoma	25%–50%	N/A	Thrombocytopenia, allergic conditions, gastrointestinal issues	Somatic mutations often affecting the second RUNX1 allele and GATA2		Brown et al. (2020), Cunningham et al. (2023), and Homan et al. (2023)
ANKRD26	ANKRD26-related thrombocytopenia	AD	MDS/AML, CML, chronic lymphocytic leukemia (CLL)	<10%	N/A	Thrombocytopenia, qualitative platelet dysfunction	Deficiency of platelet α-granules, elevated thrombopoietin levels, dysmegakaryopoiesis		Kewan et al. (2020) and Homan et al. (2023)
PAX5	N/A	AD	B-cell ALL	Unknown	N/A	Neurodevelopmental disorder with developmental delay and/or autism spectrum disorder	Somatic deletion of WT allele		Shah et al. (2013) and Gofin et al. (2022)
KDM1A	N/A	AD	Multiple myeloma (MM)	Unknown	N/A	N/A			Wei et al. (2018)

Continued

Cite this article as *Cold Spring Harb Perspect Med* doi: 10.1101/cshperspect.a041585

Table 3. *Continued*

Gene(s)	Syndrome	Inheritance pattern	Most common hematologic malignancies	Lifetime risk for hematologic malignancy	Other malignancy risks	Additional clinical features	Additional findings that support diagnosis	Additional comments	References
Cancer predisposition syndromes with risk for hematologic malignancies									
TP53	Li–Fraumeni syndrome	AD	ALL, AML, MDS, CML, lymphomas	4%–6%	Breast cancer, osteosarcoma, soft tissue sarcomas, adrenocortical carcinoma, choroid plexus carcinoma, brain tumors, and other solid tumors	N/A	Hypodiploid ALL	Consider whole body/brain magnetic resonance imaging (MRI) to evaluate for other malignancies before initiating hematopoietic stem cell transplant (HSCT); therapy-related myeloid malignancies including MDS and AML are common in patients with Li-Fraumeni syndrome (LFS) and often associated with poor prognosis with standard therapies and even allogenic HSCT; up to 20% de novo rate	Swaminathan et al. (2019) and de Andrade et al. (2021)
MBD4	MBD4 deficiency	AR	AML	Unknown, suspected high penetrance	Polyposis, uveal melanoma, and nervous system tumors	N/A	Hypermutator genomic signature, somatic *DNMT3A* mutations common		Sanders et al. (2018) and Palles et al. (2022)
MLH1, MSH2, MSH6, PMS2, EPCAM	Constitutional mismatch repair deficiency (CMMRD)	AR, majority inherited from parent with Lynch syndrome	Lymphoma (typically non-Hodgkin), ALL, AML	33%–42%	Colorectal polyposis, colorectal cancer, brain tumors, gastric, small bowel, pancreatic, endometrial, urinary tract cancers, and other solid tumors	Café-au-lait spots	Hypermutator genomic signature, MMR deficient	Consider whole body/brain MRI and colonoscopy to evaluate for other malignancies before initiating HSCT; consanguinity of the parents and/or homozygosity for a founder mutation is observed in more than 50% of individuals with CMMRD	Bakry et al. (2014), Abedalthagafi (2018), and Durno et al. (2021)

Continued

Table 3. *Continued*

Gene(s)	Syndrome	Inheritance pattern	Most common hematologic malignancies	Lifetime risk for hematologic malignancy	Other malignancy risks	Additional clinical features	Additional findings that support diagnosis	Additional comments	References
POT1	POT1 tumor predisposition	AD	CLL	3.6-fold increased risk for CLL	Melanoma, angiosarcoma (primarily cardiac), glioma, meningioma, and other solid tumors	N/A	Unknown		Speedy et al. (2016) and Herrera-Mullar et al. (2023)
Bone marrow failure syndromes									
AR: *FANCA, FANCC, FANCD1 (BRCA2), FANCD2, FANCE, FANCF, FANCG (XRCC9), FANCI, FANCJ (BRIP1), FANCL, FANCM, FANCN (PALB2), FANCO (RAD51C), FANCP (SLX4), FANCQ (ERCC4), FANCS (BRCA1), FANCT (UBE2T), FANCU (XRCC2), FANCV (REV7/MAD2L2), FANCW (RFWD3), FANCY (FAAP100)* AD: *FANCR (RAD51)* XLR: *FANCB*	Fanconi anemia	AR, AD, XLR	MDS/AML, T-ALL, CMML	>50% by age 50	Head and neck squamous cell carcinomas, anogenital squamous cell carcinoma, and liver tumors; for FANCD1/BRCA2 subtype: additional cancer risks, including CNS tumors, Wilms tumors, other solid tumors	Bone marrow failure, short stature, skin hyper-/hypopigmentation, skeletal, GU, neurologic, and heart abnormalities	Abnormal chromosome breakage; somatic *RUNX1* mutations	Can experience chemosensitivity and risks from alkylating and DNA cross-linking agents	Alter (2014)

Continued

Cite this article as *Cold Spring Harb Perspect Med* doi: 10.1101/cshperspect.a041585

Table 3. Continued

Gene(s)	Syndrome	Inheritance pattern	Most common hematologic malignancies	Lifetime risk for hematologic malignancy	Other malignancy risks	Additional clinical features	Additional findings that support diagnosis	Additional comments	References
AD: *TERC, TINF2, ZCCHC8, NAF1, RPA1, MDM4, NPM1* AR: *NOP10, NHP2, WRAP53, CTC1, STN1, DCLRE1B, USB1* AD/AR: *TERT, RTEL1, PARN, ACD, POT1* XLR: *DKC1*	Telomere biology disorders (dyskeratosis congenita)	AD, AR, XLR	MDS/AML, lymphoma	Unknown, genotype/phenotype correlations exist	Head and neck squamous cell carcinomas, gastrointestinal cancers	Bone marrow failure, leukoplakia, dysplastic nails, abnormal skin pigmentation, pulmonary fibrosis, premature graying, liver cirrhosis, arteriovenous malformation, skeletal anomalies, short stature, cytopenia, and immunodeficiency	Telomere lengths typically below tenth percentile for age in multiple subsets		Savage (2022)
AR: *SBDS, DNAJC21, EFL1* AD: *SRP54* (most are de novo)	Schwachman–Diamond syndrome	AR, AD	MDS/AML	~17%	N/A	Bone marrow failure, exocrine pancreatic dysfunction, neutropenia, short stature, skeletal anomalies (including chondrodysplasia and congenital thoracic dystrophy), hepatomegaly, cognitive and behavioral impairment	Somatic del(20)(q11) and i(7)(q10), somatic *TP53* mutations, abnormal trypsinogen, and/or isoamylase levels		Furutani et al. (2022)
AD: *RPL5, RPL9, RPL11, RPL15, RPL18, RPL26, RPL27, RPL31, RPL35, RPL35A, RPS7, RPS10, RPS15A, RPS17, RPS19, RPS24, RPS26, RPS27, RPS28, RPS29* XLR: *GATA1, TSR2*	Diamond–Blackfan anemia	AD, XLR	MDS/AML	Reduced penetrance, <15% by age 45	Osteosarcoma, colorectal cancer, other solid tumors	Bone marrow failure, short stature, cleft/lip palate, thumb, cardiac, craniofacial, and kidney anomalies	Elevated erythrocyte adenosine deaminase (eADA), macrocytic anemia with red cell aplasia		Vlachos et al. (2012, 2018)
ELANE	*ELANE*-related neutropenia	AD	MDS/AML	~15%–36% within 15 years of initiating treatment with G-CSF	N/A	Congenital neutropenia, cyclic neutropenia, severe and recurrent infections	Monosomy 7, somatic mutations in *CSF3R*, and *RUNX1*		Rosenberg et al. (2008) and Dale et al. (2022)

Continued

Table 3. *Continued*

Gene(s)	Syndrome	Inheritance pattern	Most common hematologic malignancies	Lifetime risk for hematologic malignancy	Other malignancy risks	Additional clinical features	Additional findings that support diagnosis	Additional comments	References
AR: *HAX1*, *G6PC3, CSF3R*, *VPS45, JAGN1* AD: *GFI1*, *CXCR4* XLR: *WAS*	Severe congenital neutropenia (see above for *ELANE*-related severe congenital neutropenia)	AR, AD, XLR	MDS/AML	Unknown	N/A	Congenital neutropenia, cyclic neutropenia, severe and recurrent infections, osteopenia, seizures, developmental delays, genital, and heart abnormalities	Specific features are gene dependent		Fadeel et al. (2021)
GATA2	GATA2 deficiency (also referred to as monocytopenia and mycobacterial infection (MonoMAC)/ dendritic cell, monocyte, B and NK lymphoid (DCML) deficiency	AD	MDS/AML, ALL, lymphoma, myeloproliferative neoplasm (MPN)	80% by age 40	N/A	Immunodeficiency, disseminated human papillomavirus (HPV) and mycobacterial infections, lymphedema, hearing loss, pulmonary alveolar proteinosis, and warts	Somatic monosomy 7 and trisomy 8 and somatic mutations in *SETBP1* and *ASXL1*; B-, natural killer (NK)-, dendritic cell (DC)-, and mono-cytopenia, hypocellular myelodysplasia with loss of monocytes and hematogones, megakaryocytes with separated nuclear lobes, micromegakaryocytes, and megakaryocytes with hypolobated nuclei	High-risk disease with risk for life-threatening infections, warrants consideration of timely HSCT	Wlodarski et al. (2017), McReynolds et al. (2018), and Sahoo et al. (2020)
MECOM	N/A	AD	MDS/AML	Unknown	N/A	Congenital hypomegakaryocytic thrombocytopenia, radioulnar cynostosis, recurrent bacterial and fungal infections, deafness, renal defects, and congenital heart defects	B-cell deficiency, hypogammaglobulinemia		Germeshausen et al. (2018) and Lozano Chinga et al. (2023)
SAMD9	MIRAGE	AD	MDS/AML	Reduced penetrance	N/A	MIRAGE: myelodysplasia, infection, restriction of growth, adrenal hypoplasia, genital phenotypes, enteropathy, and cytopenia	Monosomy 7	Somatic hematopoietic rescue can occur leading to spontaneous hematological remission	Sahoo et al. (2020)

Continued

Cite this article as *Cold Spring Harb Perspect Med* doi: 10.1101/cshperspect.a041585

Table 3. *Continued*

Gene(s)	Syndrome	Inheritance pattern	Most common hematologic malignancies	Lifetime risk for hematologic malignancy	Other malignancy risks	Additional clinical features	Additional findings that support diagnosis	Additional comments	References
SAMD9L	Ataxia-pancytopenia	AD	MDS/AML	Reduced penetrance	N/A	Ataxia, pancytopenia	Monosomy 7	Somatic hematopoietic rescue can occur leading to spontaneous hematological remission; germline mutations are gain of function	Tesi et al. (2017) and Sahoo et al. (2020)
ERCC6L2	N/A	AR	MDS/AML, T-ALL	Unknown, suspected high penetrance	N/A	Neurologic abnormalities reported	AML-M6, Monosomy 7, complex karyotype with somatic mutations in *TP53*	Poor prognosis warrants consideration of preemptive HSCT before progression to HM	Armes et al. (2022) and Baccelli et al. (2023)
MPL	Congenital amegakaryocytic thrombocytopenia (CAMT)	AR, AD	MDS/AML	Unknown	N/A	Thrombocytopenia, intracranial bleeds, bone marrow failure, and nervous system anomalies	Few or absent megakaryocytes	HSCT is only cure and is often recommended at a young age due to bleeding risks	Germeshausen and Ballmaier (2021)
RBM8A	Thrombocytopenia absent radius	AR	MDS/AML, ALL	Unknown, reduced penetrance	N/A	Bilateral absence of radii with the presence of both thumbs, thrombocytopenia, cow milk allergy, skeletal, heart, and genitourinary anomalies	Unknown	Unknown	Adam et al. (1993) and Jameson-Lee et al. (2018)
Other									
N/A	Trisomy 21 (Down syndrome)	N/A	Transient myeloproliferative disorder (TMD), acute megakaryoblastic leukemia (AMKL), ALL	5%–30% risk of TMD, which develops into AMKL in 20% of individuals, 7-20-fold increased risk of childhood ALL	Increased risk for germ cell tumors, and testicular tumors	Hypotonia, short stature, intellectual disability, heart defects, hearing loss, and obstructive sleep apnea	Somatic *GATA1* short and cohesin gene mutations		Xavier et al. (2009), Hasaart et al. (2021), and Osuna-Marco et al. (2021)

Continued

Table 3. *Continued*

Gene(s)	Syndrome	Inheritance pattern	Most common hematologic malignancies	Lifetime risk for hematologic malignancy	Other malignancy risks	Additional clinical features	Additional findings that support diagnosis	Additional comments	References
BLM	Bloom syndrome (sometimes referred to as Bloom–Torre–Machacek syndrome or congenital telangiectatic erythema)	AR	AML, ALL, non-Hodgkin lymphoma	Unknown	GI cancers, GU cancers, osteosarcoma, squamous cell carcinoma of skin, and other solid tumors	Extreme sun sensitivity, frequent infections, growth delay and short stature, endocrine and immune system abnormalities	Unknown	AML often occurs after chemotherapy for previous malignancy	Osuna-Marco et al. (2021)
PTPN11, NF1, CBL	RASopathies	AD	Transient myeloproliferative neoplasm in infancy, JMML.	For germline PTPN11: ~6% develop transient myeloproliferative neoplasm in infancy, ~3% JMML	Risks are gene-dependent but can include brain tumors, neuroblastoma, rhabdomyosarcoma, and other solid tumors	Short stature, heart abnormalities, skeletal abnormalities, distinct facial differences, developmental delay		Children with JMML and germline predisposition have a high rate of spontaneous resolution of JMML.	Strullu et al. (2014)

(AD) Autosomal dominant, (AR) autosomal recessive, (XLR) X-linked recessive, (AML) acute myeloid leukemia, (ALL) acute lymphoblastic leukemia, (JMML) juvenile myelomonocytic leukemia, (GI) gastrointestinal, (GU) genitourinary, (CNS) central nervous system, (G-CSF) granulocyte colony-stimulating factor, (WT) wild-type.

Cite this article as *Cold Spring Harb Perspect Med* doi: 10.1101/cshperspect.a041585

GPS, when present, to identify the optimal treatment approach (Gurnari et al. 2023). In the case of chromosomal breakage disorders like FA, ataxia–telangiectasia, and Nijmegen breakage disorder, in which genes integral to DNA repair are mutated, one may be particularly sensitive to alkylators like cyclophosphamide as well as radiation. Patients with dyskeratosis congenita and other TBDs also fare poorly with alkylator treatment (Deeg et al. 1983; Gluckman et al. 1995; Rossini et al. 2022; Hudda and Myers 2023). With knowledge of an underlying GPS, definitive treatment may entail the use of an HSCT to eradicate the aberrant population of BM cells at risk for leukemic transformation. This is largely the case when a myeloid hematopoietic malignancy is found in a child with a GPS, even if the cytomolecular risk features of their malignancy would not support HSCT otherwise. In addition, individuals with *GATA2* deficiency are at risk for serious infectious complications given their baseline immune dysfunction, prompting swift transition to HSCT for treatment consolidation when feasible, to minimize risk for infectious complications (Wlodarski et al. 2017; Gurnari et al. 2023). While HSCT is not recommended routinely in most children with a GPS and acute lymphoblastic lymphoma, such individuals may be prone to emergence of additional leukemic clones during or following completion of their initial leukemia therapy, thereby warranting close observation (McReynolds et al. 2022; Gutierrez-Rodrigues et al. 2023). Taken together, special considerations regarding treatment exist in individuals with an HM and underlying GPS and awareness of these nuances provides an opportunity for the advancement of disease-free survival with minimization of morbidity.

Toxicities following treatment may also occur at a higher frequency than the general population. Patients with TBDs are at risk for liver disease, pulmonary fibrosis, and arteriovenous malformations following treatment (Savage 2022). In addition, as a subset of GPSs are associated with both hematologic and solid tumors, the risk for secondary HMs following chemotherapy for a solid tumor may be heightened, warranting surveillance for this risk while also monitoring for solid tumor disease recurrence.

HSCT Considerations

For patients undergoing allogeneic HSCT with a related donor, knowledge of a GPS in the family can play a major role in donor selection. For a patient with a GPS, all potential family member donors should be tested for that GPS regardless of whether the family member has clinical features suggestive of the condition. When there is a strong suspicion for a GPS but no genetic diagnosis is identified, family donor evaluation may include a BM examination to assess for abnormalities. A donor with a known germline leukemia predisposition abnormality should not be used; the rationale for this stance extends beyond the obvious concern for risk for donor-derived malignancies following HSCT (Trottier and Godley 2021). For example, previous studies have demonstrated that, when related donors with deleterious germline variants such as *CEPBA* or *RUNX1* or *GATA2* are utilized, inferior HSCT outcomes are seen due to poor HSC mobilization of the donor, delayed or failed engraftment, poor immune function post-HSCT (Fogarty et al. 2003; Owen et al. 2008; Trottier and Godley 2021), as well as risk for leukemia development in the donor after stem cell mobilization/collection (Xiao et al. 2011; Kraft and Godley 2020).

In the absence of a suspicious family history in the patient undergoing a transplant, a comprehensive health history and screening CBC may be sufficient for the potential related donor. Universal NGS panel testing of donors has not become the standard of care given the ethics surrounding this approach. Moreover, widespread germline testing of donors may increase the potential for identification of VUSs that may, needlessly, delay HSCT putting the recipient at higher risk of recurrence and/or infection and create undue stress in a potential donor who has, to date, had no clinical stigmata of the GPS in consideration.

While universal donor testing is not recommended at present, treating physicians must be aware of the potential risk for the development of a donor-derived clonal population in an HSCT recipient. Two to five percent of leukemia relapses post-HSCT are donor derived (Williams et al. 2021). In this context, it is hypothesized that such conferred risk might reflect an unrecognized GPS in the donor versus donor clonal hematopoiesis that evolved a second hit in the recipient and resulted in disease. Cases of donor-derived malignancy after related and unrelated transplants suggest that pathogenic variants in genes such as *CEPBA*, *DDX41*, *GATA2*, and *RUNX1* may contribute to poor HSCT outcomes (Fogarty et al. 2003; Owen et al. 2008; Williams et al. 2021). Although universal donor testing for GPS is not felt to be cost effective or ethical, the National Marrow Donor Program 2023 policy allows donors to opt in for notification, and ideally genetic counseling, if a genetic variant in a recipient is confirmed or suspected to be donor derived (Williams et al. 2021).

In the context of patients with GPS who need HSCT, DNA damage disorders like FA or a TBD have significantly inferior outcomes to those with ribosome biology disorders or disorders of hematopoiesis (Agarwal 2023; Hudda and Myers 2023). Moreover, TRM is reported to be 10% or higher in patients with a GPS (Agarwal 2023). Subsequent advances in HSCT approaches have strived to improve these outcomes by trialing alternative reduced-intensity conditioning regimens that include fludarabine or treosulfan and/or eliminate radiation (MacMillan and Wagner 2010; Peffault de Latour et al. 2015; Ebens et al. 2017; Mehta et al. 2017; Hudda and Myers 2023). A recent prospective multicenter trial of conditioning with treosulfan, an alkylating agent with a favorable toxicity profile, fludarabine, and, in some cases, rabbit antithymocyte globulin (ATG), in 14 patients with inherited BM failure without progression to HM reported 100% engraftment and 92% survival (Burroughs et al. 2017). This study is now expanded to see whether similar results are seen in a larger cohort (NCT04965597). Efforts are also ongoing to develop a similar trial of treosulfan conditioning for patients with known BM failure syndrome and subsequent HM.

With changes in HSCT conditioning, we are optimistic that outcomes for patients with GPS and secondary HM will improve, although the need to monitor for both disease recurrence and secondary malignancy is imperative. In addition, there are a number of post-HSCT complications that are of higher risk in patients with GPS that warrant consideration. Late effects in post-HSCT survivors, including but not limited to delayed immune reconstitution, iron overload, pulmonary complications, infertility, renal impairment, and growth failure, are more prevalent and more severe in a number of GPSs (Alter 2017). Detailed discussion of these risks and their clinical features, specific to individual GPSs seen in pediatrics, are well described in the 2017 blood article by Blanche Alter (Avagyan and Shimamura 2022). A recent American Society of Hematology (ASH) education review by Hudda and Myers (2023) provides a thorough review of post-HSCT monitoring recommended for some of the most common GPSs seen in children.

ADDITIONAL CONSIDERATIONS

Disease Surveillance in Individuals with a Known GPS

The frequency of disease monitoring in patients with a GPS, both those with and without a history of HM, is not clear. Recommendations are informed by the magnitude of the malignancy risk, outcomes of malignancy, and disease-specific patterns of progression to malignancy. For conditions with poor survival, once malignancy develops, surveillance offers an opportunity to intervene with HSCT before progression to frank malignancy. For example, disease progression has been studied longitudinally in SDS, a GPS with a high risk of *TP53*-mutated myeloid malignancy and poor survival once malignancy develops (Myers et al. 2020). Although small heterozygous *TP53*-mutated clones develop early in life and may persist stably for years, the acquisition of biallelic *TP53* mutations in the marrow is a harbinger of malignancy in SDS (Kennedy et al. 2021; Reilly and Shimamura 2023).

As most GPSs lack longitudinal data to guide practice approaches, the present standard of care is largely informed by expert opinion and/or extrapolated from other hereditary hematopoietic disorders (Godley and Shimamura 2017; Porter et al. 2017; Roloff et al. 2021). A baseline BM evaluation at the time of diagnosis of a GPS is generally recommended, before HM development. Marrow examinations allow assessment for early signs of clonal evolution including dysplasia, cytogenetics, FISH, and somatic mutation analysis. Patients are then generally followed every 3–6 months with clinical evaluation and CBC, with consideration of annual BM examinations for conditions with high leukemia risk. Should cytopenias develop in one or more lineages, repeat CBC in 2–4 weeks is prudent with a low threshold to repeat BM testing if there is ongoing clinical concern. Data are lacking on the utility of molecular surveillance of the PB, although this is increasingly used as a screen for clonal progression in the absence of serial BM evaluations. If a somatic panel shows evidence of clonal evolution in the PB, additional marrow assessment may be warranted (Porter et al. 2017). If concerning clonal abnormalities or dysplasias are noted in the BM without frank malignancy, more frequent marrow testing using multiple modalities to assess for disease progression (i.e., FISH, karyotype, NGS somatic panel) is recommended (Roloff et al. 2021).

Given there are inherent differences in the natural history of acute lymphoid versus myeloid malignancies, decisions regarding the type and frequency of disease surveillance may vary. For example, due to the rapid onset of NHL, ALL, and some subsets of AML, there is limited data that PB and/or BM surveillance facilitates early detection of HM that translates into improved outcomes. However, as some heritable forms of AML and MDS can be more indolent, disease progression has been detected with serial monitoring for PB cytopenias and potentially surveillance BM evaluations (Porter et al. 2017; Förster et al. 2023). Of note, blood counts alone have been demonstrated to be an insensitive marker of impending myeloid malignancy in patients with SDS, so regular marrow examinations are recommended in that context (Myers et al. 2020; Reilly and Shimamura 2023).

Regarding risk for HM recurrence, the initial frequency of evaluations should be done as per the standard monitoring proposed for patients with de novo HM including, at a minimum, visits, and laboratory work done at regular intervals until 3 years from the end of treatment. At that time, many centers transition to yearly evaluations. However, given the underlying GPS and potential risk for a new HM, more frequent monitoring may be prudent and should be tailored to align with general surveillance guidelines for a given GPS as referenced above (Porter et al. 2017). Monitoring for disease post-HSCT can also introduce confounders. While surveillance with serial CBC is reasonable, some providers feel that molecular analysis in the setting of full donor chimerism should be avoided as such testing may result in the identification of gene variants derived from the donor, which may raise undue concern in transplanted individuals (Alter 2017; Trottier and Godley 2021). However, given the risk for recurrence of a primary HM as well as the potential for secondary hematopoietic malignancy in a patient with a GPS, awareness of the significance of donor mutations may be beneficial, and practice variations regarding disease monitoring exist as a result.

Cascade Testing and Reproductive Considerations in Patients with a GPS

Patients with GPS should be educated about the risks for their family members, including a discussion of the inheritance pattern, penetrance of HM, variability in phenotype, and surveillance that will be recommended for any family members who have the mutation. Physicians should collaborate with genetic services at their institution to coordinate cascade testing for family members or find a local certified genetic counselor specializing in hereditary cancer risks through the National Society of Genetic Counselors (NSGC) via the online genetic counselor tool at findageneticcounselor.nsgc.org.

Patients should also be made aware of various reproductive options, including natural conception with prenatal genetic testing through amniocentesis or chorionic villus sampling, natural conception with postnatal genetic testing, egg, or sperm donation, in vitro fertilization with preimplantation genetic testing (IVF with PGT), and adoption. Patients interested in learning more about PGT as well as other reproductive options should be referred to a prenatal genetic counselor or maternal–fetal medicine provider. For patients with autosomal recessive conditions, it is important to discuss the value of testing one's partner to better understand risks for offspring.

SUMMARY

In summary, as awareness of genetic predisposition to HM expands, close collaboration between families, physicians, genetic counselors, scientists, and bioinformatics experts is critical. Diagnostic evaluation of GPS integrating clinical history, family history, physical examination, laboratory testing, and consideration of germline genetic testing for individuals with an HM has been outlined. Diagnosis of a GPS impacts decisions regarding chemotherapy treatment, the role of HSCT, as well as HSCT donor considerations. Moreover, such testing may also raise awareness as to the benefit of disease surveillance in other family members with the GPS. Lastly, as more children and adults with GPS are identified, there is a growing need for evidence-based guidelines regarding disease surveillance. As health care providers integrate the diagnostic workup for a GPS into the evaluation, monitoring, and treatment of HM patients, we anticipate greater survival of HM patients with GPS and less treatment-related morbidity/mortality.

REFERENCES

Abedalthagafi M. 2018. Constitutional mismatch repair-deficiency: current problems and emerging therapeutic strategies. *Oncotarget* 9: 35458–35469. doi:10.18632/on cotarget.26249

Adam MP, Feldman J, Mirzaa GM, Pagon RA, Wallace SE, Bean LJH, Gripp KW, Amemiya A, ed. 1993. *GeneReviews* [Internet]. University of Washington, Seattle, Seattle, WA.

Agarwal S. 2023. Minimal intensity conditioning strategies for bone marrow failure: is it time for "preventative" transplants? *Hematology* 2023: 135–140. doi:10.1182/he matology.2023000470

Alter BP. 2014. Fanconi anemia and the development of leukemia. *Best Pract Res Clin Haematol* 27: 214–221. doi:10.1016/j.beha.2014.10.002

Alter BP. 2017. Inherited bone marrow failure syndromes: considerations pre- and posttransplant. *Hematology Am Soc Hematol Educ Program* 2017: 88–95. doi:10.1182/ asheducation-2017.1.88

Alter BP, Rosenberg PS, Day T, Menzel S, Giri N, Savage SA, Thein SL. 2013. Genetic regulation of fetal haemoglobin in inherited bone marrow failure syndromes. *Br J Haematol* 162: 542–546. doi:10.1111/bjh.12399

Andreasson P, Johansson B, Arheden K, Billström R, Mitelman F, Höglund M. 1997. Deletions of CDKN1B and ETV6 in acute myeloid leukemia and myelodysplastic syndromes without cytogenetic evidence of 12p abnormalities. *Genes Chromosomes Cancer* 19: 77–83. doi:10 .1002/(SICI)1098-2264(199706)19:2<77::AID-GCC2>3 .0.CO;2-X

Arber DA, Orazi A, Hasserjian RP, Borowitz MJ, Calvo KR, Kvasnicka HM, Wang SA, Bagg A, Barbui T, Branford S, et al. 2022. International Consensus Classification of Myeloid Neoplasms and Acute Leukemias: integrating morphologic, clinical, and genomic data. *Blood* 140: 1200–1228. doi:10.1182/blood.2022015850

Armes H, Bewicke-Copley F, Rio-Machin A, Di Bella D, Philippe C, Wozniak A, Tummala H, Wang J, Ezponda T, Prosper F, et al. 2022. Germline ERCC excision repair 6 like 2 (*ERCC6L2*) mutations lead to impaired erythropoiesis and reshaping of the bone marrow microenvironment. *Br J Haematol* 199: 754–764. doi:10.1111/bjh .18466

Avagyan S, Shimamura A. 2022. Lessons from pediatric MDS: approaches to germline predisposition to hematologic malignancies. *Front Oncol* 12: 813149. doi:10.3389/ fonc.2022.813149

Baccelli F, Leardini D, Cerasi S, Messelodi D, Bertuccio SN, Masetti R. 2023. ERCC6L2-related disease: a novel entity of bone marrow failure disorder with high risk of clonal evolution. *Ann Hematol* 102: 699–705. doi:10.1007/ s00277-023-05128-2

Bakry D, Aronson M, Durno C, Rimawi H, Farah R, Alharbi QK, Alharbi M, Shamvil A, Ben-Shachar S, Mistry M, et al. 2014. Genetic and clinical determinants of constitutional mismatch repair deficiency syndrome: report from the constitutional mismatch repair deficiency consortium. *Eur J Cancer* 50: 987–996. doi:10.1016/j.ejca.2013 .12.005

Brown AL, Arts P, Carmichael CL, Babic M, Dobbins J, Chong CE, Schreiber AW, Feng J, Phillips K, Wang PPS, et al. 2020. RUNX1-mutated families show phenotype heterogeneity and a somatic mutation profile unique to germline predisposed AML. *Blood Adv* 4: 1131–1144. doi:10.1182/bloodadvances.2019000901

Bruzzese A, Leardini D, Masetti R, Strocchio L, Girardi K, Algeri M, Del Baldo G, Locatelli F, Mastronuzzi A. 2020. GATA2 related conditions and predisposition to pediatric myelodysplastic syndromes. *Cancers (Basel)* 12: 2962. doi:10.3390/cancers12102962

Burroughs L, Woolfrey A, Shimamura A. 2009. Shwachman–Diamond syndrome: a review of the clinical presentation, molecular pathogenesis, diagnosis, and treatment.

Hematol Oncol Clin North Am **23**: 233–248. doi:10.1016/j
.hoc.2009.01.007

Burroughs LM, Shimamura A, Talano JA, Domm JA, Baker
KK, Delaney C, Frangoul H, Margolis DA, Baker KS,
Nemecek ER, et al. 2017. Allogeneic hematopoietic cell
transplantation using treosulfan-based conditioning for
treatment of marrow failure disorders. *Biol Blood Marrow
Transplant* **23**: 1669–1677. doi:10.1016/j.bbmt.2017.06
.002

Cerhan JR, Slager SL. 2015. Familial predisposition and genetic risk factors for lymphoma. *Blood* **126**: 2265–2273.
doi:10.1182/blood-2015-04-537498

Choi JW, Kim Y, Fujino M, Ito M. 2002. Significance of fetal
hemoglobin-containing erythroblasts (F blasts) and the F
blast/F cell ratio in myelodysplastic syndromes. *Leukemia*
16: 1478–1483. doi:10.1038/sj.leu.2402536

Cioc AM, Wagner JE, MacMillan ML, DeFor T, Hirsch B.
2010. Diagnosis of myelodysplastic syndrome among a
cohort of 119 patients with Fanconi anemia: morphologic
and cytogenetic characteristics. *Am J Clin Pathol* **133**: 92–
100. doi:10.1309/AJCP7W9VMJENZOVG

Collin M, Dickinson R, Bigley V. 2015. Haematopoietic and
immune defects associated with *GATA2* mutation. *Br J
Haematol* **169**: 173–187. doi:10.1111/bjh.13317

Cunningham L, Merguerian M, Calvo KR, Davis J, Deuitch
NT, Dulau-Florea A, Patel N, Yu K, Sacco K, Bhattacharya
S, et al. 2023. Natural history study of patients with fami-
lial platelet disorder with associated myeloid malignancy.
Blood **142**: 2146–2158. doi:10.1182/blood.2023019746

Dale DC, Bolyard AA, Shannon JA, Connelly JA, Link DC,
Bonilla MA, Newburger PE. 2022. Outcomes for patients
with severe chronic neutropenia treated with granulocyte
colony-stimulating factor. *Blood Adv* **6**: 3861–3869.
doi:10.1182/bloodadvances.2021005684

Davidsson J, Puschmann A, Tedgård U, Bryder D, Nilsson L,
Cammenga J. 2018. SAMD9 and SAMD9L in inherited
predisposition to ataxia, pancytopenia, and myeloid malignancies. *Leukemia* **32**: 1106–1115. doi:10.1038/s41375-
018-0074-4

de Andrade KC, Khincha PP, Hatton JN, Frone MN, Weg-
man-Ostrosky T, Mai PL, Best AF, Savage SA. 2021. Cancer incidence, patterns, and genotype-phenotype associations in individuals with pathogenic or likely pathogenic
germline TP53 variants: an observational cohort study.
Lancet Oncol **22**: 1787–1798. doi:10.1016/S1470-2045
(21)00580-5

Deeg HJ, Storb R, Thomas ED, Appelbaum F, Buckner CD,
Clift RA, Doney K, Johnson L, Sanders JE, Stewart P, et al.
1983. Fanconi's anemia treated by allogeneic marrow
transplantation. *Blood* **61**: 954–959. doi:10.1182/blood
.V61.5.954.954

DiNardo CD, Bannon SA, Routbort M, Franklin A, Mork M,
Armanios M, Mace EM, Orange JS, Jeff-Eke M, Churpek
JE, et al. 2016. Evaluation of patients and families with
concern for predispositions to hematologic malignancies
within the Hereditary Hematologic Malignancy Clinic
(HHMC). *Clin Lymphoma Myeloma Leuk* **16**: 417–428.
e2. doi:10.1016/j.clml.2016.04.001

Di Paola J, Porter CC. 2019. ETV6-related thrombocytopenia and leukemia predisposition. *Blood* **134**: 663–667.
doi:10.1182/blood.2019852418

Durno C, Ercan AB, Bianchi V, Edwards M, Aronson M,
Galati M, Atenafu EG, Abebe-Campino G, Al-Battashi A,
Alharbi M, et al. 2021. Survival benefit for individuals
with constitutional mismatch repair deficiency undergo-
ing surveillance. *J Clin Oncol* **39**: 2779–2790. doi:10.1200/
JCO.20.02636

Ebens CL, MacMillan ML, Wagner JE. 2017. Hematopoietic
cell transplantation in Fanconi anemia: current evidence,
challenges and recommendations. *Expert Rev Hematol*
10: 81–97. doi:10.1080/17474086.2016.1268048

Fadeel B, Garwicz D, Carlsson G, Sandstedt B, Nordenskjöld
M. 2021. Kostmann disease and other forms of severe
congenital neutropenia. *Acta Paediatr* **110**: 2912–2920.
doi:10.1111/apa.16005

Fargo JH, Kratz CP, Giri N, Savage SA, Wong C, Backer K,
Alter BP, Glader B. 2013. Erythrocyte adenosine deami-
nase: diagnostic value for Diamond–Blackfan anaemia.
Br J Haematol **160**: 547–554. doi:10.1111/bjh.12167

Fargo JH, Rochowski A, Giri N, Savage SA, Olson SB, Alter
BP. 2014. Comparison of chromosome breakage in non-
mosaic and mosaic patients with Fanconi anemia, relatives, and patients with other inherited bone marrow failure syndromes. *Cytogenet Genome Res* **144**: 15–27. doi:10
.1159/000366251

Feurstein S, Godley LA. 2017. Germline ETV6 mutations
and predisposition to hematological malignancies. *Int J
Hematol* **106**: 189–195. doi:10.1007/s12185-017-2259-4

Feurstein S, Trottier AM, Estrada-Merly N, Pozsgai M,
McNeely K, Drazer MW, Ruhle B, Sadera K, Koppayi
AL, Scott BL, et al. 2022a. Germ line predisposition variants occur in myelodysplastic syndrome patients of all
ages. *Blood* **140**: 2533–2548. doi:10.1182/blood.20220
15790

Feurstein S, Hahn CN, Mehta N, Godley LA. 2022b. A practical guide to interpreting germline variants that drive
hematopoietic malignancies, bone marrow failure, and
chronic cytopenias. *Genet Med* **24**: 931–954. doi:10
.1016/j.gim.2021.12.008

Flactif M, Lai JL, Preudhomme C, Fenaux P. 1994. Fluores-
cence in situ hybridization improves the detection of
monosomy 7 in myelodysplastic syndromes. *Leukemia*
8: 1012–1018.

Fogarty PF, Yamaguchi H, Wiestner A, Baerlocher GM,
Sloand E, Zeng WS, Read EJ, Lansdorp PM, Young NS.
2003. Late presentation of dyskeratosis congenita as apparently acquired aplastic anaemia due to mutations in
telomerase RNA. *Lancet* **362**: 1628–1630. doi:10.1016/
S0140-6736(03)14797-6

Förster A, Davenport C, Duployez N, Erlacher M, Ferster A,
Fitzgibbon J, Göhring G, Hasle H, Jongmans MC, Kole-
nova A, et al. 2023. European standard clinical practice -
key issues for the medical care of individuals with familial
leukemia. *Eur J Med Genet* **66**: 104727. doi:10.1016/j.ejmg
.2023.104727

Furutani E, Liu S, Galvin A, Steltz S, Malsch MM, Loveless
SK, Mount L, Larson JH, Queenan K, Bertuch AA, et al.
2022. Hematologic complications with age in Shwachman–Diamond syndrome. *Blood Adv* **6**: 297–306.
doi:10.1182/bloodadvances.2021005539

Germeshausen M, Ballmaier M. 2021. CAMT-MPL: congenital amegakaryocytic thrombocytopenia caused by
MPL mutations—heterogeneity of a monogenic disor-

der—a comprehensive analysis of 56 patients. *Haematologica* **106:** 2439–2448. doi:10.3324/haematol.2020.257972

Germeshausen M, Ancliff P, Estrada J, Metzler M, Ponstingl E, Rütschle H, Schwabe D, Scott RH, Unal S, Wawer A, et al. 2018. MECOM-associated syndrome: a heterogeneous inherited bone marrow failure syndrome with amegakaryocytic thrombocytopenia. *Blood Adv* **2:** 586–596. doi:10.1182/bloodadvances.2018016501

Gluckman E, Auerbach AD, Horowitz MM, Sobocinski KA, Ash RC, Bortin MM, Butturini A, Camitta BM, Champlin RE, Friedrich W, et al. 1995. Bone marrow transplantation for Fanconi anemia. *Blood* **86:** 2856–2862. doi:10.1182/blood.V86.7.2856.2856

Godley LA, Shimamura A. 2017. Genetic predisposition to hematologic malignancies: management and surveillance. *Blood* **130:** 424–432. doi:10.1182/blood-2017-02-735290

Gofin Y, Wang T, Gillentine MA, Scott TM, Berry AM, Azamian MS, Genetti C, Agrawal PB, Picker J, Wojcik MH, et al. 2022. Delineation of a novel neurodevelopmental syndrome associated with *PAX5* haploinsufficiency. *Hum Mutat* **43:** 461–470. doi:10.1002/humu.24332

Gonzalez KD, Buzin CH, Noltner KA, Gu D, Li W, Malkin D, Sommer SS. 2009. High frequency of de novo mutations in Li–Fraumeni syndrome. *J Med Genet* **46:** 689–693. doi:10.1136/jmg.2008.058958

Gross M, Hanenberg H, Lobitz S, Friedl R, Herterich S, Dietrich R, Gruhn B, Schindler D, Hoehn H. 2002. Reverse mosaicism in Fanconi anemia: natural gene therapy via molecular self-correction. *Cytogenet Genome Res* **98:** 126–135. doi:10.1159/000069805

Gurnari C, Robin M, Godley LA, Drozd-Sokołowska J, Włodarski MW, Raj K, Onida F, Worel N, Ciceri F, Carbacioglu S, et al. 2023. Germline predisposition traits in allogeneic hematopoietic stem-cell transplantation for myelodysplastic syndromes: a survey-based study and position paper on behalf of the Chronic Malignancies Working Party of the EBMT. *Lancet Haematol* **10:** e994–e1005. doi:10.1016/S2352-3026(23)00265-X

Gutierrez-Rodrigues F, Patel BA, Groarke EM. 2023. When to consider inherited marrow failure syndromes in adults. *Hematology* **2023:** 548–555. doi:10.1182/hematology.2023000488

Hasaart KAL, Bertrums EJM, Manders F, Goemans BF, van Boxtel R. 2021. Increased risk of leukaemia in children with Down syndrome: a somatic evolutionary view. *Expert Rev Mol Med* **23:** e5. doi:10.1017/erm.2021.6

Herrera-Mullar J, Fulk K, Brannan T, Yussuf A, Polfus L, Richardson ME, Horton C. 2023. Characterization of POT1 tumor predisposition syndrome: tumor prevalence in a clinically diverse hereditary cancer cohort. *Genet Med* **25:** 100937. doi:10.1016/j.gim.2023.100937

Hock H, Shimamura A. 2017. ETV6 in hematopoiesis and leukemia predisposition. *Semin Hematol* **54:** 98–104. doi:10.1053/j.seminhematol.2017.04.005

Homan CC, Scott HS, Brown AL. 2023. Hereditary platelet disorders associated with germ line variants in *RUNX1*, *ETV6*, and *ANKRD26*. *Blood* **141:** 1533–1543. doi:10.1182/blood.2022017735

Hudda Z, Myers KC. 2023. Posttransplant complications in patients with marrow failure syndromes: are we improving long-term outcomes? *Hematology* **2023:** 141–148. doi:10.1182/hematology.2023000471

Jameson-Lee M, Chen K, Ritchie E, Shore T, Al-Khattab O, Gergis U. 2018. Acute myeloid leukemia in a patient with thrombocytopenia with absent radii: a case report and review of the literature. *Hematol Oncol Stem Cell Ther* **11:** 245–247. doi:10.1016/j.hemonc.2017.02.001

Keel SB, Scott A, Sanchez-Bonilla M, Ho PA, Gulsuner S, Pritchard CC, Abkowitz JL, King MC, Walsh T, Shimamura A. 2016. Genetic features of myelodysplastic syndrome and aplastic anemia in pediatric and young adult patients. *Haematologica* **101:** 1343–1350. doi:10.3324/haematol.2016.149476

Kennedy AL, Myers KC, Bowman J, Gibson CJ, Camarda ND, Furutani E, Muscato GM, Klein RH, Ballotti K, Liu S, et al. 2021. Distinct genetic pathways define pre-malignant versus compensatory clonal hematopoiesis in Shwachman–Diamond syndrome. *Nat Commun* **12:** 1334. doi:10.1038/s41467-021-21588-4

Kewan T, Noss R, Godley LA, Rogers HJ, Carraway HE. 2020. Inherited thrombocytopenia caused by germline *ANKRD26* mutation should be considered in young patients with suspected myelodysplastic syndrome. *J Investig Med High Impact Case Rep* **8:** 2324709620938941. doi:10.1177/2324709620938941

Khoury JD, Solary E, Abla O, Akkari Y, Alaggio R, Apperley JF, Bejar R, Berti E, Busque L, Chan JKC, et al. 2022. The 5th edition of the World Health Organization classification of haematolymphoid tumours: myeloid and histiocytic/dendritic neoplasms. *Leukemia* **36:** 1703–1719. doi:10.1038/s41375-022-01613-1

Klco JM, Mullighan CG. 2021. Advances in germline predisposition to acute leukaemias and myeloid neoplasms. *Nat Rev Cancer* **21:** 122–137. doi:10.1038/s41568-020-00315-z

Kraft IL, Godley LA. 2020. Identifying potential germline variants from sequencing hematopoietic malignancies. *Blood* **136:** 2498–2506. doi:10.1182/blood.2020006910

Kudo S, Harigae H, Watanabe N, Takasawa N, Kimura J, Kameoka J, Meguro K, Imaizumi M, Kaku M, Sasaki T. 2000. Increased HbF levels in dyserythropoiesis. *Clin Chim Acta* **291:** 83–87. doi:10.1016/s0009-8981(99)00186-2

Levy B, Burnside RD. 2019. Are all chromosome microarrays the same? What clinicians need to know. *Prenat Diagn* **39:** 157–164. doi:10.1002/pd.5422

Li J, Bledsoe JR. 2023. Inherited bone marrow failure syndromes and germline predisposition to myeloid neoplasia: a practical approach for the pathologist. *Semin Diagn Pathol* **40:** 429–442. doi:10.1053/j.semdp.2023.06.006

Li P, White T, Xie W, Cui W, Peker D, Zeng G, Wang HY, Vagher J, Brown S, Williams M, et al. 2022. AML with germline DDX41 variants is a clinicopathologically distinct entity with an indolent clinical course and favorable outcome. *Leukemia* **36:** 664–674. doi:10.1038/s41375-021-01404-0

Linabery AM, Ross JA. 2008. Childhood and adolescent cancer survival in the US by race and ethnicity for the diagnostic period 1975–1999. *Cancer* **113:** 2575–2596. doi:10.1002/cncr.23866

Lozano Chinga MM, Bertuch AA, Afify Z, Dollerschell K, Hsu JI, John TD, Rao ES, Rowe RG, Sankaran VG,

Cite this article as *Cold Spring Harb Perspect Med* doi: 10.1101/cshperspect.a041585

Shimamura A, et al. 2023. Expanded phenotypic and hematologic abnormalities beyond bone marrow failure in *MECOM*-associated syndromes. *Am J Med Genet A* **191**: 1826–1835. doi:10.1002/ajmg.a.63208

MacMillan ML, Wagner JE. 2010. Haematopoeitic cell transplantation for Fanconi anaemia—when and how? *Br J Haematol* **149**: 14–21. doi:10.1111/j.1365-2141.2010.08078.x

Marquez R, Hantel A, Lorenz R, Neistadt B, Wong J, Churpek JE, Mardini NA, Shaukat I, Gurbuxani S, Miller JL, et al. 2014. A new family with a germline *ANKRD26* mutation and predisposition to myeloid malignancies. *Leuk Lymphoma* **55**: 2945–2946. doi:10.3109/10428194.2014.903476

McReynolds LJ, Calvo KR, Holland SM. 2018. Germline GATA2 mutation and bone marrow failure. *Hematol Oncol Clin North Am* **32**: 713–728. doi:10.1016/j.hoc.2018.04.004

McReynolds LJ, Rafati M, Wang Y, Ballew BJ, Kim J, Williams VV, Zhou W, Hendricks RM, Dagnall C, Freedman ND, et al. 2022. Genetic testing in severe aplastic anemia is required for optimal hematopoietic cell transplant outcomes. *Blood* **140**: 909–921. doi:10.1182/blood.2022016508

Mehta PA, Davies SM, Leemhuis T, Myers K, Kernan NA, Prockop SE, Scaradavou A, O'Reilly RJ, Williams DA, Lehmann L, et al. 2017. Radiation-free, alternative-donor HCT for Fanconi anemia patients: results from a prospective multi-institutional study. *Blood* **129**: 2308–2315. doi:10.1182/blood-2016-09-743112

Myers KC, Furutani E, Weller E, Siegele B, Galvin A, Arsenault V, Alter BP, Boulad F, Bueso-Ramos C, Burroughs L, et al. 2020. Clinical features and outcomes of patients with Shwachman–Diamond syndrome and myelodysplastic syndrome or acute myeloid leukaemia: a multicentre, retrospective, cohort study. *Lancet Haematol* **7**: e238–e246. doi:10.1016/S2352-3026(19)30206-6

Obrochta E, Godley LA. 2018. Identifying patients with genetic predisposition to acute myeloid leukemia. *Best Pract Res Clin Haematol* **31**: 373–378. doi:10.1016/j.beha.2018.09.014

Osuna-Marco MP, López-Barahona M, López-Ibor B, Tejera Á. 2021. Ten reasons why people with Down syndrome are protected from the development of most solid tumors—a review. *Front Genet* **12**: 749480. doi:10.3389/fgene.2021.749480

Owen CJ, Toze CL, Koochin A, Forrest DL, Smith CA, Stevens JM, Jackson SC, Poon MC, Sinclair GD, Leber B, et al. 2008. Five new pedigrees with inherited RUNX1 mutations causing familial platelet disorder with propensity to myeloid malignancy. *Blood* **112**: 4639–4645. doi:10.1182/blood-2008-05-156745

Padron E, Ball MC, Teer JK, Painter JS, Yoder SJ, Zhang C, Zhang L, Moscinski LC, Rollison DE, Gore SD, et al. 2018. Germ line tissues for optimal detection of somatic variants in myelodysplastic syndromes. *Blood* **131**: 2402–2405. doi:10.1182/blood-2018-01-827881

Palles C, West HD, Chew E, Galavotti S, Flensburg C, Grolleman JE, Jansen EAM, Curley H, Chegwidden L, Arbe-Barnes EH, et al. 2022. Germline MBD4 deficiency causes a multi-tumor predisposition syndrome. *Am J Hum Genet* **109**: 953–960. doi:10.1016/j.ajhg.2022.03.018

Peffault de Latour R, Soulier J. 2016. How I treat MDS and AML in Fanconi anemia. *Blood* **127**: 2971–2979. doi:10.1182/blood-2016-01-583625

Peffault de Latour R, Peters C, Gibson B, Strahm B, Lankester A, de Heredia CD, Longoni D, Fioredda F, Locatelli F, Yaniv I, et al. 2015. Recommendations on hematopoietic stem cell transplantation for inherited bone marrow failure syndromes. *Bone Marrow Transplant* **50**: 1168–1172. doi:10.1038/bmt.2015.117

Polprasert C, Schulze I, Sekeres MA, Makishima H, Przychodzen B, Hosono N, Singh J, Padgett RA, Gu X, Phillips JG, et al. 2015. Inherited and somatic defects in DDX41 in myeloid neoplasms. *Cancer Cell* **27**: 658–670. doi:10.1016/j.ccell.2015.03.017

Porter CC, Druley TE, Erez A, Kuiper RP, Onel K, Schiffman JD, Wolfe Schneider K, Scollon SR, Scott HS, Strong LC, et al. 2017. Recommendations for surveillance for children with leukemia-predisposing conditions. *Clin Cancer Res* **23**: e14–e22. doi:10.1158/1078-0432.CCR-17-0428

Reilly CR, Shimamura A. 2023. Predisposition to myeloid malignancies in Shwachman–Diamond syndrome: biological insights and clinical advances. *Blood* **141**: 1513–1523. doi:10.1182/blood.2022017739

Reinhardt D, Haase D, Schoch C, Wollenweber S, Hinkelmann E, v Heyden W, Lentini G, Wörmann B, Schröter W, Pekrun A. 1998. Hemoglobin F in myelodysplastic syndrome. *Ann Hematol* **76**: 135–138. doi:10.1007/s002770050377

Revy P, Kannengiesser C, Fischer A. 2019. Somatic genetic rescue in Mendelian haematopoietic diseases. *Nat Rev Genet* **20**: 582–598. doi:10.1038/s41576-019-0139-x

Rigolin GM, Bigoni R, Milani R, Cavazzini F, Roberti MG, Bardi A, Agostini P, Della Porta M, Tieghi A, Piva N, et al. 2001. Clinical importance of interphase cytogenetics detecting occult chromosome lesions in myelodysplastic syndromes with normal karyotype. *Leukemia* **15**: 1841–1847. doi:10.1038/sj.leu.2402293

Roloff GW, Drazer MW, Godley LA. 2021. Inherited susceptibility to hematopoietic malignancies in the era of precision oncology. *JCO Precis Oncol* **5**: 107–122. doi:10.1200/PO.20.00387

Rosenberg PS, Alter BP, Link DC, Stein S, Rodger E, Bolyard AA, Aprikyan AA, Bonilla MA, Dror Y, Kannourakis G, et al. 2008. Neutrophil elastase mutations and risk of leukaemia in severe congenital neutropenia. *Br J Haematol* **140**: 210–213. doi:10.1111/j.1365-2141.2007.06897.x

Rossini L, Durante C, Bresolin S, Opocher E, Marzollo A, Biffi A. 2022. Diagnostic strategies and algorithms for investigating cancer predisposition syndromes in children presenting with malignancy. *Cancers (Basel)* **14**: 3741. doi:10.3390/cancers14153741

Rudelius M, Weinberg OK, Niemeyer CM, Shimamura A, Calvo KR. 2023. The International Consensus Classification (ICC) of hematologic neoplasms with germline predisposition, pediatric myelodysplastic syndrome, and juvenile myelomonocytic leukemia. *Virchows Arch* **482**: 113–130. doi:10.1007/s00428-022-03447-9

Sahoo SS, Kozyra EJ, Wlodarski MW. 2020. Germline predisposition in myeloid neoplasms: unique genetic and clinical features of GATA2 deficiency and SAMD9/SAMD9L syndromes. *Best Pract Res Clin Haematol* **33**: 101197. doi:10.1016/j.beha.2020.101197

Sahoo SS, Pastor VB, Goodings C, Voss RK, Kozyra EJ, Szvetnik A, Noellke P, Dworzak M, Starý J, Locatelli F, et al. 2021. Clinical evolution, genetic landscape and trajectories of clonal hematopoiesis in SAMD9/SAMD9L syndromes. *Nat Med* **27:** 1806–1817. doi:10.1038/s41591-021-01511-6

Sanders MA, Chew E, Flensburg C, Zeilemaker A, Miller SE, Al Hinai AS, Bajel A, Luiken B, Rijken M, Mclennan T, et al. 2018. MBD4 guards against methylation damage and germ line deficiency predisposes to clonal hematopoiesis and early-onset AML. *Blood* **132:** 1526–1534. doi:10.1182/blood-2018-05-852566

Savage SA. 2022. Dyskeratosis congenita and telomere biology disorders. *Hematology* **2022:** 637–648. doi:10.1182/hematology.2022000394

Schratz KE. 2023. Clonal evolution in inherited marrow failure syndromes predicts disease progression. *Hematology* **2023:** 125–134. doi:10.1182/hematology.2023000469

Shah S, Schrader KA, Waanders E, Timms AE, Vijai J, Miething C, Wechsler J, Yang J, Hayes J, Klein RJ, et al. 2013. A recurrent germline PAX5 mutation confers susceptibility to pre-B cell acute lymphoblastic leukemia. *Nat Genet* **45:** 1226–1231. doi:10.1038/ng.2754

Sood R, Kamikubo Y, Liu P. 2017. Role of RUNX1 in hematological malignancies. *Blood* **129:** 2070–2082. doi:10.1182/blood-2016-10-687830

Speedy HE, Kinnersley B, Chubb D, Broderick P, Law PJ, Litchfield K, Jayne S, Dyer MJS, Dearden C, Follows GA, et al. 2016. Germline mutations in shelterin complex genes are associated with familial chronic lymphocytic leukemia. *Blood* **128:** 2319–2326. doi:10.1182/blood-2016-01-695692

Spinner MA, Sanchez LA, Hsu AP, Shaw PA, Zerbe CS, Calvo KR, Arthur DC, Gu W, Gould CM, Brewer CC, et al. 2014. GATA2 deficiency: a protean disorder of hematopoiesis, lymphatics, and immunity. *Blood* **123:** 809–821. doi:10.1182/blood-2013-07-515528

Strullu M, Caye A, Lachenaud J, Cassinat B, Gazal S, Fenneteau O, Pouvreau N, Pereira S, Baumann C, Contet A, et al. 2014. Juvenile myelomonocytic leukaemia and Noonan syndrome. *J Med Genet* **51:** 689–697. doi:10.1136/jmedgenet-2014-102611

Stubbins RJ, Korotev S, Godley LA. 2022. Germline CHEK2 and ATM variants in myeloid and other hematopoietic malignancies. *Curr Hematol Malig Rep* **17:** 94–104. doi:10.1007/s11899-022-00663-7

Swaminathan M, Bannon SA, Routbort M, Naqvi K, Kadia TM, Takahashi K, Alvarado Y, Ravandi-Kashani F, Patel KP, Champlin R, et al. 2019. Hematologic malignancies and Li–Fraumeni syndrome. *Cold Spring Harb Mol Case Stud* **5:** a003210. doi:10.1101/mcs.a003210

Tawana K, Wang J, Renneville A, Bödör C, Hills R, Loveday C, Savic A, Van Delft FW, Treleaven J, Georgiades P, et al. 2015. Disease evolution and outcomes in familial AML with germline CEBPA mutations. *Blood* **126:** 1214–1223. doi:10.1182/blood-2015-05-647172

Tawana K, Rio-Machin A, Preudhomme C, Fitzgibbon J. 2017. Familial CEBPA-mutated acute myeloid leukemia. *Semin Hematol* **54:** 87–93. doi:10.1053/j.seminhematol.2017.04.001

Tesi B, Davidsson J, Voss M, Rahikkala E, Holmes TD, Chiang SCC, Komulainen-Ebrahim J, Gorcenco S, Rund-berg Nilsson A, Ripperger T, et al. 2017. Gain-of-function SAMD9L mutations cause a syndrome of cytopenia, immunodeficiency, MDS, and neurological symptoms. *Blood* **129:** 2266–2279. doi:10.1182/blood-2016-10-743302

Tiong IS, Stevenson WS, Wall M, Yap YZ, Seymour JF, Kenealy M, Blombery P; Australasian Leukaemia & Lymphoma Group. 2023. Favorable outcomes of DDX41-mutated myelodysplastic syndrome and low blast count acute myeloid leukemia treated with azacitidine ± lenalidomide. *EJHaem* **4:** 1212–1215. doi:10.1002/jha2.767

Trottier AM, Godley LA. 2021. Inherited predisposition to haematopoietic malignancies: overcoming barriers and exploring opportunities. *Br J Haematol* **194:** 663–676. doi:10.1111/bjh.17247

Vlachos A, Rosenberg PS, Atsidaftos E, Alter BP, Lipton JM. 2012. Incidence of neoplasia in Diamond Blackfan anemia: a report from the Diamond Blackfan Anemia Registry. *Blood* **119:** 3815–3819. doi:10.1182/blood-2011-08-375972

Vlachos A, Rosenberg PS, Atsidaftos E, Kang J, Onel K, Sharaf RN, Alter BP, Lipton JM. 2018. Increased risk of colon cancer and osteogenic sarcoma in Diamond–Blackfan anemia. *Blood* **132:** 2205–2208. doi:10.1182/blood-2018-05-848937

Waisfisz Q, Morgan NV, Savino M, de Winter JP, van Berkel CG, Hoatlin ME, Ianzano L, Gibson RA, Arwert F, Savoia A, et al. 1999. Spontaneous functional correction of homozygous Fanconi anaemia alleles reveals novel mechanistic basis for reverse mosaicism. *Nat Genet* **22:** 379–383. doi:10.1038/11956

Wei X, Calvo-Vidal MN, Chen S, Wu G, Revuelta MV, Sun J, Zhang J, Walsh MF, Nichols KE, Joseph V, et al. 2018. Germline lysine-specific demethylase 1 (LSD1/KDM1A) mutations confer susceptibility to multiple myeloma. *Cancer Res* **78:** 2747–2759. doi:10.1158/0008-5472.CAN-17-1900

Westbrook CA, Hsu WT, Chyna B, Litvak D, Raza A, Horrigan SK. 2000. Cytogenetic and molecular diagnosis of chromosome 5 deletions in myelodysplasia. *Br J Haematol* **110:** 847–855. doi:10.1046/j.1365-2141.2000.02285.x

Williams L, Doucette K, Karp JE, Lai C. 2021. Genetics of donor cell leukemia in acute myelogenous leukemia and myelodysplastic syndrome. *Bone Marrow Transplant* **56:** 1535–1549. doi:10.1038/s41409-021-01214-z

Wlodarski MW, Hirabayashi S, Pastor V, Starý J, Hasle H, Masetti R, Dworzak M, Schmugge M, van den Heuvel-Eibrink M, Ussowicz M, et al. 2016. Prevalence, clinical characteristics, and prognosis of GATA2-related myelodysplastic syndromes in children and adolescents. *Blood* **127:** 1387–1397; quiz 1518. doi:10.1182/blood-2015-09-669937

Wlodarski MW, Collin M, Horwitz MS. 2017. GATA2 deficiency and related myeloid neoplasms. *Semin Hematol* **54:** 81–86. doi:10.1053/j.seminhematol.2017.05.002

Wlodarski MW, Sahoo SS, Niemeyer CM. 2018. Monosomy 7 in pediatric myelodysplastic syndromes. *Hematol Oncol Clin North Am* **32:** 729–743. doi:10.1016/j.hoc.2018.04.007

Wong JC, Bryant V, Lamprecht T, Ma J, Walsh M, Schwartz J, Del Pilar Alzamora M, Mulligan CG, Loh ML, Ribeiro

Cite this article as *Cold Spring Harb Perspect Med* doi: 10.1101/cshperspect.a041585

R, et al. 2018. Germline SAMD9 and SAMD9L mutations are associated with extensive genetic evolution and diverse hematologic outcomes. *JCI Insight* **3:** e121086. doi:10.1172/jci.insight.121086

Xavier AC, Ge Y, Taub JW. 2009. Down syndrome and malignancies: a unique clinical relationship: a paper from the 2008 William Beaumont Hospital Symposium on Molecular Pathology. *J Mol Diagn* **11:** 371–380. doi:10.2353/jmoldx.2009.080132

Xiao H, Shi J, Luo Y, Tan Y, He J, Xie W, Zhang L, Wang Y, Liu L, Wu K, et al. 2011. First report of multiple CEBPA mutations contributing to donor origin of leukemia relapse after allogeneic hematopoietic stem cell transplantation. *Blood* **117:** 5257–5260. doi:10.1182/blood-2010-12-326322

New Paradigms in the Clinical Management of Li–Fraumeni Syndrome

Camilla Giovino,[1,2,5] **Vallijah Subasri,**[1,2,5] **Frank Telfer,**[1,2,5] **and David Malkin**[1,2,3,4]

[1]Genetics and Genome Biology Program, The Hospital for Sick Children, Toronto, Ontario M5G 1L7, Canada

[2]Department of Medical Biophysics, Temerty Faculty of Medicine, University of Toronto, Toronto, Ontario M5G 1L7, Canada

[3]Institute of Medical Science, Temerty Faculty of Medicine, University of Toronto, Toronto, Ontario M5S 1A8, Canada

[4]Division of Hematology-Oncology, The Hospital for Sick Children, Department of Pediatrics, University of Toronto, Toronto, Ontario M5G 1X8, Canada

Correspondence: david.malkin@sickkids.ca

Approximately 8.5%–16.2% of childhood cancers are associated with a pathogenic/likely pathogenic germline variant—a prevalence that is likely to rise with improvements in phenotype recognition, sequencing, and variant validation. One highly informative, classical hereditary cancer predisposition syndrome is Li–Fraumeni syndrome (LFS), associated with germline variants in the *TP53* tumor suppressor gene, and a >90% cumulative lifetime cancer risk. In seeking to improve outcomes for young LFS patients, we must improve the specificity and sensitivity of existing cancer surveillance programs and explore how to complement early detection strategies with pharmacology-based risk-reduction interventions. Here, we describe novel precision screening technologies and clinical strategies for cancer risk reduction. In particular, we summarize the biomarkers for early diagnosis and risk stratification of LFS patients from birth, noninvasive and machine learning–based cancer screening, and drugs that have shown the potential to be repurposed for cancer prevention.

Although hereditary cancer has long been thought of as rare, germline variants associated with increased cancer risk are more prevalent than previously understood. This is especially relevant in the pediatric context with recent reports suggesting that at least 15% of childhood cancers are associated with germline pathogenic variants across several dozen genes (Mody et al. 2015; Zhang et al. 2015; ICGC PedBrain-Seq Project et al. 2018; Akhavanfard et al. 2020; Wong et al. 2020; Villani et al. 2023). One highly informative, classical hereditary cancer predisposition is Li–Fraumeni syndrome (LFS), characterized by a broad spectrum of early onset malignancies, a lifetime cancer risk approaching 75% in males and >90% in females, and a >83-fold increased risk to develop multiple primary tumors. Clinical management of cancer predisposition syndromes (CPSs) such as LFS has long emphasized risk reduction primarily with the introduction of comprehensive site-directed surveillance tools including imaging and bloodwork to great success (McBride et al. 2014; Kratz et al. 2017). However, existing screening modalities are intensive and one-size-fits-all; patients screened from the beginning of life are subject to a standard protocol, which may evolve with age as CPS-specific tumor risks change, but which does not take into account patient-unique differences that contribute to variable cancer risk. Furthermore, few interventions short of life-altering prophylactic surgeries exist for primary prevention of disease. The need for preventive strategies is further underlined by the challenges associated with the cumulative toxicities of therapy for multiple primary cancers or recurrent tumors and the potential for treatment-induced secondary malignancies.

Recent advances in sequencing and machine learning have the potential to better capture the vast clinical heterogeneity of LFS, permitting stratification of patients into appropriate, personalized surveillance programs. Clinical implementation of sophisticated diagnostic models for the early detection of cancer type would identify individuals at a high risk of certain cancers and eliminate unnecessary prophylactic measures (e.g., a patient not at a high risk of breast cancer [BC] could avoid prophylactic mastectomy). Likewise, models that facilitate the prediction of the age of cancer onset would enable the reduction of more burdensome components of the surveillance protocol, such as the annual whole-body magnetic resonance imaging (MRI) where sedation is typically necessary in a significant proportion of pediatric patients (Villani et al. 2016; Ballinger et al. 2017). Herein, we will describe two potential precision surveillance strategies: (1) early diagnosis and risk stratification from birth; and (2) cancer screening and monitoring. Early diagnosis and risk stratification from birth include the identification of inherited germline drivers and modifiers of tumor phenotypes. Cancer screening and monitoring involves real-time tracking of cancer progression using imaging and molecular correlates of tumor phenotypes.

Improvements in our understanding of the correlates of malignancy of LFS will permit a shift in the narrative of care through the complementary approach of pharmacology-based risk reduction. A number of preclinical and clinical studies have recently been undertaken to investigate the role of metabolic changes in LFS patients. The potential importance of these metabolic shifts may be in informing cancer prevention efforts. For example, numerous groups have established correlations between reductions in cellular metabolism with reductions in cancer burden. Others have observed reduced cancer rates in patients and preclinical models involving drugs that act on growth-promoting metabolic pathways. If clinical studies successfully establish cancer risk-reduction efficacy for these drugs, the impact could be profound, especially given the near-inevitability of cancer in LFS. Such interventions could extend the cancer-free period and life span for LFS patients as well as reduce the frequency and intensity of surveillance. In this review, we will describe a few key agents and other interventions that have shown cancer-preventive promise, particularly the use of metformin (glucophage) and rapamycin (sirolimus).

Implementation of this research in the clinic has the potential to reduce the physiological and psychological burdens of screening and treatment, contribute to decreased cancer incidence and mortality, and minimize the resources and costs associated with such measures (Samuel et al. 2014; Ho et al. 2020). We hope to provide an overview of these important developments, as well as potential synergies between them, and accelerate further preclinical and clinical research that would see them implemented for LFS and potentially other patients at increased risk of cancer, particularly in the pediatric and young adult contexts.

HEREDITARY PEDIATRIC CANCER

Pediatric Cancer

The genetic basis of childhood cancer differs fundamentally from that of adult-onset cancer. The vast majority of adult cancers are sporadic—they are triggered by the accumulation of somatic mutations in essential driver genes, many resulting from carcinogenic environmental exposures over a patient's lifetime. The effects of long-term carcinogen exposure on the development of somatic mutations are negligible in children. Pediatric cancer genomes tend to be "quiet," often harboring far fewer mutations—many of which are rare or absent in adult cancers. Adult cancers also generally exhibit far more gene mutations, with hypermutation (10–100 mut/Mb) prevalent in ∼1 in 6 adult cancers compared to ∼1 in 20 childhood cancers (Campbell et al. 2017). In addition, the spectrum of cancer types that commonly occur in children is different than in adults. Leukemias and lymphomas comprise nearly half of the pediatric cancers with an additional 26% being tumors of the central nervous system (CNS). Leukemias and lymphomas account for <10% of all cancers in adults, and CNS tumors comprise <1.5% (Linet et al. 1999).

Recent reports suggest that ∼8.5%–16.2% of childhood cancers are associated with a pathogenic germline variant in a cancer-associated gene (Mody et al. 2015; Zhang et al. 2015; ICGC PedBrain-Seq Project et al. 2018; Akhavanfard et al. 2020; Wong et al. 2020). While these figures are noteworthy, it is likely that they represent a significant underestimate of the true prevalence of germline cancer predisposition—and will likely continue to increase with the ever-improving sensitivity of sequencing methods, the ongoing discovery of additional cancer susceptibility genes (CSGs), better characterization of unvalidated variants of uncertain significance (VUSs), and improved recognition of CPS phenotypes. In Akhavanfard et al.'s (2020) study of a long-term follow-up clinic, although 12% of children and young adult patients with solid tumors were found to have a germline pathogenic variant in a known CSG, 61% were found to carry pathogenic variants in other genes whose association with cancer risk has not been clearly outlined, but which may potentially contribute to a predisposition.

Cancer Predisposition Syndromes

CPS is classically defined by genetically related individuals afflicted with a malignancy (of the same or differing types) in a predictable pattern. Over 100 CPSs have been described to date, the majority of which present with incomplete penetrance and an autosomal dominant pattern of inheritance. In many cases, these classical syndromes have been associated with pathogenic variants in critical tumor-suppressor genes (McGee and Nichols 2016). One highly informative disorder is LFS, a hereditary cancer predisposition associated with germline variants in the *TP53* tumor-suppressor gene. LFS is unique among classical CPS in its high penetrance and the unusually wide spectrum of cancer types with which it is associated.

THE CURRENT STATE OF MANAGEMENT FOR LFS

Introduction to LFS

The classical definition of LFS was established by Drs. Frederick Li and Joseph Fraumeni in 1988. The phenotypic characteristics of "classic" LFS, as well as the more recently defined (revised) Chompret criteria are outlined in Table 1 (Li et al. 1988; Malkin et al. 1990; Srivastava et al. 1990; Chompret et al. 2001; Tinat et al. 2009; Holmfeldt et al. 2013). In general, these criteria emphasize family history (McGee and Nichols 2016). However, the use of family history alone as a diagnostic tool for hereditary cancer is insufficient and often unreliable, especially in the pediatric context. There can be substantial difficulty in acquiring and maintaining accurate and comprehensive family medical histories. Moreover, in the absence of a strong family history of cancer, many high-risk patients are not identified due to the missed detection of de novo variants (Gonzalez et al. 2009; Renaux-Petel et al. 2018). Classical clinical criteria have also proven inadequate for other reasons, namely unrecognized variability in phenotypes and cancer penetrance across individuals (Nagy et al. 2004; Knapke et al. 2012).

Approximately 60%–80% of LFS patients diagnosed using classical criteria carry a pathogenic germline variant in *TP53* (Olivier et al. 2003). Lifetime cancer risk for LFS patients approaches 75% in males and >90% in females. While many

Table 1. Clinical criteria for identification and diagnosis of Li–Fraumeni syndrome

Classical Li–Fraumeni syndrome (Li et al. 1988)	
Proband must fulfill all of the following criteria:	Proband with a sarcoma <45 yr, AND
	Proband with a first-degree relative diagnosed with any cancer under 45, AND
	Proband with another first- or second-degree relative diagnosed with any cancer under 45, OR a sarcoma at any age
Revised Chompret criteria (Bougeard et al. 2015)	
Familial presentation	Proband with a tumor belonging to the LFS tumor spectrum (i.e., premenopausal breast cancer, soft tissue sarcoma, osteosarcoma, CNS tumor, adrenocortical carcinoma) before age 46 yr, AND at least one first- or second-degree relative with LFS tumor (except breast cancer if proband has breast cancer) before age 56 yr or with multiple tumors OR
Multiple primitive tumors	Proband with multiple tumors (except multiple breast tumors), two of which belong to the LFS tumor spectrum and the first of which occurred before age 46 yr OR
Rare tumors	Proband with adrenocortical carcinoma, choroid plexus tumor, or rhabdomyosarcoma of embryonal anaplastic subtype, irrespective of family history OR
Early onset breast cancer	Proband with breast cancer before age 31 yr

(LFS) Li–Fraumeni syndrome, (CNS) central nervous system.

cancer types have been reported in LFS, five account for the majority of observed tumors (the "core" cancers): adrenocortical carcinomas (ACCs), BCs, tumors of the CNS, osteosarcomas (OSs), and soft tissue sarcomas (STSs). LFS patients also have an increased risk for early onset malignancies and an 83-fold increased risk to develop multiple primary cancers (McBride et al. 2014; Kratz et al. 2017). Mounting evidence suggests that germline *TP53* variants may occur more frequently, and their role may be more phenotypically diverse than initially suspected. It has been reported that 80% of children with rhabdomyosarcoma with diffuse anaplasia, 50% of children with ACC, 40% of children with choroid plexus carcinoma (CPC), and up to 10% of children with OS, carry a pathogenic germline *TP53* variant, often in the absence of an obvious family history (Bougeard et al. 2015; Wasserman et al. 2015; Zhang et al. 2015; Kratz et al. 2017).

While the prevalence of various cancer types differs across age groups (and indeed across different studies), STS and OS are the most commonly diagnosed in children followed by ACC and tumors of the CNS (Gonzalez et al. 2009; Ognjanovic et al. 2012; Sorrell et al. 2013; Bougeard et al. 2015; Mai et al. 2016).

With respect to overall cancer risk, Mai et al. reported that 12%, 25%, 52%, and 80% of *TP53* variant carriers had developed cancer by the ages of 20, 30, 40, and 50 years, respectively, whereas only 0.7%, 1.0%, 2.2%, and 5.1% of noncarriers had developed cancer by the same ages. Furthermore, ~50% of patients who develop a malignancy go on to develop another within the subsequent 10 years.

Cancer Surveillance and LFS

Much of the progress in widespread, population-based screening has been limited to sporadic adult cancers. Implementation of optimal, large-scale screening programs for pediatric cancer and for hereditary cancer, in particular, remains an ongoing challenge. Moreover, 5-year overall survival for pediatric patients afflicted with metastatic disease is significantly worse than that of patients with localized disease, reinforcing the impact of early cancer detection on patient outcomes (Fineberg et al. 2020).

As previously stated, clinical management of LFS patients focuses on cancer surveillance and prevention. Historically, the development of comprehensive screening protocols was hindered by the pleiotropic nature of LFS. In recent years, improvements in imaging technologies and genetic testing have minimized some of these difficulties. The first comprehensive surveillance protocol designed specifically for LFS patients was reported in 2011 (and coined the "Toronto Protocol," depicted in Fig. 1) using frequent physical examinations, whole-body MRI and ultrasound (as well as mammograms and colonoscopies for adults), and biomarkers that were tumor-specific (e.g., adrenocortical hor-

mone profile for early ACC detection) or nonspecific (e.g., complete blood count [CBC] or lactate dehydrogenase [LDH]) at regular, frequent intervals (Villani et al. 2011). Variations of this protocol have been developed and clinically validated at multiple institutions around the world and captured in consensus guidelines developed in 2017 (McBride et al. 2014; Villani et al. 2016; Daly et al. 2017; Kratz et al. 2017). A landmark, 11-year follow-up of a prospective observational study of LFS patients who underwent

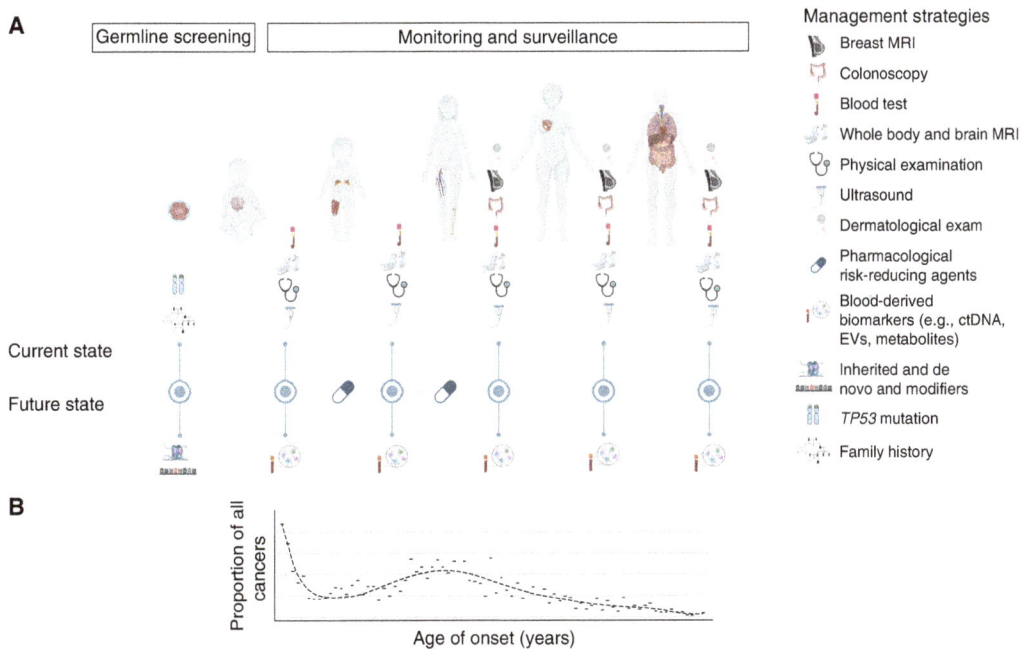

Figure 1. A novel paradigm for the management of LI–Fraumeni syndrome (LFS): pharmacological risk reduction slows tumor development, and personalized cancer screening permits early detection and intervention of specific tumors. (A) The current and future state of management for an individual with LFS is outlined across the human life span. Individuals ascertained to have a germline TP53 pathogenic variant are monitored for tumor development using intensive tumor surveillance protocols, as specified by the Toronto Protocol, starting at birth and continuing throughout life. The current state of management, as depicted on the upper timeline, includes blood tests, whole body and brain magnetic resonance imaging (MRI), physical examination, and ultrasound at regular, specific intervals based on the individual's age, as per the Toronto Surveillance Protocol (Villani et al. 2011, 2016), with the addition of breast MRI, colonoscopy, and dermatological examination in individuals over the age of 18 yr. The future state of management is depicted on the lower timeline, highlighting the potential for personalized screening to improve early tumor detection, inform risk stratification, as well as improve the ability to monitor for specific tumor types. This will entail screening for other inherited and de novo germline modifiers, in addition to detection and analysis of blood-derived biomarkers including circulating tumor DNA (ctDNA), extracellular vesicles (EVs), and metabolites at regular intervals. Future management may also involve the administration of pharmacological risk-reducing agents to delay or completely eliminate tumor development, thus influencing the frequency and intensity of surveillance. (B) Temporal patterns of tumor onset in germline TP53 mutation carriers. (B, Reprinted from Amadou et al. 2018, with permission from Wolters Kluwer, Inc., © 2017/2018.) Amadou et al. (2018) plotted the proportion of all cancers occurring by each year of age in TP53 mutation carriers. (Created with BioRender.com.)

surveillance revealed the effectiveness of such protocols in detecting asymptomatic tumors. Notably, enrollment in the protocol contributed to improved 5-year overall survival relative to patients who were not enrolled (88.8% [95% CI; 78.7–100] vs. 59.6% [47.2–75.2]) (Villani et al. 2016). Given that the prevalence of different tumor types changes with age and sex, the screening guidelines incorporate specific modalities based on these factors. For example, children with LFS are more susceptible to ACC, various types of brain and blood cancers, as well as soft tissue and bone sarcomas. The protocols are thus optimized to ensure the detection of these malignancies over other common, but later-onset LFS-associated cancer types such as BC. Children from birth (or at the time of confirmation of their mutant *TP53* carrier status) to 18 years of age receive an ultrasound of the abdomen and pelvis every 3–4 months together with blood and urine tests, as well as annual brain and whole-body MRIs (Kratz et al. 2017).

Protocol personalization could be substantially improved with the incorporation of specific genotypes and other information collected from next-generation sequencing (NGS), as we will discuss in a later section below. One group has suggested the prioritization of the *TP53* variant type in the development of a management plan. In line with this, modifications to the aforementioned consensus guidelines have been recently proposed by other groups. In particular, some have suggested a wholesale reframing of the LFS diagnosis around *TP53* variant status (The European Reference Network GENTURIS et al. 2020). These shifts may be warranted, and with further research genotype–phenotype correlations will likely form a central component of clinical care in the future.

Cancer-Preventive Interventions for LFS

Beyond screening, there are few prophylactic interventions specifically recommended to be considered for LFS patients. As a precaution against early BC development, some female LFS patients choose to undergo bilateral mastectomy in early adulthood. Similarly, the early introduction of screening colonoscopies with the accompanying removal of any detected polyps is commonly recommended for all LFS patients. Low-grade gliomas are reported to have a higher propensity to malignant transformation; therefore, some consideration for resection of these lesions if technically feasible has also been advocated for (Villani et al. 2023). Primary prevention strategies in LFS include avoidance of exposure to cancer-causing agents in the environment or dietary substances that result in molecular changes that could increase cancer risk (i.e., sun exposure, tobacco use, diagnostic and therapeutic radiation, occupational exposures to certain chemicals, and excessive alcohol use). Unfortunately, beyond general avoidance of these known carcinogens, no other risk-reduction strategies have been reported to date. Furthermore, there is a lack of specific preventive interventions for the vast majority of the LFS tumor spectrum, especially those that are more prevalent in children (Kratz et al. 2017). While precision cancer screening is essential to improving patient outcomes, a number of recent preclinical and clinical studies have suggested a complementary approach: the implementation of pharmacological cancer risk-reduction strategies, to be discussed in a later section (Comas et al. 2012; Komarova et al. 2012; Wang et al. 2017; Walcott et al. 2020).

PERSONALIZED CANCER SURVEILLANCE: MOVING BEYOND A ONE-SIZE-FITS-ALL APPROACH

Early Diagnosis and Risk Stratification from Birth

Characterizing the genetic and epigenetic underpinnings of LFS will make possible the personalization of clinical management based on individual cancer risk (Fig. 1). Traditionally, our understanding of hereditary cancer has focused on single-gene disorders (Malkin et al. 1990). However, recent evidence has suggested a high prevalence of polygenic effects through modifier events (Ballinger et al. 2016). The pleiotropism of LFS-associated cancers, insufficiently explained by the germline *TP53* variant alone, suggests this is particularly true for the LFS population. Franceschi et al. (2017) attempted to demonstrate this by performing trio sequencing of an LFS family, which revealed that the affected child and mother shared a pathogenic *TP53* variant, whereas the child and unaffected father shared a pathogenic variant in *ERCC3*. The authors hypothesized that the variant in *ERCC3*, a gene that encodes a DNA helicase involved in nucleotide excision repair, acted as a modifier in the child by further impairing DNA damage response, which could consequently explain the accelerated tumor onset in the child (Franceschi et al. 2017). To date, only a handful of modifiers have been identified in LFS, primarily associated with earlier age of tumor onset and increased tumor aggressiveness. Murine double minute 2 (*MDM2*) is a negative regulator of p53 (Fig. 2). Genetic variants altering the MDM2-TP53 autoregulatory feedback loop including the *MDM2* SNP309T > G, miR-605 A > G, and TP53 codon 72Pro > Arg are associated with accelerated tumor onset (Bougeard et al. 2015; Id Said and Malkin 2015). More broadly, genome-wide events such as telomere attrition, copy number variation, and methylation reflect underlying genomic instability in LFS patients and confer increased cancer risk (Shlien et al. 2008; Tabori et al. 2010; Samuel et al. 2016). The addition of screening protocols that incorporate these modifiers into the clinical management of LFS patients is an intuitive next step in moving toward personalized cancer surveillance programs.

Precision oncology initiatives assessing the feasibility and clinical utility of molecular features for screening and therapeutic target discovery have been increasing worldwide (Fig. 2). A summary of the major molecular profiling methods used for said precision oncology initiatives is outlined in Table 2. Several programs including TARGET (Therapeutically Applicable Research to Generate Effective Treatments, USA), PRO-FYLE (Precision Oncology for Children and Young People, Canada), and the Zero Childhood Cancer Program (Australia) have used NGS to inform clinical management for children with cancers that are rare, associated with poor outcomes, relapsed, or treatment-refractory. The Zero Childhood Cancer Program's multiplatform approach uses DNA methylation profiling in addition to NGS and found 93.7% of enrolled patients had at least one germline or somatic

aberration, 71.4% had therapeutic targets, and 5.2% necessitated a change in diagnosis (Wong et al. 2020). A growing number of studies have also performed large-scale, genome-wide association susceptibility (GWAS) studies, identifying common susceptibility loci in the germline associated with increased risk of LFS-associated cancers such as OS, acute lymphoblastic leukemia (ALL), and BC (Treviño et al. 2009; Michailidou et al. 2013; Savage et al. 2013). Of note, a study by Holmfeldt et al. (2013) identified *TP53* mutations in 91.2% of childhood cases of low-hypodiploid ALL, with these mutations also present in nontumor cells in 43.3% of those cases (Holmfeldt et al. 2013). Taken together, these findings highlight the inherited nature of these mutations and that low-hypodiploid ALL is present in LFS (Holmfeldt et al. 2013). Moreover, as alluded to in Figure 2, PRSs developed using population-level data have been shown to confer increased cancer risk, even in hereditary cancers, in the presence of a pathogenic variant in high-risk genes like *BRCA1* and *BRCA2* (Barnes et al. 2022). As a result, the construction of PRS in LFS using previously identified susceptibility loci could permit the stratification of patients into high- and low-risk strata to guide surveillance to a specific tumor site.

Traditionally, risk stratification has relied on clinicopathologic features; however, prognostic biomarkers offer a much greater depth of information and accuracy (Fig. 2). In recent years, significant progress has been made in the identification of prognostic biomarkers for risk stratification in several LFS-associated cancers including tumors of the CNS, ACC, and STS (Arnold and Barr 2017). In medulloblastoma (MB), four subgroups with distinct molecular, demographic, and clinical characteristics have been identified: WNT, sonic hedgehog (SHH), Group 3, and Group 4. The SHH-MB subtype, in particular, is strongly associated with LFS. In one study by Rausch et al. (2012), the identification of SHH-MB led clinicians to uncover five previously undiagnosed LFS cases, which permitted prospective management of these patients and their respective families. Furthermore, the identification of MB subtypes has led to the stratification of adjuvant therapy in MB clinical trials based upon molec-

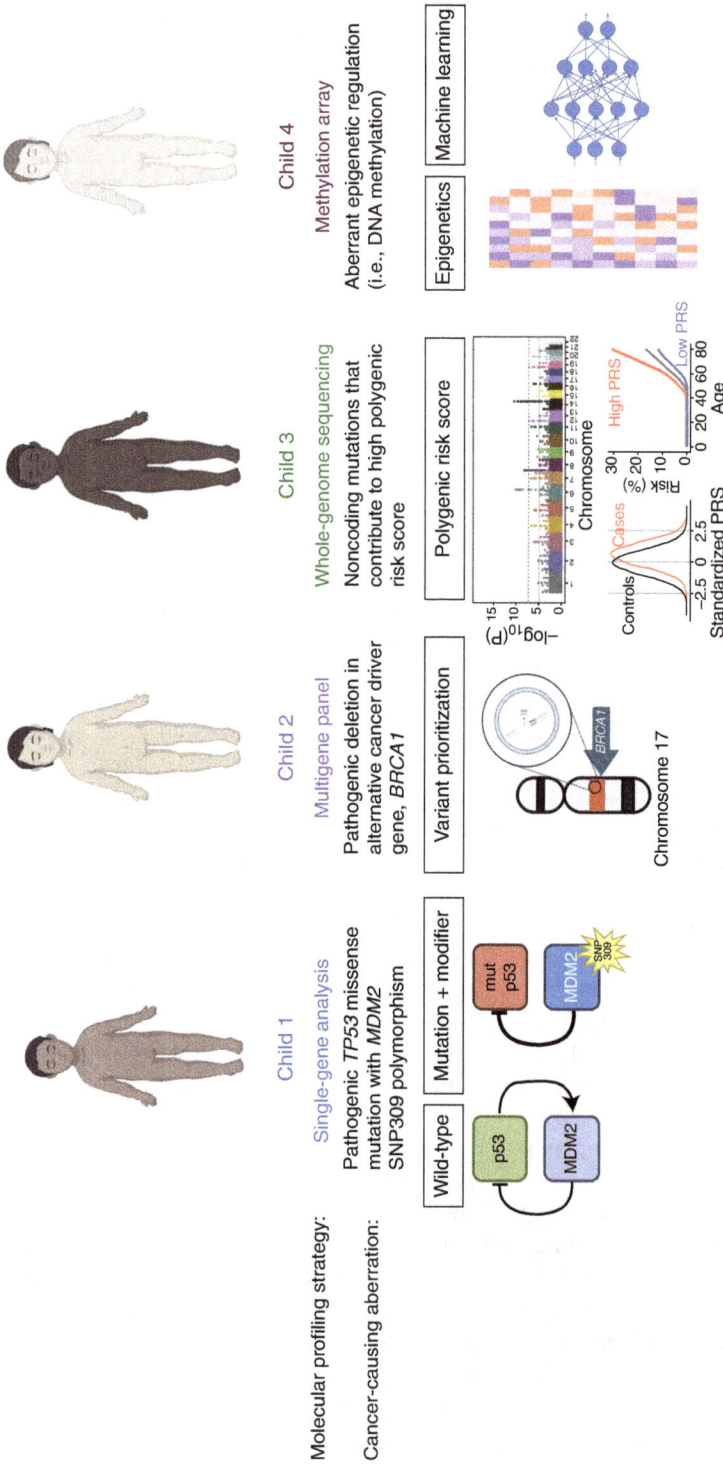

Figure 2. A multiomics, data-driven approach is required for personalized cancer risk evaluation. Four children are depicted, each harboring a different cancer-causing genetic or epigenetic aberration, highlighting the advantages of multiomics screening approaches that cover the entire genome and epigenome to facilitate personalized cancer risk screening. Child 1 harbors a pathogenic *TP53* missense mutation, as well as a single-nucleotide polymorphism (SNP) at nucleotide 309 in *MDM2* ascertained using single-gene testing. Reciprocal interactions between wild-type p53 and MDM2 are shown, as well as changes to this interaction in the context of *TP53* mutation and SNP in *MDM2*. Child 2 harbors a pathogenic deletion in an alternative cancer-causing gene, namely *BRCA1*, ascertained using multigene panel testing. Child 3 harbors a pathogenic noncoding mutation and exhibits a high polygenic risk score (PRS), depicted in the Manhattan plot and ascertained through whole-genome sequencing (WGS). Child 4 harbors pathogenic epigenetic aberrations, exhibiting a dysregulated methylation profile, ascertained through a methylation array. (Created with BioRender.com.)

Cite this article as *Cold Spring Harb Perspect Med* doi: 10.1101/cshperspect.a041584

Table 2. Molecular profiling methods with associated advantages and disadvantages

Molecular profiling method	Advantages and disadvantages
Genetic profiling	
PCR (for recurrent mutations)	*Advantages*: rapid and sensitive, with the ability to produce high product yield for sequencing, cost-effective *Disadvantages*: specificity of the generated PCR product may be altered by nonspecific primer binding
Sanger sequencing (for single-gene analysis)	*Advantages*: accurate, reliable, and easy to perform *Disadvantages*: slow, expensive, labor-intensive for large-scale projects, requires a large amount of input DNA and known target sequence
MLPA (for single-gene analysis)	*Advantages*: high-throughput and cost-effective for detecting duplications and deletions, can be multiplexed for up to 60 products per reaction, reproducible, easy to perform, and sensitive *Disadvantages*: inability to detect changes at the single-cell level, inability to detect unknown point mutations, more sensitive to contaminants compared to conventional PCR, analysis from heterogeneous samples will yield an average copy number per cell
Multigene panel	*Advantages*: parallel testing of multiple genes, cost-effective, reduces testing for individuals who meet criteria for more than one hereditary cancer syndrome *Disadvantages*: high chance of identifying an incidental finding or VUS associated with increased patient anxiety and identification of mutations in genes with limited clinical management guidelines, can be less accurate than conventional sequencing methods and thus requires confirmatory testing
WES/WGS	*Advantages*: higher throughput, lower cost, greater sensitivity, and more flexibility relative to Sanger *Disadvantages*: higher complexity, lower accuracy, and more technical challenges relative to Sanger
Epigenetic profiling	
Array/WGBS (for methylation)	*Advantages*: high-throughput and genome-wide coverage, allows for the modeling of temporal dynamics *Disadvantages*: difficulty to discern causality as methylation does not follow a Mendelian inheritance pattern, batch effects, complex and dynamic nature due to variation between tissues and cell types, and due to a variety of exposures (e.g., inherited, environmental, treatment)
ATAC-seq, DNase-seq (for open chromatin analysis)	*Advantages*: does not require prior knowledge of regulatory elements *Disadvantages*: not all chromatin regulators have an associated DNA motif, integration with ChIP-seq data is necessary to identify and differentiate protein-binding sites
ChIP-seq (for transcription factor binding)	*Advantages*: can help understand disease mechanism, integration with other genomic data can reveal regulatory landscape *Disadvantages*: high cost, requires high tissue volume, dependent on antibody quality

(PCR) Polymerase chain reaction, (MLPA) multiplex ligation-dependent probe amplification, (NGS) next-generation sequencing, (WES) whole-exome sequencing, (WGS) whole-genome sequencing, (WGBS) whole-genome bisulfite sequencing, (ATAC-seq) assay for transposase-accessible chromatin with sequencing, (DNase-seq) DNase I hypersensitive sites sequencing, (ChIP-seq) chromatin immunoprecipitation followed by sequencing, (VUS) variant of unknown significance.

ular subgroups, and in particular for SHH-MB (NCT01601184, NCT03904862, NCT02875314, NCT04023669) (Thompson et al. 2020). Similar-ly, molecular features such as *PAX-FOXO1* fusion status, "metagene" expression signatures (e.g., 5-gene metagene signature in fusion-negative tumors), and *MYOD1* variants inform rhabdomyosarcoma patient outcomes, although this has not yet been widely incorporated into clinical practice (Shern et al. 2014; Hingorani et al. 2015). In the context of ACC, Assié et al. (2014) and Zheng et al. (2016) identified prognostic subgroups according to transcriptomic, chromosome alteration, and methylome profile. Stratification based on these subgroups outperformed clinical factors such as tumor stage and proliferation index in recapitulating disease-free survival and overall survival, highlighting the prognostic value of molecular data in LFS (Assié et al. 2019).

Epigenetic data may also inform approaches for early detection and risk stratification of hereditary cancers (Fig. 1). One such potential application in LFS is the identification of constitutional epimutations in *TP53*. Constitutional epimutations are altered epigenotypes widely distributed in normal tissue that result in aberrant gene expression (Hitchins 2015). In the absence of genetic mutations, intergenerational transmission of constitutional epimutations has been observed in *MLH1* and *MSH2* in the context of another important hereditary cancer predisposition, Lynch syndrome (Dámaso et al. 2018). Similarly, multiple-case BC families devoid of pathogenic variants in known breast CSG have been shown to harbor heritable DNA methylation marks associated with elevated BC risk (kConFab et al. 2018). For LFS patients lacking genetic alterations in *TP53*, constitutional epimutations present a possible alternative cause of cancer and therefore an opportunity for early intervention.

Due to the involvement of multiple tissue sites, systemic biomarkers of risk are imperative for LFS patients. Blood is an attractive analyte as it permits noninvasive, multicancer testing—this is especially important given the diversity in ages of onset and lifetime cancer risk in LFS (Fig. 1). DNA methylation data procured from peripheral blood leukocytes (PBLs; a proxy for the germline) has demonstrated great potential as an informative biomarker of cancer risk and prognosis (Fig. 2). It has been observed that global methylation of PBL is significantly different between healthy controls and patients afflicted with a variety of cancer types (Moore et al. 2008; Choi et al. 2009; Marsit et al. 2011). In previous work, Marsit et al. (2011) analyzed the DNA methylation status of PBL to build a classifier to predict bladder cancer patients from controls, achieving an area under the receiver operator characteristic curve (auROC) of 0.76. Luo et al. (2016) identified individual gene methylation of *IRF4*, *FOXE-1*, *AOX-1*, *ADAMTS9*, and *RERG* in PBL was associated with an increased risk of colorectal carcinoma. In LFS, Samuel et al. (2016) showed that germline *TP53* variant carriers possess a distinct PBL methylation profile compared to their sporadic counterparts, suggesting applications for the development of clinical surveillance tools. Building on this work, Subasri et al. (2023) utilized WGS and DNA methylation of blood from a large ($n = 396$) LFS cohort, identifying variants in the WNT signaling pathway that were associated with decreased cancer risk. Conversely, inherited epimutations in genes including *ASXL1*, *ETV6*, and *LEF1* conferred increased cancer risk. A machine learning model developed using this methylation data predicts cancer risk with an auROC of 0.725 (Subasri et al. 2023).

Noninvasive Cancer Screening and Monitoring

Liquid biopsies have emerged as a favorable alternative to conventional tissue biopsies, permitting a noninvasive approach for the screening and monitoring of cancer (Fig. 1). In particular, ctDNA diagnostic approaches have seen increased interest for pediatric hereditary cancer screening after recently receiving FDA approval as companion diagnostic tests for several cancer types largely affecting adults (Kwapisz 2017; Center for Drug Evaluation and Research 2020). ctDNA comprises the fraction of cell-free DNA (cfDNA) released by the tumor into the blood and which can be quantified for cancer screening, monitoring treatment response, and as a prognostic biomarker (Dawson et al. 2013; Siravegna et al. 2015; Shen et al. 2018). Several programs have begun to evaluate the clinical utility of liquid biopsies on a large scale (Cohen et al. 2018; Lennon et al. 2020; Liu et al. 2020). Of note, a prospective case-control substudy from CCGA and STRIVE demonstrated the effectiveness of a targeted methylation cfDNA strategy to detect multiple cancer types, across all stages, with high specificity (Liu et al. 2020). Methylation-based cfDNA screening strategies are of great interest in the context of LFS where tumor hypermethylation of prognostic biomarkers such as miR-34A has been strongly associated with decreased overall survival (Samuel et al. 2016).

Prospective studies intended to evaluate the utility of ctDNA for cancer screening and monitoring of LFS patients are currently underway and have shown promising early results (NCT04367246, NCT04261972). For instance, ctDNA of LFS patients consistently exhibit abnormal fragmentation profiles and methylation patterns compared to non-LFS healthy controls (Wong et al. 2022). One study used ctDNA extracted from cerebrospinal fluid for early cancer detection in MB and identified the emergence of an independent primary malignancy at relapse in a patient with LFS (Escudero et al. 2020). This patient was ineligible for a biopsy of their relapsed tumor and, as such, intratumoral changes would have otherwise gone unnoticed. The most comprehensive study published to date exploring the potential of ctDNA as a viable biomarker for early cancer detection was recently published by Wong et al. (2022). This study assessed the efficacy of a multimodal liquid biopsy assay that integrates a targeted gene panel, shallow whole genome, fragmentomics, and cell-free methylated DNA immunoprecipitation sequencing in a longitudinal cohort of 89 LFS patients. This approach increased the detection rate in patients with an active cancer diagnosis over unimodal analyses. It was able to detect a cancer-associated signal in carriers before diagnosis with conventional screening (positive predictive value [PPV] = 67.6%; negative predictive value [NPV] = 96.5%). This study provides a framework for the integration of liquid biopsy into current surveillance methods for individuals with LFS. Implementation of ctDNA strategies such as that reported by Wong et al. in LFS patients will not only aid in the identification of tumors and their tissue of origin but will also help differentiate between primary and therapy-induced malignancies to implement personalized surveillance strategies. This is especially important in LFS where patients may be at elevated risk of radiation-induced secondary malignancies (Hendrickson et al. 2020).

Machine Learning for Cancer Surveillance

To diagnose LFS in individuals who harbor a de novo *TP53* variant, in the absence of a family history of cancer, it is critical to distinguish *TP53* variants as pathogenic or benign. Pathogenicity classification of *TP53* variants not only informs LFS diagnosis but permits the stratification of patients by survival outcomes. In this regard, Ben-Cohen et al. (2022) built an ML classifier to predict the impact of missense mutations in *TP53* and achieved an accuracy of 93.1% and 95.9% on two independent cohorts of LFS patients. Pedigree structure has also been leveraged to build several models to predict cancer risk in LFS patients (Shin et al. 2019, 2020a,b). One such model performed penetrance estimates for multiple primary cancers in LFS (Shin et al. 2019). Another generated penetrance estimates for both first and second primary cancer diagnoses in LFS patients from a pediatric sarcoma cohort (Shin et al. 2020a). These models demonstrate the potential for the prediction of cancer risk in LFS patients; however, as we move beyond family history and toward a genetic definition of LFS, they would be significantly enhanced with the incorporation of "omic" data. To this end, Capper et al. (2018) leveraged DNA methylation data and successfully applied a CNS tumor classification model across a broad range of CNS tumor types and age groups in a routine diagnostic setting. Brain tumors are among the most common cancer types occurring in LFS; as such, this tool and others like it will prove valuable for clinical decision-making.

The applications of ML in the classification of images obtained in cancer surveillance show great promise in improving the performance of the human user when used in tandem with standard diagnostic protocols (Ehteshami Bejnordi et al. 2017; Nam et al. 2019; Yamashita et al. 2021). Several of these artificial intelligence (AI) systems have been successfully applied in staging adult cancer patients from whole-body MRIs; however, this may prove more challenging to implement for LFS patients due to varying bone signals during growth, the limited compliance of young children during imaging, and the rarity of positive cases essential for developing models. Nonetheless, deep learning models using whole-body MRI images of LFS patients have demonstrated that state-of-the-art generative adversarial networks (GANs) are able to generate pediatric whole-body MRI images that allow for sufficient training data (Chang et al. 2020). The generated images were then utilized to build an ML model tasked with anomaly detection of cancer modules. Furthermore, improvements to the prediction performance of such models will ultimately enable automated cancer detection for LFS patients.

In addition to building robust clinical ML models for LFS and other hereditary cancer syndromes, it is important to establish a level of explainability to provide insights regarding the rationale of predictions; this will enable a greater mechanistic understanding of LFS phenotype heterogeneity and ultimately serve to build clinician confidence in AI models (Saria et al. 2018; Kelly et al. 2019). Establishing causality is also imperative as it increases our mechanistic understanding of LFS tumor development and may eventually assist in guiding the design of preventive and therapeutic strategies (Saria et al. 2018; Holzinger et al. 2019).

PHARMACOLOGICAL CANCER RISK REDUCTION: A SUPPLEMENT TO CONVENTIONAL SCREENING

Improvements to our understanding of the correlates of malignancy in LFS have suggested a potential supplemental approach to traditional cancer screening: pharmacology-based cancer risk reduction. In this section, we will discuss a few important examples of drug repurposing for cancer preventive goals. Though research remains ongoing, implementation of such protocols will likely become a major clinical focus, particularly in the context of highly penetrant syndromes such as LFS, which confer lifelong cancer risk (Fig. 1).

Targeting Metabolic Dysregulation

Despite the many pathways involved in cancer, cancer prevention is most closely associated with anabolic metabolism. Specifically, reductions in anabolic metabolism by various means have been linked to diminished cancer rates, particularly in the preclinical setting.

Systemic metabolic changes have been noted in patients with CPS, in particular LFS. One striking study that compared LFS patients harboring germline *TP53* variants to noncarrier family members and healthy volunteers found that carriers exhibited elevated oxidative phosphorylation in skeletal muscle tissue (Wang et al. 2017). Two important drugs, both converging on metabolism, have recently been investigated for cancer risk reduction in LFS: metformin and rapamycin. Administration of these drugs in murine models, as well as small clinical pilot studies of LFS patients, has suggested potential preventive efficacy (Comas et al. 2012; Komarova et al. 2012; Livi et al. 2013; Hasty et al. 2014; Popovich et al. 2014; Wang et al. 2017; Walcott et al. 2020).

Repurposing Metformin for Cancer Risk Reduction

Metformin is an oral biguanide used as a first-line therapy in the treatment of hyperglycemia for patients with type II diabetes. A number of retrospective, observational studies have suggested a separate, unexpected application for this agent. Specifically, patients with type II diabetes being treated with metformin presented with lower cancer incidence and mortality compared to healthy individuals and patients not receiving the drug (Evans et al. 2005; Libby et al. 2009; Landman et al. 2010). Subsequent prospective investigation of genetically engineered mouse models (GEMMs) of cancer predisposition further underlined the role of metformin in reducing or eliminating cancer risk (Anisimov et al. 2005; Wang et al. 2017). Of particular note is a study by Wang et al. (2017), which evaluated the efficacy of this drug in delaying cancer onset in a murine model of LFS. The authors found that the administration of metformin shortly after weaning increased the median cancer-free survival of mice homozygous for the R172H mutation in *Trp53* (homolog of the

R175H human *TP53* variant) by 27% (Wang et al. 2017).

The well-established, ideal safety profile and low cost of metformin make it an especially attractive candidate for repurposing as a cancer chemopreventive. Metformin also has a unique potential for adoption in the pediatric population over other agents of interest given its long history of use in children.

The preclinical evaluation of metformin for use in LFS has recently been complemented by the initiation of a series of clinical trials assessing the safety and tolerability of metformin in cohorts of LFS patients (NCT01981525). In a small pilot study, healthy LFS patients treated with metformin exhibited decreases in mitochondrial activity and proliferative signaling with no significant adverse effects (Wang et al. 2017). A subsequent phase I study by Walcott et al. (2020) also showed no adverse effects associated with its potential use as a cancer chemopreventive. Although these clinical studies did not assess the efficacy of metformin in decreasing cancer burden, efforts are currently underway to expand upon this work. A longitudinal, multicenter, randomized phase II clinical trial (Metformin in LFS—MILI) has been initiated out of both the National Institutes of Health and Oxford University for adults with LFS; consideration for expansion of a pediatric cohort is being actively explored.

Important remaining questions include the ideal dosing strategy and the specific age at which administration will be safe and efficacious—all likely to be central considerations in the aforementioned multicenter trial. Moreover, further investigation into the biology underlying the chemopreventive effects of metformin and the unique metabolic phenotype of LFS patients is warranted and may yield further candidate drugs for cancer risk reduction.

Repurposing Rapamycin for Cancer Risk Reduction

Originally applied as an antifungal and as an immunosuppressant for the prevention of transplant rejection, rapamycin, and its analogs, all direct inhibitors of mechanistic target of rapamycin complex 1 (mTORC1), have long been investi-gated for potential anticancer properties. Although rapamycin is generally ineffective in the treatment of established malignancies, it has been shown to reduce or, in certain cases, ablate cancer risk in GEMMs of cancer. Komorova et al. (2012) examined the cancer preventive efficacy of rapamycin in heterozygous p53 null mice, revealing delayed tumorigenesis and a 28% mean extension in life span when rapamycin was started early in life (i.e., before 5 months of age). A second study of heterozygous p53 null mice by Christy et al. (2015) identified a modest, but significant increase in life span associated with rapamycin treatment. Comas et al. (2012) found that a nanoformulated preparation of rapamycin, called Rapatar, delayed tumorigenesis and extended mean life span of homozygous p53 null mice by 30%.

While rapamycin has shown promise in murine models, considerable obstacles may preclude its use in the cancer clinic as a preventive agent, particularly in the pediatric context. As already mentioned, rapamycin exerts substantial immunosuppressive activity—chronic use of this drug is likely to bring about significant adverse effects. Rapamycin also has the potential to induce growth inhibition, which especially complicates its use in children. It may be possible to combat some of these limitations by using short and/or intermittent dosing strategies, which have shown efficacy in the preclinical context. Additionally, it could potentially be used in high-risk patients. Specifically, with an improved understanding of genotype–phenotype correlations in LFS, cancer preventive pharmacology may be better personalized for affected patients. Exploring the role of p53-targeted agents such as APR-246 and p53 conformation activating peptides (CAPs) (agents that restore p53's wild-type conformation/function) may also be worthwhile (Bykov et al. 2005; Tal et al. 2016). These are currently being investigated for therapeutic purposes, but they may also have a role in the context of cancer predisposition.

CONCLUDING REMARKS

The management of CPS, such as LFS, focuses primarily on prevention and early tumor detection. This emphasis has substantially improved patient outcomes and continues to be an important area of research. In this review, we have summarized critical technological and basic science developments that have the potential to further reshape the clinical management of hereditary cancer. In the future, a suspected LFS diagnosis will be confirmed immediately with genetic testing of both the *TP53* genotype and modifiers of risk. Depending on the suspected severity of their case as indicated by modeling of their specific disease course, this patient will be enrolled in a tailored surveillance protocol emphasizing age- and tumor-type appropriate screening interventions based on genomic features and precise predictive multifactorial modeling (Fig. 1). Such an approach would lead to reduced burden of current intensive surveillance protocols on the patient and economic burdens on health care systems while at the same time reducing cancer risk. For a patient at higher risk where more invasive measures may have once been warranted, they may instead be offered primary pharmacologic prevention. This strategy would leverage previous successes in risk-informed care, while acknowledging the need to incorporate individualized genetic information in the management of a genetic disease, reframing the narrative of care for patients at extremely high risk of cancer.

ACKNOWLEDGMENTS

This work is supported in part by a New Frontiers Program Project Grant (#1081) from the Terry Fox Research Institute with funds from the Terry Fox Foundation, and a Foundation Scheme Grant (#143234) from the Canadian Institutes for Health Research (D.M.). D.M. holds the CIBC Children's Foundation Chair in Child Health Research at The Hospital for Sick Children.

REFERENCES

Akhavanfard S, Padmanabhan R, Yehia L, Cheng F, Eng C. 2020. Comprehensive germline genomic profiles of children, adolescents and young adults with solid tumors. *Nat Commun* **11:** 2206. doi:10.1038/s41467-020-16067-1

Amadou A, Waddington Achatz MI, Hainaut P. 2018. Revising tumor patterns and penetrance in germline TP53 mutations carriers: temporal phases of Li–Fraumeni syndrome. *Curr Opin Oncol* **30:** 23–29.

Anisimov VN, Berstein LM, Egormin PA, Piskunova TS, Popovich IG, Zabezhinski MA, Kovalenko IG, Poroshina TE, Semenchenko AV, Provinciali M, et al. 2005. Effect of metformin on life span and on the development of spontaneous mammary tumors in HER-2/neu transgenic mice. *Exp Gerontol* **40:** 685–693. doi:10.1016/j.exger.2005.07.007

Arnold MA, Barr FG. 2017. Molecular diagnostics in the management of rhabdomyosarcoma. *Expert Rev Mol Diagn* **17:** 189–194. doi:10.1080/14737159.2017.1275965

Assié G, Letouzé E, Fassnacht M, Jouinot A, Luscap W, Barreau O, Omeiri H, Rodriguez S, Perlemoine K, René-Corail F, et al. 2014. Integrated genomic characterization of adrenocortical carcinoma. *Nat Genet* **46:** 607–612. doi:10.1038/ng.2953

Assié G, Jouinot A, Fassnacht M, Libé R, Garinet S, Jacob L, Hamzaoui N, Neou M, Sakat J, de La Villéon B, et al. 2019. Value of molecular classification for prognostic assessment of adrenocortical carcinoma. *JAMA Oncol* **5:** 1440.

Ballinger ML, Goode DL, Ray-Coquard I, James PA, Mitchell G, Niedermayr E, Puri A, Schiffman JD, Dite GS, Cipponi A, et al. 2016. Monogenic and polygenic determinants of sarcoma risk: an international genetic study. *Lancet Oncol* **17:** 1261–1271. doi:10.1016/S1470-2045(16)30147-4

Ballinger ML, Best A, Mai PL, Khincha PP, Loud JT, Peters JA, Achatz MI, Chojniak R, da Costa A B, Santiago KM, et al. 2017. Baseline surveillance in Li-Fraumeni syndrome using whole-body magnetic resonance imaging: a meta-analysis. *JAMA Oncol* **3:** 1634. doi:10.1001/jamaoncol.2017.1968

Barnes DR, Silvestri V, Leslie G, McGuffog L, Dennis J, Yang X, Adlard J, Agnarsson BA, Ahmed M, Aittomäki K, et al. 2022. Breast and prostate cancer risks for male *BRCA1* and *BRCA2* pathogenic variant carriers using polygenic risk scores. *J Natl Cancer Inst* **114:** 109–122. doi:10.1093/jnci/djab147

Ben-Cohen G, Doffe F, Devir M, Leroy B, Soussi T, Rosenberg S. 2022. TP53_PROF: a machine learning model to predict impact of missense mutations in *TP53*. *Brief Bioinform* **23:** bbab524. 10.1093/bib/bbab524

Bougeard G, Renaux-Petel M, Flaman J-M, Charbonnier C, Fermey P, Belotti M, Gauthier-Villars M, Stoppa-Lyonnet D, Consolino E, Brugières L, et al. 2015. Revisiting Li-Fraumeni syndrome from *TP53* mutation carriers. *J Clin Oncol* **33:** 2345–2352. doi:10.1200/JCO.2014.59.5728

Bykov VJN, Zache N, Stridh H, Westman J, Bergman J, Selivanova G, Wiman KG. 2005. PRIMA-1MET synergizes with cisplatin to induce tumor cell apoptosis. *Oncogene* **24:** 3484–3491. doi:10.1038/sj.onc.1208419

Campbell BB, Light N, Fabrizio D, Zatzman M, Fuligni F, de Borja R, Davidson S, Edwards M, Elvin JA, Hodel KP, et al. 2017. Comprehensive analysis of hypermutation in human cancer. *Cell* **171:** 1042–1056.e10. doi:10.1016/j.cell.2017.09.048

Capper D, Jones DTW, Sill M, Hovestadt V, Schrimpf D, Sturm D, Koelsche C, Sahm F, Chavez L, Reuss DE, et al. 2018. DNA methylation-based classification of central nervous system tumours. *Nature* **555:** 469–474. doi:10.1038/nature26000

Center for Drug Evaluation Research. 2020. FDA approves liquid biopsy NGS companion diagnostic test for multiple cancers and biomarkers. *US Food and Drug Administration*. https://www.fda.gov/drugs/resources-information-approved-drugs/fda-approves-liquid-biopsy-ngs-companion-diagnostic-test-multiple-cancers-and-biomarkers (Accessed July 12, 2022).

Chang A, Suriyakumar VM, Moturu A, Tewattanarat N, Doria A, Goldenberg A. 2020. Using generative models for pediatric wbMRI. http://arxiv.org/abs/2006.00727

Choi JY, James SR, Link PA, McCann SE, Hong CC, Davis W, Nesline MK, Ambrosone CB, Karpf AR. 2009. Association between global DNA hypomethylation in leukocytes and risk of breast cancer. *Carcinogenesis* 30: 1889–1897. doi:10.1093/carcin/bgp143

Chompret A, Abel A, Stoppa-Lyonnet D, Brugieres L, Pages S, Feunteun J, Bonaiti-Pellie C. 2001. Sensitivity and predictive value of criteria for p53germline mutation screening. *J Med Genet* 38: 43–47. doi:10.1136/jmg.38.1.43

Christy B, Demaria M, Campisi J, Huang J, Jones D, Dodds SG, Williams C, Hubbard G, Livi CB, Gao X, et al. 2015. P53 and rapamycin are additive. *Oncotarget* 6: 15802–15813. doi:10.18632/oncotarget.4602

Cohen JD, Li L, Wang Y, Thoburn C, Afsari B, Danilova L, Douville C, Javed AA, Wong F, Mattox A, et al. 2018. Detection and localization of surgically resectable cancers with a multi-analyte blood test. *Science* 359: 926–930. doi:10.1126/science.aar3247

Comas M, Toshkov I, Kuropatwinski KK, Chernova OB, Polinsky A, Blagosklonny MV, Gudkov AV, Antoch MP. 2012. New nanoformulation of rapamycin Rapatar extends lifespan in homozygous $p53^{-/-}$ mice by delaying carcinogenesis. *Aging* 4: 715–722. doi:10.18632/aging.100496

Daly MB, Pilarski R, Berry M, Buys SS, Farmer M, Friedman S, Garber JE, Kauff ND, Khan S, Klein C, et al. 2017. NCCN guidelines insights: genetic/familial high-risk assessment: breast and ovarian, version 2.2017. *J Natl Compr Canc Netw* 15: 9–20. doi:10.6004/jnccn.2017.0003

Dámaso E, Castillejo A, del M Arias M, Canet-Hermida J, Navarro M, del Valle J, Campos O, Fernández A, Marín F, Turchetti D, et al. 2018. Primary constitutional MLH1 epimutations: a focal epigenetic event. *Br J Cancer* 119: 978–987. doi:10.1038/s41416-018-0019-8

Dawson SJ, Tsui DWY, Murtaza M, Biggs H, Rueda OM, Chin SF, Dunning MJ, Gale D, Forshew T, Mahler-Araujo B, et al. 2013. Analysis of circulating tumor DNA to monitor metastatic breast cancer. *N Engl J Med* 368: 1199–1209. doi:10.1056/NEJMoa1213261

Ehteshami Bejnordi B, Veta M, Johannes van Diest P, van Ginneken B, Karssemeijer N, Litjens G, van der Laak JAWM, the CAMELYON16 Consortium, Hermsen M, Manson QF, et al. 2017. Diagnostic assessment of deep learning algorithms for detection of lymph node metastases in women with breast cancer. *J Am Med Assoc* 318: 2199–2210.

Escudero L, Llort A, Arias A, Diaz-Navarro A, Martínez-Ricarte F, Rubio-Perez C, Mayor R, Caratù G, Martínez-Sáez E, Vázquez-Méndez É, et al. 2020. Circulating tumour DNA from the cerebrospinal fluid allows the characterisation and monitoring of medulloblastoma. *Nat Commun* 11: 5376. 10.1038/s41467-020-19175-0

Evans JMM, Donnelly LA, Emslie-Smith AM, Alessi DR, Morris AD. 2005. Metformin and reduced risk of cancer in diabetic patients. *Br Med J* 330: 1304–1305. doi:10.1136/bmj.38415.708634.F7

Fineberg R, Zahedi S, Eguchi M, Hart M, Cockburn M, Green AL. 2020. Population-based analysis of demographic and socioeconomic disparities in pediatric CNS cancer survival in the United States. *Sci Rep* 10: 4588. doi:10.1038/s41598-020-61237-2

Franceschi S, Spugnesi L, Aretini P, Lessi F, Scarpitta R, Galli A, Congregati C, Caligo MA, Mazzanti CM. 2017. Whole-exome analysis of a Li–Fraumeni family trio with a novel TP53 PRD mutation and anticipation profile. *Carcinogenesis* 38: 938–943. doi:10.1093/carcin/bgx069

Gonzalez KD, Noltner KA, Buzin CH, Gu D, Wen-Fong CY, Nguyen VQ, Han JH, Lowstuter K, Longmate J, Sommer SS, et al. 2009. Beyond Li Fraumeni syndrome: clinical characteristics of families with *p53* germline mutations. *J Clin Oncol* 27: 1250–1256. doi:10.1200/JCO.2008.16.6959

Hasty P, Livi CB, Dodds SG, Jones D, Strong R, Javors M, Fischer KE, Sloane L, Murthy K, Hubbard G, et al. 2014. Erapa restores a normal life span in a FAP mouse model. *Cancer Prev Res* 7: 169–178. doi:10.1158/1940-6207.CAPR-13-0299

Hendrickson PG, Luo Y, Kohlmann W, Schiffman J, Maese L, Bishop AJ, Lloyd S, Kokeny KE, Hitchcock YJ, Poppe MM, et al. 2020. Radiation therapy and secondary malignancy in Li-Fraumeni syndrome: a hereditary cancer registry study. *Cancer Med* 9: 7954–7963. doi:10.1002/cam4.3427

Hingorani P, Missiaglia E, Shipley J, Anderson JR, Triche TJ, Delorenzi M, Gastier-Foster J, Wing M, Hawkins DS, Skapek SX. 2015. Clinical application of prognostic gene expression signature in fusion gene–negative rhabdomyosarcoma: a report from the Children's Oncology Group. *Clin Cancer Res* 21: 4733–4739. doi:10.1158/1078-0432.CCR-14-3326

Hitchins MP. 2015. Constitutional epimutation as a mechanism for cancer causality and heritability? *Nat Rev Cancer* 15: 625–634. doi:10.1038/nrc4001

Ho D, Quake SR, McCabe ERB, Chng WJ, Chow EK, Ding X, Gelb BD, Ginsburg GS, Hassenstab J, Ho CM, et al. 2020. Enabling technologies for personalized and precision medicine. *Trends Biotechnol* 38: 497–518. doi:10.1016/j.tibtech.2019.12.021

Holmfeldt L, Wei L, Diaz-Flores E, Walsh M, Zhang J, Ding L, Payne-Turner D, Churchman M, Andersson A, Chen SC, et al. 2013. The genomic landscape of hypodiploid acute lymphoblastic leukemia. *Nat Genet* 45: 242–252. doi:10.1038/ng.2532

Holzinger A, Langs G, Denk H, Zatloukal K, Müller H. 2019. Causability and explainability of artificial intelligence in medicine. *Wiley Interdiscip Rev Data Min Knowl Discov* 9: e1312. doi:10.1002/widm.1312

ICGC PedBrain-Seq Project; ICGC MMML-Seq Project; Gröbner SN, Worst BC, Weischenfeldt J, Buchhalter I, Kleinheinz K, Rudneva VA, Johann PD, Balasubramanian GP, et al. 2018. The landscape of genomic alterations across childhood cancers. *Nature* 555: 321–327. doi:10.1038/nature25480

Id Said B, Malkin D. 2015. A functional variant in miR-605 modifies the age of onset in Li-Fraumeni syndrome. *Can-*

cer Genet **208**: 47–51. doi:10.1016/j.cancergen.2014.12
.003

kConFab; Joo JE, Dowty JG, Milne RL, Wong EM, Dugué P-
A, English D, Hopper JL, Goldgar DE, Giles GG, et al.
2018. Heritable DNA methylation marks associated with
susceptibility to breast cancer. *Nat Commun* **9**: 867.

Kelly CJ, Karthikesalingam A, Suleyman M, Corrado G, King
D. 2019. Key challenges for delivering clinical impact with
artificial intelligence. *BMC Med* **17**: 195. doi:10.1186/
s12916-019-1426-2

Knapke S, Zelley K, Nichols KE, Kohlmann W, Schiffman JD.
2012. Identification, management, and evaluation of children with cancer-predisposition syndromes. *Am Soc Clin
Oncol Educ Book* **32**: 576–584. doi:10.14694/EdBook_AM
.2012.32.8

Komarova EA, Antoch MP, Novototskaya LR, Chernova OB,
Paszkiewicz G, Leontieva OV, Blagosklonny MV, Gudkov
AV. 2012. Rapamycin extends lifespan and delays tumorigenesis in heterozygous p53$^{+/-}$ mice. *Aging* **4**: 709–714.
doi:10.18632/aging.100498

Kratz CP, Achatz MI, Brugières L, Frebourg T, Garber JE,
Greer MLC, Hansford JR, Janeway KA, Kohlmann WK,
McGee R, et al. 2017. Cancer screening recommendations
for individuals with Li-Fraumeni syndrome. *Clin Cancer
Res* **23**: e38–e45. doi:10.1158/1078-0432.CCR-17-0408

Kwapisz D. 2017. The first liquid biopsy test approved. Is it a
new era of mutation testing for non-small cell lung cancer?
Ann Transl Med **5**: 46. doi:10.21037/atm.2017.01.32

Landman GWD, Kleefstra N, van Hateren KJJ, Groenier KH,
Gans ROB, Bilo HJG. 2010. Metformin associated with
lower cancer mortality in type 2 diabetes: ZODIAC-16.
Diabetes Care **33**: 322–326.

Lennon AM, Buchanan AH, Kinde I, Warren A, Honushef-
sky A, Cohain AT, Ledbetter DH, Sanfilippo F, Sheridan K,
Rosica D, et al. 2020. Feasibility of blood testing combined
with PET-CT to screen for cancer and guide intervention.
Science **369**: eabb9601. doi:10.1126/science.abb9601

Li FP, Fraumeni JF, Mulvihill JJ, Blattner WA, Dreyfus MG,
Tucker MA, Miller RW. A cancer family syndrome in
twenty-four kindreds. *Cancer Res* **48**: 5368–5362.

Libby G, Donnelly LA, Donnan PT, Alessi DR, Morris AD,
Evans JMM. 2009. New users of metformin are at low risk
of incident cancer: a cohort study among people with type
2 diabetes. *Diabetes Care* **32**: 1620–1625.

Linet MS, Ries LAG, Smith MA, Tarone RE, Devesa SS. 1999.
Cancer surveillance series: recent trends in childhood can-
cer incidence and mortality in the United States. *J Natl
Cancer Inst* **91**: 1051–1058. doi:10.1093/jnci/91.12.1051

Liu MC, Oxnard GR, Klein EA, Swanton C, Seiden MV,
Cummings SR, Absalan F, Alexander G, Allen B, Amini
H, et al. 2020. Sensitive and specific multi-cancer detection
and localization using methylation signatures in cell-free
DNA. *Ann Oncol* **31**: 745–759.

Livi CB, Hardman RL, Christy BA, Dodds SG, Jones D, Williams C, Strong R, Bokov A, Javors MA, Ikeno Y, et al. 2013.
Rapamycin extends life span of Rb1$^{+/-}$ mice by inhibiting
neuroendocrine tumors. *Aging* **5**: 100–110. doi:10.18632/
aging.100533

Luo X, Huang R, Sun H, Liu Y, Bi H, Li J, Yu H, Sun J, Lin S, Cui
B, et al. 2016. Methylation of a panel of genes in peripheral
blood leukocytes is associated with colorectal cancer. *Sci
Rep* **6**: 29922. doi:10.1038/srep29922

Mai PL, Best AF, Peters JA, DeCastro RM, Khincha PP, Loud
JT, Bremer RC, Rosenberg PS, Savage SA. 2016. Risks of
first and subsequent cancers among *TP53* mutation carriers in the national cancer institute Li-Fraumeni syndrome
cohort: cancer risk in *TP53* mutation carriers. *Cancer* **122**:
3673–3681.

Malkin D, Li F, Strong L, Fraumeni J, Nelson C, Kim D, Kassel
J, Gryka M, Bischoff F, Tainsky M, et al. 1990. Germ line
p53 mutations in a familial syndrome of breast cancer,
sarcomas, and other neoplasms. *Science* **250**: 1233–1238.
doi:10.1126/science.1978757

Marsit CJ, Koestler DC, Christensen BC, Karagas MR,
Houseman EA, Kelsey KT. 2011. DNA methylation array
analysis identifies profiles of blood-derived DNA methylation associated with bladder cancer. *J Clin Oncol* **29**:
1133–1139. doi:10.1200/JCO.2010.31.3577

McBride KA, Ballinger ML, Killick E, Kirk J, Tattersall MHN,
Eeles RA, Thomas DM, Mitchell G. 2014. Li-Fraumeni
syndrome: cancer risk assessment and clinical management. *Nat Rev Clin Oncol* **11**: 260–271. doi:10.1038/nrcli
nonc.2014.41

McGee RB, Nichols KE. 2016. Introduction to cancer genetic
susceptibility syndromes. *Hematology* **2016**: 293–301.
doi:10.1182/asheducation-2016.1.293

Michailidou K, Hall P, Gonzalez-Neira A, Ghoussaini M,
Dennis J, Milne RL, Schmidt MK, Chang-Claude J, Boje-
sen SE, Bolla MK, et al. 2013. Large-scale genotyping identifies 41 new loci associated with breast cancer risk. *Nat
Genet* **45**: 353–361. doi:10.1038/ng.2563

Mody RJ, Wu Y-M, Lonigro RJ, Cao X, Roychowdhury S, Vats
P, Frank KM, Prensner JR, Asangani I, Palanisamy N, et al.
2015. Integrative clinical sequencing in the management
of refractory or relapsed cancer in youth. *J Am Med Assoc*
314: 913. doi:10.1001/jama.2015.10080

Moore LE, Pfeiffer RM, Poscablo C, Real FX, Kogevinas M,
Silverman D, García-Closas R, Chanock S, Tardón A, Serra
C, et al. 2008. Genomic DNA hypomethylation as a bio-
marker for bladder cancer susceptibility in the Spanish
Bladder Cancer Study: a case-control study. *Lancet Oncol*
9: 359–366. doi:10.1016/S1470-2045(08)70038-X

Nagy R, Sweet K, Eng C. 2004. Highly penetrant hereditary
cancer syndromes. *Oncogene* **23**: 6445–6470. doi:10.1038/
sj.onc.1207714

Nam JG, Park S, Hwang EJ, Lee JH, Jin KN, Lim KY, Vu TH,
Sohn JH, Hwang S, Goo JM, et al. 2019. Development and
validation of deep learning-based automatic detection al-
gorithm for malignant pulmonary nodules on chest radio-
graphs. *Radiology* **290**: 218–228. doi:10.1148/radiol.20
18180237

Ognjanovic S, Olivier M, Bergemann TL, Hainaut P. 2012.
Sarcomas in TP53 germline mutation carriers: a review of
the IARC TP53 database. *Cancer* **118**: 1387–1396.

Olivier M, Goldgar DE, Sodha N, Ohgaki H, Kleihues P,
Hainaut P, Eeles RA. 2003. Li-Fraumeni and related syndromes: correlation between tumor type, family structure,
and TP53 genotype. *Cancer Res* **63**: 6643–6650.

Popovich IG, Anisimov VN, Zabezhinski MA, Semenchenko
AV, Tyndyk ML, Yurova MN, Blagosklonny MV. 2014.
Lifespan extension and cancer prevention in HER-2/neu
transgenic mice treated with low intermittent doses of
rapamycin. *Cancer Biol Ther* **15**: 586–592.

Cite this article as *Cold Spring Harb Perspect Med* doi: 10.1101/cshperspect.a041584

Rausch T, Jones DTW, Zapatka M, Stütz AM, Zichner T, Weischenfeldt J, Jäger N, Remke M, Shih D, Northcott PA, et al. 2012. Genome sequencing of pediatric medulloblastoma links catastrophic DNA rearrangements with TP53 mutations. *Cell* **148:** 59–71.

Renaux-Petel M, Charbonnier F, Théry J-C, Fermey P, Lienard G, Bou J, Coutant S, Vezain M, Kasper E, Fourneaux S, et al. 2018. Contribution of de novo and mosaic *TP53* mutations to Li-Fraumeni syndrome. *J Med Genet* **55:** 173–180. doi:10.1136/jmedgenet-2017-104976

Samuel N, Villani A, Fernandez CV, Malkin D. 2014. Management of familial cancer: sequencing, surveillance and society. *Nat Rev Clin Oncol* **11:** 723–731. doi:10.1038/nrclinonc.2014.169

Samuel N, Wilson G, Lemire M, Id Said B, Lou Y, Li W, Merino D, Novokmet A, Tran J, Nichols KE, et al. 2016. Genome-wide DNA methylation analysis reveals epigenetic dysregulation of microRNA-34A in *TP53*-associated cancer susceptibility. *J Clin Oncol* **34:** 3697–3704. doi:10.1200/JCO.2016.67.6940

Saria S, Butte A, Sheikh A. 2018. Better medicine through machine learning: what's real, and what's artificial? *PLoS Med* **15:** e1002721. doi:10.1371/journal.pmed.1002721

Savage SA, Mirabello L, Wang Z, Gastier-Foster JM, Gorlick R, Khanna C, Flanagan AM, Tirabosco R, Andrulis IL, Wunder JS, et al. 2013. Genome-wide association study identifies two susceptibility loci for osteosarcoma. *Nat Genet* **45:** 799–803. doi:10.1038/ng.2645

Shen SY, Singhania R, Fehringer G, Chakravarthy A, Roehrl MHA, Chadwick D, Zuzarte PC, Borgida A, Wang TT, Li T, et al. 2018. Sensitive tumour detection and classification using plasma cell-free DNA methylomes. *Nature* **563:** 579–583. doi:10.1038/s41586-018-0703-0

Shern JF, Chen L, Chmielecki J, Wei JS, Patidar R, Rosenberg M, Ambrogio L, Auclair D, Wang J, Song YK, et al. 2014. Comprehensive genomic analysis of rhabdomyosarcoma reveals a landscape of alterations affecting a common genetic axis in fusion-positive and fusion-negative tumors. *Cancer Discov* **4:** 216–231. doi:10.1158/2159-8290.CD-13-0639

Shin SJ, Yuan Y, Strong LC, Bojadzieva J, Wang W. 2019. Bayesian semiparametric estimation of cancer-specific age-at-onset penetrance with application to Li-Fraumeni syndrome. *J Am Stat Assoc* **114:** 541–552. doi:10.1080/01621459.2018.1482749

Shin SJ, Dodd-Eaton EB, Gao F, Bojadzieva J, Chen J, Kong X, Amos CI, Ning J, Strong LC, Wang W. 2020a. Penetrance estimates over time to first and second primary cancer diagnosis in families with Li-Fraumeni syndrome: a single institution perspective. *Cancer Res* **80:** 347–353.

Shin SJ, Dodd-Eaton EB, Peng G, Bojadzieva J, Chen J, Amos CI, Frone MN, Khincha PP, Mai PL, Savage SA, et al. 2020b. Penetrance of different cancer types in families with Li-Fraumeni syndrome: a validation study using multicenter cohorts. *Cancer Res* **80:** 354–360. doi:10.1158/0008-5472.CAN-19-0728

Shlien A, Tabori U, Marshall CR, Pienkowska M, Feuk L, Novokmet A, Nanda S, Druker H, Scherer SW, Malkin D. 2008. Excessive genomic DNA copy number variation in the Li-Fraumeni cancer predisposition syndrome. *Proc Natl Acad Sci* **105:** 11264–11269. doi:10.1073/pnas.0802970105

Siravegna G, Mussolin B, Buscarino M, Corti G, Cassingena A, Crisafulli G, Ponzetti A, Cremolini C, Amatu A, Lauricella C, et al. 2015. Clonal evolution and resistance to EGFR blockade in the blood of colorectal cancer patients. *Nat Med* **21:** 795–801.

Sorrell AD, Espenschied CR, Culver JO, Weitzel JN. 2013. Tumor protein p53 (TP53) testing and Li-Fraumeni syndrome: current status of clinical applications and future directions. *Mol Diagn Ther* **17:** 31–47. doi:10.1007/s40291-013-0020-0

Srivastava S, Zou Z, Pirollo K, Blattner W, Chang EH. 1990. Germ-line transmission of a mutated p53 gene in a cancer-prone family with Li–Fraumeni syndrome. *Nature* **348:** 747–749. doi:10.1038/348747a0

Subasri V, Light N, Kanwar N, Brzezinski J, Luo P, Hansford JR, Cairney E, Portwine C, Elser C, Finlay JL, et al. 2023. Multiple germline events contribute to cancer development in patients with Li-Fraumeni syndrome. *Cancer Res Commun* **3:** 738–754. doi:10.1158/2767-9764.CRC-22-0402

Tabori U, Shlien A, Baskin B, Levitt S, Ray P, Alon N, Hawkins C, Bouffet E, Pienkowska M, Lafay-Cousin L, et al. 2010. TP53 alterations determine clinical subgroups and survival of patients with choroid plexus tumors. *J Clin Oncol* **28:** 1995–2001. doi:10.1200/JCO.2009.26.8169

Tal P, Eizenberger S, Cohen E, Goldfinger N, Pietrokovski S, Oren M, Rotter V. 2016. Cancer therapeutic approach based on conformational stabilization of mutant p53 protein by small peptides. *Oncotarget* **7:** 11817–11837.

The European Reference Network GENTURIS; Frebourg T, Bajalica Lagercrantz S, Oliveira C, Magenheim R, Evans DG. 2020. Guidelines for the Li–Fraumeni and heritable TP53-related cancer syndromes. *Eur J Hum Genet* **28:** 1379–1386. doi:10.1038/s41431-020-0638-4

Thompson EM, Ashley D, Landi D. 2020. Current medulloblastoma subgroup specific clinical trials. *Transl Pediatr* **9:** 157–162.

Tinat J, Bougeard G, Baert-Desurmont S, Vasseur S, Martin C, Bouvignies E, Caron O, Bressac-de Paillerets B, Berthet P, Dugast C, et al. 2009. 2009 version of the Chompret criteria for Li Fraumeni syndrome. *J Clin Oncol* **27:** e108–e109. doi:10.1200/JCO.2009.22.7967

Treviño LR, Yang W, French D, Hunger SP, Carroll WL, Devidas M, Willman C, Neale G, Downing J, Raimondi SC, et al. 2009. Germline genomic variants associated with childhood acute lymphoblastic leukemia. *Nat Genet* **41:** 1001–1005. doi:10.1038/ng.432

Villani A, Tabori U, Schiffman J, Shlien A, Beyene J, Druker H, Novokmet A, Finlay J, Malkin D. 2011. Biochemical and imaging surveillance in germline TP53 mutation carriers with Li-Fraumeni syndrome: a prospective observational study. *Lancet Oncol* **12:** 559–567.

Villani A, Shore A, Wasserman JD, Stephens D, Kim RH, Druker H, Gallinger B, Naumer A, Kohlmann W, Novokmet A, et al. 2016. Biochemical and imaging surveillance in germline TP53 mutation carriers with Li-Fraumeni syndrome: 11 year follow-up of a prospective observational study. *Lancet Oncol* **17:** 1295–1305. doi:10.1016/S1470-2045(16)30249-2

Villani A, Davidson S, Kanwar N, Lo WW, Li Y, Cohen-Gogo S, Fuligni F, Edward L-M, Light N, Layeghifard M, et al. 2023. The clinical utility of integrative genomics in childhood cancer extends beyond targetable mutations. *Nat Cancer* **4**: 203–221. doi:10.1038/s43018-022-00474-y

Walcott FL, Wang P-Y, Bryla CM, Huffstutler RD, Singh N, Pollak MN, Khincha PP, Savage SA, Mai PL, Dodd KW, et al. 2020. Pilot study assessing tolerability and metabolic effects of metformin in patients with Li-Fraumeni syndrome. *JNCI Cancer Spectr* **4**: kaa063. doi:10.1093/jncics/pkaa063

Wang PY, Li J, Walcott FL, Kang JG, Starost MF, Talagala SL, Zhuang J, Park JH, Huffstutler RD, Bryla CM, et al. 2017. Inhibiting mitochondrial respiration prevents cancer in a mouse model of Li-Fraumeni syndrome. *J Clin Invest* **127**: 132–136. doi:10.1172/JCI88668

Wasserman JD, Novokmet A, Eichler-Jonsson C, Ribeiro RC, Rodriguez-Galindo C, Zambetti GP, Malkin D. 2015. Prevalence and functional consequence of *TP53* mutations in pediatric adrenocortical carcinoma: a Children's Oncology Group Study. *J Clin Oncol* **33**: 602–609. doi:10.1200/JCO.2013.52.6863

Wong M, Mayoh C, Lau LMS, Khuong-Quang D-A, Pinese M, Kumar A, Barahona P, Wilkie EE, Sullivan P, Bowen-James R, et al. 2020. Whole genome, transcriptome and methylome profiling enhances actionable target discovery

in high-risk pediatric cancer. *Nat Med* **26**: 1742–1753. doi:10.1038/s41591-020-1072-4

Wong D, Znassi N, Luo P, Oldfield LE, Bruce J, Danesh A, Prokopec S, Basra P, Pederson S, Wellum J, et al. 2022. OP015: multi-omic analysis of circulating tumour DNA for the early detection of cancer in patients with Li-Fraumeni syndrome. *Genet Med* **24**: S346–S347. doi:10.1016/j.gim.2022.01.609

Yamashita R, Long J, Longacre T, Peng L, Berry G, Martin B, Higgins J, Rubin DL, Shen J. 2021. Deep learning model for the prediction of microsatellite instability in colorectal cancer: a diagnostic study. *Lancet Oncol* **22**: 132–141. doi:10.1016/S1470-2045(20)30535-0

Zhang J, Walsh MF, Wu G, Edmonson MN, Gruber TA, Easton J, Hedges D, Ma X, Zhou X, Yergeau DA, et al. 2015. Germline mutations in predisposition genes in pediatric cancer. *N Engl J Med* **373**: 2336–2346. doi:10.1056/NEJMoa1508054

Zheng S, Cherniack AD, Dewal N, Moffitt RA, Danilova L, Murray BA, Lerario AM, Else T, Knijnenburg TA, Ciriello G, et al. 2016. Comprehensive pan-genomic characterization of adrenocortical carcinoma. *Cancer Cell* **29**: 723–736. doi:10.1016/j.ccell.2016.04.002

Developmental Dysregulation of Childhood Cancer

Thomas R.W. Oliver[1,2] and Sam Behjati[2,3,4]

[1]Department of Histopathology and Cytology, Cambridge University Hospitals NHS Foundation Trust, Cambridge, Cambridgeshire CB2 0QQ, United Kingdom

[2]Wellcome Sanger Institute, Hinxton, Cambridgeshire CB10 1RQ, United Kingdom

[3]Department of Paediatrics, University of Cambridge, Cambridge, Cambridgeshire CB2 0QQ, United Kingdom

[4]Department of Paediatric Haematology and Oncology, Cambridge University Hospitals NHS Foundation Trust, Cambridge, Cambridgeshire CB2 0QQ, United Kingdom

Correspondence: to3@sanger.ac.uk; sb31@sanger.ac.uk

Most childhood cancers possess distinct clinicopathological profiles from those seen in adulthood, reflecting their divergent mechanisms of carcinogenesis. Rather than depending on the decades-long, stepwise accumulation of changes within a mature cell that defines adult carcinomas, many pediatric malignancies emerge rapidly as the consequence of random errors during development. These errors—whether they be genetic, epigenetic, or microenvironmental—characteristically block maturation, resulting in phenotypically primitive neoplasms. Only an event that falls within a narrow set of spatiotemporal parameters will forge a malignant clone; if it occurs too soon then the event might be lethal, or negatively selected against, while if it is too late or in an incorrectly primed precursor cell then the necessary intracellular conditions for transformation will not be met. The precise characterization of these changes, through the study of normal tissues and tumors from patients and model systems, will be essential if we are to develop new strategies to diagnose, treat, and perhaps even prevent childhood cancer.

Traditional models of cancer development describe a stepwise accumulation of cancer-causing (driver) mutations in a differentiated cell, or established stem cell, that ultimately culminates in malignant transformation. Much of the focus in cancer research for over half a century has been on quantifying these steps and identifying the driver events that underpin them. The earliest efforts to address the former question used epidemiological data of cancer death rates by age group (Nordling 1953; Armitage and Doll 1954). For epithelial tumors, it was suggested that approximately six successive driver events were necessary for cancer formation, close to the estimated four to five critical mutations derived from large-scale sequencing experiments decades later (Martincorena et al. 2017; ICGC/TCGA Pan-Cancer Analysis of Whole Genomes Consortium 2020).

Cancers of children and young adults are inconsistent with these models, however. They occur too frequently to be the result of the same multistep process seen in carcinomas of older individuals. For a small number of cases that share a morphomolecular profile with their adult counterparts, the answer often lies in an inherited cancer predisposition. The underlying germline mutation the individual possesses can take one of two forms: either it can represent the first step in a defined pathway to malignancy or it makes cells constitutively prone to mutation, thus accelerating the rate of driver acquisition. A good example here would be colorectal cancer (CRC). Some younger patients have an inherited loss-of-function mutation in the *APC* gene, a classic CRC driver event, while those with constitutional mismatch repair deficiency lack this first step and instead have an elevated point mutation rate (Fearon and Vogelstein 1990; Sanders et al. 2021). In both cases, the resultant tumor has progressed through the same multistage model as older cases, simply in a shortened time frame.

This cannot explain why the young usually develop tumors seldom seen at other ages, however. Leukemias, central nervous system tumors, and sarcomas are proportionally much more frequent cancers in children than they are in older individuals and possess distinct histological and/ or genetic profiles compared to adult variants of

these cancers (Pfister et al. 2022). Other cancers that commonly affect children, such as Wilms tumor, almost never occur in adults (Spreafico et al. 2021). These observations suggest that different models of carcinogenesis are required.

There are now credible developmental pathways described for many pediatric tumors, but critical to each of them is the notion that cancer emerges as a consequence of maturation failure within a developing cell. This does not only account for the histological and demographic differences we have highlighted, but a great deal else we shall discuss further (Table 1). In this review, we will present the case for childhood cancer as dysregulation of development and outline contributing genetic, epigenetic, and microenvironmental factors. We present a unifying model of pediatric oncogenesis in Figure 1.

HISTORIC CLINICOPATHOLOGICAL OBSERVATION

If many tumors that occur in childhood are seldom found in older patients, it follows that they emerge within cell states that only transiently exist during development and thus are unavailable to undergo malignant transformation later on. Each pediatric cancer has its own restricted

Table 1. Features that reflect the developmental origins of childhood cancer

Clinical
Some present in utero/at birth
Almost exclusively occur within a narrow time frame of childhood/young adulthood
Capacity of some precursor lesions and tumors to regress spontaneously
No established epidemiological link to environmental factors
Some associated with congenital malformation syndromes
Histopathological
Histological similarity to corresponding fetal tissue
Precursor lesions found during infant autopsies
Maturation of some cancers observed upon exposure to systemic therapy
Genomic/transcriptomic/epigenetic
Low overall and driver mutation burdens
Cell-intrinsic mutagenic processes dominate mutational landscapes
Driver mutations/epimutations typically perturb the pathways that regulate the epigenome or fetal growth and differentiation
Driver mutations/epimutations found to pervade large areas of adjacent normal tissue
Transcriptomes and epigenomes closely correlate to fetal cell populations

     Cite this article as *Cold Spring Harb Perspect Med* doi: 10.1101/cshperspect.a041580

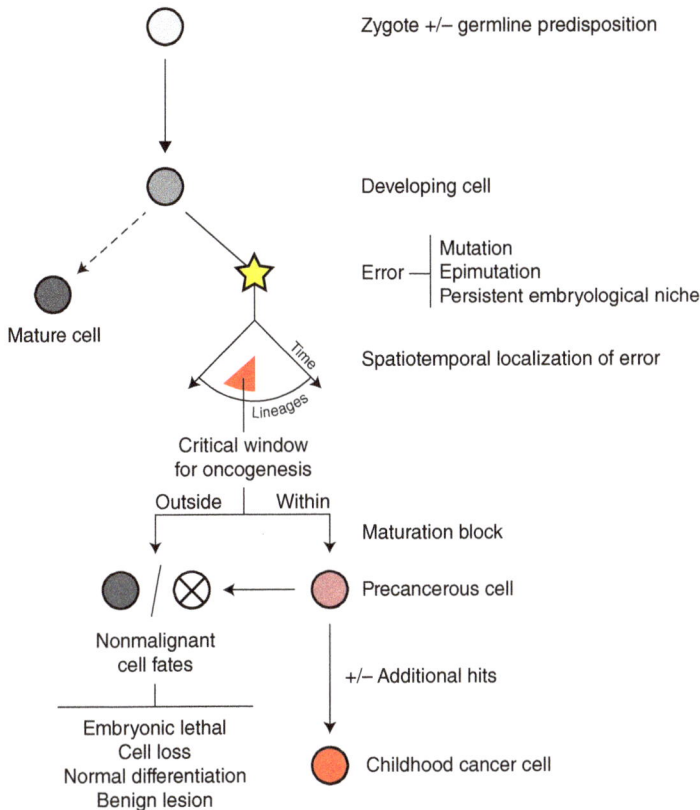

Figure 1. A developmental model of pediatric oncogenesis. A catastrophic error within a developing cell derived from a specific lineage at a critical time point may prevent it from differentiating any further. Further "hits" may be necessary for this persistently primitive cell to continue to proliferate and evade cell death upon withdrawal of its transient microenvironment. Failure to overcome additional barriers to cancer development, or events that fall outside of the necessary spatiotemporal window, may be lethal to the individual or clone. Alternatively, the affected cell may differentiate either into normal, mature tissue or into a benign lesion. All of these steps may or may not occur in the context of an inherited mutation that predisposes the child to developing cancer.

time frame in which to emerge. For example, those of the viscera tend to present in infants or young children, while osteosarcomas are seen in adolescents (Siegel et al. 2023). This, it has been suggested, reflects each tissue's period of greatest growth, allowing for delay in presentation due to precursor intermediates or subclinical disease (Scotting et al. 2005).

Consistent with the idea that these tumors must progress within a critical developmental window is the unusual capacity of some to involute or mature spontaneously. The classic example here is neuroblastoma, the most common extracranial solid tumor of childhood (Siegel et al. 2023), a tumor arising in a precursor to the sympathetic nervous system and most commonly found within the adrenal medulla. The estimated prevalence of this tumor is 0.3%–1%, based on autopsies of infants, around 40 times more frequent than would be expected based on the incidence of clinically evident, invasive disease (Beckwith and Perrin 1963). Even if it does manifest in infancy, it will often regress without intervention, even in the presence of metastases to specific sites such as the skin or liver (Nuchtern et al. 2012; Tas et al. 2020). Similar observations have also been made in the kidney. Persistent foci of blastema, termed nephrogenic rests, have been reported in 1% of infant kidneys (Beckwith 1993), again much more frequently than their malignant counterpart, the Wilms tumor (nephroblastoma). The fact that these rests are seen in up to 40% of cases of unilateral Wilms and almost 100% of bilateral cases would suggest that they at least share a root perturbation (Vujanić et al. 2017), whether or not they represent true in situ (preinvasive) disease. Collectively, these insights suggest that development offers a substrate for pediatric malignancy, but additional steps are necessary for lesional cells to avoid elimination or maturation over time. A similar phenomenon is seen in normal embryogenesis; organs are populated by more embryonic cells than they need to develop, and the surplus undergoes apoptosis upon withdrawal of trophic factors (Marshall et al. 2014).

Perhaps the oldest observations that reveal the early origins of childhood cancers have been made through light microscopy. Some neoplasms, such as Wilms tumor and neuroblastoma, are referred to as embryonal tumors or "blastomas," which reflects the histological similarity they share with tissues usually only found during in utero life. Max Wilms first noted that his eponymous tumor was a facsimile of the fetal kidney over a century ago (Wilms 1899). These data provide further compelling, albeit indirect, evidence of a maturation block at the center of tumorigenesis.

Collectively, while imprecise and often qualitative, clinicopathological studies have formed the essential foundations on which the developmental theory of pediatric oncogenesis was built. As we will show throughout the remainder of this article, the tools, patient tissues, and model systems are now available for us to quantify the underlying perturbation more accurately.

GENETIC FACTORS

Few Driver Mutations Are Required for Malignant Transformation

Large-scale, pan-pediatric cancer whole exome and genome sequencing experiments have revealed that, despite being vastly diverse, these tumors possess recurring features that may be a consequence of their developmental origins. Mostly prominently, in contrast to the long, stepwise accumulation of driver events seen in tumors of adulthood, childhood cancers require very few oncogenic mutations to emerge. Of course, short models of pediatric tumor development are not new; Knudson notably proposed his two-hit theory for retinoblastoma years before the era of next-generation sequencing (Knudson 1971). Nonetheless, we now know that this holds true for many other pediatric tumors: most have only one or no driver point mutation, compared to adult cancers that have around four (Martincorena et al. 2017; Gröbner et al. 2018).

Not only are fewer driver mutations necessary but the genes implicated are different too, frequently being pathognomonic to a single disease entity. In one experiment, only 45% of probable driver genes identified were previously reported in adult cancer (Ma et al. 2018). Pediatric tumors instead often sustain driver events in the components of epigenetic regulation, such as histones or chromatin remodeling complexes, or signaling pathways that predominantly operate during in utero life, like the neurotropic tyrosine receptor kinase or NTRK family of proto-oncogenes (Filbin and Monje 2019; Zhao et al. 2021). Mutations in the common cell survival or proliferation pathways seen in adulthood, on the other hand, play a relatively minor role. For example, while PI3K pathway mutations are the most commonly reported in adult cancers (31%), they comprise only 3% of drivers in younger patients (Kandoth et al. 2013; Gröbner et al. 2018). Taken together, these findings provide compelling evidence for why these tumors manifest so early on and for their root within development; namely, the primed precursor is vulnerable to as little as a single mutation in pathways tightly regulating growth and differentiation that have a diminished role in later life.

Tumors Emerge from Errors in Normal, Cell-Intrinsic Processes during Development

The wider repertoire of mutations found within a tumor (beyond the handful of driver mutations we have considered so far) can also tell us about the broader mutational processes, both intrinsic

Cite this article as *Cold Spring Harb Perspect Med* doi: 10.1101/cshperspect.a041580

and extrinsic, operating within it and its precursor. This is because different mutagens will inflict unique patterns of mutation, also known as mutational signatures, which facilitate carcinogenesis (Alexandrov et al. 2020). Among adult cancers, we now know the mutational signatures for several well-known environmental agents implicated in their development, such as ultraviolet light, smoking, and aristolochic acid. The role of the environment in pediatric cancer has long been debated. The only proven causal links are to prior ionizing radiation and chemotherapy exposure (Spector et al. 2015). Examination of the mutational processes in childhood tumors confirms how minor a role the environment appears to play in shaping their genomes. Most mutations can be attributed to intrinsic, clocklike processes, such as the deamination of 5-methylcytosine, oxidative stress, or errors of DNA repair (Thatikonda et al. 2023). The rare exceptions to this are therapy-induced mutations, consistent with the aforementioned clinical data, which may provoke the emergence of a second malignancy, and the intriguing detection of two ultraviolet light signatures (SBS7A, SBS7B) in hypodiploid B-cell acute lymphoblastic leukemia (B-ALL) (Ma et al. 2018). The origin of this latter signature activity is unclear, and the possibility remains that it is derived from an unknown mutational process that mirrors that of ultraviolet light. Nevertheless, these data would be consistent overall with treatment-naive pediatric cancer emerging as a random event due to failure of intrinsic, ubiquitous cellular processes that occur during development or in early postnatal life.

Mutation Reveals the Relationship between Organogenesis and Tumorigenesis

High-penetrance, germline mutations are now recognized to play a critical role in the formation of many pediatric tumors. Overall, the proportion with an underlying cancer predisposition syndrome is probably ~10%, although this varies by tumor type, rising to 50% in adrenal cortical carcinoma (Zhang et al. 2015; Gröbner et al. 2018). While some of these will be in DNA repair pathway genes, thus simply accelerating the acquisition of more traditional, adult-type tumors,

like the constitution mismatch repair deficiency example in CRC we gave earlier, others are more specific to pediatric-type tumors. Many of these syndromes will be covered in detail elsewhere in this collection. They are, however, worth highlighting here because of the developmental link they allow us to draw between organogenesis and pediatric tumorigenesis. For example, several of the syndromes that predispose a child to Wilms tumor are associated with congenital malformations, especially of the urogenital tract, such as WAGR syndrome and Denys–Drash syndrome, or result from conditions typified by generalized organ overgrowth (Beckwith–Wiedemann syndrome) (Treger et al. 2019).

This link between malformation and mutation is not confined to inherited syndromes, however; somatic mutation in classic cancer driver genes is now recognized as the cause of isolated malformations in nonsyndromic children. For instance, some congenital pulmonary airway malformations are the consequence of hotspot KRAS missense mutations, while the neurodevelopmental anomaly focal cortical dysplasia can result from somatic mutation in the mTOR signaling pathway (Hermelijn et al. 2020; Jesus-Ribeiro et al. 2021). Such findings show that the developmental consequence of mutation is context dependent, varying according to the cell type affected, the developmental stage, and perhaps additional microenvironmental cues (discussed more later). Pediatric cancer can be thought of as existing on a spectrum with more benign developmental conditions.

Mutation Disrupts a Multitude of Pathways that Regulate Development

Each type of developing cell depends upon a combination of generic and more specialized pathways of growth and differentiation. It is impossible to observe and test the effects of perturbation of these pathways in humans prospectively; however, by examining the genetic alterations in pediatric cancer, we can infer some of the most critical pathways for its survival. Around 10%–20% of Wilms tumors, for example, harbor driver events in the WT1 gene (Treger et al. 2019). WT1 is a tumor suppressor gene and transcriptional regulator, which, crucially, has long been recognized to play an important role in genitourinary development (Pritchard-Jones et al. 1990). WT1 protein is thought to facilitate the mesenchymal-to-epithelial transition of the metanephric mesenchyme toward nephron differentiation (Hastie 2017). It is notable then that Wilms tumors—which may comprise a varying combination of blastemal, epithelial, and stromal histological elements—favor a mesenchymal or stromal appearance when *WT1* is mutated (Schumacher et al. 2003). Furthermore, it is recognized that the stromal component may give rise to heterologous elements, including skeletal and smooth muscle, and it has been reported that *WT1* expression during development inhibits the expression of myogenic genes (Miyagawa et al. 1998). These genes are highly expressed in *WT1*-mutant tumors. These data suggest then that the loss of *WT1* in the precursor to Wilms tumor may dysregulate development in two ways: inhibition of kidney development and potentiation of the aberrant pursuit of other mesodermal cell lineages. Some Wilms tumor driver mutations are also found in other genes critical to nephrogenesis, like the transcription factors *SIX1* and *SIX2* (Wegert et al. 2015).

Another example of a driver event specific to human development is amplification of the *MYCN* gene. In neuroblastoma, for example, *MYCN* amplification may be sufficient alone to permit malignant transformation, as supported by evidence from clinical samples and cell lines derived from the murine neural crest (Olsen et al. 2017; Brady et al. 2020). Occurring in 20%–30% of cases, it accounts for ~50% of high-risk cases at diagnosis but is not seen as a late event in other molecular subtypes, further supporting its role as only an initiator of the disease (Brodeur et al. 1984; Seeger et al. 1985; Brady et al. 2020; Otte et al. 2021). Like other members of the MYC gene family, *MYCN* has a diverse range of cellular functions. In the developing sympathetic nervous system, however, studies in model systems indicate that its expression plays a central role in determining neural fate while preventing terminal differentiation (Knoepfler et al. 2002; Huang and Weiss 2013; Ponzoni et al. 2022). Ordinarily, *MYCN* expression would also provoke p53-mediated apoptosis. In neuroblastoma and its

precursors, however, additional antiapoptotic factors appear to mitigate this, since *TP53* mutations are rare in untreated tumors (Brady et al. 2020; Otte et al. 2021). For example, *MYCN* expression also promotes *MDM2* transcription (Slack et al. 2005), a gene that codes for an E3 ubiquitin ligase that facilitates the degradation of p53. Coactivation of *ALK* has also been proposed as an alternative mechanism, based on data from transgenic mice and zebrafish models (Berry et al. 2012; Zhu et al. 2012) through activation of the PI3K and MAPK pathways. This is especially compelling as the co-occurrence of *MYCN* and *ALK* mutations confers an especially poor prognosis (Brady et al. 2020). These data represent plausible pathways for the developmental emergence of aggressive neuroblastoma.

Mutations within the genes responsible for regulating the epigenome are an especially important group of pediatric cancer drivers. Cell differentiation depends upon careful and coordinated changes to the epigenome and single base errors in the machinery maintaining it can have catastrophic, genome-wide consequences to gene expression (Filbin and Monje 2019). Driver variants in the genes coding for histone H3, so-called oncohistone mutations, are probably the best recognized of these in pediatric cancer, affecting residues in the amino-terminal tail that are usually subject to posttranslational modification (PTM). Disruption to histone PTMs is known to have wide-ranging epigenetic consequences, including disruption to histone reading, writing, and erasing, as well as alteration to the histone variants comprising the octamer complex and the function of chromatin remodelers (Feinberg et al. 2016; Amatori et al. 2021). It is striking that different missense hotspots in H3 genes exhibit exquisite spatiotemporal and cancer-type specificity in a variety of tumors of children and young adults: K27M/I in diffuse midline gliomas, G34R/V in diffuse hemispheric gliomas, G34W/L in giant cell tumors of the bone, and K36M in most chondroblastomas (Khuong-Quang et al. 2012; Schwartzentruber et al. 2012; Wu et al. 2012; Behjati et al. 2013).

How the oncohistones exert their effects is becoming clearer. We now know that K27M, the first oncohistone mutation to be discovered, provokes a global loss of H3-K27 tri- and dimethylation through interference with the methyltransferase activity of PRC2, a multiprotein complex that is responsible for transcriptional repression of key developmental genes (Chammas et al. 2020; Amatori et al. 2021). It is this derepression that facilitates tumorigenesis and a similar mechanism likely underpins other H3 mutations too; namely, mutation prevents methylation of the residue or its neighbor in the wild-type H3 protein pool through interaction with histone-modifying enzymes (Amatori et al. 2021), permitting expression of genes that promote primitive, developmental cell states. Such is the strong, dominant-negative effect of these mutations that K27M has been shown to potentiate its neoplastic effects when found in as little as 3% of H3 histones (Lewis and Allis 2013).

The evidence for the developmental origins of many oncohistones is compelling. K27M/I mutant gliomas tend to occur in the thalamus and brainstem of younger children while G34R/V mutant gliomas are found in the cerebral cortex of older children and younger adults (Filbin and Monje 2019; Amatori et al. 2021). Both groups are seldom seen in older adults and the underlying mutations are almost invariably clonal, early events (Salloum et al. 2017; Hoffman et al. 2019; Arunachalam et al. 2022). H3 mutation must occur within a highly constrained window of development to initiate gliomagenesis. Data from preclinical studies would support this. K27M mutations in mouse zygotes are embryonic lethal while those induced in the neonatal pup fail to form tumors, even with loss of *ATRX* and *Trp53* (Pathania et al. 2017). Targeting the neural progenitor cells (NPCs; H3.3-K27M and *Trp53* loss) in the E12.5–E13.5 mouse embryo, however, is sufficient for malignancy. Similar observations have been made in vitro using human embryonic stem cell–derived NPCs, but with *TP53* loss and *PDGFRA* inactivation (Funato et al. 2014). Note that transformation in both of these models necessitated mutations in additional, classic high-grade glioma genes. In fact, withdrawal of growth factor from the NPC cell cultures in the latter experiment demonstrated that apoptosis was only significantly up-regulated in

cells that possessed a solitary H3.3 K27M mutation, similar to the physiological embryonic cells we described earlier. This suggests H3-K27M invokes a maturation block but depends on additional mutations for neoplasia. Curiously, while H3-K27M diffuse midline gliomas are invariably fatal (Lam et al. 2015), there are rare instances of pilocytic astrocytoma, a low-grade glioma, that possess the same mutation (Schwartzentruber et al. 2012; Orillac et al. 2016) in addition to their usual MAPK pathway variants (Jones et al. 2013). The divergent fate of both gliomas must be dependent on other factors then, such as their cell of origin and/or their co-occurring driver mutations.

EPIGENETIC AND MICROENVIRONMENTAL FACTORS

Finding the Gaps between Genetic Alteration and Transformation

For much of this review, the focus has been on genetic variation as a means of dysregulating development. Mutation alone, however, is only part of the story, as shown by the discovery of pediatric cell populations that carry their tissue-specific driver mutations and yet remain phenotypically normal. This has been most thoroughly demonstrated in B-ALL. Early data indicated that monozygotic twins who concurrently developed this disease shared the same underlying clonal somatic rearrangement, likely due to intraplacental spread from one twin to the other (Ford et al. 1993). B-ALL's prenatal origins would later be confirmed in singleton children too through detection of the causative rearrangement in their neonatal blood spot samples (Gale et al. 1997). Applying this approach to unselected neonates though, revealed an astonishing ~1% carried a recognized fusion oncogene driver of childhood leukemia in their blood (Mori et al. 2002). This is about 100 times greater than the cumulative lifetime risk of acquiring the related tumors. Other factors must facilitate tumorigenesis: mutation is not enough.

Other drivers of childhood cancer have also been identified in their corresponding normal tissues in recent years. Studies examining children with Wilms tumors have identified hypermethylation of the parentally imprinted gene (imprinting discussed in greater detail later) for the long noncoding RNA *H19*, an established driver event, in both the tumor and the background kidney (Moulton et al. 1994; Steenman et al. 1994; Coorens et al. 2019). One experiment found unusually long, shared genetic ancestry between the two tissues, with the epimutation pervading potentially large (>10%) numbers of normal kidney cells (Coorens et al. 2019). Together, these data place *H19* hypermethylation in a common precursor that is responsible for both the tumor and the cells that form the normal, anatomically complex kidney, perhaps pervading several mature cell lineages. Similar insights have been made in a malignant rhabdoid tumor (MRT) of the kidney as well, where biallelic loss of *SMARCB1*, the canonical driver event, has been found in both the tumor and histologically normal hilar nerve tissue (Custers et al. 2021).

It might be tempting here to draw parallels between these premalignant, embryologically derived expansions and the vast array of precancerous clones now identified in healthy adult epithelial tissues. Such comparisons, however, are misguided. In both the Wilms tumor and MRT studies we describe, the driver events found in the adjacent normal tissues should be sufficient to permit cancer formation, unlike the premalignant expansions in adult tissues that likely lack the number and combination of driver events necessary for transformation. Cure rates for MRT remain poor (Tomlinson et al. 2005; Reinhard et al. 2008; van den Heuvel-Eibrink et al. 2011), especially metastatic disease, which is rarely treatable, so it is remarkable that the same driver repertoire can exist in totally innocuous tissue with a shared lineage. Furthermore, these precursors do not carry their cancer risk into adult life. Some may regress while it is plausible that others are forever trapped within a differentiated cell phenotype. As deeper, multiomic studies are performed on normal pediatric tissues, it is likely many more genetic precursors will be found. The next, more challenging step will be to better understand what governs their fate.

Major Role Played by Epigenetic and Microenvironmental Factors in Pediatric Oncogenesis

Beyond the impact of mutations in the genes regulating the epigenome, there is a wider role for the epigenome in the development of pediatric cancers. In utero or germline perturbation of genomic imprinting is one such example. Genomic imprinting is the mechanism by which ~200 genes are expressed in a parent-of-origin manner in mammals through the use of methylation (Tucci et al. 2019). Many of the genes that are controlled in this way are essential regulators of the growth of the embryo and placenta. Intriguingly, growth promoters are expressed on the paternal allele and suppressors on the maternal one (Ishida and Moore 2013; Barlow and Bartolomei 2014). A popular explanation for this division of labor is termed "conflict theory" (Moore and Haig 1991), where paternally expressed genes seek to maximize extraction of maternal resources while maternally expressed genes favor resource conservation, possibly to allocate to later offspring. It stands to reason that disruption to the balance of expression of these genes may have a profound impact on fetal growth and, indeed, this is well known to be the case. One well-studied imprinted locus is 11p15, which contains *IGF2* and *H19*; *IGF2* is progrowth and expressed on the paternal allele, whereas *H19* is expressed on the maternal allele and seeks to antagonize *IGF2*'s function (Ishida and Moore 2013). Constitutional epigenetic silencing of *H19* leads to *IGF2* overexpression and overgrowth (Beckwith–Wiedemann syndrome) and vice versa results in growth restriction (Silver–Russell syndrome). Somatic aberrant methylation at this locus, resulting in *IGF2* overexpression, is the most common driver of Wilms tumor but is also seen in other pediatric malignancies, such as hepatoblastoma (Honda et al. 2008; Spreafico et al. 2021).

Germ cell tumors (GCTs) are perhaps the clearest example of a pediatric tumor where nongenetic factors are known to disturb development and facilitate their pathogenesis, as recently reviewed in detail by Oosterhuis and Looijenga (2019). GCTs are a diverse collection of tumors of varying malignant potential and histological

Cite this article as *Cold Spring Harb Perspect Med* doi: 10.1101/cshperspect.a041580

composition. Most malignant GCTs originate from a primordial germ cell that has undergone aberrant reprogramming, due to a failure of cell-intrinsic and niche signaling factors to maintain its phenotype. Primordial germ cells may be particularly susceptible to neoplastic transformation, even without mutation, in view of their expression of telomerase (Wright et al. 1996; Hiyama and Hiyama 2007), which ensures their replicative immortality.

The evidence placing the origins of GCTs in utero is substantial. For example, a subset of tumors is located in midline structures, rather than the gonads, reflecting the migratory pathway primordial germ cells take in early embryogenesis (Oosterhuis et al. 2007). These tumors are genetically, epigenetically, and histologically similar to their gonadal counterparts (Schneider et al. 2001; Wang et al. 2014; Fukushima et al. 2017; Shen et al. 2018; Van Nieuwenhuysen et al. 2018; Oliver et al. 2022), supporting the notion that they are derived from the same cell type. Ordinarily, germ cells that fail to migrate to the gonad will eventually apoptose due to loss of KITLG-KIT signaling, so they must either be reprogrammed or establish themselves in a microenvironment with KITLG-secreting niche cells, such as the thymus (Oosterhuis et al. 2007). Furthermore, GCT global methylation profiles match those of primordial germ cells, including loss of genomic imprinting. Imprinting is normally lost in germ cells in utero before it is reestablished in a sex-specific manner (Ishida and Moore 2013; Shen et al. 2018). Lastly, genomic analyses that have timed the first copy number changes in these tumors predict that they took place in utero in many instances (Oliver et al. 2022).

Studying the Maturation Block

Central to everything we have discussed in this review is the notion that it is the arrest of maturation that forms the foundation for pediatric cancer. Historically, this would have been observed and measured in comparison to an approximate counterpart in the normal fetus on the basis of morphology and immunohistochemistry. Now, however, the widespread availability of single-cell RNA sequencing has enabled us to

profile the transcriptomes of fetal tissues extensively, thus providing a library of cell-specific readouts to quantifiably match against a tumor transcriptome (Coorens and Behjati 2022).

Many studies have been published now highlighting the transcriptional similarities between pediatric cancers and prenatal cell populations (Young et al. 2018, 2021; Jessa et al. 2019; Kildisiute et al. 2021). This has been seen as a way of determining the cell-of-origin of pediatric cancers, on the basis that the tumor will still retain at least some transcriptional similarity with its precursor cell that underwent the maturation block (Behjati et al. 2021). While an appealing idea, it is fraught with complications. Crucially, the transcriptome is dynamic, perhaps nowhere more so than in the embryonic precursors of many pediatric cancers that possess great plasticity. This is best exemplified within the aforementioned GCTs, which are able to recapitulate a vast array of embryonic and extraembryonic tissues, all of which are transcriptionally distinct and may bear no relationship to their cell-of-origin (Oliver et al. 2022). Even if one discounts tumors with innate morphological heterogeneity, many others are studied after treatment exposure. The effects of this are not always inconsiderable. Pediatric tumors, like GCTs and neuroblastoma, are well recognized to mature on exposure to cytotoxic chemotherapy. This will inevitably perturb any cellular signals inherited from the precursor on malignant transformation, as has been seen in treated Wilms tumor where its resemblance to the early fetal nephron is attenuated (Young et al. 2021). Nonetheless, such experiments hold great possible therapeutic value. Sequencing data from fetal tissues have revealed the differentiation trajectories that govern development. By understanding these, we can look to manipulate these pathways in their malignant counterparts. This is the basis of differentiation therapy, a topic that will be covered in greater detail later in other articles in this collection.

CONCLUDING REMARKS

Throughout this review, we have presented the case for developmental dysregulation as the mechanism by which many childhood cancers occur. Random mutation and/or epimutation in a developing cell is the necessary first step, assuming that there is no inherited cancer predisposition. This will usually affect either specific pathways of embryogenesis and fetal growth, or the machinery of epigenetic regulation. The exact timing of this is crucial; some oncogenic mutations will be lethal to the cell or embryo if they happen too early and those that occur later may lack the primed, primitive state they need to transform. Furthermore, those that successfully disrupt embryogenesis may instead form benign malformations, again depending on the cell state and developmental window they affect. The aberrant cell that manages to avoid all of these nonmalignant fates will need to achieve a stable maturation block. This means they must retain their pro-proliferative, early phenotype while avoiding the cell death that normally follows once the associated physiological developmental niche, including stimulation by trophic factors, is withdrawn. Lastly, some tumors will require additional hits, genetic or otherwise, to provide the necessary fitness for their persistence, expansion, and invasion.

While our understanding of pediatric oncogenesis has advanced significantly in recent years, many gaps remain. Filling these gaps should be a research priority: complete models of tumor development are essential to the discovery of new diagnostic, therapeutic, and preventative strategies for patients. We must not limit ourselves to the study of tumor tissue either, but rather pursue a deep characterization of the genotypes and phenotypes of fetal and pediatric tissues as well if we are to understand the developmental origins of these malignancies fully.

ACKNOWLEDGMENTS

We thank Drs. Joseph Christopher and Henry Lee-Six for their review of this manuscript.

REFERENCES

Alexandrov LB, Kim J, Haradhvala NJ, Huang MN, Tian Ng AW, Wu Y, Boot A, Covington KR, Gordenin DA, Bergstrom EN, et al. 2020. The repertoire of mutational signatures in human cancer. *Nature* **578:** 94–101. doi:10.1038/s41586-020-1943-3

Amatori S, Tavolaro S, Gambardella S, Fanelli M. 2021. The dark side of histones: genomic organization and role of oncohistones in cancer. *Clin Epigenetics* **13:** 71. doi:10.1186/s13148-021-01057-x

Armitage P, Doll R. 1954. The age distribution of cancer and a multi-stage theory of carcinogenesis. *Br J Cancer* **8:** 1–12. doi:10.1038/bjc.1954.1

Arunachalam S, Szlachta K, Brady SW, Ma X, Ju B, Shaner B, Mulder HL, Easton J, Raphael BJ, Myers M, et al. 2022. Convergent evolution and multi-wave clonal invasion in H3 K27-altered diffuse midline gliomas treated with a PDGFR inhibitor. *Acta Neuropathol Commun* **10:** 80. doi:10.1186/s40478-022-01381-0

Barlow DP, Bartolomei MS. 2014. Genomic imprinting in mammals. *Cold Spring Harb Perspect Biol* **6:** a018382. doi:10.1101/cshperspect.a018382

Beckwith JB. 1993. Precursor lesions of Wilms tumor: clinical and biological implications. *Med Pediatr Oncol* **21:** 158–168. doi:10.1002/mpo.2950210303

Beckwith JB, Perrin EV. 1963. In situ neuroblastomas: a contribution to the natural history of neural crest tumors. *Am J Pathol* **43:** 1089–1104.

Behjati S, Tarpey PS, Presneau N, Scheipl S, Pillay N, Van Loo P, Wedge DC, Cooke SL, Gundem G, Davies H, et al. 2013. Distinct H3F3A and H3F3B driver mutations define chondroblastoma and giant cell tumor of bone. *Nat Genet* **45:** 1479–1482. doi:10.1038/ng.2814

Behjati S, Gilbertson RJ, Pfister SM. 2021. Maturation block in childhood cancer. *Cancer Discov* **11:** 542–544. doi:10.1158/2159-8290.CD-20-0926

Berry T, Luther W, Bhatnagar N, Jamin Y, Poon E, Sanda T, Pei D, Sharma B, Vetharoy WR, Hallsworth A, et al. 2012. The ALK(F1174L) mutation potentiates the oncogenic activity of MYCN in neuroblastoma. *Cancer Cell* **22:** 117–130. doi:10.1016/j.ccr.2012.06.001

Brady SW, Liu Y, Ma X, Gout AM, Hagiwara K, Zhou X, Wang J, Macias M, Chen X, Easton J, et al. 2020. Pan-neuroblastoma analysis reveals age- and signature-associated driver alterations. *Nat Commun* **11:** 5183. doi:10.1038/s41467-020-18987-4

Brodeur GM, Seeger RC, Schwab M, Varmus HE, Bishop JM. 1984. Amplification of N-*myc* in untreated human neuroblastomas correlates with advanced disease stage. *Science* **224:** 1121–1124. doi:10.1126/science.6719137

Chammas P, Mocavini I, Di Croce L. 2020. Engaging chromatin: PRC2 structure meets function. *Br J Cancer* **122:** 315–328. doi:10.1038/s41416-019-0615-2

Coorens THH, Behjati S. 2022. Tracing and targeting the origins of childhood cancer. *Annu Rev Cancer Biol* **6:** 35–47. doi:10.1146/annurev-cancerbio-070620-091632

Coorens TH, Treger TD, Al-Saadi R, Moore L, Tran MG, Mitchell TJ, Tugnait S, Thevanesan C, Young MD, Oliver TR, et al. 2019. Embryonal precursors of Wilms tumor. *Science* **366:** 1247–1251. doi:10.1126/science.aax1323

Custers L, Khabirova E, Coorens THH, Oliver TRW, Calandrini C, Young MD, Vieira Braga FA, Ellis P, Mamanova L, Segers H, et al. 2021. Somatic mutations and single-cell transcriptomes reveal the root of malignant rhabdoid tumours. *Nat Commun* **12:** 1407. doi:10.1038/s41467-021-21675-6

Cite this article as *Cold Spring Harb Perspect Med* doi: 10.1101/cshperspect.a041580

Fearon ER, Vogelstein B. 1990. A genetic model for colorectal tumorigenesis. *Cell* **61:** 759–767. doi:10.1016/0092-8674 (90)90186-I

Feinberg AP, Koldobskiy MA, Göndör A. 2016. Epigenetic modulators, modifiers and mediators in cancer aetiology and progression. *Nat Rev Genet* **17:** 284–299. doi:10.1038/nrg.2016.13

Filbin M, Monje M. 2019. Developmental origins and emerging therapeutic opportunities for childhood cancer. *Nat Med* **25:** 367–376. doi:10.1038/s41591-019-0383-9

Ford AM, Ridge SA, Cabrera ME, Mahmoud H, Steel CM, Chan LC, Greaves M. 1993. In utero rearrangements in the trithorax-related oncogene in infant leukaemias. *Nature* **363:** 358–360. doi:10.1038/363358a0

Fukushima S, Yamashita S, Kobayashi H, Takami H, Fukuoka K, Nakamura T, Yamasaki K, Matsushita Y, Nakamura H, Totoki Y, et al. 2017. Genome-wide methylation profiles in primary intracranial germ cell tumors indicate a primordial germ cell origin for germinomas. *Acta Neuropathol* **133:** 445–462. doi:10.1007/s00401-017-1673-2

Funato K, Major T, Lewis PW, Allis CD, Tabar V. 2014. Use of human embryonic stem cells to model pediatric gliomas with H3.3K27M histone mutation. *Science* **346:** 1529–1533. doi:10.1126/science.1253799

Gale KB, Ford AM, Repp R, Borkhardt A, Keller C, Eden OB, Greaves MF. 1997. Backtracking leukemia to birth: identification of clonotypic gene fusion sequences in neonatal blood spots. *Proc Natl Acad Sci* **94:** 13950–13954. doi:10.1073/pnas.94.25.13950

Gröbner SN, Worst BC, Weischenfeldt J, Buchhalter I, Kleinheinz K, Rudneva VA, Johann PD, Balasubramanian GP, Segura-Wang M, Brabetz S, et al. 2018. The landscape of genomic alterations across childhood cancers. *Nature* **555:** 321–327. doi:10.1038/nature25480

Hastie ND. 2017. Wilms' tumour 1 (WT1) in development, homeostasis and disease. *Development* **144:** 2862–2872. doi:10.1242/dev.153163

Hermelijn SM, Wolf JL, Dorine den Toom T, Wijnen RMH, Rottier RJ, Schnater JM, von der Thüsen JH. 2020. Early KRAS oncogenic driver mutations in nonmucinous tissue of congenital pulmonary airway malformations as an indicator of potential malignant behavior. *Hum Pathol* **103:** 95–106. doi:10.1016/j.humpath.2020.07.015

Hiyama E, Hiyama K. 2007. Telomere and telomerase in stem cells. *Br J Cancer* **96:** 1020–1024. doi:10.1038/sj.bjc.6603671

Hoffman M, Gillmor AH, Kunz DJ, Johnston M, Nikolic A, Narta K, Zarrei M, King J, Ellestad K, Dang NH, et al. 2019. Intratumoral genetic and functional heterogeneity in pediatric glioblastoma. *Cancer Res* **79:** 2111–2123. doi:10.1158/0008-5472.CAN-18-3441

Honda S, Arai Y, Haruta M, Sasaki F, Ohira M, Yamaoka H, Horie H, Nakagawara A, Hiyama E, Todo S, et al. 2008. Loss of imprinting of IGF2 correlates with hypermethylation of the H19 differentially methylated region in hepatoblastoma. *Br J Cancer* **99:** 1891–1899. doi:10.1038/sj.bjc.6604754

Huang M, Weiss WA. 2013. Neuroblastoma and MYCN. *Cold Spring Harb Perspect Med* **3:** a014415. doi:10.1101/cshperspect.a014415

ICGC/TCGA Pan-Cancer Analysis of Whole Genomes Consortium. 2020. Pan-cancer analysis of whole genomes. *Nature* **578:** 82–93. doi:10.1038/s41586-020-1969-6

Ishida M, Moore GE. 2013. The role of imprinted genes in humans. *Mol Aspects Med* **34:** 826–840. doi:10.1016/j.mam.2012.06.009

Jessa S, Blanchet-Cohen A, Krug B, Vladoiu M, Coutelier M, Faury D, Poreau B, De Jay N, Hébert S, Monlong J, et al. 2019. Stalled developmental programs at the root of pediatric brain tumors. *Nat Genet* **51:** 1702–1713. doi:10.1038/s41588-019-0531-7

Jesus-Ribeiro J, Pires LM, Melo JD, Ribeiro IP, Rebelo O, Sales F, Freire A, Melo JB. 2021. Genomic and epigenetic advances in focal cortical dysplasia types I and II: a scoping review. *Front Neurosci* **14:** 580357. doi:10.3389/fnins.2020.580357

Jones DTW, Hutter B, Jäger N, Korshunov A, Kool M, Warnatz H-J, Zichner T, Lambert SR, Ryzhova M, Quang DAK, et al. 2013. Recurrent somatic alterations of FGFR1 and NTRK2 in pilocytic astrocytoma. *Nat Genet* **45:** 927–932. doi:10.1038/ng.2682

Kandoth C, McLellan MD, Vandin F, Ye K, Niu B, Lu C, Xie M, Zhang Q, McMichael JF, Wyczalkowski MA, et al. 2013. Mutational landscape and significance across 12 major cancer types. *Nature* **502:** 333–339. doi:10.1038/nature12634

Khuong-Quang D-A, Buczkowicz P, Rakopoulos P, Liu X-Y, Fontebasso AM, Bouffet E, Bartels U, Albrecht S, Schwartzentruber J, Letourneau L, et al. 2012. K27m mutation in histone H3.3 defines clinically and biologically distinct subgroups of pediatric diffuse intrinsic pontine gliomas. *Acta Neuropathol* **124:** 439–447. doi:10.1007/s00401-012-0998-0

Kildisiute G, Kholosy WM, Young MD, Roberts K, Elmentaite R, van Hooff SR, Pacyna CN, Khabirova E, Piapi A, Thevanesan C, et al. 2021. Tumor to normal single-cell mRNA comparisons reveal a pan-neuroblastoma cancer cell. *Sci Adv* **7:** eabd3311. doi:10.1126/sciadv.abd3311

Knoepfler PS, Cheng PF, Eisenman RN. 2002. N-myc is essential during neurogenesis for the rapid expansion of progenitor cell populations and the inhibition of neuronal differentiation. *Genes Dev* **16:** 2699–2712. doi:10.1101/gad.1021202

Knudson AG. 1971. Mutation and cancer: statistical study of retinoblastoma. *Proc Natl Acad Sci* **68:** 820–823. doi:10.1073/pnas.68.4.820

Lam S, Lin Y, Auffinger B, Melkonian S. 2015. Analysis of survival in pediatric high-grade brainstem gliomas: a population-based study. *J Pediatr Neurosci* **10:** 199–206. doi:10.4103/1817-1745.165656

Lewis PW, Allis CD. 2013. Poisoning the "histone code" in pediatric gliomagenesis. *Cell Cycle* **12:** 3241–3242. doi:10.4161/cc.26356

Ma X, Liu Y, Liu Y, Alexandrov LB, Edmonson MN, Gawad C, Zhou X, Li Y, Rusch MC, Easton J, et al. 2018. Pan-cancer genome and transcriptome analyses of 1,699 paediatric leukaemias and solid tumours. *Nature* **555:** 371–376. doi:10.1038/nature25795

Marshall GM, Carter DR, Cheung BB, Liu T, Mateos MK, Meyerowitz JG, Weiss WA. 2014. The prenatal origins of cancer. *Nat Rev Cancer* **14:** 277–289. doi:10.1038/nrc3679

Martincorena I, Raine KM, Gerstung M, Dawson KJ, Haase K, Van Loo P, Davies H, Stratton MR, Campbell PJ. 2017. Universal patterns of selection in cancer and somatic tissues. *Cell* 171: 1029–1041.e21. doi:10.1016/j.cell.2017.09 .042

Miyagawa K, Kent J, Moore A, Charlieu JP, Little MH, Williamson KA, Kelsey A, Brown KW, Hassam S, Briner J, et al. 1998. Loss of WT1 function leads to ectopic myogenesis in Wilms' tumour. *Nat Genet* 18: 15–17. doi:10.1038/ ng0198-15

Moore T, Haig D. 1991. Genomic imprinting in mammalian development: a parental tug-of-war. *Trends Genet* 7: 45–49. doi:10.1016/0168-9525(91)90040-W

Mori H, Colman SM, Xiao Z, Ford AM, Healy LE, Donaldson C, Hows JM, Navarrete C, Greaves M. 2002. Chromosome translocations and covert leukemic clones are generated during normal fetal development. *Proc Natl Acad Sci* 99: 8242–8247. doi:10.1073/pnas.112218799

Moulton T, Crenshaw T, Hao Y, Moosikasuwan J, Lin N, Dembitzer F, Hensle T, Weiss L, McMorrow L, Loew T, et al. 1994. Epigenetic lesions at the H19 locus in Wilms' tumour patients. *Nat Genet* 7: 440–447. doi:10.1038/ ng0794-440

Nordling CO. 1953. A new theory on the cancer-inducing mechanism. *Br J Cancer* 7: 68–72. doi:10.1038/bjc.1953.8

Nuchtern JG, London WB, Barnewolt CE, Naranjo A, McGrady PW, Geiger JD, Diller L, Schmidt ML, Maris JM, Cohn SL, et al. 2012. A prospective study of expectant observation as primary therapy for neuroblastoma in young infants: a Children's Oncology Group study. *Ann Surg* 256: 573–580. doi:10.1097/SLA.0b013e31826cbbbd

Oliver TRW, Chappell L, Sanghvi R, Deighton L, Ansari-Pour N, Dentro SC, Young MD, Coorens THH, Jung H, Butler T, et al. 2022. Clonal diversification and histogenesis of malignant germ cell tumours. *Nat Commun* 13: 4272. doi:10 .1038/s41467-022-31375-4

Olsen RR, Otero JH, García-López J, Wallace K, Finkelstein D, Rehg JE, Yin Z, Wang YD, Freeman KW. 2017. MYCN induces neuroblastoma in primary neural crest cells. *Oncogene* 36: 5075–5082. doi:10.1038/onc.2017.128

Oosterhuis JW, Looijenga LHJ. 2019. Human germ cell tumours from a developmental perspective. *Nat Rev Cancer* 19: 522–537. doi:10.1038/s41568-019-0178-9

Oosterhuis JW, Stoop H, Honecker F, Looijenga LHJ. 2007. Why human extragonadal germ cell tumours occur in the midline of the body: old concepts, new perspectives. *Int J Androl* 30: 256–263; discussion 263–264. doi:10.1111/j .1365-2605.2007.00793.x

Orillac C, Thomas C, Dastagirzada Y, Hidalgo ET, Golfinos JG, Zagzag D, Wisoff JH, Karajannis MA, Snuderl M. 2016. Pilocytic astrocytoma and glioneuronal tumor with histone H3 K27M mutation. *Acta Neuropathol Commun* 4: 84. doi:10.1186/s40478-016-0361-0

Otte J, Dyberg C, Pepich A, Johnsen JI. 2021. MYCN function in neuroblastoma development. *Front Oncol* 10: 624079. doi:10.3389/fonc.2020.624079

Pathania M, De Jay N, Maestro N, Harutyunyan AS, Nitarska J, Pahlavan P, Henderson S, Mikael LG, Richard-Londt A, Zhang Y, et al. 2017. H3.3K27M cooperates with Trp53 loss and PDGFRA gain in mouse embryonic neural progenitor cells to induce invasive high-grade gliomas. *Cancer Cell* 32: 684–700.e9. doi:10.1016/j.ccell.2017.09.014

Pfister SM, Reyes-Múgica M, Chan JKC, Hasle H, Lazar AJ, Rossi S, Ferrari A, Jarzembowski JA, Pritchard-Jones K, Hill DA, et al. 2022. A summary of the inaugural WHO classification of pediatric tumors: transitioning from the optical into the molecular era. *Cancer Discov* 12: 331–355. doi:10.1158/2159-8290.CD-21-1094

Ponzoni M, Bachetti T, Corrias MV, Brignole C, Pastorino F, Calarco E, Bensa V, Giusto E, Ceccherini I, Perri P. 2022. Recent advances in the developmental origin of neuroblastoma: an overview. *J Exp Clin Cancer Res* 41: 92. doi:10 .1186/s13046-022-02281-w

Pritchard-Jones K, Fleming S, Davidson D, Bickmore W, Porteous D, Gosden C, Bard J, Buckler A, Pelletier J, Housman D. 1990. The candidate Wilms' tumour gene is involved in genitourinary development. *Nature* 346: 194–197. doi:10 .1038/346194a0

Reinhard H, Reinert J, Beier R, Furtwängler R, Alkasser M, Rutkowski S, Frühwald M, Koscielniak E, Leuschner I, Kaatsch P, et al. 2008. Rhabdoid tumors in children: prognostic factors in 70 patients diagnosed in Germany. *Oncol Rep* 19: 819–823.

Salloum R, McConechy MK, Mikael LG, Fuller C, Drissi R, DeWire M, Nikbakht H, De Jay N, Yang X, Boue D, et al. 2017. Characterizing temporal genomic heterogeneity in pediatric high-grade gliomas. *Acta Neuropathol Commun* 5: 78. doi:10.1186/s40478-017-0479-8

Sanders MA, Vöhringer H, Forster VJ, Moore L, Campbell BB, Hooks Y, Edwards M, Bianchi V, Coorens THH, Butler TM, et al. 2021. Life without mismatch repair. bioRxiv doi:10.1101/2021.04.14.437578

Schneider DT, Schuster AE, Fritsch MK, Hu J, Olson T, Lauer S, Göbel U, Perlman EJ. 2001. Multipoint imprinting analysis indicates a common precursor cell for gonadal and nongonadal pediatric germ cell tumors. *Cancer Res* 61: 7268–7276.

Schumacher V, Schuhen S, Sonner S, Weirich A, Leuschner I, Harms D, Licht J, Roberts S, Royer-Pokora B. 2003. Two molecular subgroups of Wilms' tumors with or without WT1 mutations. *Clin Cancer Res* 9: 2005–2014.

Schwartzentruber J, Korshunov A, Liu X-Y, Jones DTW, Pfaff E, Jacob K, Sturm D, Fontebasso AM, Quang D-AK, Tönjes M, et al. 2012. Driver mutations in histone H3.3 and chromatin remodelling genes in paediatric glioblastoma. *Nature* 482: 226–231. doi:10.1038/nature10833

Scotting PJ, Walker DA, Perilongo G. 2005. Childhood solid tumours: a developmental disorder. *Nat Rev Cancer* 5: 481–488. doi:10.1038/nrc1633

Seeger RC, Brodeur GM, Sather H, Dalton A, Siegel SE, Wong KY, Hammond D. 1985. Association of multiple copies of the N-*myc* oncogene with rapid progression of neuroblastomas. *N Engl J Med* 313: 1111–1116. doi:10.1056/ NEJM198510313131802

Shen H, Shih J, Hollern DP, Wang L, Bowlby R, Tickoo SK, Thorsson V, Mungall AJ, Newton Y, Hegde AM, et al. 2018. Integrated molecular characterization of testicular germ cell tumors. *Cell Rep* 23: 3392–3406. doi:10.1016/j.celrep .2018.05.039

Siegel DA, King JB, Lupo PJ, Durbin EB, Tai E, Mills K, Van Dyne E, Buchanan Lunsford N, Henley SJ, Wilson RJ. 2023. Counts, incidence rates, and trends of pediatric cancer in the United States, 2003–2019. *J Natl Cancer Inst* 115: 1337–1354. doi:10.1093/jnci/djad115

Cite this article as *Cold Spring Harb Perspect Med* doi: 10.1101/cshperspect.a041580

Slack A, Chen Z, Tonelli R, Pule M, Hunt L, Pession A, Shohet JM. 2005. The p53 regulatory gene *MDM2* is a direct transcriptional target of MYCN in neuroblastoma. *Proc Natl Acad Sci* 102: 731–736. doi:10.1073/pnas.0405495102

Spector LG, Pankratz N, Marcotte EL. 2015. Genetic and nongenetic risk factors for childhood cancer. *Pediatr Clin North Am* 62: 11–25. doi:10.1016/j.pcl.2014.09.013

Spreafico F, Fernandez CV, Brok J, Nakata K, Vujanic G, Geller JI, Gessler M, Maschietto M, Behjati S, Polanco A, et al. 2021. Wilms tumour. *Nat Rev Dis Primers* 7: 1–21. doi:10.1038/s41572-021-00308-8

Steenman MJ, Rainier S, Dobry CJ, Grundy P, Horon IL, Feinberg AP. 1994. Loss of imprinting of IGF2 is linked to reduced expression and abnormal methylation of H19 in Wilms' tumour. *Nat Genet* 7: 433–439. doi:10.1038/ng0794-433

Tas ML, Nagtegaal M, Kraal KCJM, Tytgat GAM, Abeling NGGM, Koster J, Pluijm SMF, Zwaan CM, de Keizer B, Molenaar JJ, et al. 2020. Neuroblastoma stage 4S: tumor regression rate and risk factors of progressive disease. *Pediatr Blood Cancer* 67: e28061. doi:10.1002/pbc.28061

Thatikonda V, Islam SMA, Autry RJ, Jones BC, Gröbner SN, Warsow G, Hutter B, Huebschmann D, Fröhling S, Kool M, et al. 2023. Comprehensive analysis of mutational signatures reveals distinct patterns and molecular processes across 27 pediatric cancers. *Nat Cancer* 4: 276–289. doi:10.1038/s43018-022-00509-4

Tomlinson GE, Breslow NE, Dome J, Guthrie KA, Norkool P, Li S, Thomas PRM, Perlman E, Beckwith JB, D'Angio GJ, et al. 2005. Rhabdoid tumor of the kidney in the National Wilms' Tumor Study: age at diagnosis as a prognostic factor. *J Clin Oncol* 23: 7641–7645. doi:10.1200/JCO.2004.00.8110

Treger TD, Chowdhury T, Pritchard-Jones K, Behjati S. 2019. The genetic changes of Wilms tumour. *Nat Rev Nephrol* 15: 240–251. doi:10.1038/s41581-019-0112-0

Tucci V, Isles AR, Kelsey G, Ferguson-Smith AC, Tucci V, Bartolomei MS, Benvenisty N, Bourc'his D, Charalambous M, Dulac C, et al. 2019. Genomic imprinting and physiological processes in mammals. *Cell* 176: 952–965. doi:10.1016/j.cell.2019.01.043

van den Heuvel-Eibrink MM, van Tinteren H, Rehorst H, Coulombe A, Patte C, de Camargo B, de Kraker J, Leuschner I, Lugtenberg R, Pritchard-Jones K, et al. 2011. Malignant rhabdoid tumours of the kidney (MRTKs), registered on recent SIOP protocols from 1993 to 2005: a report of the SIOP Renal Tumour Study Group. *Pediatr Blood Cancer* 56: 733–737. doi:10.1002/pbc.22922

Van Nieuwenhuysen E, Busschaert P, Neven P, Han SN, Moerman P, Liontos M, Papaspirou M, Kupryjanczyk J, Hogdall C, Hogdall E, et al. 2018. The genetic landscape of 87 ovarian germ cell tumors. *Gynecol Oncol* 151: 61–68. doi:10.1016/j.ygyno.2018.08.013

Vujanić GM, Apps JR, Moroz V, Ceroni F, Williams RD, Sebire NJ, Pritchard-Jones K. 2017. Nephrogenic rests in Wilms tumors treated with preoperative chemotherapy: the UK SIOP Wilms Tumor 2001 Trial experience. *Pediatr Blood Cancer* 64. doi:10.1002/pbc.26547

Wang L, Yamaguchi S, Burstein MD, Terashima K, Chang K, Ng HK, Nakamura H, He Z, Doddapaneni H, Lewis L, et al. 2014. Novel somatic and germline mutations in intracranial germ cell tumours. *Nature* 511: 241–245. doi:10.1038/nature13296

Wegert J, Ishaque N, Vardapour R, Geörg C, Gu Z, Bieg M, Ziegler B, Bausenwein S, Nourkami N, Ludwig N, et al. 2015. Mutations in the SIX1/2 pathway and the DROSHA/DGCR8 miRNA microprocessor complex underlie high-risk blastemal type Wilms tumors. *Cancer Cell* 27: 298–311. doi:10.1016/j.ccell.2015.01.002

Wilms M. 1899. *Die Mischgeschwülste der Niere.* Verlag von Arthur Georgi, Leipzig, Germany.

Wright WE, Piatyszek MA, Rainey WE, Byrd W, Shay JW. 1996. Telomerase activity in human germline and embryonic tissues and cells. *Dev Genet* 18: 173–179. doi:10.1002/(SICI)1520-6408(1996)18:2<173::AID-DVG10>3.0.CO;2-3

Wu G, Broniscer A, McEachron TA, Lu C, Paugh BS, Becksfort J, Qu C, Ding L, Huether R, Parker M, et al. 2012. Somatic histone H3 alterations in pediatric diffuse intrinsic pontine gliomas and non-brainstem glioblastomas. *Nat Genet* 44: 251–253. doi:10.1038/ng.1102

Young MD, Mitchell TJ, Vieira Braga FA, Tran MG, Stewart BJ, Ferdinand JR, Collord G, Botting RA, Popescu DM, Loudon KW, et al. 2018. Single cell transcriptomes from human kidneys reveal the cellular identity of renal tumors. *Science* 361: 594–599. doi:10.1126/science.aat1699

Young MD, Mitchell TJ, Custers L, Margaritis T, Morales-Rodriguez F, Kwakwa K, Khabirova E, Kildisiute G, Oliver TRW, de Krijger RR, et al. 2021. Single cell derived mRNA signals across human kidney tumors. *Nat Commun* 12: 3896. doi:10.1038/s41467-021-23949-5

Zhang J, Walsh MF, Wu G, Edmonson MN, Gruber TA, Easton J, Hedges D, Ma X, Zhou X, Yergeau DA, et al. 2015. Germline mutations in predisposition genes in pediatric cancer. *N Engl J Med* 373: 2336–2346. doi:10.1056/NEJMoa1508054

Zhao X, Kotch C, Fox E, Surrey LF, Wertheim GB, Baloch ZW, Lin F, Pillai V, Luo M, Kreiger PA, et al. 2021. NTRK fusions identified in pediatric tumors: the frequency, fusion partners, and clinical outcome. *JCO Precis Oncol* 1: PO.20.00250. doi:10.1200/PO.20.00250

Zhu S, Lee J-S, Guo F, Shin J, Perez-Atayde AR, Kutok JL, Rodig SJ, Neuberg DS, Helman D, Feng H, et al. 2012. Activated ALK collaborates with MYCN in neuroblastoma pathogenesis. *Cancer Cell* 21: 362. doi:10.1016/j.ccr.2012.02.010

Developmental Modeling of Childhood Cancers

Kosuke Funato[1] and Viviane Tabar[2]

[1]Center for Molecular Medicine, Department of Biochemistry and Molecular Biology, University of Georgia, Athens, Georgia 30602-4712, USA

[2]Department of Neurosurgery, Cancer Biology and Genetics Program, Memorial Sloan Kettering Cancer Center, New York, New York 10065, USA

Correspondence: kosuke.funato@uga.edu

Growing evidence indicates that childhood cancer is a developmental disease and the oncogenic impact of mutations depends on spatiotemporal developmental contexts. This dependency leads to distinct molecular, genetic, and clinical characteristics across various cancer (sub)types. However, the underlying molecular mechanisms of tumorigenesis are not fully understood, and the development of precision medicine for childhood cancers is still an ongoing effort, partially due to their relative rarity. Therefore, it is crucial to develop and use "developmental models" that replicate both mutations and specific developmental contexts that determine their impact. In this review, we summarize recent advances in the growing field of developmental modeling of childhood cancers, which enhance our understanding of the pathogenic mechanisms and pave the way for the development of new therapeutic approaches.

In recent years, we have gained significant and growing understanding of the molecular and genomic landscape of childhood cancer. Efforts to translate this knowledge into novel therapies are ongoing but face significant hurdles, particularly in view of the relative rarity of childhood cancers. Globally, it is estimated that 20 million adults are diagnosed with cancer each year (Bray et al. 2024); however, the number goes down significantly to 400,000 for children aged 0–19 (Ward et al. 2019). This rarity makes it difficult to conduct clinical trials and collect clinical specimens to study cancer biology. Moreover, currently available patient-derived cell lines are primarily maintained in two-dimensional (2D) culture and thus do not fully preserve the cellular and molecular characteristics, intratumor heterogeneity, and three-dimensional (3D) architecture of original tumors. In some hard-to-detect cancer types, their specimens and cell lines often represent mid- to late stages of cancer development, offering limited insight into the mechanisms of early tumorigenesis of childhood cancer. Growing evidence supports the concept of childhood cancer as a developmental disease, in which dysregulated developmental programs lead to uncontrolled proliferation and impaired differentiation, and that space (cell type) and time (developmental stage) are key elements of the oncogenic impact of driver mutations (Filbin and Monje 2019). Therefore, to elucidate the pathogenic mechanisms of childhood cancers, it is crucial to create "developmental models" that replicate not only the genetic features of patient tumors, but also the time and space context that allows mutations to co-opt developmental programs built into the cells of origin. In this section, we outline recent endeavors to enhance, expand, and diversify models of pediatric tumors that provide us with a deeper understanding of the role of developmental programs in the onset and progression of childhood cancers.

MAJOR TYPES OF CHILDHOOD CANCER MODELS

Patient-Derived Cells

Since the first establishment of an immortal human cell line in 1951, patient-derived cell lines have remained a cornerstone for cancer research, including childhood brain cancers. However, the success rate of derivation varies widely and is nearly nil for some cancer types. Furthermore, in vitro culture conditions exert selective pressures on these cells, leading to a loss of the molecular and genetic characteristics of their original cancer (Gillet et al. 2013; Ben-David et al. 2018). While primary culture cells and low-passage cell lines maintain a closer resemblance to the original cancer, they may exhibit reproducibility issues due to selection and genetic drift. Another limitation of patient-derived cell lines is their inability to maintain structural organization and crucial components like vasculature and immune cells. To mitigate these limitations, organotypic and organoid culture methods can be employed to better preserve the complex architecture and heterogeneity of tumor tissue (Shimizu et al. 2011; Veninga and Voest 2021).

Stem Cell–Based Models

Recent strides in stem cell technologies, such as the establishment and controlled differentiation of pluripotent stem cells (PSCs), including embryonic stem cells (ESCs) and induced pluripotent stem cells (iPSCs), have opened doors to their use in cancer modeling (Smith and Tabar 2019). ESC is a pluripotent cell line derived from an inner cell mass of blastocysts (Evans and Kaufman 1981; Martin 1981; Thomson et al. 1998). Later, Shinya Yamanaka's group estab-lished iPSCs by introducing Oct3/4, Sox2, Myc, and Klf4 (so-called Yamanaka factors) into somatic cells (Takahashi and Yamanaka 2006). Implantation of PSCs into immunodeficient mice leads to the formation of teratoma, which contains cells from all three germ layers, and is widely used as a proof of pluripotency but rarely used as a tool to study cancer biology. Several research groups, including our own, have further advanced this concept and successfully developed childhood cancer models based on human PSCs (Smith and Tabar 2019). The advantages of using PSCs in developmental oncology are substantial. They provide access to cell types in early developmental stages that are otherwise technically and ethically challenging to obtain. Thanks to their pluripotency, these cells can differentiate into various cell types within our bodies, enabling a systematic comparison of the effects of oncogenic mutations in different cell types. PSCs are amenable to monoclonal culture, affording a precise system for studying cancer biology, compared to patient-derived tissues and cell lines that display combinatorial effects of driver mutations, passenger mutations, and original genetic background of each patient. Additionally, PSC cultures are highly scalable, making them suitable for large-scale assays, including pharmacological and genetic screenings. Studies by our group and others have demonstrated the utility of PSCs in investigating the cell-type-specific effects of driver mutations and the role of developmental programs in the formation of childhood cancers.

Animal Models

The earliest cancer model was established by Katsusaburo Yamagiwa's group by repeated application of coal tar in rabbits' ears (Yamagiwa and Ichikawa 1918). Since then, many animal models have been created through chemical exposure, transplantation of human cancer cells, and genetic engineering (Onaciu et al. 2020). Before the advent of cell culture systems, xenotransplantation was the sole method to sustain cancer cells outside a patient's body. However, the immune response of host animals posed a substantial challenge to the growth of transplanted tumor cells.

The derivation of mouse ESCs and the development of genetically engineered mice paved the way for genetically engineered mouse models (GEMMs). Subsequent derivation of immunodeficient mouse strains facilitated more efficient engraftment of cancer cells (Flanagan 1966; Bosma et al. 1983). However, immunodeficient animals, which lack functional immune cells, are unsuitable for studying the role of immunity in cancer progression. Beyond laboratory mice, several other species, including fruit flies and zebrafish, have been employed in cancer modeling. Recent breakthroughs in genome editing technologies, such as CRISPR-Cas9, have further streamlined the creation of new cancer models in mice and other species.

MODELING OF CHILDHOOD BRAIN TUMORS

Cell-Based Models of Childhood Brain Tumors

In the past decade, large-scale genomic studies of childhood brain tumors have identified several new genetic mutations that can define molecularly and genetically distinct new subtypes (Khuong-Quang et al. 2012; Schwartzentruber et al. 2012; Sturm et al. 2012, 2016; Wu et al. 2012a; Mackay et al. 2017). These findings led to a major update of the WHO Classification of Tumors of the Central Nervous System, revolutionizing the diagnosis of childhood brain tumors (Louis et al. 2021). To better understand subtype-specific tumor biology, a great amount of effort has been made to establish patient-derived cell lines from different tumor histologies (Monje et al. 2011; Hashizume et al. 2012; Caretti et al. 2014; Xu et al. 2015; Ijaz et al. 2020). These cell lines have been used for multiomics analysis as well as genetic and pharmacological screens (He et al. 2021; Sun et al. 2023). However, the cell-line-based approach needs to be complemented with a bottom-up approach to fully elucidate the pathogenic mechanisms, especially the precise molecular role of driver mutations in specific developmental contexts.

Among the newly identified driver mutations, recurrent mutations in histone genes (so-called oncohistone mutations) have gained particular attention as they are the first recurrent mutations found in histone genes and highly prevalent in some childhood brain tumors and other tumor types (described below). For instance, H3K27M mutations are predominantly found in diffuse midline glioma (DMG), which includes tumors formally categorized as diffuse intrinsic pontine glioma (DIPG). An elegant study by David Allis's group showed that H3K27M mutations, including H3.3K27M and H3.1K27M mutations, induce a global reduction of histone H3K27 trimethylation by inhibiting the Polycomb Repressive Complex 2 (PRC2) (Lewis et al. 2013). However, expression of H3K27M in Nestin-positive progenitors of the mouse brain failed to induce a tumor, suggesting the context-dependent effect of the mutation. This idea is further supported by the other type of histone mutation H3.3G34R/V. In contrast to H3K27M mutations that are confined to the brainstem, H3.3G34R/V mutations are mostly found in cortical high-grade glioma. Additionally, high-grade glioma with H3.3G34R/V mutations exhibits totally distinct transcriptional, epigenetic, and molecular features and characteristics compared to H3K27M-mutant DMG. These observations underscore the cell-type-specific effects of these driver mutations. Thus, to replicate the oncogenic impact of these mutations, it is essential to create a model based on cells that represent the originating cell type and developmental stage. On the flip side, if a driver mutation exhibits its oncogenic phenotype solely in a particular cell type, it is highly likely that this cell type represents the origin. Owing to the development of robust differentiation protocols that recapitulate key developmental stages and chronology ex vivo, PSCs offer a potent platform for systematically exploring the cell-type-specific effects of mutations. In the case of brain tumor modeling, different protocols are available that can drive PSCs into neuroectoderm lineage. Initially, the embryoid body-based approach was commonly used; however, the purity of the neuroectoderm lineage was often suboptimal as embryoid bodies contain cells from other lineages. A significant breakthrough came with the Dual-Smad inhibition protocol developed by Lorenz Studer's group

(Chambers et al. 2009). The simplicity and robustness of this protocol have paved the way for the use of human PSCs in various neurological research and cell-replacement therapy.

Using this approach, our group pioneered the creation of the first genetically engineered PSC-based brain tumor model (Funato et al. 2014). In this study, we focused on H3.3K27M-mutant DMG, the most aggressive form of brain cancer in children, with a median survival of less than a year. Our findings show that the H3.3K27M mutation transforms neural progenitor cells (NPCs) but not astrocytes or fibroblasts, supporting cell-type-specific effects of the mutation. Upon transplantation into the brainstem of immunodeficient mice, these transformed cells gave rise to tumors closely resembling those in DMG patients. Our data also indicated that the H3.3K27M mutation holds cells in an undifferentiated state by maintaining the expression of developmental genes such as LIN28B and PLAGL1, both of which are highly expressed in the early stages of brain development, in particular at neuroepithelial cells, but not at later stages. Furthermore, our research also highlighted that the PSC-based model serves as a robust platform for drug screening, leading to the identification of a potential therapeutic agent. In a parallel effort, Stefan Pfister's group engineered iPSCs to carry an inducible H3.3-K27M allele at the endogenous locus (Haag et al. 2021). Consistent with our findings, their data demonstrated synergistic effects of H3.3K27M and p53 loss on proliferation, apoptosis, and in vivo tumorigenicity in iPSC-derived neural stem cells, but not in astroglial precursors. Their results also showed that iPSC-derived oligodendrocyte progenitor cells (OPCs) responded to the mutations in vitro but failed to form tumors in vivo, suggesting that H3.3K27M-mutant DMGs originate from a specific cell type at a particular developmental stage.

Several other groups have used PSCs to model different brain tumor types. Juan Carlos Izpisua Belmonte's group generated an iPSC-based glioma model for the first time by genetically manipulating p53 and receptor tyrosine kinase signaling pathways (Sancho-Martinez et al. 2016). The transformed NPCs showed in vivo tumorigenicity and metabolic reprogramming,

which can potentially be targeted by small molecule inhibitors. Yasuhiro Yamada's group successfully established an iPSC-based model of atypical teratoid/rhabdoid tumors (AT/RT) by knocking out the SMARCB1 and TP53 genes (Terada et al. 2019). The transformed NPCs developed tumors in vivo displaying rhabdoid histology, a hallmark of AT/RT. Frank Furnari's group also employed an iPSC-based model and showed the role of SMARCB1 in neuronal differentiation (Parisian et al. 2020). Our group and others used human ESC (hESC)-derived NPCs to demonstrate the oncogenic role of FOXR2 in childhood brain tumors (Tsai et al. 2022; Royston et al. 2024). Luca Tiberi's group developed a model of group 3 medulloblastoma by introducing Otx2 and c-Myc into iPSCs (Ballabio et al. 2020). Additionally, several independent groups reported the development of iPSC-based models of the Sonic Hedgehog (SHH) group medulloblastoma (Čančer et al. 2019; Huang et al. 2019; Susanto et al. 2020; van Essen et al. 2024). Two of them employed iPSCs derived from Gorlin syndrome patients harboring a germline mutation in the SHH receptor PTCH1. Additionally, patient-derived iPSCs are useful for investigating genetic predisposition in childhood cancers such as Neurofibromatosis and Li–Fraumeni syndrome (Mazuelas et al. 2022).

In recent years, advances in 3D culture systems, including cerebral organoid culture, have allowed the modeling of the complex architecture and cellular heterogeneity of brain tumor tissues. Several groups have developed organoid culture systems for brain tumor cells, including patient specimens (Hubert et al. 2016; Jacob et al. 2020; Lago et al. 2023). These models recapitulate the original tumor architecture, including hypoxic gradients and cellular heterogeneity, and maintain in vivo tumorigenicity. Howard Fine's group cocultured adult glioblastoma cells with hESC-derived cerebral organoids and showed an invasive phenotype and the formation of a tumor microtube network (Linkous et al. 2019). In addition to the patient-derived brain tumor organoid models, several groups have developed models by introducing oncogenic mutations into PSCs or PSC-derived cerebral organoids (Bian et al. 2018;

Ogawa et al. 2018; Funato et al. 2021). Jürgen Knoblich's group systematically investigated the impact of multiple oncogenes found in both adult and childhood brain cancers in hESC-derived cerebral organoids. While histone mutations did not yield striking phenotypes, this model revealed the overgrowth of cells bearing MYC amplification or combinations of recurrent mutations seen in adult glioblastoma, such as EGFRvIII, CDKN2A deletion, and PTEN deletion. Our group also employed region-specific brain spheroids to demonstrate the region-specific effects of H3.3G34R and concomitant mutations in *ATRX* and *TP53* (Funato et al. 2021). While the introduction of these mutations in interneuronal progenitors of the ventral forebrain led to tumor formation in vivo, ventral hindbrain progenitors with the same set of mutations failed to form tumors. Our PSC-based model provided the first experimental evidence of the interneuronal origin of H3.3G34R-mutant cortical high-grade gliomas, which was corroborated by bioinformatics analyses of patient tumors (Chen et al. 2020a; Liu et al. 2024). Our study also demonstrated that the 3D culture system is particularly useful for investigating cell-contact-based signaling pathways such as the Notch signaling pathway. In addition to these studies, cerebral organoid culture systems have been used in some of the aforementioned studies (Ballabio et al. 2020; Parisian et al. 2020). The organoid-based model holds significant promise for incorporating other cellular components, such as endothelial cells and immune cells (Abud et al. 2017; Pham et al. 2018; Cakir et al. 2019; Schafer et al. 2023). The more complex organoid-based approach is suitable for analyzing complex tumor architecture and interactions among heterogenous cell populations.

In addition to PSCs, primary cultured cells from the mouse or human brain have been used to model brain cancers. Kristian Helin's group developed a H3.3K27M-mutant DMG model by introducing H3.3K27M mutation and platelet-derived growth factor (PDGF) into mouse neural stem cells and showed that EZH2 is a potential therapeutic target for DMG (Mohammad et al. 2017). Steven Pollard's group established a

collection of primary NPC lines derived from various locations in the human fetal brain, revealing region-specific effects of histone H3.3G34R mutation (Bressan et al. 2021). His group also used hindbrain NPCs to create an H3K27M-mutant DMG model and unveiled the underlying epigenetic mechanisms (Brien et al. 2021). Paul Knoepfler's group used CRISPR-Cas9 genome editing technology to introduce H3.3K27M and G34R mutations into human astrocytes and showed the activation of the Notch signaling pathway (Chen et al. 2020b).

Mouse Models of Childhood Brain Tumors

Mouse models, including xenograft models and GEMMs, are widely employed in the field. In addition to conventional knockout and knockin mice, various gene transfer and engineering technologies have been developed and used. For instance, the RCAS-based system, developed by Eric Holland and Harold Varmus, enables spatial–temporal control of transgene introduction (Holland and Varmus 1998). The Sleeping Beauty transposon system, initially used by John Ohlfest's group, along with other transposon-based systems, has dramatically expedited the development of various brain tumor models, including those for H3.3K27M-mutant DMG and H3.3G34R-mutant cortical high-grade glioma (Wiesner et al. 2009; Pathania et al. 2017; Haase et al. 2022). In addition, Sleeping Beauty–based mutagenesis has been used for forward genetic screening to identify new oncogenes and tumor suppressor genes in childhood brain tumors (Wu et al. 2012b; Genovesi et al. 2013; Beckmann et al. 2019).

In line with observations from stem cell–based models, mouse models of brain tumors have also demonstrated cell-type-specific effects of driver mutations, as well as the synergistic effects of multiple mutations. For example, initial studies indicate that RCAS-based transduction of K27M-mutant histone H3.3 is insufficient to induce full-blown tumors in the mouse brain (Lewis et al. 2013). Oren Becher's group developed mouse models of H3.3K27M-mutant DMG by combining H3.3K27M with other oncogenic mutations, including the overexpression of

PDGF ligand and p53 knockout (Cordero et al. 2017; Tomita et al. 2022). His group and others demonstrated that their DMG mouse model serves as a valuable platform to assess the efficacy of therapies (Deland et al. 2021; Watanabe et al. 2024). Although overexpressing K27M-mutant histone H3.3 successfully triggers tumor development when combined with other mutations, the dosage of wild-type and mutant histones remains a subject of debate. To address this concern, Suzanne Baker's group created a conditional knockin mouse model in which K27M-mutant histone H3.3 is expressed from the endogenous *H3f3a* locus following Cre recombinase-mediated excision of a loxP-flanked transcriptional STOP cassette (Larson et al. 2019). Consistent with our study, the addition of p53 knockout and activation of platelet-derived growth factor receptor α (PDGFRα) accelerated tumorigenesis in the brainstem. Another group introduced K27M-mutant histone H3.3 into embryonic precursor cells in the fetal brain (E12.5–E13.5) through in utero electroporation (IUE) of piggyBac transposon-based vectors (Pathania et al. 2017). This study demonstrated that postnatal introduction of K27M-mutant histone H3.3 and p53 knockout failed to induce tumor formation, suggesting an embryonic origin of H3K27M-mutant DMG. The same study also reported the establishment of a mouse model for the H3.3G34 subtype. Mouse models for H3.3G34R-mutant cortical high-grade glioma have been independently reported by several groups (Kim et al. 2019; Haase et al. 2022; Abdallah et al. 2023). All of them showed the synergistic effects of H3.3G34R mutation, Atrx loss, and p53 loss, which is consistent with the cancer genomics studies that showed the frequent co-occurrence of these mutations in patient tumors (Schwartzentruber et al. 2012; Sturm et al. 2012; Mackay et al. 2017).

GEMMs are also widely used to study medulloblastoma, one of the most prevalent types of brain cancer in young children. It is widely accepted that there are four subgroups of medulloblastoma (WNT, SHH, group 3, and group 4). Significant efforts have been made to create mouse models that represent each subgroup. One of the most frequently used models is the SHH subgroup model using *Ptch1*$^{+/-}$ mice, first

reported in 1997 by Matthew Scott's group (Goodrich et al. 1997). Since then, more than 50 medulloblastoma GEMMs have been generated by using various technologies, providing valuable insight into subtype-specific pathogenic mechanisms, including the cell types of origins (Roussel and Stripay 2020). For example, conditional knockout of *Ptch1* as well as conditional expression of an activated allele of Smoothened (SmoM2) resulted in the formation of an SHH subgroup from granule neuron precursors and stem cells (Schüller et al. 2008; Yang et al. 2008). Daniel Fults' group employed the RCAS system to overexpress Shh or Myc to generate the SHH subgroup and group 3 subgroup, respectively (Rao et al. 2004; Jenkins et al. 2016). A WNT subtype model can be achieved by activating the Wnt signaling pathway and inhibiting the p53 pathway in the floor of the fourth ventricle (Gibson et al. 2010). This study indicates that the Wnt subtype medulloblastoma originates from the embryonic dorsal brainstem, setting it apart from other medulloblastoma subtypes that are derived from the cerebellum. A group 3 model can be created by overexpressing either N-Myc or Myc oncogene, alone or in combination with other mutations (Swartling et al. 2010; Kawauchi et al. 2012; Pei et al. 2012; Jenkins et al. 2016). Regarding group 4, the first GEMMs were established by introducing a dominant negative form of Trp53 (DNp53) and a constitutively active form of SRC into the nuclear transitory zone (NTZ) progenitors of the developing cerebellum (Forget et al. 2018).

In addition to these medulloblastoma models, GEMMs for other brain tumor subtypes have been developed. For example, Eric Holland's group created a new ependymoma mouse model by introducing the C11orf95–RELA fusion gene by the RCAS system (Ozawa et al. 2018). Sila Ultanir's group created another ependymoma model by expressing active nuclear YAP1 under NEX/NeuroD6-Cre (Eder et al. 2020). Their study also showed that conditional double knockout of the YAP1 suppressors LATS1 and LATS2, resulted in the formation of tumors that display histological features of ependymoma. Tom Curran's group showed that conditional double knockout of Snf5/Smarcb1 and p53 led to the

Cite this article as *Cold Spring Harb Perspect Med* doi: 10.1101/cshperspect.a041711

formation of highly aggressive brain tumors displaying hallmarks of CNS AT/RT (Ng et al. 2015).

In sum, these mouse models clearly demonstrated the cell-type-specific effects of oncogenic mutations and the crucial role of developmental programs in the formation of childhood brain tumors. Also, mouse models can be used for identifying and assessing new therapeutic approaches. Furthermore, GEMMs and syngeneic mouse models provide powerful platforms to study the tumor microenvironment, including tumor–immune interactions, as they have an intact immune system (Messiaen et al. 2023).

MODELING OF BLOOD CANCER AND MDS

Animal Models of Blood Cancers

Blood cancers, including leukemia, lymphoma, and myeloma, are the most prevalent types of cancer in children, with an incidence of ∼7.4 cases per 100,000 children. In the 1950s, researchers established mouse models for blood cancers by introducing chemical carcinogens or oncogenic viruses into mice. These models allowed for the serial transplantation of cancerous cells into irradiated mice, enabling their long-term maintenance. The discovery of the nude mouse, an immunocompromised strain carrying the Foxn1 mutation, marked a significant breakthrough (Flanagan 1966). This strain and subsequent highly immunocompromised mouse models greatly improved the engraftment of human normal hematopoietic cells and blood cancer cells, leading to groundbreaking findings, such as the identification of cancer stem cells by John Dick's group (Kamel-Reid and Dick 1988; Lapidot et al. 1994). This discovery underscored the presence of a cellular hierarchy within the blood cancer population, resembling the normal hematopoietic system. However, a limitation of using immunodeficient animals is the absence of immune system interaction. To address this, researchers developed "humanized mice" by transplanting human hematopoietic stem cells (HSCs) into irradiated immunocompromised mice. Another approach to creating mouse models for blood cancers is via genetic engineering. One no-

table model is Eμ-Myc mice, which express the c-myc oncogene under the IgH enhancer (Adams et al. 1985), mimicking the IGH-MYC translocation commonly found in diffuse large B-cell lymphoma (DLBCL) and Burkitt lymphoma. These mice develop B-cell leukemia/lymphoma with a 100% penetrance rate. Similarly, models expressing BCL6 (B-cell lymphoma 6) under the endogenous Iμ promoter were created based on the findings that BCL6 is essential for germinal center (GC) formation and inhibits GC B-cell differentiation (Cattoretti et al. 2005). These models demonstrated dysregulated GC B-cell proliferation and differentiation due to driver mutations. Following the development of lymphoma models, extensive efforts have been directed toward creating GEMMs for leukemias, particularly acute myeloid leukemia (AML), which has a poor prognosis with a 5-year survival rate of ∼30% (Kurtz et al. 2022). Researchers have tested various methods, including the overexpression of oncogenic fusion genes, to induce leukemia in murine bone marrow. These studies have highlighted the importance of identifying downstream genes of driver genes and therapeutic strategies using GEMMs. In the quest to develop GEMMs that more accurately resemble the pathogenesis of human leukemia, Terence Rabbitts's group introduced MLL–AF9 into the endogenous *Mll1* locus, successfully creating an AML model (Corral et al. 1996). Although MLL1 is ubiquitously expressed in various tissues, the transgenic mice only developed AML, suggesting the lineage-specific role of the fusion gene in cancer development. Gary Gilliland's group created a series of GEMMs by expressing FLT3 with internal tandem duplication mutation (FLT3–ITD), one of the most common mutations in AML (Kelly et al. 2002; Lee et al. 2005, 2007). Although the expression of FLT3–ITD alone resulted in a myeloproliferative disorder, not AML, the combination of FLT3–ITD with other mutations led to the development of fatal AML (Zorko et al. 2012).

Cell-Based Models of Blood Cancers

Over 100 blood cancer cell lines have been established, covering all major subtypes of lymphoma and leukemia. Thanks to the development of immunocompromised mice as described above, these cell lines have been used to dissect the molecular mechanisms of cancer formation as well as develop and evaluate new therapeutic approaches. We will not discuss the details as they are well described elsewhere (Quentmeier et al. 2019).

In 2001, James Thomson's group reported the derivation of hematopoietic cells from human ESCs by coculturing them with murine bone marrow cell line S17 or the yolk sac endothelial cell line C166 (Kaufman et al. 2001). Subsequently, more efficient, robust, and chemically defined protocols were developed (Demirci et al. 2020). Although the derivation of long-term HSCs remains challenging, hematopoietic progenitor cells of all lineages and their progeny can be derived from ESCs and iPSCs. Notably, many genes critical for hematopoietic development and differentiation are mutated in blood cancers, indicating a close link between hematopoietic development and the initiation of blood cancer. Four years after the groundbreaking discovery of iPSCs, which were initially generated from fibroblasts, reprogramming of peripheral blood cells into iPSCs was achieved. This breakthrough allowed for the derivation of iPSCs from blood cancer patients. Eirini Papapetrou's group was the first to use iPSCs to model Myelodysplastic syndromes (MDS), a condition where immature blood cells in the bone marrow cannot properly mature (Kotini et al. 2015). They derived iPSC lines from MDS patients carrying chromosome 7q deletion, a characteristic of MDS, and demonstrated that MDS iPSCs have diminished hematopoietic differentiation potential. Subsequently, her group and Ravindra Majeti's group successfully created models of AML (Chao et al. 2017; Kotini et al. 2017). These models replicated impaired differentiation and maturation and could induce AML when transplanted into immunocompromised mice. Similar approaches have been employed to model other types of blood cancers, such as chronic myeloid leukemia (CML) (Kumano et al. 2012; Telliam et al. 2023) and Juvenile myelomonocytic leukemia (JMML) (Gandre-Babbe et al. 2013; Mulero-Navarro et al. 2015; Shigemura et al. 2019; Tasian et al. 2019). These models can be further genetically modified to simulate disease progression or

correct driver mutations. Additionally, a model of acute lymphoblastic leukemia (ALL) was created by introducing the ETV6–RUNX1 fusion gene into human PSCs (Böiers et al. 2018). Collectively, these studies have demonstrated that blood cancers are caused by dysregulated developmental programs that impede the maturation of hematopoietic stem/progenitor cells.

DEVELOPMENTAL MODELS FOR OTHER CANCER TYPES

Modeling of Bone and Soft Tissue Tumors

Bone and soft tissue tumors share similarities with childhood brain cancers in terms of recurrent mutations in histone genes. Specifically, the H3.3K36M mutation is nearly ubiquitous in chondroblastoma, suggesting its significance in the cell type of origin. When expressed in a mesenchymal stem cell line, H3.3K36M suppresses differentiation and generates tumors in immunodeficient mice (Lu et al. 2016). However, H3.3K36M knockin mice display aberrations in histone marks but do not develop tumors (Abe et al. 2021). These findings indicate the presence of either another oncogenic alteration or potential species-specific effects of the mutation.

Alveolar rhabdomyosarcoma (ARMS) is a prevalent form of pediatric sarcoma driven by the PAX3–FOXO1 (previously known as PAX3–FKHR) fusion gene. The expression of the PAX3–FOXO1 fusion gene has the capability to transform cells, promoting anchorage-independent growth. However, additional cooperating mutations are required to initiate in vivo tumor formation (Scheidler et al. 1996). Knockin mice expressing the PAX3–FOXO1 fusion protein in neural crest and muscle precursor cells only develop tumors when combined with other oncogenic mutations in either p53 or Ink4a/Arf (Keller et al. 2004). Another study showed that PAX3–FOXO1 fusion protein reprograms endothelial progenitors to functional myogenic stem cells, suggesting that endothelial progenitors can be the cell of origin of ARMS (Searcy et al. 2023).

Ewing sarcoma, aggressive bone sarcomas predominantly affecting children, is primarily

Cite this article as *Cold Spring Harb Perspect Med* doi: 10.1101/cshperspect.a041711

caused by chromosomal translocations that produce the EWS–FLI1 fusion gene. More than a dozen cell lines have been derived from tumors from Ewing sarcoma patients and have been used for the research. Development of GEMMs of Ewing sarcoma is challenging because overexpression of EWS–FLT1 fusion oncogene tends to induce cell death in mouse ESCs and other embryonic cell types (Minas et al. 2017). Even in permissive cell types, such as mesenchymal stem cells, neural crest stem cells, and embryonic osteochondrogenic progenitor cells, no tumor formation has been observed. Consequently, several research groups have explored alternative approaches to create a preclinical model of Ewing sarcoma. For example, James Amatruda's group has developed a zebrafish model of Ewing sarcoma by overexpressing the EWSR1–FLT1 fusion protein (Leacock et al. 2012; Vasileva et al. 2022).

Peripheral nerve sheath tumors, including schwannomas and neurofibromas, are a type of soft tissue tumors. Although most peripheral nerve sheath tumors are benign, some become malignant referred to as malignant peripheral nerve sheath tumors (MPNSTs). Patient-derived cell lines and patient-derived xenograft (PDX) models have been used to elucidate the molecular mechanisms underlying the development of MPNST and to identify therapeutic approaches (Miller et al. 2006; Pollard et al. 2020; Larsson et al. 2023; Magallón-Lorenz et al. 2023). Additionally, multiple GEMMs of MPNST have been created and used for investigating druggable downstream pathways, the time and lineage dependency of the tumorigenesis, and stem cell–like populations in MPNST (Wu et al. 2018; Inoue et al. 2021; Sun et al. 2021).

Modeling of Neuroblastoma

Neuroblastoma is a childhood cancer found in various parts of the body including the adrenal gland, neck, chest, and spinal cord. Currently, more than 100 neuroblastoma cell lines of both human and rodent origin are available, thanks in part to collaborative efforts led by the Children's Oncology Group. Although cell lines serve as a useful platform for dissecting tumor biology and

developing new therapeutic strategies, they are not ideal for investigating the developmental origins of neuroblastoma, which remains a topic of debate. To address this question, GEMMs and human stem cell–based models have been created. Several mouse studies showed that overexpression of MYCN and/or mutant ALK in sympathoadrenal cells led to the formation of neuroblastoma (Berry et al. 2012; Heukamp et al. 2012; Cazes et al. 2014). Similarly, MYCN overexpression transforms human ESC-derived neural crest cells (Weng et al. 2022). The transformed cells exhibited enhanced proliferation, impaired differentiation, and in vivo tumorigenicity. Additionally, this study identified CD55 as a novel cancer stem cell marker as well as therapeutic target. Other models used directly reprogrammed human NPCs and sympathetic neuroblasts from chick embryos (Kramer et al. 2016; Sitnikov et al. 2023). These studies collectively indicate the sympathoadrenal origin of neuroblastoma, which is consistent with a recent single-cell transcriptomic study (Dong et al. 2020). Recent studies have developed protocols for differentiating PSCs into various cell types associated with neuroblastoma, including sympathetic neurons and their progenitors (Saxena et al. 2013; Carr-Wilkinson et al. 2018; Wu et al. 2024). These advances hold promise for further elucidating the cellular and molecular mechanisms underlying neuroblastoma formation.

Modeling of Retinoblastoma

Retinoblastoma is a malignant tumor of the developing retina that mostly affects young children.

Mouse models of retinoblastoma have been established by combining retinoblastoma (RB1) knockout with the deletion of additional tumor suppressor genes, such as p107 and p130 (Chen et al. 2004; MacPherson et al. 2004). These mouse models suggest that a double knockout results in cell death in ganglion precursors but not amacrine precursors, thereby supporting the concept of cell-type-specific tumorigenesis. Retinal organoids have also been developed from PSCs, and several studies have investigated retinoblastoma initiation in vitro (Eiraku et al. 2011; Meyer et al.

2011; Nakano et al. 2012). Michael Dyer's group used RB1 mutant human patient–derived iPSCs and created a stem cell–based retinoblastoma model (Norrie et al. 2021). Even though the prognosis of retinoblastoma patients is generally good with the overall survival rate of 95% or higher, some patients with high-risk retinoblastoma undergo eye removal surgery and, in rare cases, succumb to the disease. Thus, there is still a clinical need to further advance retinoblastoma research and treatment.

CONCLUSION

Dysregulated developmental programs are widely observed in childhood cancers. Recent scientific and technological advances in various fields, including gene transfer, genome editing, and stem cell biology, have enabled the development of models that can recapitulate the genetic and molecular characteristics of patient tumors and experimental investigation of the precise cellular and molecular mechanisms underlying the development of cancers. More recently, the rapid development of single-cell transcriptomic, proteomic, epigenetic, and genomic technologies has led to additional insight into the complexity of cellular heterogeneity and lineage trajectories, aiding the development of new models. Together, these approaches have led strong support to the concept that childhood cancer is a developmental disease. Developmental programs are precisely regulated in utero, and these regulations are cell type/lineage dependent. The behavior of each cell type/lineage is programmed by the interrelation between the gene regulatory network and the epigenetic landscape in specific developmental contexts of time and space. This can explain why driver mutations exhibit cell type/lineage-specific effects in many models where they dysregulate developmental programs that control stemness, differentiation, and proliferation in specific cells of origin. Ongoing investigations will provide more insight into the mechanisms of developmental dysregulation and perhaps how best to block them. Additionally, these models have been used to identify and assess new therapeutic approaches. Most childhood cancers are rare diseases that also exhibit significant hetero-

geneity as shown in multiomics studies. Complementing other approaches, developmental models can serve as powerful and practical tools to advance our understanding of childhood cancers and achieve precision medicine for children suffering from the devastating disease.

REFERENCES

Abdallah AS, Cardona HJ, Gadd SL, Brat DJ, Powla PP, Alruwalli WS, Shen C, Picketts DJ, Li X-N, Becher OJ. 2023. Novel genetically engineered H3.3G34R model reveals cooperation with ATRX loss in upregulation of *Hoxa* cluster genes and promotion of neuronal lineage. *Neurooncol Adv* 5: vdad003. doi:10.1093/noajnl/vdad003

Abe S, Nagatomo H, Sasaki H, Ishiuchi T. 2021. A histone H3.3K36M mutation in mice causes an imbalance of histone modifications and defects in chondrocyte differentiation. *Epigenetics* 16: 1123–1134. doi:10.1080/15592294.2020.1841873

Abud EM, Ramirez RN, Martinez ES, Healy LM, Nguyen CHH, Newman SA, Yeromin AV, Scarfone VM, Marsh SE, Fimbres C, et al. 2017. iPSC-derived human microglia-like cells to study neurological diseases. *Neuron* 94: 278–293.e9. doi:10.1016/j.neuron.2017.03.042

Adams JM, Harris AW, Pinkert CA, Corcoran LM, Alexander WS, Cory S, Palmiter RD, Brinster RL. 1985. The c-myc oncogene driven by immunoglobulin enhancers induces lymphoid malignancy in transgenic mice. *Nature* 318: 533–538. doi:10.1038/318533a0

Ballabio C, Anderle M, Gianesello M, Lago C, Miele E, Cardano M, Aiello G, Piazza S, Caron D, Gianno F, et al. 2020. Modeling medulloblastoma in vivo and with human cerebellar organoids. *Nat Commun* 11: 583. doi:10.1038/s41467-019-13989-3

Beckmann PJ, Larson JD, Larsson AT, Ostergaard JP, Wagner S, Rahrmann EP, Shamsan GA, Otto GM, Williams RL, Wang J, et al. 2019. *Sleeping Beauty* insertional mutagenesis reveals important genetic drivers of central nervous system embryonal tumors. *Cancer Res* 79: 905–917. doi:10.1158/0008-5472.CAN-18-1261

Ben-David U, Siranosian B, Ha G, Tang H, Oren Y, Hinohara K, Strathdee CA, Dempster J, Lyons NJ, Burns R, et al. 2018. Genetic and transcriptional evolution alters cancer cell line drug response. *Nature* 560: 325–330. doi:10.1038/s41586-018-0409-3

Berry T, Luther W, Bhatnagar N, Jamin Y, Poon E, Sanda T, Pei D, Sharma B, Vetharoy WR, Hallsworth A, et al. 2012. The ALKF1174L mutation potentiates the oncogenic activity of MYCN in neuroblastoma. *Cancer Cell* 22: 117–130. doi:10.1016/j.ccr.2012.06.001

Bian S, Repic M, Guo Z, Kavirayani A, Burkard T, Bagley JA, Krauditsch C, Knoblich JA. 2018. Genetically engineered cerebral organoids model brain tumor formation. *Nat Methods* 15: 631–639. doi:10.1038/s41592-018-0070-7

Böiers C, Richardson SE, Laycock E, Zriwil A, Turati VA, Brown J, Wray JP, Wang D, James C, Herrero J, et al. 2018. A human IPS model implicates embryonic B-myeloid fate restriction as developmental susceptibility to B acute lym-

Cite this article as *Cold Spring Harb Perspect Med* doi: 10.1101/cshperspect.a041711

phoblastic leukemia-associated ETV6-RUNX1. *Dev Cell* **44:** 362–377.e7. doi:10.1016/j.devcel.2017.12.005

Bosma GC, Custer RP, Bosma MJ. 1983. A severe combined immunodeficiency mutation in the mouse. *Nature* **301:** 527–530. doi:10.1038/301527a0

Bray F, Laversanne M, Sung H, Ferlay J, Siegel RL, Soerjomataram I, Jemal A. 2024. Global cancer statistics 2022: GLOBOCAN estimates of incidence and mortality worldwide for 36 cancers in 185 countries. *CA Cancer J Clin* **74:** 229–263. doi:10.3322/caac.21834

Bressan RB, Southgate B, Ferguson KM, Blin C, Grant V, Alfazema N, Wills JC, Marques-Torrejon MA, Morrison GM, Ashmore J, et al. 2021. Regional identity of human neural stem cells determines oncogenic responses to histone H3.3 mutants. *Cell Stem Cell* **28:** 877–893.e9. doi:10.1016/j.stem.2021.01.016

Brien GL, Bressan RB, Monger C, Gannon D, Lagan E, Doherty AM, Healy E, Neikes H, Fitzpatrick DJ, Deevy O, et al. 2021. Simultaneous disruption of PRC2 and enhancer function underlies histone H3-K27M oncogenic activity in human hindbrain neural stem cells. *Nat Genet* **53:** 1221–1232. doi:10.1038/s41588-021-00897-w

Cakir B, Xiang Y, Tanaka Y, Kural MH, Parent M, Kang Y-J, Chapeton K, Patterson B, Yuan Y, He C-S, et al. 2019. Engineering of human brain organoids with a functional vascular-like system. *Nat Methods* **16:** 1169–1175. doi:10.1038/s41592-019-0586-5

Čančer M, Hutter S, Holmberg KO, Rosén G, Sundström A, Tailor J, Bergström T, Garancher A, Essand M, Wechsler-Reya RJ, et al. 2019. Humanized stem cell models of pediatric medulloblastoma reveal an Oct4/mTOR axis that promotes malignancy. *Cell Stem Cell* **25:** 855–870.e11. doi:10.1016/j.stem.2019.10.005

Caretti V, Sewing ACP, Lagerweij T, Schellen P, Bugiani M, Jansen MHA, van Vuurden DG, Navis AC, Horsman I, Vandertop WP, et al. 2014. Human pontine glioma cells can induce murine tumors. *Acta Neuropathol* **127:** 897–909. doi:10.1007/s00401-014-1272-4

Carr-Wilkinson J, Prathalingam N, Pal D, Moad M, Lee N, Sundaresh A, Forgham H, James P, Herbert M, Lako M, et al. 2018. Differentiation of human embryonic stem cells to sympathetic neurons: a potential model for understanding neuroblastoma pathogenesis. *Stem Cells Int* **2018:** 1–12. doi:10.1155/2018/4391641

Cattoretti G, Pasqualucci L, Ballon G, Tam W, Nandula SV, Shen Q, Mo T, Murty VV, Dalla-Favera R. 2005. Deregulated BCL6 expression recapitulates the pathogenesis of human diffuse large B cell lymphomas in mice. *Cancer Cell* **7:** 445–455. doi:10.1016/j.ccr.2005.03.037

Cazes A, Lopez-Delisle L, Tsarovina K, Pierre-Eugène C, De Preter K, Peuchmaur M, Nicolas A, Provost C, Louis-Brennetot C, Daveau R, et al. 2014. Activated Alk triggers prolonged neurogenesis and Ret upregulation providing a therapeutic target in ALK-mutated neuroblastoma. *Oncotarget* **5:** 2688–2702. doi:10.18632/oncotarget.1883

Chambers SM, Fasano CA, Papapetrou EP, Tomishima M, Sadelain M, Studer L. 2009. Highly efficient neural conversion of human ES and iPS cells by dual inhibition of SMAD signaling. *Nat Biotechnol* **27:** 275–280. doi:10.1038/nbt.1529

Chao MP, Gentles AJ, Chatterjee S, Lan F, Reinisch A, Corces MR, Xavy S, Shen J, Haag D, Chanda S, et al. 2017. Human AML-iPSCs reacquire leukemic properties after differentiation and model clonal variation of disease. *Cell Stem Cell* **20:** 329–344.e7. doi:10.1016/j.stem.2016.11.018

Chen D, Livne-bar I, Vanderluit JL, Slack RS, Agochiya M, Bremner R. 2004. Cell-specific effects of RB or RB/p107 loss on retinal development implicate an intrinsically death-resistant cell-of-origin in retinoblastoma. *Cancer Cell* **5:** 539–551. doi:10.1016/j.ccr.2004.05.025

Chen CCL, Deshmukh S, Jessa S, Hadjadj D, Lisi V, Andrade AF, Faury D, Jawhar W, Dali R, Suzuki H, et al. 2020a. Histone H3.3G34-mutant interneuron progenitors co-opt PDGFRA for gliomagenesis. *Cell* **183:** 1617–1633. e22. doi:10.1016/j.cell.2020.11.012

Chen K-Y, Bush K, Klein RH, Cervantes V, Lewis N, Naqvi A, Carcaboso AM, Lechpammer M, Knoepfler PS. 2020b. Reciprocal H3.3 gene editing identifies K27M and G34R mechanisms in pediatric glioma including NOTCH signaling. *Commun Biol* **3:** 363. doi:10.1038/s42003-020-1076-0

Cordero FJ, Huang Z, Grenier C, He X, Hu G, McLendon RE, Murphy SK, Hashizume R, Becher OJ. 2017. Histone H3.3K27M represses *p16* to accelerate gliomagenesis in a murine model of DIPG. *Mol Cancer Res* **15:** 1243–1254. doi:10.1158/1541-7786.MCR-16-0389

Corral J, Lavenir I, Impey H, Warren AJ, Forster A, Larson TA, Bell S, McKenzie ANJ, King G, Rabbitts TH. 1996. An Mll-AF9 fusion gene made by homologous recombination causes acute leukemia in chimeric mice: a method to create fusion oncogenes. *Cell* **85:** 853–861. doi:10.1016/S0092-8674(00)81269-6

Deland K, Starr BF, Mercer JS, Byemerwa J, Crabtree DM, Williams NT, Luo L, Ma Y, Chen M, Becher OJ, et al. 2021. Tumor genotype dictates radiosensitization after Atm deletion in primary brainstem glioma models. *J Clin Invest* **131:** e142158. doi:10.1172/JCI142158

Demirci S, Leonard A, Tisdale JF. 2020. Hematopoietic stem cells from pluripotent stem cells: clinical potential, challenges, and future perspectives. *Stem Cells Transl Med* **9:** 1549–1557. doi:10.1002/sctm.20-0247

Dong R, Yang R, Zhan Y, Lai H-D, Ye C-J, Yao X-Y, Luo W-Q, Cheng X-M, Miao J-J, Wang J-F, et al. 2020. Single-cell characterization of malignant phenotypes and developmental trajectories of adrenal neuroblastoma. *Cancer Cell* **38:** 716–733.e6. doi:10.1016/j.ccell.2020.08.014

Eder N, Roncaroli F, Domart M-C, Horswell S, Andreiuolo F, Flynn HR, Lopes AT, Claxton S, Kilday J-P, Collinson L, et al. 2020. YAP1/TAZ drives ependymoma-like tumour formation in mice. *Nat Commun* **11:** 2380. doi:10.1038/s41467-020-16167-y

Eiraku M, Takata N, Ishibashi H, Kawada M, Sakakura E, Okuda S, Sekiguchi K, Adachi T, Sasai Y. 2011. Self-organizing optic-cup morphogenesis in three-dimensional culture. *Nature* **472:** 51–56. doi:10.1038/nature09941

Evans MJ, Kaufman MH. 1981. Establishment in culture of pluripotential cells from mouse embryos. *Nature* **292:** 154–156. doi:10.1038/292154a0

Filbin M, Monje M. 2019. Developmental origins and emerging therapeutic opportunities for childhood cancer. *Nat Med* **25:** 367–376. doi:10.1038/s41591-019-0383-9

Flanagan SP. 1966. "Nude," a new hairless gene with pleiotropic effects in the mouse. *Genet Res* **8:** 295–309. doi:10.1017/S0016672300010168

Forget A, Martignetti L, Puget S, Calzone L, Brabetz S, Picard D, Montagud A, Liva S, Sta A, Dingli F, et al. 2018. Aberrant ERBB4-SRC signaling as a hallmark of group 4 medulloblastoma revealed by integrative phosphoproteomic profiling. *Cancer Cell* **34**: 379–395.e7. doi:10.1016/j .ccell.2018.08.002

Funato K, Major T, Lewis PW, Allis CD, Tabar V. 2014. Use of human embryonic stem cells to model pediatric gliomas with H3.3K27M histone mutation. *Science* **346**: 1529–1533. doi:10.1126/science.1253799

Funato K, Smith RC, Saito Y, Tabar V. 2021. Dissecting the impact of regional identity and the oncogenic role of human-specific NOTCH2NL in an hESC model of H3.3G34R-mutant glioma. *Cell Stem Cell* **28**: 894–905. e7. doi:10.1016/j.stem.2021.02.003

Gandre-Babbe S, Paluru P, Aribeana C, Chou ST, Bresolin S, Lu L, Sullivan SK, Tasian SK, Weng J, Favre H, et al. 2013. Patient-derived induced pluripotent stem cells recapitulate hematopoietic abnormalities of juvenile myelomonocytic leukemia. *Blood* **121**: 4925–4929. doi:10.1182/ blood-2013-01-478412

Genovesi LA, Ng CG, Davis MJ, Remke M, Taylor MD, Adams DJ, Rust AG, Ward JM, Ban KH, Jenkins NA, et al. 2013. *Sleeping Beauty* mutagenesis in a mouse medulloblastoma model defines networks that discriminate between human molecular subgroups. *Proc Natl Acad Sci* **110**: E4325–E4334. doi:10.1073/pnas.1318639110

Gibson P, Tong Y, Robinson G, Thompson MC, Currle DS, Eden C, Kranenburg TA, Hogg T, Poppleton H, Martin J, et al. 2010. Subtypes of medulloblastoma have distinct developmental origins. *Nature* **468**: 1095–1099. doi:10 .1038/nature09587

Gillet J-P, Varma S, Gottesman MM. 2013. The clinical relevance of cancer cell lines. *J Natl Cancer Inst* **105**: 452–458. doi:10.1093/jnci/djt007

Goodrich LV, Milenković L, Higgins KM, Scott MP. 1997. Altered neural cell fates and medulloblastoma in mouse *patched* mutants. *Science* **277**: 1109–1113. doi:10.1126/ science.277.5329.1109

Haag D, Mack N, Benites Goncalves da Silva P, Statz B, Clark J, Tanabe K, Sharma T, Jäger N, Jones DTW, Kawauchi D, et al. 2021. H3.3-K27M drives neural stem cell-specific gliomagenesis in a human iPSC-derived model. *Cancer Cell* **39**: 407–422.e13. doi:10.1016/j.ccell.2021.01.005

Haase S, Banerjee K, Mujeeb AA, Hartlage CS, Núñez FM, Núñez FJ, Alghamri MS, Kadiyala P, Carney S, Barissi MN, et al. 2022. H3.3-G34 mutations impair DNA repair and promote cGAS/STING-mediated immune responses in pediatric high-grade glioma models. *J Clin Invest* **132**: e154229. doi:10.1172/JCI154229

Hashimoto R, Smirnov I, Liu S, Phillips JJ, Hyer J, McKnight TR, Wendland M, Prados M, Banerjee A, Nicolaides T, et al. 2012. Characterization of a diffuse intrinsic pontine glioma cell line: implications for future investigations and treatment. *J Neurooncol* **110**: 305–313. doi:10.1007/ s11060-012-0973-6

He C, Xu K, Zhu X, Dunphy PS, Gudenas B, Lin W, Twarog N, Hover LD, Kwon C-H, Kasper LH, et al. 2021. Patient-derived models recapitulate heterogeneity of molecular signatures and drug response in pediatric high-grade glioma. *Nat Commun* **12**: 4089. doi:10.1038/s41467-021-24168-8

Heukamp LC, Thor T, Schramm A, De Preter K, Kumps C, De Wilde B, Odersky A, Peifer M, Lindner S, Spruessel A, et al. 2012. Targeted expression of mutated ALK induces neuroblastoma in transgenic mice. *Sci Transl Med* **4**: 141ra9. doi:10.1126/scitranslmed.3003967

Holland EC, Varmus HE. 1998. Basic fibroblast growth factor induces cell migration and proliferation after glia-specific gene transfer in mice. *Proc Natl Acad Sci* **95**: 1218–1223. doi:10.1073/pnas.95.3.1218

Huang M, Tailor J, Zhen Q, Gillmor AH, Miller ML, Weishaupt H, Chen J, Zheng T, Nash EK, McHenry LK, et al. 2019. Engineering genetic predisposition in human neuroepithelial stem cells recapitulates medulloblastoma tumorigenesis. *Cell Stem Cell* **25**: 433–446.e7. doi:10.1016/j .stem.2019.05.013

Hubert CG, Rivera M, Spangler LC, Wu Q, Mack SC, Prager BC, Couce M, McLendon RE, Sloan AE, Rich JN. 2016. A three-dimensional organoid culture system derived from human glioblastomas recapitulates the hypoxic gradients and cancer stem cell heterogeneity of tumors found in vivo. *Cancer Res* **76**: 2465–2477. doi:10.1158/0008-5472 .CAN-15-2402

Ijaz H, Koptyra M, Gaonkar KS, Rokita JL, Baubet VP, Tauhid L, Zhu Y, Brown M, Lopez G, Zhang B, et al. 2020. Pediatric high-grade glioma resources from the Children's Brain Tumor Tissue Consortium. *Neuro Oncol* **22**: 163–165. doi:10.1093/neuonc/noz192

Inoue A, Janke LJ, Gudenas BL, Jin H, Fan Y, Paré J, Clay MR, Northcott PA, Hirbe AC, Cao X. 2021. A genetic mouse model with postnatal *Nf1* and *p53* loss recapitulates the histology and transcriptome of human malignant peripheral nerve sheath tumor. *Neurooncol Adv* vdab129. doi:10.1093/noajnl/vdab129

Jacob F, Salinas RD, Zhang DY, Nguyen PTT, Schnoll JG, Wong SZH, Thokala R, Sheikh S, Saxena D, Prokop S, et al. 2020. A patient-derived glioblastoma organoid model and biobank recapitulates inter- and intra-tumoral heterogeneity. *Cell* **180**: 188–204.e22. doi:10.1016/j.cell.2019 .11.036

Jenkins NC, Rao G, Eberhart CG, Pedone CA, Dubuc AM, Fults DW. 2016. Somatic cell transfer of c-Myc and Bcl-2 induces large-cell anaplastic medulloblastomas in mice. *J Neurooncol* **126**: 415–424. doi:10.1007/s11060-015-1985-9

Kamel-Reid S, Dick JE. 1988. Engraftment of immune-deficient mice with human hematopoietic stem cells. *Science* **242**: 1706–1709. doi:10.1126/science.2904703

Kaufman DS, Hanson ET, Lewis RL, Auerbach R, Thomson JA. 2001. Hematopoietic colony-forming cells derived from human embryonic stem cells. *Proc Natl Acad Sci* **98**: 10716–10721. doi:10.1073/pnas.191362598

Kawauchi D, Robinson G, Uziel T, Gibson P, Rehg J, Gao C, Finkelstein D, Qu C, Pounds S, Ellison DW, et al. 2012. A mouse model of the most aggressive subgroup of human medulloblastoma. *Cancer Cell* **21**: 168–180. doi:10.1016/j .ccr.2011.12.023

Keller C, Arenkiel BR, Coffin CM, El-Bardeesy N, DePinho RA, Capecchi MR. 2004. Alveolar rhabdomyosarcomas in conditional *Pax3:Fkhr* mice: cooperativity of Ink4a/ARF and Trp53 loss of function. *Genes Dev* **18**: 2614–2626. doi:10.1101/gad.1244004

Cite this article as *Cold Spring Harb Perspect Med* doi: 10.1101/cshperspect.a041711

Kelly LM, Liu Q, Kutok JL, Williams IR, Boulton CL, Gilliland DG. 2002. FLT3 internal tandem duplication mutations associated with human acute myeloid leukemias induce myeloproliferative disease in a murine bone marrow transplant model. *Blood* **99:** 310–318. doi:10.1182/blood.V99.1.310

Khuong-Quang D-A, Buczkowicz P, Rakopoulos P, Liu X-Y, Fontebasso AM, Bouffet E, Bartels U, Albrecht S, Schwartzentruber J, Letourneau L, et al. 2012. K27m mutation in histone H3.3 defines clinically and biologically distinct subgroups of pediatric diffuse intrinsic pontine gliomas. *Acta Neuropathol* **124:** 439–447. doi:10.1007/s00401-012-0998-0

Kim GB, Rincon Fernandez Pacheco D, Saxon D, Yang A, Sabet S, Dutra-Clarke M, Levy R, Watkins A, Park H, Abbasi Akhtar A, et al. 2019. Rapid generation of somatic mouse mosaics with locus-specific, stably integrated transgenic elements. *Cell* **179:** 251–267.e24. doi:10.1016/j.cell.2019.08.013

Kotini AG, Chang C-J, Boussaad I, Delrow JJ, Dolezal EK, Nagulapally AB, Perna F, Fishbein GA, Klimek VM, Hawkins RD, et al. 2015. Functional analysis of a chromosomal deletion associated with myelodysplastic syndromes using isogenic human induced pluripotent stem cells. *Nat Biotechnol* **33:** 646–655. doi:10.1038/nbt.3178

Kotini AG, Chang C-J, Chow A, Yuan H, Ho T-C, Wang T, Vora S, Solovyov A, Husser C, Olszewska M, et al. 2017. Stage-specific human induced pluripotent stem cells map the progression of myeloid transformation to transplantable leukemia. *Cell Stem Cell* **20:** 315–328.e7. doi:10.1016/j.stem.2017.01.009

Kramer M, Ribeiro D, Arsenian-Henriksson M, Deller T, Rohrer H. 2016. Proliferation and survival of embryonic sympathetic neuroblasts by MYCN and activated ALK signaling. *J Neurosci* **36:** 10425–10439. doi:10.1523/JNEUROSCI.0183-16.2016

Kumano K, Arai S, Hosoi M, Taoka K, Takayama N, Otsu M, Nagae G, Ueda K, Nakazaki K, Kamikubo Y, et al. 2012. Generation of induced pluripotent stem cells from primary chronic myelogenous leukemia patient samples. *Blood* **119:** 6234–6242. doi:10.1182/blood-2011-07-367441

Kurtz KJ, Conneely SE, O'Keefe M, Wohlan K, Rau RE. 2022. Murine models of acute myeloid leukemia. *Front Oncol* **12:** 854973. doi:10.3389/fonc.2022.854973

Lago C, Federico A, Leva G, Mack NL, Schwalm B, Ballabio C, Gianesello M, Abballe L, Giovannoni I, Reddel S, et al. 2023. Patient- and xenograft-derived organoids recapitulate pediatric brain tumor features and patient treatments. *EMBO Mol Med* **15:** e18199. doi:10.15252/emmm.202318199

Lapidot T, Sirard C, Vormoor J, Murdoch B, Hoang T, Caceres-Cortes J, Minden M, Paterson B, Caligiuri MA, Dick JE. 1994. A cell initiating human acute myeloid leukaemia after transplantation into SCID mice. *Nature* **367:** 645–648. doi:10.1038/367645a0

Larson JD, Kasper LH, Paugh BS, Jin H, Wu G, Kwon CH, Fan Y, Shaw TI, Silveira AB, Qu C, et al. 2019. Histone H3.3 K27M accelerates spontaneous brainstem glioma and drives restricted changes in bivalent gene expression. *Cancer Cell* **35:** 140–155.e7. doi:10.1016/j.ccell.2018.11.015

Larsson AT, Bhatia H, Calizo A, Pollard K, Zhang X, Conniff E, Tibbitts JF, Rono E, Cummins K, Osum SH, et al. 2023. Ex vivo to in vivo model of malignant peripheral nerve sheath tumors for precision oncology. *Neuro Oncol* **25:** 2044–2057. doi:10.1093/neuonc/noad097

Leacock SW, Basse AN, Chandler GL, Kirk AM, Rakheja D, Amatruda JF. 2012. A zebrafish transgenic model of Ewing's sarcoma reveals conserved mediators of EWS-FLI1 tumorigenesis. *Dis Model Mech* **5:** 95–106. doi:10.1242/dmm.007401

Lee BH, Williams IR, Anastasiadou E, Boulton CL, Joseph SW, Amaral SM, Curley DP, Duclos N, Huntly BJP, Fabbro D, et al. 2005. FLT3 internal tandem duplication mutations induce myeloproliferative or lymphoid disease in a transgenic mouse model. *Oncogene* **24:** 7882–7892. doi:10.1038/sj.onc.1208933

Lee BH, Tothova Z, Levine RL, Anderson K, Buza-Vidas N, Cullen DE, McDowell EP, Adelsperger J, Fröhling S, Huntly BJP, et al. 2007. FLT3 mutations confer enhanced proliferation and survival properties to multipotent progenitors in a murine model of chronic myelomonocytic leukemia. *Cancer Cell* **12:** 367–380. doi:10.1016/j.ccr.2007.08.031

Lewis PW, Müller MM, Koletsky MS, Cordero F, Lin S, Banaszynski L, Garcia B, Muir TW, Becher OJ, Allis CD. 2013. Inhibition of PRC2 activity by a gain-of-function H3 mutation found in pediatric glioblastoma. *Science* **340:** 857–861. doi:10.1126/science.1232245

Linkous A, Balamatsias D, Snuderl M, Edwards L, Miyaguchi K, Milner T, Reich B, Cohen-Gould L, Storaska A, Nakayama Y, et al. 2019. Modeling patient-derived glioblastoma with cerebral organoids. *Cell Rep* **26:** 3203–3211.e5. doi:10.1016/j.celrep.2019.02.063

Liu I, Alencastro Veiga Cruzeiro G, Bjerke L, Rogers RF, Grabovska Y, Beck A, Mackay A, Barron T, Hack OA, Quezada MA, et al. 2024. GABAergic neuronal lineage development determines clinically actionable targets in diffuse hemispheric glioma, H3G34-mutant. *Cancer Cell* **42:** 1528–1548.e17. doi:10.1016/j.ccell.2024.08.006

Louis DN, Perry A, Wesseling P, Brat DJ, Cree IA, Figarella-Branger D, Hawkins C, Ng HK, Pfister SM, Reifenberger G, et al. 2021. The 2021 WHO classification of tumors of the central nervous system: a summary. *Neuro Oncol* **23:** 1231–1251. doi:10.1093/neuonc/noab106

Lu C, Jain SU, Hoelper D, Bechet D, Molden RC, Ran L, Murphy D, Venneti S, Hameed M, Pawel BR, et al. 2016. Histone H3K36 mutations promote sarcomagenesis through altered histone methylation landscape. *Science* **352:** 844–849. doi:10.1126/science.aac7272

Mackay A, Burford A, Carvalho D, Izquierdo E, Fazal-Salom J, Taylor KR, Bjerke L, Clarke M, Vinci M, Nandhabalan M, et al. 2017. Integrated molecular meta-analysis of 1,000 pediatric high-grade and diffuse intrinsic pontine glioma. *Cancer Cell* **32:** 520–537.e5. doi:10.1016/j.ccell.2017.08.017

MacPherson D, Sage J, Kim T, Ho D, McLaughlin ME, Jacks T. 2004. Cell type-specific effects of *Rb* deletion in the murine retina. *Genes Dev* **18:** 1681–1694. doi:10.1101/gad.1203304

Magallón-Lorenz M, Terribas E, Ortega-Bertran S, Creus-Bachiller E, Fernández M, Requena G, Rosas I, Mazuelas H, Uriarte-Arrazola I, Negro A, et al. 2023. Deep genomic analysis of malignant peripheral nerve sheath tumor cell lines challenges current malignant peripheral nerve sheath tumor diagnosis. *iScience* **26**: 106096. doi:10.1016/j.isci.2023.106096

Martin GR. 1981. Isolation of a pluripotent cell line from early mouse embryos cultured in medium conditioned by teratocarcinoma stem cells. *Proc Natl Acad Sci* **78**: 7634–7638. doi:10.1073/pnas.78.12.7634

Mazuelas H, Magallón-Lorenz M, Fernández-Rodríguez J, Uriarte-Arrazola I, Richaud-Patin Y, Terribas E, Villanueva A, Castellanos E, Blanco I, Raya Á, et al. 2022. Modeling iPSC-derived human neurofibroma-like tumors in mice uncovers the heterogeneity of Schwann cells within plexiform neurofibromas. *Cell Rep* **38**: 110385. doi:10.1016/j.celrep.2022.110385

Messiaen J, Jacobs SA, De Smet F. 2023. The tumor microenvironment in pediatric glioma: friend or foe? *Front Immunol* **14**: 1227126. doi:10.3389/fimmu.2023.1227126

Meyer JS, Howden SE, Wallace KA, Verhoeven AD, Wright LS, Capowski EE, Pinilla I, Martin JM, Tian S, Stewart R, et al. 2011. Optic vesicle-like structures derived from human pluripotent stem cells facilitate a customized approach to retinal disease treatment. *Stem Cells* **29**: 1206–1218. doi:10.1002/stem.674

Miller SJ, Rangwala F, Williams J, Ackerman P, Kong S, Jegga AG, Kaiser S, Aronow BJ, Frahm S, Kluwe L, et al. 2006. Large-scale molecular comparison of human Schwann cells to malignant peripheral nerve sheath tumor cell lines and tissues. *Cancer Res* **66**: 2584–2591. doi:10.1158/0008-5472.CAN-05-3330

Minas TZ, Surdez D, Javaheri T, Tanaka M, Howarth M, Kang H-J, Han J, Han Z-Y, Sax B, Kream BE, et al. 2017. Combined experience of six independent laboratories attempting to create an Ewing sarcoma mouse model. *Oncotarget* **8**: 34141–34163. doi:10.18632/oncotarget.9388

Mohammad F, Weissmann S, Leblanc B, Pandey DP, Højfeldt JW, Comet I, Zheng C, Johansen JV, Rapin N, Porse BT, et al. 2017. EZH2 is a potential therapeutic target for H3K27M-mutant pediatric gliomas. *Nat Med* **23**: 483–492. doi:10.1038/nm.4293

Monje M, Mitra SS, Freret ME, Raveh TB, Kim J, Masek M, Attema JL, Li G, Haddix T, Edwards MSB, et al. 2011. Hedgehog-responsive candidate cell of origin for diffuse intrinsic pontine glioma. *Proc Natl Acad Sci* **108**: 4453–4458. doi:10.1073/pnas.1101657108

Mulero-Navarro S, Sevilla A, Roman AC, Lee D-F, D'Souza SL, Pardo S, Riess I, Su J, Cohen N, Schaniel C, et al. 2015. Myeloid dysregulation in a human induced pluripotent stem cell model of PTPN11-associated juvenile myelomonocytic leukemia. *Cell Rep* **13**: 504–515. doi:10.1016/j.celrep.2015.09.019

Nakano T, Ando S, Takata N, Kawada M, Muguruma K, Sekiguchi K, Saito K, Yonemura S, Eiraku M, Sasai Y. 2012. Self-formation of optic cups and storable stratified neural retina from human ESCs. *Cell Stem Cell* **10**: 771–785. doi:10.1016/j.stem.2012.05.009

Ng JMY, Martinez D, Marsh ED, Zhang Z, Rappaport E, Santi M, Curran T. 2015. Generation of a mouse model of atypical teratoid/rhabdoid tumor of the central nervous system through combined deletion of Snf5 and p53. *Cancer Res* **75**: 4629–4639. doi:10.1158/0008-5472.CAN-15-0874

Norrie JL, Nityanandam A, Lai K, Chen X, Wilson M, Stewart E, Griffiths L, Jin H, Wu G, Orr B, et al. 2021. Retinoblastoma from human stem cell-derived retinal organoids. *Nat Commun* **12**: 4535. doi:10.1038/s41467-021-24781-7

Ogawa J, Pao GM, Shokhirev MN, Verma IM. 2018. Glioblastoma model using human cerebral organoids. *Cell Rep* **23**: 1220–1229. doi:10.1016/j.celrep.2018.03.105

Onaciu A, Munteanu R, Munteanu VC, Gulei D, Raduly L, Feder R-I, Pirlog R, Atanasov AG, Korban SS, Irimie A, et al. 2020. Spontaneous and induced animal models for cancer research. *Diagnostics* **10**: 660. doi:10.3390/diagnostics10090660

Ozawa T, Arora S, Szulzewsky F, Juric-Sekhar G, Miyajima Y, Bolouri H, Yasui Y, Barber J, Kupp R, Dalton J, et al. 2018. A de novo mouse model of C11orf95-RELA fusion-driven ependymoma identifies driver functions in addition to NF-κB. *Cell Rep* **23**: 3787–3797. doi:10.1016/j.celrep.2018.04.099

Parisian AD, Koga T, Miki S, Johann PD, Kool M, Crawford JR, Furnari FB. 2020. SMARCB1 loss interacts with neuronal differentiation state to block maturation and impact cell stability. *Genes Dev* **34**: 1316–1329. doi:10.1101/gad.339978.120

Pathania M, De Jay N, Maestro N, Harutyunyan AS, Nitarska J, Pahlavan P, Henderson S, Mikael LG, Richard-Londt A, Zhang Y, et al. 2017. H3.3K27M cooperates with Trp53 loss and PDGFRA gain in mouse embryonic neural progenitor cells to induce invasive high-grade gliomas. *Cancer Cell* **32**: 684–700.e9. doi:10.1016/j.ccell.2017.09.014

Pei Y, Moore CE, Wang J, Tewari AK, Eroshkin A, Cho Y-J, Witt H, Korshunov A, Read T-A, Sun JL, et al. 2012. An animal model of MYC-driven medulloblastoma. *Cancer Cell* **21**: 155–167. doi:10.1016/j.ccr.2011.12.021

Pham MT, Pollock KM, Rose MD, Cary WA, Stewart HR, Zhou P, Nolta JA, Waldau B. 2018. Generation of human vascularized brain organoids. *Neuroreport* **29**: 588–593. doi:10.1097/WNR.0000000000001014

Pollard K, Banerjee J, Doan X, Wang J, Guo X, Allaway R, Langmead S, Slobogean B, Meyer CF, Loeb DM, et al. 2020. A clinically and genomically annotated nerve sheath tumor biospecimen repository. *Sci Data* **7**: 184. doi:10.1038/s41597-020-0508-5

Quentmeier H, Pommerenke C, Dirks WG, Eberth S, Koeppel M, MacLeod RAF, Nagel S, Steube K, Uphoff CC, Drexler HG. 2019. The LL-100 panel: 100 cell lines for blood cancer studies. *Sci Rep* **9**: 8218. doi:10.1038/s41598-019-44491-x

Rao G, Pedone CA, Del Valle L, Reiss K, Holland EC, Fults DW. 2004. Sonic hedgehog and insulin-like growth factor signaling synergize to induce medulloblastoma formation from nestin-expressing neural progenitors in mice. *Oncogene* **23**: 6156–6162. doi:10.1038/sj.onc.1207818

Roussel MF, Stripay JL. 2020. Modeling pediatric medulloblastoma. *Brain Pathol* **30**: 703–712. doi:10.1111/bpa.12803

Royston HN, Hampton AB, Bhagat D, Pinto EF, Emerson MD, Funato K. 2024. A human embryonic stem cell-based model reveals the cell of origin of FOXR2-activated

Cite this article as *Cold Spring Harb Perspect Med* doi: 10.1101/cshperspect.a041711

CNS neuroblastoma. *Neurooncol Adv* **6**: vdae144. doi:10.1093/noajnl/vdae144

Sancho-Martinez I, Nivet E, Xia Y, Hishida T, Aguirre A, Ocampo A, Ma L, Morey R, Krause MN, Zembrzycki A, et al. 2016. Establishment of human iPSC-based models for the study and targeting of glioma initiating cells. *Nat Commun* **7**: 10743. doi:10.1038/ncomms10743

Saxena S, Wahl J, Huber-Lang MS, Stadel D, Braubach P, Debatin K-M, Beltinger C. 2013. Generation of murine sympathoadrenergic progenitor-like cells from embryonic stem cells and postnatal adrenal glands. *PLoS ONE* **8**: e64454. doi:10.1371/journal.pone.0064454

Schafer ST, Mansour AA, Schlachetzki JCM, Pena M, Ghassemzadeh S, Mitchell L, Mar A, Quang D, Stumpf S, Ortiz IS, et al. 2023. An in vivo neuroimmune organoid model to study human microglia phenotypes. *Cell* **186**: 2111–2126.e20. doi:10.1016/j.cell.2023.04.022

Scheidler S, Fredericks WJ, Rauscher FJ, Barr FG, Vogt PK. 1996. The hybrid PAX3-FKHR fusion protein of alveolar rhabdomyosarcoma transforms fibroblasts in culture. *Proc Natl Acad Sci* **93**: 9805–9809. doi:10.1073/pnas.93.18.9805

Schüller U, Heine VM, Mao J, Kho AT, Dillon AK, Han Y-G, Huillard E, Sun T, Ligon AH, Qian Y, et al. 2008. Acquisition of granule neuron precursor identity is a critical determinant of progenitor cell competence to form Shh-induced medulloblastoma. *Cancer Cell* **14**: 123–134. doi:10.1016/j.ccr.2008.07.005

Schwartzentruber J, Korshunov A, Liu X-Y, Jones DTW, Pfaff E, Jacob K, Sturm D, Fontebasso AM, Quang D-AK, Tönjes M, et al. 2012. Driver mutations in histone H3.3 and chromatin remodelling genes in paediatric glioblastoma. *Nature* **482**: 226–231. doi:10.1038/nature10833

Searcy MB, Larsen RK, Stevens BT, Zhang Y, Jin H, Drummond CJ, Langdon CG, Gadek KE, Vuong K, Reed KB, et al. 2023. PAX3-FOXO1 dictates myogenic reprogramming and rhabdomyosarcoma identity in endothelial progenitors. *Nat Commun* **14**: 7291. doi:10.1038/s41467-023-43044-1

Shigemura T, Matsuda K, Kurata T, Sakashita K, Okuno Y, Muramatsu H, Yue F, Ebihara Y, Tsuji K, Sasaki K, et al. 2019. Essential role of *PTPN11* mutation in enhanced haematopoietic differentiation potential of induced pluripotent stem cells of juvenile myelomonocytic leukaemia. *Br J Haematol* **187**: 163–173. doi:10.1111/bjh.16060

Shimizu F, Hovinga KE, Metzner M, Soulet D, Tabar V. 2011. Organotypic explant culture of glioblastoma multiforme and subsequent single-cell suspension. *Curr Protoc Stem Cell Biol* doi:10.1002/9780470151808.sc0305s19

Sitnikov D, Revkova V, Ilina I, Shatalova R, Komarov P, Struleva E, Konoplyannikov M, Kalsin V, Baklaushev V. 2023. Sensitivity of neuroblastoma and induced neural progenitor cells to high-intensity THz radiation. *Int J Mol Sci* **24**: 6558. doi:10.3390/ijms24076558

Smith RC, Tabar V. 2019. Constructing and deconstructing cancers using human pluripotent stem cells and organoids. *Cell Stem Cell* **24**: 12–24. doi:10.1016/j.stem.2018.11.012

Sturm D, Witt H, Hovestadt V, Khuong-Quang D-A, Jones DTW, Konermann C, Pfaff E, Tönjes M, Sill M, Bender S, et al. 2012. Hotspot mutations in H3F3A and IDH1 define distinct epigenetic and biological subgroups of glioblas-toma. *Cancer Cell* **22**: 425–437. doi:10.1016/j.ccr.2012.08.024

Sturm D, Orr BA, Toprak UH, Hovestadt V, Jones DTW, Capper D, Sill M, Buchhalter I, Northcott PA, Leis I, et al. 2016. New brain tumor entities emerge from molecular classification of CNS-PNETs. *Cell* **164**: 1060–1072. doi:10.1016/j.cell.2016.01.015

Sun D, Xie XP, Zhang X, Wang Z, Sait SF, Iyer SV, Chen Y-J, Brown R, Laks DR, Chipman ME, et al. 2021. Stem-like cells drive NF1-associated MPNST functional heterogeneity and tumor progression. *Cell Stem Cell* **28**: 1397–1410.e4. doi:10.1016/j.stem.2021.04.029

Sun CX, Daniel P, Bradshaw G, Shi H, Loi M, Chew N, Parackal S, Tsui V, Liang Y, Koptyra M, et al. 2023. Generation and multi-dimensional profiling of a childhood cancer cell line atlas defines new therapeutic opportunities. *Cancer Cell* **41**: 660–677.e7. doi:10.1016/j.ccell.2023.03.007

Susanto E, Marin Navarro A, Zhou L, Sundström A, van Bree N, Stantic M, Moslem M, Tailor J, Rietdijk J, Zubillaga V, et al. 2020. Modeling SHH-driven medulloblastoma with patient iPS cell-derived neural stem cells. *Proc Natl Acad Sci* **117**: 20127–20138. doi:10.1073/pnas.1920521117

Swartling FJ, Grimmer MR, Hackett CS, Northcott PA, Fan Q-W, Goldenberg DD, Lau J, Masic S, Nguyen K, Yakovenko S, et al. 2010. Pleiotropic role for *MYCN* in medulloblastoma. *Genes Dev* **24**: 1059–1072. doi:10.1101/gad.1907510

Takahashi K, Yamanaka S. 2006. Induction of pluripotent stem cells from mouse embryonic and adult fibroblast cultures by defined factors. *Cell* **126**: 663–676. doi:10.1016/j.cell.2006.07.024

Tasian SK, Casas JA, Posocco D, Gandre-Babbe S, Gagne AL, Liang G, Loh ML, Weiss MJ, French DL, Chou ST. 2019. Mutation-specific signaling profiles and kinase inhibitor sensitivities of juvenile myelomonocytic leukemia revealed by induced pluripotent stem cells. *Leukemia* **33**: 181–190. doi:10.1038/s41375-018-0169-y

Telliam G, Desterke C, Imeri J, M'kacher R, Oudrhiri N, Balducci E, Fontaine-Arnoux M, Acloque H, Bennaceur-Griscelli A, Turhan AG. 2023. Modeling global genomic instability in chronic myeloid leukemia (CML) using patient-derived induced pluripotent stem cells (iPSCs). *Cancers (Basel)* **15**: 2594. doi:10.3390/cancers15092594

Terada Y, Jo N, Arakawa Y, Sakakura M, Yamada Y, Ukai T, Kabata M, Mitsunaga K, Mineharu Y, Ohta S, et al. 2019. Human pluripotent stem cell-derived tumor model uncovers the embryonic stem cell signature as a key driver in atypical teratoid/rhabdoid tumor. *Cell Rep* **26**: 2608–2621.e6. doi:10.1016/j.celrep.2019.02.009

Thomson JA, Itskovitz-Eldor J, Shapiro SS, Waknitz MA, Swiergiel JJ, Marshall VS, Jones JM. 1998. Embryonic stem cell lines derived from human blastocysts. *Science* **282**: 1145–1147. doi:10.1126/science.282.5391.1145

Tomita Y, Shimazu Y, Somasundaram A, Tanaka Y, Takata N, Ishi Y, Gadd S, Hashizume R, Angione A, Pinero G, et al. 2022. A novel mouse model of diffuse midline glioma initiated in neonatal oligodendrocyte progenitor cells highlights cell-of-origin dependent effects of H3K27M. *Glia* **70**: 1681–1698. doi:10.1002/glia.24189

Tsai JW, Cejas P, Wang DK, Patel S, Wu DW, Arounleut P, Wei X, Zhou N, Syamala S, Dubois FPB, et al. 2022. FOXR2 is an epigenetically regulated pan-cancer oncogene that activates ETS transcriptional circuits. *Cancer Res* **82:** 2980–3001. doi:10.1158/0008-5472.CAN-22-0671

van Essen MJ, Apsley EJ, Riepsaame J, Xu R, Northcott PA, Cowley SA, Jacob J, Becker EBE. 2024. *PTCH1*-mutant human cerebellar organoids exhibit altered neural development and recapitulate early medulloblastoma tumorigenesis. *Dis Model Mech* **17:** 5009. doi:10.1242/dmm.050323

Vasileva E, Warren M, Triche TJ, Amatruda JF. 2022. Dysregulated heparan sulfate proteoglycan metabolism promotes Ewing sarcoma tumor growth. *eLife* **11:** e69734. doi:10.7554/eLife.69734

Veninga V, Voest EE. 2021. Tumor organoids: opportunities and challenges to guide precision medicine. *Cancer Cell* **39:** 1190–1201. doi:10.1016/j.ccell.2021.07.020

Ward ZJ, Yeh JM, Bhakta N, Frazier AL, Atun R. 2019. Estimating the total incidence of global childhood cancer: a simulation-based analysis. *Lancet Oncol* **20:** 483–493. doi:10.1016/S1470-2045(18)30909-4

Watanabe J, Clutter MR, Gullette MJ, Sasaki T, Uchida E, Kaur S, Mo Y, Abe K, Ishi Y, Takata N, et al. 2024. BET bromodomain inhibition potentiates radiosensitivity in models of H3K27-altered diffuse midline glioma. *J Clin Invest* **134:** e174794. doi:10.1172/JCI174794

Weng Z, Lin J, He J, Gao L, Lin S, Tsang LL, Zhang H, He X, Wang G, Yang X, et al. 2022. Human embryonic stem cell-derived neural crest model unveils CD55 as a cancer stem cell regulator for therapeutic targeting in *MYCN*-amplified neuroblastoma. *Neuro Oncol* **24:** 872–885. doi:10.1093/neuonc/noab241

Wiesner SM, Decker SA, Larson JD, Ericson K, Forster C, Gallardo JL, Long C, Demorest ZL, Zamora EA, Low WC, et al. 2009. *De novo* induction of genetically engineered brain tumors in mice using plasmid DNA. *Cancer Res* **69:** 431–439. doi:10.1158/0008-5472.CAN-08-1800

Wu G, Broniscer A, McEachron TA, Lu C, Paugh BS, Becksfort J, Qu C, Ding L, Huether R, Parker M, et al. 2012a. Somatic histone H3 alterations in pediatric diffuse intrinsic pontine gliomas and non-brainstem glioblastomas. *Nat Genet* **44:** 251–253. doi:10.1038/ng.1102

Wu X, Northcott PA, Dubuc A, Dupuy AJ, Shih DJH, Witt H, Croul S, Bouffet E, Fults DW, Eberhart CG, et al. 2012b. Clonal selection drives genetic divergence of metastatic medulloblastoma. *Nature* **482:** 529–533. doi:10.1038/nature10825

Wu LMN, Deng Y, Wang J, Zhao C, Wang J, Rao R, Xu L, Zhou W, Choi K, Rizvi TA, et al. 2018. Programming of Schwann cells by Lats1/2-TAZ/YAP signaling drives malignant peripheral nerve sheath tumorigenesis. *Cancer Cell* **33:** 292–308.e7. doi:10.1016/j.ccell.2018.01.005

Wu H-F, Art J, Saini T, Zeltner N. 2024. Protocol for generating postganglionic sympathetic neurons using human pluripotent stem cells for electrophysiological and functional assessments. *STAR Protoc* **5:** 102970. doi:10.1016/j.xpro.2024.102970

Xu J, Margol A, Asgharzadeh S, Erdreich-Epstein A. 2015. Pediatric brain tumor cell lines. *J Cell Biochem* **116:** 218–224. doi:10.1002/jcb.24976

Yamagiwa K, Ichikawa K. 1918. Experimental study of the pathogenesis of carcinoma. *J Cancer Res* **3:** 1–29.

Yang Z-J, Ellis T, Markant SL, Read T-A, Kessler JD, Bourboulas M, Schüller U, Machold R, Fishell G, Rowitch DH, et al. 2008. Medulloblastoma can be initiated by deletion of patched in lineage-restricted progenitors or stem cells. *Cancer Cell* **14:** 135–145. doi:10.1016/j.ccr.2008.07.003

Zorko NA, Bernot KM, Whitman SP, Siebenaler RF, Ahmed EH, Marcucci GG, Yanes DA, McConnell KK, Mao C, Kalu C, et al. 2012. Mll partial tandem duplication and Flt3 internal tandem duplication in a double knock-in mouse recapitulates features of counterpart human acute myeloid leukemias. *Blood* **120:** 1130–1136. doi:10.1182/blood-2012-03-415067

Cite this article as *Cold Spring Harb Perspect Med* doi: 10.1101/cshperspect.a041711

Developmental Heterogeneity of Rhabdomyosarcoma

Bradley T. Stevens[1,2] and Mark E. Hatley[1]

[1]Department of Oncology, St. Jude Children's Research Hospital, Memphis, Tennessee 38105, USA

[2]St. Jude Graduate School of Biomedical Sciences, Memphis, Tennessee 38105, USA

Correspondence: mark.hatley@stjude.org

Rhabdomyosarcoma (RMS) is a pediatric embryonal solid tumor and the most common pediatric soft tissue sarcoma. The histology and transcriptome of RMS resemble skeletal muscle progenitor cells that have failed to terminally differentiate. Thus, RMS is typically thought to arise from corrupted skeletal muscle progenitor cells during development. However, RMS can occur in body regions devoid of skeletal muscle, suggesting the potential for nonmyogenic cells of origin. Here, we discuss the interplay between RMS driver mutations and cell(s) of origin with an emphasis on driving location specificity. Additionally, we discuss the mechanisms governing RMS transformation events and tumor heterogeneity through the lens of transcriptional networks and epigenetic control. Finally, we reimagine Waddington's developmental landscape to include a plane of transformation connecting distinct lineage landscapes to more accurately reflect the phenomena observed in pediatric cancers.

Children experience different and unique exposures compared to adults, making their dynamic developmental physiology important to consider when contemplating the origins and biology of pediatric cancer. Pediatric cancers in general harbor markedly fewer coding mutations than adult cancers (Gröbner et al. 2018), and temporal models of tumorigenesis are incompatible with the acquisition time of childhood cancers. Many models of adult cancer explain tumorigenesis as a long process spanning decades and requiring the serial acquisition of multiple mutations (Vogelstein et al. 2013), whereas some childhood malignancies can present at or within months after birth. Thus, it is important to contextualize that the dynamic process of human development offers unique opportunities and vulnerabilities for cell transformation. The active developmental process does not stop at birth but continues with the expansive growth through adolescence.

Many pediatric malignancies have histology and transcriptomes that recapitulate developmental progenitors that fail to execute their respective normal developmental trajectory and are collectively termed pediatric embryonal tumors. Pediatric embryonal solid tumors can occur within the central nervous system (CNS) and as solid tumors outside the CNS. Pediatric embryonal tumors of the CNS include medulloblastoma, atypical teratoid/rhabdoid tumors, embryonal tumors with multilayered rosettes,

Cite this article as *Cold Spring Harb Perspect Med* doi: 10.1101/cshperspect.a041583

pineoblastoma, and others (Torp et al. 2022). Pediatric solid tumors outside the CNS appear to arise from developmental progenitor cells and include rhabdomyosarcoma (RMS), neuroblastoma, Wilms' tumor, hepatoblastoma, and retinoblastoma. Pediatric solid tumors seem to hijack these dynamic processes arresting differentiation for tumorigenesis. For example, during development, there exists an abundance of multipotent progenitors present, and many proto-oncogenes and tumor suppressors play important roles in normal development (Chen et al. 2015b; Puisieux et al. 2018). Activation of oncogenes and/or silencing of tumor suppressors is not tolerated in every cell, but the developmental competence of multipotent progenitors may uniquely allow for transformation or even alter the developmental trajectory of a given progenitor. Furthermore, as embryos develop from a single-cell zygote and undergo morphogenesis into a highly polarized organism, the specificity that dictates the anatomical location of cell types is as important as developmental competence. The intersection of developmental cell state, spatial patterning and morphogenesis, and specific genetic perturbations in pediatric embryonal tumors remains poorly understood. This work aims to discuss this intersectionality with a focus on the pediatric embryonal solid tumor RMS.

RHABDOMYOSARCOMA

Sarcomas are rare tumors that develop in supporting and connective tissues and are broadly categorized as either soft tissue or bone sarcomas. Soft tissue sarcomas develop in mesenchymal tissues, including skeletal and smooth muscle, adipose, cartilage, blood vessels, and other support tissue. RMS is the most common pediatric soft tissue sarcoma (Skapek et al. 2019; Gartrell and Pappo 2020). Children with RMS often present with the disease younger than 10 years old with males affected slightly more than females (Ognjanovic et al. 2009). RMS cells histologically resemble skeletal muscle progenitor cells that have failed to terminally differentiate, which is confirmed through the expression of the skeletal muscle intermediate filament DESMIN and characteristic myogenic determination transcription factors (TFs)

MYOD1 and MYOGENIN (Rudzinski et al. 2021). Primary RMS tumors occur throughout the body, including in the head and neck, extremities, and abdominal wall. Furthermore, RMS can arise in body areas devoid of skeletal muscle, such as the male and female genitourinary tract, omentum, salivary gland, gallbladder, and pulmonary artery (Amer et al. 2019). RMS is staged by tumor features, including location and whether it is localized or metastatic disease and grouped by the extent of disease following surgery (Meza et al. 2006; Crane et al. 2022). The current standard of care includes combinations of chemotherapy, surgery, and radiation (Gartrell and Pappo 2020). Importantly, this treatment regimen has remained the same for decades, underscoring the importance of understanding the basic biology driving RMS so that new therapeutic targets can be discovered and exploited.

RMS is further classified into four main subtypes based on its histologic characteristics and molecular drivers (Fig. 1). The four main subtypes of RMS are alveolar RMS (ARMS), embryonal RMS (ERMS), spindle cell/sclerosing RMS, and pleiomorphic RMS (Yohe et al. 2019; Dehner et al. 2023). ARMS accounts for ~22% of RMS cases and has a poor prognosis with a 35% 10-year survival (Amer et al. 2019). ARMS cells have a high nuclear-to-cytoplasm ratio and appear loosely dispersed in areas such that the cells resemble an alveolar pattern. The majority of ARMS tumors harbor either t(2;13)(q35;q14) or t(1;13)(q36;q14) chromosomal translocations that result in PAX3::FOXO1 or PAX7::FOXO1 fusion oncoproteins, respectively (Barr et al. 1993; Galili et al. 1993; Shapiro et al. 1993; Davis et al. 1994). Children with alveolar histology tumors harboring *PAX3/7::FOXO1* gene fusions have a much worse prognosis than children whose tumors lack these translocations whose survival is similar to ERMS (Williamson et al. 2010). Thus, the current convention parses patients as fusion-positive RMS (FP-RMS) and fusion-negative RMS (FN-RMS) based on the presence or absence of the *PAX3/7:: FOXO1* translocations. Genetically, FP-RMS tumors harbor few secondary perturbations with an average of 6.4 somatic mutations, indicating the potency of oncogenic fusion proteins (Chen et al. 2013, 2015a; Shern et al. 2014). Histologically,

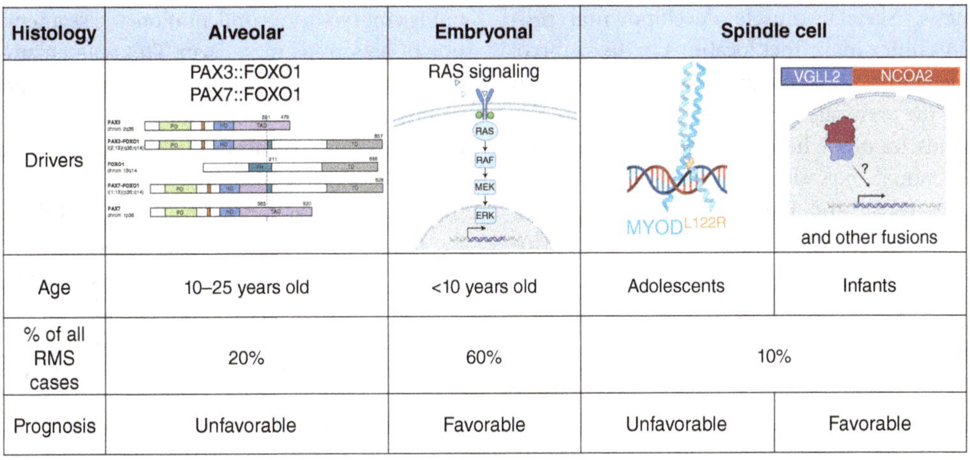

Histology	Alveolar	Embryonal	Spindle cell	
Drivers	PAX3::FOXO1 PAX7::FOXO1	RAS signaling	MYOD^{L122R}	VGLL2 / NCOA2 and other fusions
Age	10–25 years old	<10 years old	Adolescents	Infants
% of all RMS cases	20%	60%	10%	
Prognosis	Unfavorable	Favorable	Unfavorable	Favorable

Figure 1. Rhabdomyosarcoma (RMS) subtypes. The subtypes of RMS are subdivided based on histological features and each has unique driving mutations. Alveolar RMS (ARMS) is driven through the oncofusion proteins PAX3::FOXO1 or PAX7::FOXO1 and designated fusion-positive RMS, typically occurring in adolescents 10–25 years old. ARMS has an unfavorable prognosis and accounts for 20% of the cases. Embryonal RMS (ERMS) is heterogeneous with many tumors being driven by mutant RAS signaling, typically occurring in children <10 years old. ERMS accounts for 60% of RMS cases and has a more favorable prognosis. Spindle cell RMS accounts for 10% of all RMS cases affecting adolescents and infants. Adolescents typically present with spindle cell RMS driven by MYOD^{L122R} mutations and have an unfavorable prognosis. Infants present with spindle cell RMS driven by other fusions such as VGLL2::NCOA2 and have a favorable prognosis.

ERMS resembles small round blue cells, and comprises the majority of FN-RMS cases. FN-RMS is genetically heterogenous with no unifying genetic perturbation, but alterations in the *RAS* family, *FGFR4*, *PIK3CA*, *NF1*, *FBXW7*, *TP53*, and *DICER1* are all seen in patients (Chen et al. 2013; Shern et al. 2014). Spindle cell/sclerosing RMS is parsed into two groups based on age and genetic mutation. Spindle cell occurring in adolescents and young adults typically harbor MYOD1^{L122R} mutations and has a particularly poor prognosis (Agaram et al. 2014, 2019; Kohsaka et al. 2014) while spindle cell RMS in infants has gene fusions with *VGLL2*, *TEAD1*, and *NCOA2* and has an excellent prognosis (Alaggio et al. 2016). Pleiomorphic RMS is a subtype that occurs exclusively in adults (Sultan et al. 2009; Rudzinski et al. 2015). The clinical, pathological, biological, and molecular features of RMS have been more extensively reviewed elsewhere (Abraham et al. 2014; Skapek et al. 2019; Yohe et al. 2019; Pomella et al. 2023). Given that the histopathologic features and transcriptome of RMS cells are strikingly like that of immature skeletal muscle myoblasts, RMS has

been viewed through the prism of normal skeletal muscle development.

MUSCLE DEVELOPMENT

This section is not designed to be a comprehensive guide to muscle development, but rather to give an overview of the complexity of muscle development and introduce the distinct transcriptional regulators driving muscle development in different locations. A more in-depth guide to muscle development can be found in contemporary reviews (Buckingham 2017; Chal and Pourquie 2017; Esteves de Lima and Relaix 2021).

Myogenesis is an expansive dynamic developmental process that spans from generating embryonic musculature through adolescent muscle growth. RMS resembles skeletal muscle progenitors that have failed to terminally differentiate thus rationalizing embryonic muscle progenitors or muscle stem cells (satellite cells) as the perceived cells of origin for RMS. An understanding of normal skeletal muscle development is essential to discussing rhabdomyosarcoma-genesis. Skeletal muscle developmental programs differ in distinct locations in the embryo. Skeletal muscles of the limbs and trunk develop from the paraxial mesoderm, whereas the skeletal muscles of the head and neck originate from the cranial mesoderm within the branchial arches (Chal and Pourquie 2017; Vyas et al. 2020). Importantly, distinct myogenic transcriptional networks drive the development and differentiation into mature muscle cells in different anatomic locations (Fig. 2).

Much of our understanding of muscle development centers on that of trunk and limb skeletal muscle through somitogenesis (McDaniel et al. 2024). The paraxial mesoderm forms in the devel-

oping embryo from condensation and segmentation of presomitic mesoderm. This segmentation occurs at a wavefront of opposing morphogen gradients of FGF/WNT from the tailbud and retinoic acid (RA) from the head. The timing of the segmentation clock is dictated by molecular oscillatory cycles of Notch, BMP, FGF, and Wnt signaling to segment into individual somites (Naiche et al. 2011; Bentzinger et al. 2012; Chal and Pourquie 2017). High levels of FGF/WNT signaling, such as near the tailbud, maintain the somitic cells in an undifferentiated/stem-like state until compartmentalization can occur. Declining FGF/WNT gradients and increased RA signaling cause the cells to compartmentalize and begin commit-

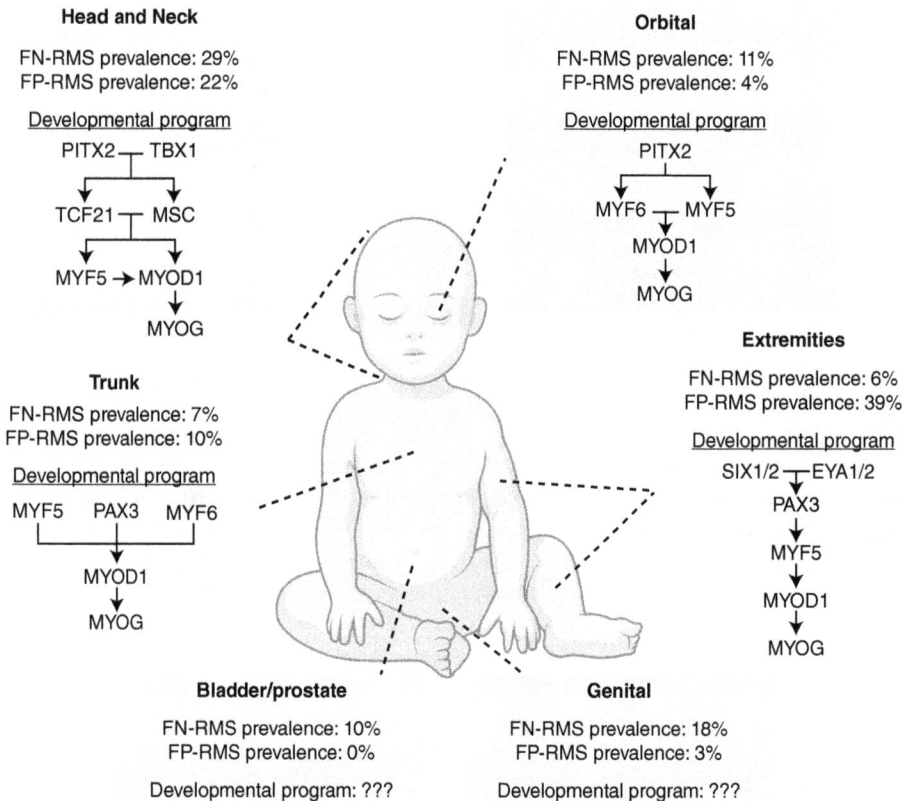

Figure 2. Rhabdomyosarcoma (RMS) prevalence is different in distinct body locations. Fusion-negative RMS (FN-RMS) and fusion-positive RMS (FP-RMS) occur in different distinct body locations at different prevalence. Discrete transcriptional networks drive skeletal muscle development in different anatomic locations. The prevalence of RMS and correlated survival differences at different anatomic locations could be due to the cell of origin and the ability of driver mutations to corrupt the different developmental programs. Furthermore, RMS can occur in body locations devoid of skeletal muscle calling into question whether an alternative cell(s) of origin exists for RMS in those sites.

    Cite this article as *Cold Spring Harb Perspect Med* doi: 10.1101/cshperspect.a041583

ment (Aulehla and Pourquié 2010). These somites contain the cells that undergo specification toward muscle progenitor cells, but they are not yet committed to a specific cell fate (Aoyama and Asamoto 1988). After compartmentalization and segmentation, the somite subdivides into the dermomyotome, myotome, and sclerotome (Ordahl and Le Douarin 1992). Remarkably, somites along the entire body axis follow the same developmental trajectory after their independent compartmentalization (Ibarra-Soria et al. 2023).

The dermomyotome cells will ultimately establish the primary myofiber scaffold, termed primary myogenesis, and are identified by their expression of the TF PAX3 (Williams and Ordahl 1994). The dermomyotome is classically divided into the dorsomedial lip, central dermomyotome, and ventrolateral lip. During primary myogenesis, a subpopulation of PAX3-positive progenitor cells found in the dorsomedial lip start to express MYF5 and migrate ventrally to form the myotome (Ott et al. 1991; Bober et al. 1994; Daston et al. 1996; Kassar-Duchossoy et al. 2005). Here, the cells down-regulate PAX3 expression and differentiate into myocytes expressing cytoskeletal proteins essential to create the primary myofiber scaffold (Denetclaw et al. 1997; Gros et al. 2009). As this process continues, more cells from the dorsomedial lip and eventually the central dermomyotome migrate and differentiate to add to the templated primary myofiber. During myofiber addition and expansion, a subset of PAX3-positive cells begin to express PAX7, exit the cell cycle, and establish the quiescent muscle stem cell compartment (Seale et al. 2000; Gros et al. 2005; Lepper and Fan 2010). This phenomenon has been recapitulated in vitro with mouse embryonic stem cells (Chal et al. 2015).

Cells that give rise to muscles of the head and neck share no clonal relation to somite-derived cells (Lescroart et al. 2015) and thus have a distinct developmental origin (Buckingham 2017; Swedlund and Lescroart 2020; Grimaldi and Tajbakhsh 2021). In conjunction with somitogenesis, pharyngeal pouches, or branchial arches, begin to protrude from the ventral cephalic region of the embryo. The pharyngeal/branchial arches contain overlapping layers of endoderm and then mesoderm followed by ectoderm-derived

cells. The cranial mesoderm in these branchial arches gives rise to the head and neck muscles with each branchial arch giving rise to a different set of muscles. Muscle precursors from branchial arch mesoderm move en masse guided by neuro-ectodermal-derived connective tissue to their terminal site of differentiation (Noden and Trainor 2005; Noden and Francis-West 2006; Heude et al. 2018). In contrast to somitic-derived muscle, these migrating masses begin to express myogenic determination genes on the way to their destination. Ultimately, all myogenic cells require MYOD1 for commitment and MYOG for differentiation, but the upstream transcriptional control differs dramatically based on developmental location (Fig. 2).

The process from mesodermal precursor to myofiber is regulated by a hierarchical network of core TFs that differs by location of muscle development. For all locations, the family of myogenic regulatory factors (MRFs), including MYOD1, MYF5, MYOG, and MYF6 plays a central role in skeletal muscle cell determination and differentiation. At different locations, the TFs acting in specification upstream of MRFs and the specific cascade of MRF expression differ (Fig. 2). For limb muscles, SIX1/2 and EYA1/2 work cooperatively to drive PAX3 expression in the presomitic mesoderm (Giordani et al. 2007). Subsequently, PAX3 expression activates PAX7, and initiates a wave of expression of MRFs MYF5 and MYOD1. MYF5 and MYOD1 in turn activate muscle differentiation TFs MYOGENIN and MYF6 (aka MRF4) as well as repress the expression of PAX3/7. MYOGENIN and MYF6 activate the expression of myotube-specific genes. In trunk muscles, PAX3, MYF5, and MYF6 can activate MYOD1 expression.

The muscles of the head and neck have distinct developmental programs. This is shown by $Myf5^{-/-};Pax3^{-/-}$ compound mutant mice having no trunk skeletal muscle while having normal head and neck muscle (Tajbakhsh et al. 1997). Muscles of the head and neck rely on TFs other than PAX3 to govern the skeletal muscle specification hierarchy. TFs PITX2 and TBX1 activate TCF21 and MSC that in turn activate MYF5 and MYOD1. Mice lacking both *Msc* and *Tcf21* fail to develop first branchial arch-derived facial muscles (Lu et al. 2002). Extraorbital muscles seem to only rely on PITX2 for activation of MYF5/6 preceding MYOD1 activation and subsequent skeletal muscle determination (Dong et al. 2006; Braun and Gautel 2011; Buckingham and Rigby 2014; Buckingham 2017). These same TFs are up-regulated in RMS, suggesting these core TFs are important for RMS cell identity (Gryder et al. 2019a), but their importance in tumor initiation has yet to be determined. Further investigation into the expression of these muscle specification TFs could reveal insights into the distinct biology of RMS occurring in different locations.

Mesodermal cells give rise to other cell types other than muscle, such as adipocytes, bone, and endothelium. The regulation of these fate choices is determined by levels of competing fate-defining TFs. Progenitor cells must skew the transcriptional balance to select a specific cell fate. Thus, a progenitor cell expressing both myogenic specification factors as well as other fate-defining TFs could be the competent cell of origin, forcibly shifting the cell fate to myogenic-like depending on the genetic perturbation driving oncogenesis. Furthermore, the cells of origin could be entirely dependent on primary tumor location. More insights into developmental competency and transforming events will likely elucidate multiple cells-of-origin for RMS partially dictated by location.

MODELS OF RMS

Similarly with muscle development, this section does not aim to provide an exhaustive list of RMS models, but instead seeks to use many of the models to show that (1) there are likely multiple cells of origin for RMS, (2) similar genetic perturbations can result in RMS regardless of the cell of origin, and (3) cell of origin can impact anatomical location. Further examination of RMS models from cells to organisms is reviewed elsewhere (Kashi et al. 2015; Dehner et al. 2021).

Being that RMS is perceived to originate from skeletal muscle, many RMS models, both human and animal, have been designed by activating oncogenes and silencing tumor suppressors in muscle progenitor cells. There is a spectrum of models of RMS spanning from genetically modified primary cells; genetically engineered *Drosophila*, zebrafish, and mice; and cell lines, patient-derived xenografts, and organoids from children's tumors. Each of these models has utilities and limitations (Kashi et al. 2015; Yohe et al. 2019). Many ARMS and ERMS cell lines derived from children's tumors are available through the American Type Culture Collection (ATCC), the Childhood Cancer Repository from the Children's Oncology Group and Alex's Lemonade Stand Foundation (www.cccells.org), and have been extensively reviewed previously (Hinson et al. 2013; Sokolowski et al. 2014). Directly akin to dealing with primary tumors, both mouse orthotopic pediatric-derived xenografts (PDXs) and zebrafish xenotransplants have been developed that have aided in the rapid testing of clinically relevant compounds (Chen et al. 2013; Stewart et al. 2017, 2018; Yan et al. 2019). The Childhood Solid Tumor Network provides an extensive repository of freely available RMS PDX models (Stewart et al. 2017). While cell lines and xenografts allow for insightful investigations in RMS, it is difficult to gain insights into the developmental corruption causative to RMS initiation and determination of cell(s) of origin or how non-cell-autonomous interactions from the immune system contribute to RMS biology.

Illustrative of the capacity for skeletal muscle progenitors to form RMS through genetic corruption, human skeletal muscle myoblasts form embryonal-like RMS when injected into immunocompromised mice upon overexpression of T/t-antigen, hTERT, and HRAS^{G12V} (Linardic et al. 2005) and alveolar-like RMS when injected into mice upon overexpression of PAX3::FOXO1 (P3F), MYCN, and hTERT (Linardic et al. 2007; Naini et al. 2008). These models have defined the minimal essential genetic perturbations necessary to drive RMS transformation in skeletal muscle myoblasts. As this system genetically manipulates an already determined myoblast, it remains to be seen whether mesodermal progenitor cells before muscle determination with ectopically expressed MYOD1 and MYF5 could likewise transform into RMS.

Given that many RMS tumors are found in tissues devoid of skeletal muscle, it is tempting to speculate that transdifferentiation from cells of mesodermal lineages other than skeletal muscle, as well as cells from endodermal or ectodermal lineages could serve as an origin for RMS. Indeed, expression of P3F in ovo converts chick neural progenitor cells to cells with FP-RMS molecular features showing the potential for P3F to drive FP-RMS from entirely unrelated germ layers (Gonzalez Curto et al. 2020). Directed differentiation of human-induced pluripotent stem cells (iPSCs) directly models the cell states traversed during normal development and would allow for investigation of transformation-competent cell states. Recently, our laboratory has developed an iPSC-derived model of FP-RMS through the expression of P3F in endothelial-directed differentiating human iPSCs with *TP53* deletion, which we have termed induced ARMS (iARMS). In this system, the introduction of P3F into endothelial progenitor cells obtained from directed differentiation of iPSCs yields cells that fully transdifferentiate and transform with a gene expression profile similar to that of human FP-RMS (Searcy et al. 2023). Interrogation of tumorigenesis using this directed differentiation system could provide insights into the interplay between cellular developmental state, plasticity, and key RMS oncogenic determinants.

To better understand the developmental etiology of RMS, animal models provide the unique ability to investigate RMS drivers in vivo in various cellular contexts and differentiation states. Like genetically manipulated cell models, many animal models have focused RMS development from the corruption of cells along the spectrum of skeletal muscle differentiation by leveraging Cre recombinase with expression dictated by *cis*-regulatory regions of MRFs. Compound mutant mice with conditional expression of *Pax3::Foxo1* and deletion of either *Cdkn2a* (*Ink4a/Arf*) or *Trp53* in *Myf6-Cre* expressing cells resulted in mice developing FP-RMS tumors in the extremities, trunk, and the head and neck (Keller et al. 2004a; Abraham et al. 2014). This model conditionally converts the *Pax3* locus into expressing *Pax3::Foxo1* (*Pax3^{P3Fm}*) in skeletal muscle cells that express *Myf6*. *Myf6-Cre* is expressed in the somite from E9.0 to E11.5 and then reappears in the differentiating muscle fibers at E16.5 (Rhodes

and Konieczny 1989; Sambasivan et al. 2013). Using a series of distinct muscle lineage Cre alleles to drive *Pax3::Foxo1* expression, fetal myoblasts were identified as the most susceptible cells driving ARMS. Surprisingly, the adult muscle stem cells or satellite cells were the least likely to drive ARMS formation (Abraham et al. 2014). Dictating the expression of PAX3::FOXO1 during prenatal myogenesis with *MCre* (hypaxial restricted *Pax3-Cre*) also gave rise to ARMS. ARMS from recombination with *MCre* only occurred in the extremities while ARMS from the *Myf6-Cre* occurred in the head and neck, trunk, and extremity. These findings show that distinct cell states are permissive for PAX3::FOXO1-induced ARMS transformation and have varied potential for transformation in different locations.

Similar findings are seen in FN-RMS mouse models. *Myf6-Cre;Trp53^{−/−}* mice with no fusion protein expression preferentially develop FN-RMS (Rubin et al. 2011). Conditional expression of *Kras^{G12D}* and deletion of *Trp53* in adult mouse satellite cells through tamoxifen-inducible *Pax7-CreER* resulted in FN-RMS (Blum et al. 2013). To explore FN-RMS transformation during embryo development, *Myod1-Cre*-driven expression of *Kras^{G12D}* and mutant *p53^{R172H}* resulted in mice that develop tumors resembling FN-RMS (Nakahata et al. 2022). In contrast, *Myod1-CreER*-mediated expression of *Kras^{G12D}* and loss of *p53* in adult mice resulted exclusively in undifferentiated pleomorphic sarcoma when injected with tamoxifen after 6 weeks of age (Blum et al. 2013). These models suggest that MYOD1-positive embryonic progenitors can develop FN-RMS, but not adult myoblasts in the context of KRAS^{G12D} expression and *Trp53* loss. Taken together, these models suggest that both early myogenic progenitors arising in the somites as well as adult muscle stem cells possess the capability to transform into RMS, but that it is highly dependent on the cocktail of activated oncogenes and silenced tumor suppressors.

The aforementioned models of RMS arise from myogenic origins in skeletal muscle progenitors or stem cells, which explain the ability for myogenic progenitor cells to become genetically corrupted resulting in arrested differentiation. However, this does not address the fact that in children RMS can arise in multiple body regions that are devoid of skeletal muscle. This suggests that RMS must arise from cells of origin other than skeletal muscle progenitors. In zebrafish, driving expression of KRAS[G12D] with the *rag2* promoter (*rag2-kRAS[G12D]*) that was expressed in T and B cells, olfactory rosettes, and sperm injected at the 1-cell stage of development resulted in FN-RMS tumors (Langenau et al. 2007). This showed the potential of nonmuscle cells as a potential origin of FN-RMS. The coupling of genetic mutations to the tumor location specificity of FN-RMS development is evident from two studies with either *Fos* germline deletion or *Erbb2* activation coupled with *Trp53* germline deletion in GEMMs. *Trp53[KO];Fos[KO]* mice get ERMS in the face and orbit with 93% penetrance by 6 months (Fleischmann et al. 2003), while all male mice with heterozygous germline deletion of *Trp53* and an activating *Erbb2* transgene develop prostate ERMS (Nanni et al. 2003). It remains unclear how the coupling of genetic mutation and cells of origin dictate the location specificity seen in RMS.

Expression of a human oncogenic *Smoothened* variant (*Smo[W535L]* or *Smo[M2]*) found in basal cell carcinoma results in constitutively active sonic hedgehog (SHH) signaling (Xie et al. 1998). A conditional *R26-Smo[M2]* mouse allele expressing rat *Smo[W539L]* allows for constitutive activation in cells dictated by Cre recombination (Jeong et al. 2004). Ubiquitous expression of SMO[M2] in *CAGGS-CreER;R26-Smo[M2]* compound mutant mice with either tamoxifen administered at day of life 10 (P10) or without tamoxifen relying on spontaneous leaky Cre expression resulted in FN-RMS in all mice. Whereas all mice in the spontaneous leaky group had FN-RMS localized to the rear thigh and abdominal wall, mice that received tamoxifen at P10 developed FN-RMS in the head and neck, tongue, and paratestis region (Mao et al. 2006). This suggests in addition to specific genetic perturbations driving location specificity that there is a temporal window dictating FN-RMS transformation occurring in specific locations.

Originally interested in studying the effect of SHH signaling in adipogenesis, a transgenic mouse with the *cis*-regulatory region of adipose protein 2 fused to Cre recombinase (*aP2*-Cre) was bred to the *R26-Smo[M2]* with the intention of dictating SHH signaling in developing adipocytes. Surprisingly, the adipose tissues in these mice were normal but nearly all *aP2-Cre;Smo[M2]* compound mutant mice developed tumors localized exclusively to the head and neck region histologically resembling human well-differentiated RMS (Hatley et al. 2012). As *aP2-Cre* was not expressed in skeletal muscle, interrogation of this model offers an avenue to explore the cell(s) of origin of FN-RMS other than that of the skeletal muscle. Lineage tracing experiments of *aP2-Cre; R26-tdTomato* compound mutant mice identified Tomato expressing endothelial cells within muscle interstitium between myofibers in all skeletal muscles. Through tracking the Tomato-positive cells in *aP2-Cre;R26[SmoM2/tdTom]* compound mutant mice through embryonic development, proliferations and expansions of Tomato-positive/MYOD1-negative cells in the skeletal muscle interstitium were observed at E15.5 specifically in the head and neck. At E17.5, these Tomato-positive muscle interstitial cells expressed MYOD1. Importantly, the adult *aP2-Cre* endothelial cells do not express SHH target genes despite SMO[M2] expression. This suggests that RMS transformation via transdifferentiation and cell reprogramming must occur in endothelial progenitors before terminal differentiation (Drummond et al. 2018b). As the *aP2-Cre;Smo[M2]* FN-RMS mouse model does not originate from skeletal muscle cells, this model provides a unique opportunity to investigate the essential drivers of RMS tumorigenesis, location specificity, and determinates of RMS cell fate.

The simplicity and rapid tumor onset in the *aP2-Cre;Smo[M2]* FN-RMS mouse model provides a means to explore the function of cooperating genetic perturbations. For example, the *PTEN* promoter is hypermethylated in the majority of FN-RMS leading to decreased *PTEN* expression (Seki et al. 2015) and more than 25% of FN-RMS cases have copy number loss of *PTEN* (Xu et al. 2018). This suggests a strong selective pressure to attenuate PTEN function in FN-RMS. To define the role of PTEN loss in FN-RMS, Langdon et al. conditionally deleted *Pten* in *aP2-Cre;Smo[M2]* mice, which resulted in faster tumor onset, increased proliferation and penetrance, and tumor cells that were less differentiated more closely resembling human ERMS compared to *Pten* WT tumors. Interestingly, the phenotype of *Pten* loss tumors appeared not to be driven by regulation of PI3K/AKT/mTOR signaling but by loss of PTEN in the nucleus, implicating transcriptional and epigenetic mechanisms as the causes for increased tumorigenesis. This phenotype was specific for PTEN and not simply generic tumor suppressor loss as deletion of *Trp53*, *Cdkn2a*, and *Rb1* did not phenocopy *Pten* loss. PAX7 was identified a key TF in these tumors and dual *Pten* and *Pax7* conditional deletion blocked the effect of PTEN loss as well as shifted tumor cell fate to resemble leiomyosarcoma (Langdon et al. 2021). The synthetic essential interaction between PTEN and PAX7 highlights how the cell of origin, transcriptional programs, and mutational spectrum intersect in pediatric embryonal tumors.

Similar to FN-RMS, FP-RMS can occur in sites devoid of skeletal muscle and without an identifiable primary tumor with either widespread disease or only with bone marrow involvement (Sandberg et al. 2001; Affinita et al. 2022). As discussed above, FP-RMS is driven by PAX3-FOXO1 or PAX7-FOXO1 oncofusion proteins. Elegant work has shown that in embryogenesis, a common progenitor in the somite possesses both endothelial and myoblast potential based on a transcriptional equilibrium between FOXC2 and PAX3, respectively (Lagha et al. 2009). The $3'$ *cis*-regulatory region of both mouse and human *FOXO1* contains a strongly conserved endothelial enhancer that is preserved in the *PAX3::FOXO1* rearrangement suggesting a potential to express *PAX3::FOXO1* in endothelial cells (Vicente-García et al. 2017). Furthermore, PAX3::FOXO1 possesses pioneering activity capable of binding and opening condensed chromatin that could enable the activation of muscle transcriptional programs (Sunkel et al. 2021). To test the capacity of P3F to direct reprogramming and transformation of endothelial cells, Searcy et al. conditionally drove *P3F* expression and *Cdkn2a* loss in aP2-Cre cells. This resulted in reprogramming aP2-Cre expressing endothelial cells in the muscle interstitium into PAX7-positive cells capable of contributing to muscle regeneration following cardiotoxin injury only in the head and neck. No FP-RMS tu-

mors arose in the *aP2-Cre;Pax3*^{P3Fm}*;Cdkn2a*^{Flox} mice; however, FP-RMS tumors were observed when an independently derived *aP2-Cre* (*Fabp4-Cre*) was used. Conditional expression of *P3F* and deletion of *Cdkn2a* in the more endothelial-restricted *Tek-Cre* expressing cells resulted in FP-RMS anatomically restricted to the snout. Genomic analysis of these tumors compared to the *Myf6-Cre;Pax3*^{P3Fm}*;Cdkn2a*^{flox} tumors revealed almost identical gene expression, enhancer landscape, and 3D chromatin architecture without any identifiable endothelial cell features solidifying the notion that cell of origin cannot be determined from tumor cell–specific characteristics (Searcy et al. 2023). In addition, tumors occurring in the head and neck were indistinguishable from those in the extremities and trunk showing that the cell of origin dictates the location without affecting the features of the tumor cell.

Taken together, there are many interesting relationships that can be drawn from RMS model systems. First, tumor location and thus patient survival could be dictated by the cell of origin. Second, the cell of origin seems to be highly interchangeable with the same genetic drivers. For instance, expression of KRAS^{G12D} in somitic muscle progenitors, satellite cells, and *rag2*-positive progenitors all result in tumors resembling human FN-RMS. Furthermore, oncogenic SHH signaling results in FN-RMS from endothelial progenitors expressing *aP2-Cre* as well as from *Myog-Cre* expressing muscle progenitor cells (Hatley et al. 2012). This indicates the third point, that the developmental pliancy of the cells of origin and the ability of oncogenic drivers to enact a specific transcriptional program is just as important as the proliferative growth advantage provided by the genetic drivers. In some cases, the growth advantage seems to be mediated entirely through transcriptional and epigenetic changes, highlighted by the synthetic essential relationship between *Pten* loss and *Pax7* in FN-RMS (Langdon et al. 2021). These insights can also be seen in FP-RMS, where expression of P3F in either endothelial or myogenic progenitors results in tumors resembling FP-RMS. In the latter, loss of either *Trp53* or *Cdkn2a* accelerated onset, but both resulted in FP-RMS tumor development. Thus, P3F is a potent oncogenic TF with the ability solely to transform cells specifically through transcriptional and epigenetic changes. This is shown by the full transformation of *TP53* KO iPSC-derived endothelial progenitors upon merely the introduction of P3F. While an exhaustive list of potential cells of origin for RMS remains elusive, each model discussed provides an excellent foundation for future exploration.

Furthermore, the models discussed above develop RMS in a multitude of locations, with some models displaying anatomically restricted tumors. Comparisons of these cells of origin would reveal insights into developmental competency and potentially uncover a link between the cell of origin and primary tumor location. For instance, *aP2-Cre* expressing cells in the head and neck skeletal muscle interstitium derived from the branchial arches can uniquely give rise to FN-RMS specifically located in the head and neck. However, *aP2-Cre* expressing cells in muscle interstitium in muscle derived from somites fail to transform (Hatley et al. 2012; Drummond et al. 2018a). Additionally, *Tek-Cre* cells expressing P3F form FP-RMS tumors anatomically restricted to the snout (Searcy et al. 2023) while FP-RMS in *Myf6-Cre* cells expressing P3F develop primarily in the extremities and trunk and less frequently in the head and neck (Keller et al. 2004; Abraham et al. 2014), showing that distinct cells of origin with the same genetic drivers can lead to tumors in divergent locations. Direct comparisons between location-specific cells of origin will provide further insight into the clinical features of RMS. Furthermore, understanding how the cells of origin progress from normal development to RMS will provide insights into designing model systems that more closely recapitulate the children's disease providing a better platform to understand the biology and identify novel therapeutic targets that could be location dependent.

RHABDOMYOSARCOMA CELL FATE CONTROL THROUGH TF BALANCE, CIRCUITRY, AND EPIGENETIC REMODELING

The most recent update of the hallmarks of cancer was aimed at unifying characteristics seen in all cancers (Hanahan 2022). The newest additions of "unlocking phenotypic plasticity" and "nonmutational epigenetic reprogramming" describe the phenomena seen in pediatric cancers well. Since pediatric tumors harbor very low mutational burden, they rely on epigenetic reprogramming to maintain high proliferation while stalling differentiation. These cancers are largely based on the high expression of fate-determining master TFs, where many of those TFs have been determined to be dependencies through genome-wide CRISPR screens (Tsherniak et al. 2017; Dharia et al. 2021). The expression of fate-defining TFs and their ability to co-opt chromatin remodelers allows pediatric cancer cells to unlock plasticity and navigate multiple cell state transitions instead of being locked into one specific state. These cell states are largely thought to be characterized by feedforward transcription loops termed core-regulatory circuits (CRCs).

CRCs are self-sustaining fate-defining networks of TFs defined by the following criteria: (1) TF autoregulation, (2) TF regulation of other CRC members that result in, (3) feedforward loops, and (4) removal of one member results in circuit ablation (Chen et al. 2020). CRCs are dominated by cell-type-specific enhancers with binding sequences of known master TFs (Neph et al. 2012; Whyte et al. 2013; Zhang et al. 2014; Stewart et al. 2018; Gryder et al. 2019a). In many cases, these master TFs are found across the spectrum of normal development, but ultimately are all connected in a malignant circuit. For instance, during normal muscle development PAX3/7, MYOD1, and MYOG overlap expression for very brief times during differentiation (Bentzinger et al. 2012; Almeida et al. 2016), yet are maintained at high levels together in an RMS cell state (Gryder et al. 2019a). This is likely due to miswired enhancers that lock RMS cells into a state where high expression of these master TFs enforces a myogenic phenotype (Gryder et al. 2019a, 2020; Hsu et al. 2022). With the advent of induced proximity methods such as proteolysis targeting chimeras (PROTACs) and molecular glues, targeting these TFs is of great interest for differentiation therapy (Dale et al. 2021). While the targeted removal of one TF should be enough to break the circuit and induce differentiation or cell death, RMS cells are highly plastic and can recur without the targeted member and/or change fate while remaining malignant. For example, in a human myoblast model with doxycycline-inducible P3F, xenografted cells form tumors resembling FP-RMS in mice. Upon doxycycline withdrawal, these tumors regress only to paradoxically recur in the absence of doxycycline. Furthermore, 60% of the recurrent FP-RMS tumors did not express P3F despite bearing the mutation (Pandey et al. 2017), indicating that P3F is either not a part of the FP-RMS CRC following tumor initiation or that cellular plasticity allows for circuit compensation in relapse. Furthermore, FN-RMS is dependent on PAX7 for its proliferative drive and myogenic fate, but deletion of PAX7 in a mouse model shifted tumor cells to become leiomyosarcoma-like (Hanna et al. 2016; Langdon et al. 2021). Together these data indicate that RMS core TFs are essential for cell identity initiation as well as maintenance. However, the TFs do not seem to describe the entire situation. While defined by core TFs, RMS tumors have a high degree of cell heterogeneity that could explain their capability to rewire transcriptional circuitry upon targeted removal of core TFs. Thus, a deeper understanding of the biology of these developmental transcriptional circuits is needed before therapeutically targeting TFs with proximity ligation methods.

Technological advances have allowed for the profiling of patient-derived tumors at single-cell resolution to identify the different heterogeneous cell populations in RMS. Multiple studies reveal insights into developmental origins, cell fate trajectories, and potential therapeutic targets (Patel et al. 2022; Wei et al. 2022; Danielli et al. 2023; DeMartino et al. 2023; Sara et al. 2023). Indeed, these investigations revealed marked heterogeneity within both FP-RMS and FN-RMS tumors with indications that a malignant stem cell–like population is responsible for treatment escape and ultimate repopulation of the tumors in relapsed disease. Furthermore, when compared to embryonic myogenesis, RMS tumors displayed characteristic markers across the entire myogenic development spectrum resolidifying the notion of RMS origins from hijacked-skeletal muscle development or rather that the tumor cells recapitulate a muscle-like cell. These studies provide insight into RMS cell states and their therapeutic implications, but it remains

unclear exactly how RMS transformation occurs and from which cell type in children. Further examination of these data will likely illuminate candidate mechanisms that can be explored in model systems. Interestingly, a subpopulation of FP-RMS cells expressed neural pathway genes, confirming the previously identified connections between neural stem cells and muscle stem cells (Galli et al. 2000; Raghavan et al. 2019; Lee et al. 2020). Importantly, this data could indicate that the plastic stem cell populations sample cell fates other than myogenic states. PAX3 and PAX7 are well-known regulators of neural crest and neural tube development highlighting the potential of these master TFs to alter cell state in these highly plastic cell populations (Boudjadi et al. 2018; Thompson et al. 2021). Furthermore, PAX proteins are known to be involved in epigenetic remodeling to alter cell state and some members demonstrate pioneering activity (Mayran et al. 2015, 2018, 2019; Pelletier et al. 2021) indicating that cell state sampling of nonmyogenic states in RMS could occur through PAX3/7 recognition and remodeling around nonmyogenic genes. With the knowledge that RMS is defined by genes of cells derived from mesoderm and can express genes related to ectoderm, it is curious what the molecular mechanisms of fate hijacking are and what cells could be excluded as cells of origin. Investigations into RMS utilization of chromatin remodelers, pioneer TFs, and iPSC reprogramming events will be essential to dissect these mechanisms.

RMS cells rely on chromatin remodeling complexes to facilitate tumorigenic gene expression and 3D chromatin architecture. Chromatin remodelers can be divided into four main subfamilies: imitation switch (ISWI), chromodomain helicase DNA binding (CHD), switch/sucrose nonfermentable (SWI/SNF), and the INO80 subfamilies. Together, chromatin remodelers use ATP to assemble nucleosomes, alter nucleosome spacing for gene regulation, and edit nucleosomes with histone variants. Their mechanisms of action and specific functions have been reviewed previously (Clapier et al. 2017) and their roles in RMS tumorigenesis are under active investigation (Pomella et al. 2023).

FN-RMS is genetically heterogeneous lacking a unifying protein to drive the disease. Instead, FN-RMS cells rely on growth and proliferation-promoting signaling from activation of the RAS pathway and developmental morphogens such as NOTCH and SHH. In addition to its role in suppressing MYOD1 binding at enhancers (Pomella et al. 2021), SNAI2 has been shown to interact with CTCF to maintain a subtopologically associated domain boundary that directly maintains high NOTCH1 expression (Fig. 3; Sreenivas et al. 2023). This phenomenon appears to be related to mutant RAS signaling (Yohe et al. 2018) and shows the power of manipulating 3D genome to maintain a tumorigenic state (Wang et al. 2023). Maintaining a tumorigenic state, however, is multifactorial and requires constant reinforcement. To this end, TWIST2 directly competes with shared MYOD1-binding sites to

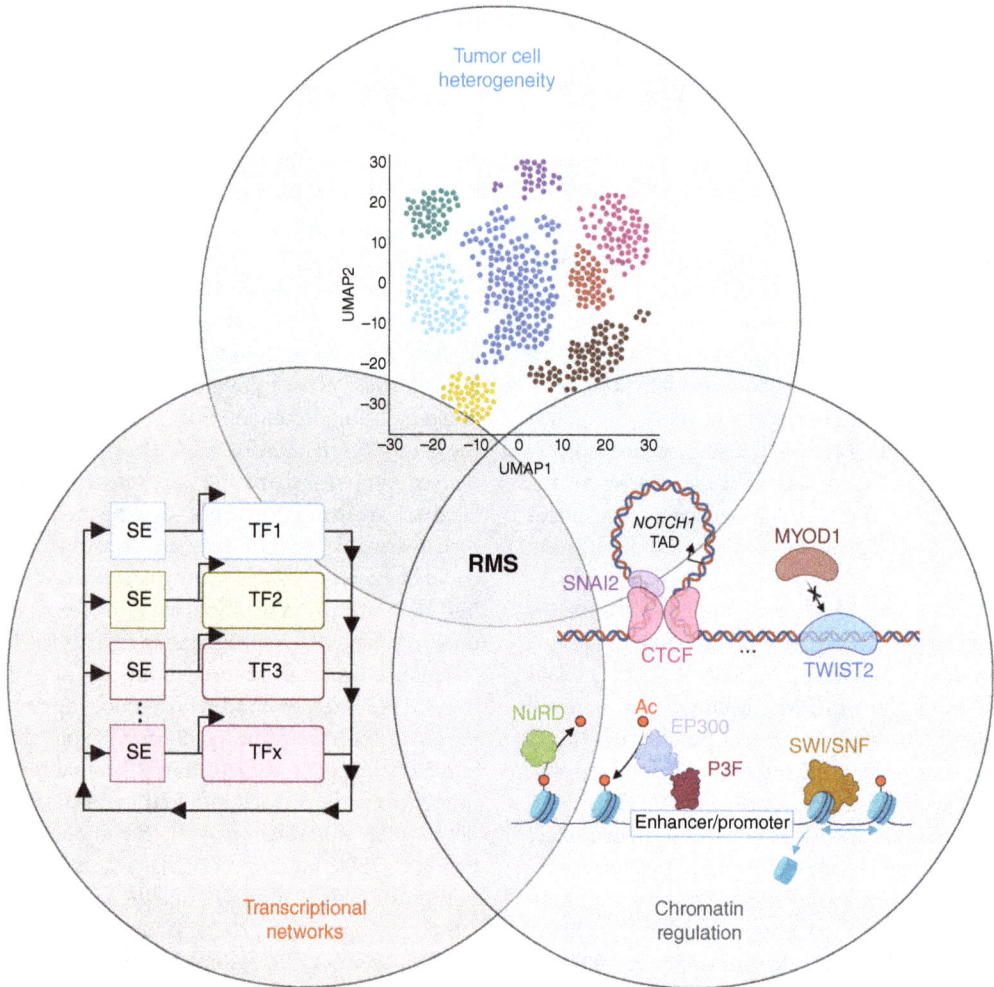

Figure 3. Rhabdomyosarcoma (RMS) tumors are dictated by tumor cell heterogeneity, chromatin regulation, and transcriptional networks. The Venn diagram displays the overlap of mechanistic features used by RMS to maintain a tumorigenic state. Tumor cell heterogeneity comprises populations spanning from mesenchymal stem cell–like to more differentiated muscle-like. To maintain these cell fates, fusion-negative RMS (FN-RMS) and fusion-positive RMS (FP-RMS) use different transcriptional mechanisms. FN-RMS has been shown to maintain high NOTCH1 signaling through corrupted 3D genome architecture while stalling differentiation signals by blocking MYOD1-binding sites. FP-RMS has been shown to maintain discrete enhancer and promoter boundaries with oncofusion P3F recruiting chromatin-modifying enzymes to sites of interest. These mechanisms are used to maintain feedforward transcriptional networks that reinforce cell fate.

Cite this article as *Cold Spring Harb Perspect Med* doi: 10.1101/cshperspect.a041583

prevent differentiation (Li et al. 2019) and activates enhancers controlling growth and proliferation genes by relocating SWI/SNF and CHD3 remodeler subunits (Fig. 3; Shah et al. 2023). In accordance with this data, silencing of SWI/SNF member BAF53a induces FN-RMS cell differentiation (Taulli et al. 2014). Future investigations into the interplay between the chromatin remodeling complexes will be greatly beneficial to understanding the complexity of FN-RMS.

FP-RMS is driven by oncofusion proteins, and thus research on chromatin regulation is largely focused on the effects mediated by either P3/7F to maintain a tumorigenic state. P3F is a strong activating TF that is able to bind heterochromatin (Sunkel et al. 2021) and establish active enhancers (Zhang et al. 2022) through direct interactions with the histone acetyltransferase EP300, BRD4, and SWI/SNF members (Fig. 3; Gryder et al. 2017; Laubscher et al. 2021). Enhancer maintenance is dependent on chromatin remodelers due to their ability to clear nucleosomes from stretches of DNA, and enhancer maintenance is critical to maintaining a tumorigenic state. When expression of SWI/SNF subunits BAF53a and/or SMARCA4 is decreased, FP-RMS cells differentiate and have reduced long-term survival (Taulli et al. 2014; Bharathy et al. 2022). This phenotype is also observed upon SMARCA4 inhibition (Laubscher et al. 2021). Importantly, P3F-bound enhancers are frequently cobound and coregulated by subunits of the Nucleosome Remodeling and Deacetylase (NuRD, CHD4 containing) complex (Böhm et al. 2016). Paradoxically, the NuRD complex also contains a histone deacetylase 2 (HDAC2) subunit that would function to remove active acetylation marks at P3F-bound enhancers (Fig. 3). CDH4 depletion removed HDAC2 from chromatin, resulting in spreading of acetylation (Marques et al. 2020), a phenomenon observed previously when RMS cells were treated with HDAC inhibitors (Gryder et al. 2019a,b). Spreading acetylation precedes the removal of RNA polymerase II from core RMS TF genetic elements (Gryder et al. 2019a; Marques et al. 2020), indicating that not only the presence of, but defined boundaries of epigenetic regulatory elements such as enhancers/super-enhancers are essential for RMS maintenance. The molecular mechanisms governing the initiation of these events and their essentiality in transformation are still unknown and may reveal important cooperating targets or competent cell states.

CONCLUDING REMARKS

Waddington's epigenetic landscape, described in 1957, sets the stage for the analysis of cell fate decisions (Waddington 1957). As the iconic marble rolls down the valleys and navigates bifurcations, pliancy and proliferation decreases and the cell ultimately achieves terminal differentiation. While this model is pivotal in normal development, it needs to be adapted to accurately describe RMS and pediatric cancers as a whole. The unadapted model assumes a stem cell state is unstable and that dedifferentiation must take the same path back to a common progenitor. However, many pediatric cancers resemble arrested states of development, suggesting that there exists a detour from the normal Waddington trajectory to a novel plane of transformation caused by oncogenic perturbation. The transformation plane is flat and perpendicular to the Waddington landscape. The plane represents the cell fate sampling space that gives rise to tumor cell heterogeneity. The plane is resident to a stable cancer stem cell population that navigates the plane through expression of fate-defining TFs and epigenetic changes. As cancer stem cells sample fates, accompanying changes in chromatin architecture, the epigenome, and gene expression can reinforce stable fates or help transition to alternative fates. Low mutational burdens necessitating the use of master TFs and epigenetic regulators would allow for sampling across multiple cell fates and could be indicative of many cells of origin. Divots in the transformation plane, representing more committed fates than the cancer stem cell, are dispersed throughout the plane. As cancer stem cells sample different cell fates with their master TFs, they may acquire expression of specification and/or determination TFs resulting in gene expression programs similar to normal cells that have transited further down their normal Waddington landscape. On the transformation plane, these cells would be resident in the divots. However, unlike their untransformed counterparts, these tumor cells do not continue toward terminal differentiation and continue to proliferate adding to tumor mass.

Utilizing this framework to summarize what we know about RMS etiology and disease we generated a 3D model seen in Figure 4. Each side represents the normal differentiation of a known cell of origin (Fig. 4). One side shows a path myogenic progenitors take to reach terminal differentiation and the other shows a path taken by endothelial cell progenitors. Both progenitors can enter the transformation plane through oncogenic insults (acquisition of P3F, RAS mutation, constitutive SHH signaling) and result in the same tumor subtype (FN vs. FP) heterogeneity regardless of the cell of origin. A particular oncogenic insult can result in progenitor cells entering the transformation plane from either side. After entering the transformation plane, the transformed cell results in the same heterogeneity of cell states, occupying identical locations within the transformation plane. RMS cancer stem cells reside on the plane and sample different cell fates guided by chromatin remodelers and master TFs (PAX members, MYOD1, SWI/SNF, NuRD, etc.). Cells that acquire expression of MYOD1 and/or MYOG fall into a divot close to the myogenic Waddington landscape and continue to proliferate without terminally differentiating. Divots further away from the myogenic and endothelial Waddington landscapes represent seemingly unrelated fates that are sampled, such as the neural population seen in FP-RMS tumors. The recent single-cell analyses of RMS have delineated distinct cell states

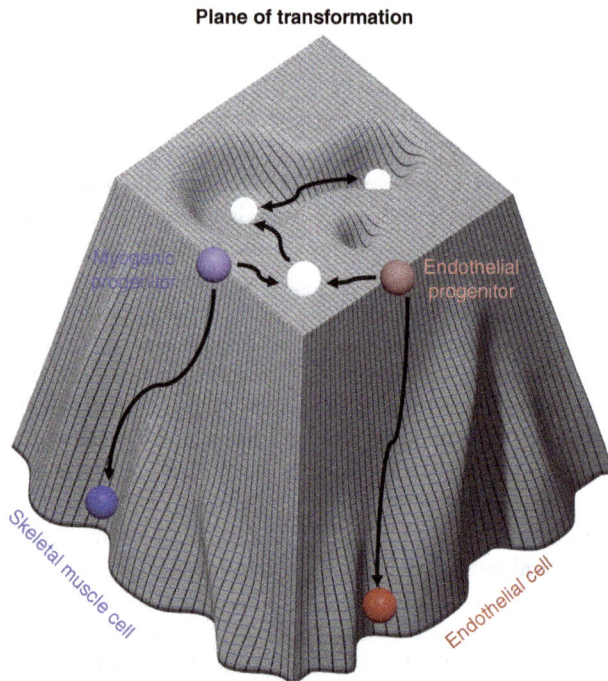

Figure 4. The plane of transformation. The normal Waddington landscape does not accurately describe the landscape of pediatric embryonal tumors. Here, we have reimagined the Waddington landscape to include a new plane of transformation that can be entered by normally differentiating cells through cell reprogramming secondary to an oncogenic insult. The plane of transformation is a cell fate sampling space resident to cancer stem cells that can completely establish tumor cell heterogeneity by navigating the plane. Navigating the plane is driven by master transcription factors and accompanying epigenome changes that either reinforce a cell fate decision or help guide the next fate sampled. Overlayed onto this 3D model is myogenic (blue) and endothelial (red) cell differentiation with oncogenic insult resulting in rhabdomyosarcoma (RMS) tumors (white).

spanning from mesenchymal-like precursors to more differentiated skeletal muscle cells (Patel et al. 2022; Wei et al. 2022; Danielli et al. 2023; DeMartino et al. 2023; Sara et al. 2023). In support of our model, Wei et al. isolated the mesenchymal-like precursors from FN-RMS cell lines (RD cells) and PDXs and showed their capacity to repopulate each of the other cell states and form tumors when engrafted in mice. Further investigations will test and refine this model, but this model provides a theoretical framework to explain the many cells of origin and marked heterogeneity seen in RMS tumors.

Normal development has taught us that TF and epigenetic balance are essential for making timely fate decisions to establish the correct body plan. Furthermore, similar cell states can be reached by different TF circuits and epigenetic modulations that occur in locations anatomically isolated from each other. Patients with RMS display a broad range of clinical heterogeneity where RMS can occur throughout the body, including sites devoid of skeletal muscle, and these locations of the primary tumor are indicative of the children's survival. RMS occurring at any of these locations could be derived from completely different cells of origin. Understanding how cell(s) of origin and oncogenic driving perturbations impact tumor formation at distinct locations could elucidate the underling biology that drives childhood RMS survival. Differentiation therapy to force terminal differentiation and exit of the cell cycle is a compelling idea in pediatric embryonal tumors. However, the ability of RMS to derive from multiple nonskeletal muscle cells could limit the capacity for this approach. RMS from cells outside of the skeletal muscle lineage may lack the epigenetic priming necessary to fully execute the skeletal muscle differentiation program. Furthermore, targeting RMS essential TFs is full of therapeutic possibilities. Current therapies do not target fate-defining TFs and thus, there is no selective pressure for the tumors to recur without them, yet the possibility warrants further investigations. Ultimately, studies dissecting the developmental origins of RMS are required to elucidate tumorigenic mechanisms and therapeutic targets to increase the survival rates of children.

ACKNOWLEDGMENTS

We thank the Hatley laboratory members for advice, support, and helpful comments on the manuscript. We thank Trevor Penix in aiding in the construction of our reimagined Waddington landscape and transformation plane. Research reported in this publication was supported by the National Cancer Institute of the National Institutes of Health under award numbers R01CA216344, R01CA251436, R01CA266600 (M.E.H.), and F31CA281254 (B.T.S.). The content is solely the responsibility of the authors and does not necessarily represent the official views of the National Institutes of Health. The Hatley laboratory is also supported by grants from the V Foundation for Cancer Research, the Rally Foundation for Childhood Cancer Research and Open Hands Overflowing Hearts award number 20IC23 (M.E.H.), St. Jude Cancer Center support grant (P30 CA21765), American Lebanese Syrian Associated Charities of St. Jude Children's Research Hospital, and the St. Jude Graduate School of Biomedical Sciences. M.E.H. has served on the advisory board for Servier. We sincerely apologize to our colleagues whose work we could not cite secondary to space constraints.

REFERENCES

Abraham J, Nuñez-Álvarez Y, Hettmer S, Carrió E, Chen HI, Nishijo K, Huang ET, Prajapati SI, Walker RL, Davis S, et al. 2014. Lineage of origin in rhabdomyosarcoma informs pharmacological response. *Genes Dev* 28: 1578–1591. doi:10.1101/gad.238733.114

Affinita MC, Merks JHM, Chisholm JC, Haouy S, Rome A, Rabusin M, Brennan B, Bisogno G. 2022. Rhabdomyosarcoma with unknown primary tumor site: a report from European Pediatric Soft Tissue Sarcoma Study Group (EpSSG). *Pediatr Blood Cancer* 69: e29967. doi:10.1002/pbc.29967

Agaram NP, Chen CL, Zhang L, LaQuaglia MP, Wexler L, Antonescu CR. 2014. Recurrent *MYOD1* mutations in pediatric and adult sclerosing and spindle cell rhabdomyosarcomas: evidence for a common pathogenesis. *Genes Chromosomes Cancer* 53: 779–787. doi:10.1002/gcc.22187

Agaram NP, LaQuaglia MP, Alaggio R, Zhang L, Fujisawa Y, Ladanyi M, Wexler L, Antonescu C. 2019. MYOD1-mutant spindle cell and sclerosing rhabdomyosarcoma: an aggressive subtype irrespective of age. A reappraisal for molecular classification and risk stratification. *Mod Pathol* 32: 27–36. doi:10.1038/s41379-018-0120-9

Alaggio R, Zhang L, Sung YS, Huang SC, Chen CL, Bisogno G, Zin A, Agaram NP, LaQuaglia MP, Wexler L, et al. 2016. A molecular study of pediatric spindle and sclerosing rhabdomyosarcoma: identification of novel and recurrent VGLL2-related fusions in infantile cases. *Am J Surg Pathol* **40:** 224–235. doi:10.1097/PAS.0000000000000538

Almeida CF, Fernandes SA, Ribeiro Junior AF, Keith Okamoto O, Vainzof M. 2016. Muscle satellite cells: exploring the basic biology to rule them. *Stem Cells Int* **2016:** 1078686. doi:10.1155/2016/1078686

Amer KM, Thomson JE, Congiusta D, Dobitsch A, Chaudhry A, Li M, Chaudhry A, Bozzo A, Siracuse B, Aytekin MN, et al. 2019. Epidemiology, incidence, and survival of rhabdomyosarcoma subtypes: SEER and ICES database analysis. *J Orthop Res* **37:** 2226–2230. doi:10.1002/jor.24387

Aoyama H, Asamoto K. 1988. Determination of somite cells: independence of cell differentiation and morphogenesis. *Development* **104:** 15–28. doi:10.1242/dev.104.1.15

Aulehla A, Pourquié O. 2010. Signaling gradients during paraxial mesoderm development. *Cold Spring Harb Perspect Biol* **2:** a000869. doi:10.1101/cshperspect.a000869

Barr FG, Galili N, Holick J, Biegel JA, Rovera G, Emanuel BS. 1993. Rearrangement of the PAX3 paired box gene in the paediatric solid tumour alveolar rhabdomyosarcoma. *Nat Genet* **3:** 113–117. doi:10.1038/ng0293-113

Bentzinger CF, Wang YX, Rudnicki MA. 2012. Building muscle: molecular regulation of myogenesis. *Cold Spring Harb Perspect Biol* **4:** a008342. doi:10.1101/cshperspect.a008342

Bharathy N, Cleary MM, Kim JA, Nagamori K, Crawford KA, Wang E, Saha D, Settelmeyer TP, Purohit R, Skopelitis D, et al. 2022. SMARCA4 biology in alveolar rhabdomyosarcoma. *Oncogene* **41:** 1647–1656. doi:10.1038/s41388-022-02205-0

Blum JM, Añó L, Li Z, Van Mater D, Bennett BD, Sachdeva M, Lagutina I, Zhang M, Mito JK, Dodd LG, et al. 2013. Distinct and overlapping sarcoma subtypes initiated from muscle stem and progenitor cells. *Cell Rep* **5:** 933–940. doi:10.1016/j.celrep.2013.10.020

Bober E, Franz T, Arnold HH, Gruss P, Tremblay P. 1994. *Pax-3* is required for the development of limb muscles: a possible role for the migration of dermomyotomal muscle progenitor cells. *Development* **120:** 603–612. doi:10.1242/dev.120.3.603

Böhm M, Wachtel M, Marques JG, Streiff N, Laubscher D, Nanni P, Mamchaoui K, Santoro R, Schäfer BW. 2016. Helicase CHD4 is an epigenetic coregulator of PAX3-FOXO1 in alveolar rhabdomyosarcoma. *J Clin Invest* **126:** 4237–4249. doi:10.1172/JCI85057

Boudjadi S, Chatterjee B, Sun W, Vemu P, Barr FG. 2018. The expression and function of PAX3 in development and disease. *Gene* **666:** 145–157. doi:10.1016/j.gene.2018.04.087

Braun T, Gautel M. 2011. Transcriptional mechanisms regulating skeletal muscle differentiation, growth and homeostasis. *Nat Rev Mol Cell Biol* **12:** 349–361. doi:10.1038/nrm3118

Buckingham M. 2017. Gene regulatory networks and cell lineages that underlie the formation of skeletal muscle. *Proc Natl Acad Sci* **114:** 5830–5837. doi:10.1073/pnas.1610605114

Buckingham M, Rigby PW. 2014. Gene regulatory networks and transcriptional mechanisms that control myogenesis. *Dev Cell* **28:** 225–238. doi:10.1016/j.devcel.2013.12.020

Chal J, Pourquié O. 2017. Making muscle: skeletal myogenesis in vivo and in vitro. *Development* **144:** 2104–2122. doi:10.1242/dev.151035

Chal J, Oginuma M, Al Tanoury Z, Gobert B, Sumara O, Hick A, Bousson F, Zidouni Y, Mursch C, Moncuquet P, et al. 2015. Differentiation of pluripotent stem cells to muscle fiber to model Duchenne muscular dystrophy. *Nat Biotechnol* **33:** 962–969. doi:10.1038/nbt.3297

Chen X, Stewart E, Shelat A, Qu C, Bahrami A, Hatley M, Wu G, Bradley C, McEvoy J, Pappo A, et al. 2013. Targeting oxidative stress in embryonal rhabdomyosarcoma. *Cancer Cell* **24:** 710–724. doi:10.1016/j.ccr.2013.11.002

Chen L, Shern JF, Wei JS, Yohe ME, Song YK, Hurd L, Liao H, Catchpoole D, Skapek SX, Barr FG, et al. 2015a. Clonality and evolutionary history of rhabdomyosarcoma. *PLoS Genet* **11:** e1005075. doi:10.1371/journal.pgen.1005075

Chen X, Pappo A, Dyer MA. 2015b. Pediatric solid tumor genomics and developmental pliancy. *Oncogene* **34:** 5207–5215. doi:10.1038/onc.2014.474

Chen Y, Xu L, Lin RY, Müschen M, Koeffler HP. 2020. Core transcriptional regulatory circuitries in cancer. *Oncogene* **39:** 6633–6646. doi:10.1038/s41388-020-01459-w

Clapier CR, Iwasa J, Cairns BR, Peterson CL. 2017. Mechanisms of action and regulation of ATP-dependent chromatin-remodelling complexes. *Nat Rev Mol Cell Biol* **18:** 407–422. doi:10.1038/nrm.2017.26

Crane JN, Xue W, Qumseya A, Gao Z, Arndt CAS, Donaldson SS, Harrison DJ, Hawkins DS, Linardic CM, Mascarenhas L, et al. 2022. Clinical group and modified TNM stage for rhabdomyosarcoma: a review from the Children's Oncology Group. *Pediatr Blood Cancer* **69:** e29644. doi:10.1002/pbc.29644

Dale B, Cheng M, Park KS, Kaniskan HU, Xiong Y, Jin J. 2021. Advancing targeted protein degradation for cancer therapy. *Nat Rev Cancer* **21:** 638–654. doi:10.1038/s41568-021-00365-x

Danielli SG, Porpiglia E, De Micheli AJ, Navarro N, Zellinger MJ, Bechtold I, Kisele S, Volken L, Marques JG, Kasper S, et al. 2023. Single-cell profiling of alveolar rhabdomyosarcoma reveals RAS pathway inhibitors as cell-fate hijackers with therapeutic relevance. *Sci Adv* **9:** eade9238. doi:10.1126/sciadv.ade9238

Daston G, Lamar E, Olivier M, Goulding M. 1996. *Pax-3* is necessary for migration but not differentiation of limb muscle precursors in the mouse. *Development* **122:** 1017–1027. doi:10.1242/dev.122.3.1017

Davis RJ, D'Cruz CM, Lovell MA, Biegel JA, Barr FG. 1994. Fusion of PAX7 to FKHR by the variant t(1;13)(p36;q14) translocation in alveolar rhabdomyosarcoma. *Cancer Res* **54:** 2869–2872.

Dehner CA, Armstrong AE, Yohe M, Shern JF, Hirbe AC. 2021. Genetic characterization, current model systems and prognostic stratification in PAX fusion-negative vs. PAX fusion-positive rhabdomyosarcoma. *Genes (Basel)* **12:** 1500. doi:10.3390/genes12101500

Dehner CA, Rudzinski ER, Davis JL. 2023. Rhabdomyosarcoma: updates on classification and the necessity of molecular testing beyond immunohistochemistry. *Hum Pathol* doi:10.1016/j.humpath.2023.12.004

DeMartino J, Meister MT, Visser LL, Brok M, Groot Koerkamp MJA, Wezenaar AKL, Hiemcke-Jiwa LS, de Souza T, Merks JHM, Rios AC, et al. 2023. Single-cell transcriptomics reveals immune suppression and cell states predictive of patient outcomes in rhabdomyosarcoma. *Nat Commun* **14:** 3074. doi:10.1038/s41467-023-38886-8

Denetclaw WF Jr, Christ B, Ordahl CP. 1997. Location and growth of epaxial myotome precursor cells. *Development* **124:** 1601–1610. doi:10.1242/dev.124.8.1601

Dharia NV, Kugener G, Guenther LM, Malone CF, Durbin AD, Hong AL, Howard TP, Bandopadhayay P, Wechsler CS, Fung I, et al. 2021. A first-generation pediatric cancer dependency map. *Nat Genet* **53:** 529–538. doi:10.1038/s41588-021-00819-w

Dong F, Sun X, Liu W, Ai D, Klysik E, Lu MF, Hadley J, Antoni L, Chen L, Baldini A, et al. 2006. *Pitx2* promotes development of splanchnic mesoderm-derived branchiomeric muscle. *Development* **133:** 4891–4899. doi:10.1242/dev.02693

Drummond CJ, Hanna JA, Garcia MR, Devine D, Peters J, Frohlich V, Finkelstein D, Hatley ME. 2018a. Fusion-negative rhabdomyosarcoma originating from endothelial progenitors. *Clin Cancer Res* **24:** 43–44. doi:10.1016/j.ccell.2017.12.001

Drummond CJ, Hanna JA, Garcia MR, Devine DJ, Heyrana AJ, Finkelstein D, Rehg JE, Hatley ME. 2018b. Hedgehog pathway drives fusion-negative rhabdomyosarcoma initiated from non-myogenic endothelial progenitors. *Cancer Cell* **33:** 108–124.e5. doi:10.1016/j.ccell.2017.12.001

Esteves de Lima J, Relaix F. 2021. Master regulators of skeletal muscle lineage development and pluripotent stem cells differentiation. *Cell Regen* **10:** 31. doi:10.1186/s13619-021-00093-5

Fleischmann A, Jochum W, Eferl R, Witowsky J, Wagner EF. 2003. Rhabdomyosarcoma development in mice lacking Trp53 and Fos: tumor suppression by the Fos protooncogene. *Cancer Cell* **4:** 477–482. doi:10.1016/s1535-6108(03)00280-0

Galili N, Davis RJ, Fredericks WJ, Mukhopadhyay S, Rauscher FJ III, Emanuel BS, Rovera G, Barr FG. 1993. Fusion of a fork head domain gene to PAX3 in the solid tumour alveolar rhabdomyosarcoma. *Nat Genet* **5:** 230–235. doi:10.1038/ng1193-230

Galli R, Borello U, Gritti A, Minasi MG, Bjornson C, Coletta M, Mora M, De Angelis MG, Fiocco R, Cossu G, et al. 2000. Skeletal myogenic potential of human and mouse neural stem cells. *Nat Neurosci* **3:** 986–991. doi:10.1038/79924

Gartrell J, Pappo A. 2020. Recent advances in understanding and managing pediatric rhabdomyosarcoma. *F1000Res* **9:** 685. doi:10.12688/f1000research.22451.1

Giordani J, Bajard L, Demignon J, Daubas P, Buckingham M, Maire P. 2007. Six proteins regulate the activation of *Myf5* expression in embryonic mouse limbs. *Proc Natl Acad Sci* **104:** 11310–11315. doi:10.1073/pnas.0611299104

Gonzalez Curto G, Der Vartanian A, Frarma YEM, Manceau L, Baldi L, Prisco S, Elarouci N, Causeret F, Korenkov D, Rigolet M, et al. 2020. The PAX-FOXO1s trigger fast trans-differentiation of chick embryonic neural cells into alveolar rhabdomyosarcoma with tissue invasive properties limited by S phase entry inhibition. *PLoS Genet* **16:** e1009164. doi:10.1371/journal.pgen.1009164

Grimaldi A, Tajbakhsh S. 2021. Diversity in cranial muscles: origins and developmental programs. *Curr Opin Cell Biol* **73:** 110–116. doi:10.1016/j.ceb.2021.06.005

Gröbner SN, Worst BC, Weischenfeldt J, Buchhalter I, Kleinheinz K, Rudneva VA, Johann PD, Balasubramanian GP, Segura-Wang M, Brabetz S, et al. 2018. The landscape of genomic alterations across childhood cancers. *Nature* **555:** 321–327. doi:10.1038/nature25480

Gros J, Manceau M, Thomé V, Marcelle C. 2005. A common somitic origin for embryonic muscle progenitors and satellite cells. *Nature* **435:** 954–958. doi:10.1038/nature03572

Gros J, Serralbo O, Marcelle C. 2009. WNT11 acts as a directional cue to organize the elongation of early muscle fibres. *Nature* **457:** 589–593. doi:10.1038/nature07564

Gryder BE, Yohe ME, Chou HC, Zhang X, Marques J, Wachtel M, Schaefer B, Sen N, Song Y, Gualtieri A, et al. 2017. PAX3-FOXO1 establishes myogenic super enhancers and confers BET bromodomain vulnerability. *Cancer Discov* **7:** 884–899. doi:10.1158/2159-8290.CD-16-1297

Gryder BE, Pomella S, Sayers C, Wu XS, Song Y, Chiarella AM, Bagchi S, Chou HC, Sinniah RS, Walton A, et al. 2019a. Histone hyperacetylation disrupts core gene regulatory architecture in rhabdomyosarcoma. *Nat Genet* **51:** 1714–1722. doi:10.1038/s41588-019-0534-4

Gryder BE, Wu L, Woldemichael GM, Pomella S, Quinn TR, Park PMC, Cleveland A, Stanton BZ, Song Y, Rota R, et al. 2019b. Chemical genomics reveals histone deacetylases are required for core gene regulatory transcription. *Nat Commun* **10:** 3004. doi:10.1038/s41467-019-11046-7

Gryder BE, Wachtel M, Chang K, El Demerdash O, Aboreden NG, Mohammed W, Ewert W, Pomella S, Rota R, Wei JS, et al. 2020. Miswired enhancer logic drives a cancer of the muscle lineage. *iScience* **23:** 101103. doi:10.1016/j.isci.2020.101103

Hanahan D. 2022. Hallmarks of cancer: new dimensions. *Cancer Discov* **12:** 31–46. doi:10.1158/2159-8290.CD-21-1059

Hanna JA, Garcia MR, Go JC, Finkelstein D, Kodali K, Pagala V, Wang X, Peng J, Hatley ME. 2016. PAX7 is a required target for microRNA-206-induced differentiation of fusion-negative rhabdomyosarcoma. *Cell Death Dis* **7:** e2256. doi:10.1038/cddis.2016.159

Hatley ME, Tang W, Garcia MR, Finkelstein D, Millay DP, Liu N, Graff J, Galindo RL, Olson EN. 2012. A mouse model of rhabdomyosarcoma originating from the adipocyte lineage. *Cancer Cell* **22:** 536–546. doi:10.1016/j.ccr.2012.09.004

Heude E, Tesarova M, Sefton EM, Jullian E, Adachi N, Grimaldi A, Zikmund T, Kaiser J, Kardon G, Kelly RG, et al. 2018. Unique morphogenetic signatures define mammalian neck muscles and associated connective tissues. *eLife* **7:** e40179. doi:10.7554/eLife.40179

Hinson AR, Jones R, Crose LE, Belyea BC, Barr FG, Linardic CM. 2013. Human rhabdomyosarcoma cell lines for rhabdomyosarcoma research: utility and pitfalls. *Front Oncol* **3:** 183. doi:10.3389/fonc.2013.00183

Hsu JY, Danis EP, Nance S, O'Brien JH, Gustafson AL, Wessells VM, Goodspeed AE, Talbot JC, Amacher SL, Jedlicka P, et al. 2022. SIX1 reprograms myogenic transcription factors to maintain the rhabdomyosarcoma undifferentiated state. *Cell Rep* **38:** 110323. doi:10.1016/j.celrep.2022.110323

Ibarra-Soria X, Thierion E, Mok GF, Münsterberg AE, Odom DT, Marioni JC. 2023. A transcriptional and regulatory map of mouse somite maturation. *Dev Cell* **58:** 1983–1995. e7. doi:10.1016/j.devcel.2023.07.003

Jeong J, Mao J, Tenzen T, Kottmann AH, McMahon AP. 2004. Hedgehog signaling in the neural crest cells regulates the patterning and growth of facial primordia. *Genes Dev* **18:** 937–951. doi:10.1101/gad.1190304

Kashi VP, Hatley ME, Galindo RL. 2015. Probing for a deeper understanding of rhabdomyosarcoma: insights from complementary model systems. *Nat Rev Cancer* **15:** 426–439. doi:10.1038/nrc3961

Kassar-Duchossoy L, Giacone E, Gayraud-Morel B, Jory A, Gomès D, Tajbakhsh S. 2005. Pax3/Pax7 mark a novel population of primitive myogenic cells during development. *Genes Dev* **19:** 1426–1431. doi:10.1101/gad.345505

Keller C, Arenkiel BR, Coffin CM, El-Bardeesy N, DePinho RA, Capecchi MR. 2004. Alveolar rhabdomyosarcomas in conditional *Pax3:Fkhr* mice: cooperativity of Ink4a/ARF and Trp53 loss of function. *Genes Dev* **18:** 2614–2626. doi:10.1101/gad.1244004

Kohsaka S, Shukla N, Ameur N, Ito T, Ng CK, Wang L, Lim D, Marchetti A, Viale A, Pirun M, et al. 2014. A recurrent neomorphic mutation in MYOD1 defines a clinically aggressive subset of embryonal rhabdomyosarcoma associated with PI3K-AKT pathway mutations. *Nat Genet* **46:** 595–600. doi:10.1038/ng.2969

Lagha M, Brunelli S, Messina G, Cumano A, Kume T, Relaix F, Buckingham ME. 2009. Pax3:Foxc2 reciprocal repression in the somite modulates muscular versus vascular cell fate choice in multipotent progenitors. *Dev Cell* **17:** 892–899. doi:10.1016/j.devcel.2009.10.021

Langdon CG, Gadek KE, Garcia MR, Evans MK, Reed KB, Bush M, Hanna JA, Drummond CJ, Maguire MC, Leavey PJ, et al. 2021. Synthetic essentiality between PTEN and core dependency factor PAX7 dictates rhabdomyosarcoma identity. *Nat Commun* **12:** 5520. doi:10.1038/s41467-021-25829-4

Langenau DM, Keefe MD, Storer NY, Guyon JR, Kutok JL, Le X, Goessling W, Neuberg DS, Kunkel LM, Zon LI. 2007. Effects of RAS on the genesis of embryonal rhabdomyosarcoma. *Genes Dev* **21:** 1382–1395. doi:10.1101/gad.1545007

Laubscher D, Gryder BE, Sunkel BD, Andresson T, Wachtel M, Das S, Roschitzki B, Wolski W, Wu XS, Chou HC, et al. 2021. BAF complexes drive proliferation and block myogenic differentiation in fusion-positive rhabdomyosarcoma. *Nat Commun* **12:** 6924. doi:10.1038/s41467-021-27176-w

Lee QY, Mall M, Chanda S, Zhou B, Sharma KS, Schaukowitch K, Adrian-Segarra JM, Grieder SD, Kareta MS, Wapinski OL, et al. 2020. Pro-neuronal activity of Myod1 due to promiscuous binding to neuronal genes. *Nat Cell Biol* **22:** 401–411. doi:10.1038/s41556-020-0490-3

Lepper C, Fan CM. 2010. Inducible lineage tracing of Pax7-descendant cells reveals embryonic origin of adult satellite cells. *Genesis* **48:** 424–436. doi:10.1002/dvg.20630

Lescroart F, Hamou W, Francou A, Théveniau-Ruissy M, Kelly RG, Buckingham M. 2015. Clonal analysis reveals a common origin between nonsomite-derived neck muscles and heart myocardium. *Proc Natl Acad Sci* **112:** 1446–1451. doi:10.1073/pnas.1424538112

Li S, Chen K, Zhang Y, Barnes SD, Jaichander P, Zheng Y, Hassan M, Malladi VS, Skapek SX, Xu L, et al. 2019. Twist2

amplification in rhabdomyosarcoma represses myogenesis and promotes oncogenesis by redirecting MyoD DNA binding. *Genes Dev* **33:** 626–640. doi:10.1101/gad.324467.119

Linardic CM, Downie DL, Qualman S, Bentley RC, Counter CM. 2005. Genetic modeling of human rhabdomyosarcoma. *Cancer Res* **65:** 4490–4495. doi:10.1158/0008-5472.CAN-04-3194

Linardic CM, Naini S, Herndon JE II, Kesserwan C, Qualman SJ, Counter CM. 2007. The *PAX3-FKHR* fusion gene of rhabdomyosarcoma cooperates with loss of p16INK4A to promote bypass of cellular senescence. *Cancer Res* **67:** 6691–6699. doi:10.1158/0008-5472.CAN-06-3210

Lu JR, Bassel-Duby R, Hawkins A, Chang P, Valdez R, Wu H, Gan L, Shelton JM, Richardson JA, Olson EN. 2002. Control of facial muscle development by MyoR and capsulin. *Science* **298:** 2378–2381. doi:10.1126/science.1078273

Mao J, Ligon KL, Rakhlin EY, Thayer SP, Bronson R, Rowitch D, McMahon AP. 2006. A novel somatic mouse model to survey tumorigenic potential applied to the Hedgehog pathway. *Cancer Res* **66:** 10171–10178. doi:10.1158/0008-5472.CAN-06-0657

Marques JG, Gryder BE, Pavlovic B, Chung Y, Ngo QA, Frommelt F, Gstaiger M, Song Y, Benischke K, Laubscher D, et al. 2020. NuRD subunit CHD4 regulates super-enhancer accessibility in rhabdomyosarcoma and represents a general tumor dependency. *eLife* **9:** e54993. doi:10.7554/eLife.54993

Mayran A, Pelletier A, Drouin J. 2015. Pax factors in transcription and epigenetic remodelling. *Semin Cell Dev Biol* **44:** 135–144. doi:10.1016/j.semcdb.2015.07.007

Mayran A, Khetchoumian K, Hariri F, Pastinen T, Gauthier Y, Balsalobre A, Drouin J. 2018. Pioneer factor Pax7 deploys a stable enhancer repertoire for specification of cell fate. *Nat Genet* **50:** 259–269. doi:10.1038/s41588-017-0035-2

Mayran A, Sochodolsky K, Khetchoumian K, Harris J, Gauthier Y, Bemmo A, Balsalobre A, Drouin J. 2019. Pioneer and nonpioneer factor cooperation drives lineage specific chromatin opening. *Nat Commun* **10:** 3807. doi:10.1038/s41467-019-11791-9

McDaniel C, Simsek MF, Chandel AS, Özbudak EM. 2024. Spatiotemporal control of pattern formation during somitogenesis. *Sci Adv* **10:** eadk8937. doi:10.1126/sciadv.adk8937

Meza JL, Anderson J, Pappo AS, Meyer WH; Children's Oncology Group. 2006. Analysis of prognostic factors in patients with nonmetastatic rhabdomyosarcoma treated on intergroup rhabdomyosarcoma studies III and IV: the Children's Oncology Group. *J Clin Oncol* **24:** 3844–3851. doi:10.1200/JCO.2005.05.3801

Naiche L, Holder N, Lewandoski M. 2011. FGF4 and FGF8 comprise the wavefront activity that controls somitogenesis. *Proc Natl Acad Sci* **108:** 4018–4023. doi:10.1073/pnas.1007417108

Naini S, Etheridge KT, Adam SJ, Qualman SJ, Bentley RC, Counter CM, Linardic CM. 2008. Defining the cooperative genetic changes that temporally drive alveolar rhabdomyosarcoma. *Cancer Res* **68:** 9583–9588. doi:10.1158/0008-5472.CAN-07-6178

Nakahata K, Simons BW, Pozzo E, Shuck R, Kurenbekova L, Prudowsky Z, Dholakia K, Coarfa C, Patel TD, Donehower LA, et al. 2022. K-Ras and p53 mouse model with molecular characteristics of human rhabdomyosarcoma and translational applications. *Dis Model Mech* **15**: dmm 049004. doi:10.1242/dmm.049004

Nanni P, Nicoletti G, De Giovanni C, Croci S, Astolfi A, Landuzzi L, Di Carlo E, Iezzi M, Musiani P, Lollini PL. 2003. Development of rhabdomyosarcoma in HER-2/neu transgenic p53 mutant mice. *Cancer Res* **63**: 2728–2732.

Neph S, Stergachis AB, Reynolds A, Sandstrom R, Borenstein E, Stamatoyannopoulos JA. 2012. Circuitry and dynamics of human transcription factor regulatory networks. *Cell* **150**: 1274–1286. doi:10.1016/j.cell.2012.04.040

Noden DM, Francis-West P. 2006. The differentiation and morphogenesis of craniofacial muscles. *Dev Dyn* **235**: 1194–1218. doi:10.1002/dvdy.20697

Noden DM, Trainor PA. 2005. Relations and interactions between cranial mesoderm and neural crest populations. *J Anat* **207**: 575–601. doi:10.1111/j.1469-7580.2005.00473.x

Ognjanovic S, Linabery AM, Charbonneau B, Ross JA. 2009. Trends in childhood rhabdomyosarcoma incidence and survival in the United States, 1975–2005. *Cancer* **115**: 4218–4226. doi:10.1002/cncr.24465

Ordahl CP, Le Douarin NM. 1992. Two myogenic lineages within the developing somite. *Development* **114**: 339–353. doi:10.1242/dev.114.2.339

Ott MO, Bober E, Lyons G, Arnold H, Buckingham M. 1991. Early expression of the myogenic regulatory gene, *myf-5*, in precursor cells of skeletal muscle in the mouse embryo. *Development* **111**: 1097–1107. doi:10.1242/dev.111.4.1097

Pandey PR, Chatterjee B, Olanich ME, Khan J, Miettinen MM, Hewitt SM, Barr FG. 2017. PAX3-FOXO1 is essential for tumour initiation and maintenance but not recurrence in a human myoblast model of rhabdomyosarcoma. *J Pathol* **241**: 626–637. doi:10.1002/path.4867

Patel AG, Chen X, Huang X, Clay MR, Komorova N, Krasin MJ, Pappo A, Tillman H, Orr BA, McEvoy J, et al. 2022. The myogenesis program drives clonal selection and drug resistance in rhabdomyosarcoma. *Dev Cell* **57**: 1226–1240. e8. doi:10.1016/j.devcel.2022.04.003

Pelletier A, Mayran A, Gouhier A, Omichinski JG, Balsalobre A, Drouin J. 2021. Pax7 pioneer factor action requires both paired and homeo DNA binding domains. *Nucleic Acids Res* **49**: 7424–7436. doi:10.1093/nar/gkab561

Pomella S, Sreenivas P, Gryder BE, Wang L, Milewski D, Cassandri M, Baxi K, Hensch NR, Carcarino E, Song Y, et al. 2021. Interaction between SNAI2 and MYOD enhances oncogenesis and suppresses differentiation in fusion negative rhabdomyosarcoma. *Nat Commun* **12**: 192. doi:10.1038/s41467-020-20386-8

Pomella S, Danielli SG, Alaggio R, Breunis WB, Hamed E, Selfe J, Wachtel M, Walters ZS, Schäfer BW, Rota R, et al. 2023. Genomic and epigenetic changes drive aberrant skeletal muscle differentiation in rhabdomyosarcoma. *Cancers (Basel)* **15**: 2823. doi:10.3390/cancers15102823

Puisieux A, Pommier RM, Morel AP, Lavial F. 2018. Cellular pliancy and the multistep process of tumorigenesis. *Cancer Cell* **33**: 164–172. doi:10.1016/j.ccell.2018.01.007

Raghavan SS, Mooney KL, Folpe AL, Charville GW. 2019. OLIG2 is a marker of the fusion protein-driven neurodevelopmental transcriptional signature in alveolar rhabdomyosarcoma. *Hum Pathol* **91**: 77–85. doi:10.1016/j.humpath.2019.07.003

Rhodes SJ, Konieczny SF. 1989. Identification of MRF4: a new member of the muscle regulatory factor gene family. *Genes Dev* **3**: 2050–2061. doi:10.1101/gad.3.12b.2050

Rubin BP, Nishijo K, Chen HI, Yi X, Schuetze DP, Pal R, Prajapati SI, Abraham J, Arenkiel BR, Chen QR, et al. 2011. Evidence for an unanticipated relationship between undifferentiated pleomorphic sarcoma and embryonal rhabdomyosarcoma. *Cancer Cell* **19**: 177–191. doi:10.1016/j.ccr.2010.12.023

Rudzinski ER, Anderson JR, Hawkins DS, Skapek SX, Parham DM, Teot LA. 2015. The World Health Organization classification of skeletal muscle tumors in pediatric rhabdomyosarcoma: a report from the Children's Oncology Group. *Arch Pathol Lab Med* **139**: 1281–1287. doi:10.5858/arpa.2014-0475-OA

Rudzinski ER, Kelsey A, Vokuhl C, Linardic CM, Shipley J, Hettmer S, Koscielniak E, Hawkins DS, Bisogno G. 2021. Pathology of childhood rhabdomyosarcoma: a consensus opinion document from the Children's Oncology Group, European Paediatric Soft Tissue Sarcoma Study Group, and the Cooperative Weichteilsarkom Studiengruppe. *Pediatr Blood Cancer* **68**: e28798. doi:10.1002/pbc.28798

Sambasivan R, Comai G, Le Roux I, Gomès D, Konge J, Dumas G, Cimper C, Tajbakhsh S. 2013. Embryonic founders of adult muscle stem cells are primed by the determination gene Mrf4. *Dev Biol* **381**: 241–255. doi:10.1016/j.ydbio.2013.04.018

Sandberg AA, Stone JF, Czarnecki L, Cohen JD. 2001. Hematologic masquerade of rhabdomyosarcoma. *Am J Hematol* **68**: 51–57. doi:10.1002/ajh.1148

Sara GD, Yun W, Michael AD, Elizabeth S, Marco W, Beat WS, Anand GP, David ML. 2023. Single cell transcriptomic profiling identifies tumor-acquired and therapy-resistant cell states in pediatric rhabdomyosarcoma. bioRxiv doi:10.1101/2023.10.13.562224

Seale P, Sabourin LA, Girgis-Gabardo A, Mansouri A, Gruss P, Rudnicki MA. 2000. Pax7 is required for the specification of myogenic satellite cells. *Cell* **102**: 777–786. doi:10.1016/s0092-8674(00)00066-0

Searcy MB, Larsen RK, Stevens BT, Zhang Y, Jin H, Drummond CJ, Langdon CG, Gadek KE, Vuong K, Reed KB, et al. 2023. PAX3-FOXO1 dictates myogenic reprogramming and rhabdomyosarcoma identity in endothelial progenitors. *Nat Commun* **14**: 7291. doi:10.1038/s41467-023-43044-1

Seki M, Nishimura R, Yoshida K, Shimamura T, Shiraishi Y, Sato Y, Kato M, Chiba K, Tanaka H, Hoshino N, et al. 2015. Integrated genetic and epigenetic analysis defines novel molecular subgroups in rhabdomyosarcoma. *Nat Commun* **6**: 7557. doi:10.1038/ncomms8557

Shah AM, Guo L, Morales MG, Jaichander P, Chen K, Huang H, Cano Hernandez K, Xu L, Bassel-Duby R, Olson EN, et al. 2023. TWIST2-mediated chromatin remodeling promotes fusion-negative rhabdomyosarcoma. *Sci Adv* **9**: eade8184. doi:10.1126/sciadv.ade8184

Shapiro DN, Sublett JE, Li B, Downing JR, Naeve CW. 1993. Fusion of PAX3 to a member of the forkhead family of transcription factors in human alveolar rhabdomyosarcoma. *Cancer Res* **53**: 5108–5112.

Shern JF, Chen L, Chmielecki J, Wei JS, Patidar R, Rosenberg M, Ambrogio L, Auclair D, Wang J, Song YK, et al. 2014. Comprehensive genomic analysis of rhabdomyosarcoma reveals a landscape of alterations affecting a common genetic axis in fusion-positive and fusion-negative tumors. *Cancer Discov* **4:** 216–231. doi:10.1158/2159-8290.CD-13-0639

Skapek SX, Ferrari A, Gupta AA, Lupo PJ, Butler E, Shipley J, Barr FG, Hawkins DS. 2019. Rhabdomyosarcoma. *Nat Rev Dis Primers* **5:** 1. doi:10.1038/s41572-018-0051-2

Sokolowski E, Turina CB, Kikuchi K, Langenau DM, Keller C. 2014. Proof-of-concept rare cancers in drug development: the case for rhabdomyosarcoma. *Oncogene* **33:** 1877–1889. doi:10.1038/onc.2013.129

Sreenivas P, Wang L, Wang M, Challa A, Modi P, Hensch NR, Gryder B, Chou HC, Zhao XR, Sunkel B, et al. 2023. A SNAI2/CTCF interaction is required for *NOTCH1* expression in rhabdomyosarcoma. *Mol Cell Biol* **43:** 547–565. doi:10.1080/10985549.2023.2256640

Stewart E, Federico SM, Chen X, Shelat AA, Bradley C, Gordon B, Karlstrom A, Twarog NR, Clay MR, Bahrami A, et al. 2017. Orthotopic patient-derived xenografts of paediatric solid tumours. *Nature* **549:** 96–100. doi:10.1038/nature23647

Stewart E, McEvoy J, Wang H, Chen X, Honnell V, Ocarz M, Gordon B, Dapper J, Blankenship K, Yang Y, et al. 2018. Identification of therapeutic targets in rhabdomyosarcoma through integrated genomic, epigenomic, and proteomic analyses. *Cancer Cell* **34:** 411–426.e19. doi:10.1016/j.ccell.2018.07.012

Sultan I, Qaddoumi I, Yaser S, Rodriguez-Galindo C, Ferrari A. 2009. Comparing adult and pediatric rhabdomyosarcoma in the surveillance, epidemiology and end results program, 1973 to 2005: an analysis of 2,600 patients. *J Clin Oncol* **27:** 3391–3397. doi:10.1200/JCO.2008.19.7483

Sunkel BD, Wang M, LaHaye S, Kelley BJ, Fitch JR, Barr FG, White P, Stanton BZ. 2021. Evidence of pioneer factor activity of an oncogenic fusion transcription factor. *iScience* **24:** 102867. doi:10.1016/j.isci.2021.102867

Swedlund B, Lescroart F. 2020. Cardiopharyngeal progenitor specification: multiple roads to the heart and head muscles. *Cold Spring Harb Perspect Biol* **12:** a036731. doi:10.1101/cshperspect.a036731

Tajbakhsh S, Rocancourt D, Cossu G, Buckingham M. 1997. Redefining the genetic hierarchies controlling skeletal myogenesis: Pax-3 and Myf-5 act upstream of MyoD. *Cell* **89:** 127–138. doi:10.1016/s0092-8674(00)80189-0

Taulli R, Foglizzo V, Morena D, Coda DM, Ala U, Bersani F, Maestro N, Ponzetto C. 2014. Failure to downregulate the BAF53a subunit of the SWI/SNF chromatin remodeling complex contributes to the differentiation block in rhabdomyosarcoma. *Oncogene* **33:** 2354–2362. doi:10.1038/onc.2013.188

Thompson B, Davidson EA, Liu W, Nebert DW, Bruford EA, Zhao H, Dermitzakis ET, Thompson DC, Vasiliou V. 2021. Overview of PAX gene family: analysis of human tissue-specific variant expression and involvement in human disease. *Hum Genet* **140:** 381–400. doi:10.1007/s00439-020-02212-9

Torp SH, Solheim O, Skjulsvik AJ. 2022. The WHO 2021 classification of central nervous system tumours: a practical update on what neurosurgeons need to know—a mini-review. *Acta Neurochir* **164:** 2453–2464. doi:10.1007/s00701-022-05301-y

Tsherniak A, Vazquez F, Montgomery PG, Weir BA, Kryukov G, Cowley GS, Gill S, Harrington WF, Pantel S, Krill-Burger JM, et al. 2017. Defining a cancer dependency map. *Cell* **170:** 564–576.e16. doi:10.1016/j.cell.2017.06.010

Vicente-García C, Villarejo-Balcells B, Irastorza-Azcárate I, Naranjo S, Acemel RD, Tena JJ, Rigby PWJ, Devos DP, Gómez-Skarmeta JL, Carvajal JJ. 2017. Regulatory landscape fusion in rhabdomyosarcoma through interactions between the PAX3 promoter and FOXO1 regulatory elements. *Genome Biol* **18:** 106. doi:10.1186/s13059-017-1225-z

Vogelstein B, Papadopoulos N, Velculescu VE, Zhou S, Diaz LA Jr, Kinzler KW. 2013. Cancer genome landscapes. *Science* **339:** 1546–1558. doi:10.1126/science.1235122

Vyas B, Nandkishore N, Sambasivan R. 2020. Vertebrate cranial mesoderm: developmental trajectory and evolutionary origin. *Cell Mol Life Sci* **77:** 1933–1945. doi:10.1007/s00018-019-03373-1

Waddington CH. 1957. *The strategy of the genes.* Allen and Unwin, London.

Wang M, Sreenivas P, Sunkel BD, Wang L, Ignatius M, Stanton BZ. 2023. The 3D chromatin landscape of rhabdomyosarcoma. *NAR Cancer* **5:** zcad028. doi:10.1093/narcan/zcad028

Wei Y, Qin Q, Yan C, Hayes MN, Garcia SP, Xi H, Do D, Jin AH, Eng TC, McCarthy KM, et al. 2022. Single-cell analysis and functional characterization uncover the stem cell hierarchies and developmental origins of rhabdomyosarcoma. *Nat Cancer* **3:** 961–975. doi:10.1038/s43018-022-00414-w

Whyte WA, Orlando DA, Hnisz D, Abraham BJ, Lin CY, Kagey MH, Rahl PB, Lee TI, Young RA. 2013. Master transcription factors and mediator establish super-enhancers at key cell identity genes. *Cell* **153:** 307–319. doi:10.1016/j.cell.2013.03.035

Williams BA, Ordahl CP. 1994. *Pax-3* expression in segmental mesoderm marks early stages in myogenic cell specification. *Development* **120:** 785–796. doi:10.1242/dev.120.4.785

Williamson D, Missiaglia E, de Reyniès A, Pierron G, Thuille B, Palenzuela G, Thway K, Orbach D, Laé M, Fréneaux P, et al. 2010. Fusion gene-negative alveolar rhabdomyosarcoma is clinically and molecularly indistinguishable from embryonal rhabdomyosarcoma. *J Clin Oncol* **28:** 2151–2158. doi:10.1200/JCO.2009.26.3814

Xie J, Murone M, Luoh SM, Ryan A, Gu Q, Zhang C, Bonifas JM, Lam CW, Hynes M, Goddard A, et al. 1998. Activating Smoothened mutations in sporadic basal-cell carcinoma. *Nature* **391:** 90–92. doi:10.1038/34201

Xu L, Zheng Y, Liu J, Rakheja D, Singleterry S, Laetsch TW, Shern JF, Khan J, Triche TJ, Hawkins DS, et al. 2018. Integrative Bayesian analysis identifies rhabdomyosarcoma disease genes. *Cell Rep* **24:** 238–251. doi:10.1016/j.celrep.2018.06.006

Yan C, Brunson DC, Tang Q, Do D, Iftimia NA, Moore JC, Hayes MN, Welker AM, Garcia EG, Dubash TD, et al. 2019. Visualizing engrafted human cancer and therapy responses in immunodeficient zebrafish. *Cell* **177:** 1903–1914.e14. doi:10.1016/j.cell.2019.04.004

Yohe ME, Gryder BE, Shern JF, Song YK, Chou HC, Sindiri S, Mendoza A, Patidar R, Zhang X, Guha R, et al.

 Cite this article as *Cold Spring Harb Perspect Med* doi: 10.1101/cshperspect.a041583

2018. MEK inhibition induces MYOG and remodels super-enhancers in RAS-driven rhabdomyosarcoma. *Sci Transl Med* 10: eaan4470. doi:10.1126/scitranslmed.aan 4470

Yohe ME, Heske CM, Stewart E, Adamson PC, Ahmed N, Antonescu CR, Chen E, Collins N, Ehrlich A, Galindo RL, et al. 2019. Insights into pediatric rhabdomyosarcoma research: challenges and goals. *Pediatr Blood Cancer* 66: e27869. doi:10.1002/pbc.27869

Zhang S, Tian D, Tran NH, Choi KP, Zhang L. 2014. Profiling the transcription factor regulatory networks of human cell types. *Nucleic Acids Res* 42: 12380–12387. doi:10.1093/nar/gku923

Zhang S, Wang J, Lui Q, McDonald HW, Bomer ML, Layden HM, Ellis J, Boristein SC, Hiebert SW, Stengel KR. 2022. PAX3-FOXO1 coordinates enhancer architecture, eRNA transcription, and RNA polymerase pause release at select gene targets. *Mol Cell* 82: 4428–4442.e7. doi:10.1016/j.molcel.2022.10.025

Targeting Hyperactive Ras Signaling in Pediatric Cancer

Anya Levinson, Kevin Shannon, and Benjamin J. Huang

Department of Pediatrics; Helen Diller Family Comprehensive Cancer Center, University of California San Francisco, San Francisco, California 94158, USA

Correspondence: ben.huang@ucsf.edu

Somatic *RAS* mutations are among the most frequent drivers in pediatric and adult cancers. Somatic *KRAS*, *NRAS*, and *HRAS* mutations exhibit distinct tissue-specific predilections. Germline *NF1* and *RAS* mutations in children with neurofibromatosis type 1 and other RASopathy developmental disorders have provided new insights into Ras biology. In many cases, these germline mutations are associated with increased cancer risk. Promising targeted therapeutic strategies for pediatric cancers and neoplasms with *NF1* or *RAS* mutations include inhibition of downstream Ras effector pathways, directly inhibiting the signal output of oncogenic Ras proteins and associated pathway members, and therapeutically targeting Ras posttranslational modifications and intracellular trafficking. Acquired drug resistance to targeted drugs remains a significant challenge but, increasingly, rational drug combination approaches have shown promise in overcoming resistance. Developing predictive preclinical models of childhood cancers for drug testing is a high priority for the field of pediatric oncology.

RAS MUTATIONS IN HUMAN CANCERS

The *HRAS*, *NRAS*, and *KRAS* genes encode four highly homologous proteins (H-Ras, N-Ras, K-Ras4a, and K-Ras4b) (Malumbres and Barbacid 2003; Schubbert et al. 2007). *RAS* genes are preferentially mutated in distinct tumor types. *KRAS* mutations are highly prevalent in epithelial cancers such as pancreatic, lung, and colorectal cancer. In contrast, *HRAS* mutations are found in the bladder, squamous carcinomas, and some sarcomas, while *NRAS* mutations predominate in melanoma and hematologic malignancies, particularly juvenile myelomonocytic leukemia (JMML) and acute myeloid leukemia

(AML) (Table 1; Malumbres and Barbacid 2003; Schubbert et al. 2007; Ward et al. 2012). The skewed distribution of somatic *RAS* mutations in different tissue contexts is associated with a higher proportion of *NRAS* mutations in pediatric versus adult cancers.

Structural and Functional Properties of the Ras GTPase Switch

Ras proteins regulate cell fate decisions by cycling between active guanosine triphosphate (GTP)-bound and inactive guanosine diphosphate (GDP)-bound states (Ras-GTP and Ras-

Table 1. Ras/MAP kinase pathway mutations in pediatric cancers

Cancer	NRAS (%)	KRAS (%)	HRAS (%)	NF1 (%)	NF1/RAS (%)	RAS/MAPK (%)	Additional Ras/MAPK pathway gene alterations	References
B-ALL	15	12	0	5	32	35	PTPN11	Holmfeldt et al. 2013; Brady et al. 2022
Low hypodiploid B-ALL	0	0	0	9	9	9	—	Brady et al. 2022
Near haploid B-ALL	17	3	0	44	64	65	PTPN11	
T-ALL	7	3	0	1	11	11	—	
AML	31	12	0	2	45	57	PTPN11, CBL	Bolouri et al. 2018
JMML	24	10	0	27	61	88	CBL, PTPN11	Stieglitz et al. 2015
Langerhans cell histiocytosis	0	0	0	0	0	40–70	BRAF, MAP2K1	Badalian-Very et al. 2010; Brown et al. 2014
RMS	6	4	3	3	16	16	—	Shern et al. 2014
FOXO1 fusion-positive RMS	0	0	0	0	0	0	—	
Fusion-negative RMS	10	6	4	4	24	25	BRAF	
Malignant ectomesenchymoma	0	0	86	0	86	86	—	Huang et al. 2016
MPNST	0	0	0	87	87	87	—	Brohl et al. 2017
Low-grade glioma	0	1	0	16[a]	17	68	BRAF	Lau et al. 2000; Jacob et al. 2009; Ryall et al. 2020
High-grade glioma	1	1	0	1	3	3	—	Gröbner et al. 2018

(B-ALL) B-cell acute lymphoblastic leukemia, (T-ALL) T-cell acute lymphoblastic leukemia, (AML) acute myeloid leukemia, (JMML) juvenile myelomonocytic leukemia, (RMS) rhabdomyosarcoma, (MPNST) malignant peripheral nerve sheath tumor, (NF1) neurofibromatosis type 1.

[a]Predominantly NF1-associated optic gliomas.

Cite this article as Cold Spring Harb Perspect Med doi: 10.1101/cshperspect.a041572

GDP) (Vetter and Wittinghofer 2001; Bos et al. 2007; Schubbert et al. 2007; Stephen et al. 2014). SOS1 and other guanine nucleotide exchange factors (GEFs) respond to growth factor receptor activation and other stimuli by catalyzing nucleotide exchange on Ras, which then rebinds to free GTP or GDP in the cell. Nucleotide exchange preferentially increases Ras-GTP levels due to the much higher intracellular concentrations of GTP than GDP. Upon GTP binding, Ras interacts productively with downstream effectors, which include Raf, phosphatidylinositol-3 kinase (PI3K), and Ral-GDS, to activate kinase signaling cascades that regulate cell proliferation, differentiation, and survival (Fig. 1). Compelling data from analyses of human tumors, cancer cell lines, and mouse models strongly implicate deregulated signaling through the PI3K/Akt and Raf/MEK/ERK (mitogen-activated protein kinase [MAPK]) cascades in tumor initiation and maintenance (Prior et al. 2012). Recent cocrystallographic and functional studies support a model whereby single Ras-GTP molecules re-cruit Raf to the plasma membrane (PM), followed by lateral assembly to form an active signaling complex comprised of two Raf and two Ras proteins (Tran et al. 2021; Chen et al. 2023; Simanshu et al. 2023). Ras signal output is terminated when Ras-GTP is hydrolyzed to Ras-GDP. The slow intrinsic Ras GTPase activity is augmented thousands-of-fold by GTPase activating proteins (GAPs). Accordingly, the competing activities of GEFs and GAPs determine cellular Ras-GTP levels. As discussed below, the *NF1* tumor suppressor gene encodes neurofibromin, a GAP for Ras (Bos et al. 2007). Additional recurrently mutated genes in pediatric cancers encode proteins that function upstream of Ras by activating GEF activity to increase guanine nucleotide exchange on normal Ras proteins (e.g., mutant Flt3 and SHP2). Other mutations in pediatric cancers, such as *PTEN* deletions in T lineage acute lymphoblastic leukemia and *BRAF* alterations in glioma, drive cancer growth by constitutively activating effector pathways downstream from Ras-GTP.

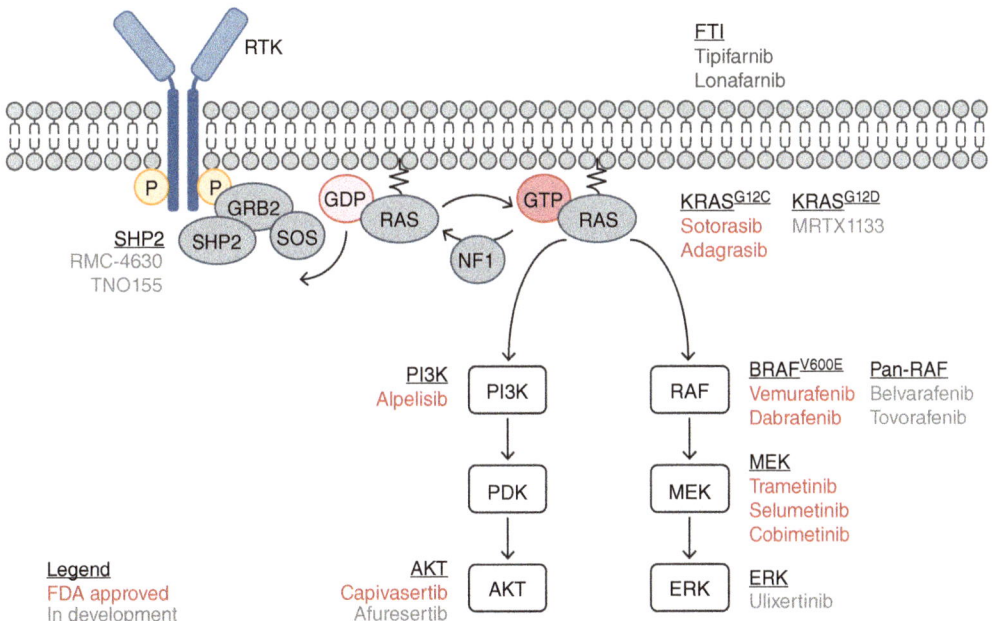

Figure 1. Ras/NF1, effector pathways, and clinically relevant inhibitors. Ras and effector pathway inhibitors that are Food and Drug Administration (FDA)-approved (red) or in development (gray). (FTI) Farnesyltransferase inhibitor, (RTK) receptor tyrosine kinase, (GTP) guanosine triphosphate, (GDP) guanosine diphosphate.

Ras Isoforms, Posttranslational Modifications, and Intracellular Trafficking

The first 85 amino acids are identical in all four Ras isoforms. This "G" domain includes the P loop that interacts with the γ phosphate of GTP and the switch I and II domains, which are stabilized by GTP binding and interact with Raf, PI3K, and other Ras effector molecules (Schubbert et al. 2007). Ras proteins share 85% identity over the next 80 amino acids, and only diverge substantially in the carboxy-terminal "hypervariable region" (HVR) (Fig. 2A). The HVR contains amino acid residues that specify posttrans-

lational modifications required for proper subcellular localization (Omerovic et al. 2007). The HVR of all Ras isoforms terminates with a CAAX motif, where the cysteine is prenylated by farnesyltransferase (FTase). This lipid modification provides weak membrane binding affinity that is stabilized by a second signal motif. For K-Ras4b, this stabilization is conferred by a polybasic lysine domain within the HVR. In contrast, H-, N-, and K-Ras4a are palmitoylated at cysteine(s) adjacent to the CAAX motif (Fig. 2B). These modifications are essential for correct trafficking and subcellular localization (Hancock et al. 1990; Omerovic et al. 2007). In particular,

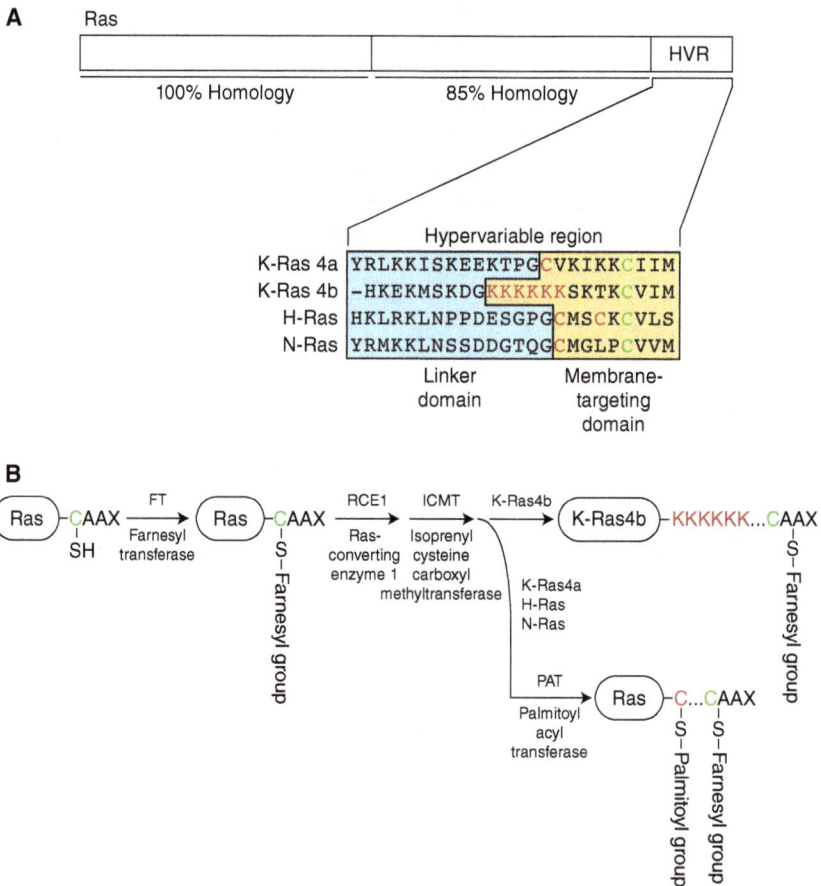

Figure 2. Hypervariable region (HVR) of Ras proteins and posttranslational modifications. (A) K-Ras4a, K-Ras4b, N-Ras, and H-Ras share homology across the G domain. The carboxy-terminal HVR contains amino acids that specify posttranslational modifications required for proper subcellular localization. (B) A farnesyl group is covalently attached to the CAAX cysteine (green). K-Ras4a, H-Ras, and N-Ras are palmitoylated at cysteine(s) adjacent to the CAAX motif (red). K-Ras4b is unique among Ras isoforms in that a polybasic lysine domain (red) stabilizes membrane binding without undergoing palmitoylation.

whereas H-, N-, and K-Ras4a are directed from the endoplasmic reticulum (ER) to the Golgi en route to the PM through the exocytotic secretory pathway, K-Ras4b is diverted before it reaches the Golgi and localizes to the PM through mechanism(s) that remain uncertain. H- and N-Ras also reside and signal in the Golgi and ER (Mor and Philips 2006). A dynamic cycle of palmitoylation/depalmitoylation mediated by palmitoyl acyl transferase (PAT) and serine hydrolase (SH) enzymes regulates intracellular H- and N-Ras trafficking across the Golgi, ER, and PM. Elegant studies have demonstrated that most N-Ras protein resides in the cytoplasm in association with the chaperone protein VPS35. However, palmitoylated N-Ras mainly localizes at the PM (Zhou et al. 2016; Zhou and Philips 2019).

Oncogenic Ras Proteins and Therapeutic Implications

The recurring somatic *RAS* mutations in adult and pediatric cancers encode missense substitutions at amino acids 12, 13, and 61 that impair intrinsic GTP hydrolysis and cause resistance to the biochemical activity of GAP proteins (Boguski and McCormick 1993; Donovan et al. 2002; Bos et al. 2007; Schubbert et al. 2007). These mutations result in constitutively elevated Ras-GTP levels, which, in turn, trigger potent negative feedback mechanisms that buffer the activation of downstream kinase effectors. This is particularly true of the Ras/MAPK signaling module, which has evolved from yeast to regulate cell decisions in all higher organisms. The existence of complex regulatory mechanisms for modulating Ras signal output has important implications for interpreting experimental data. Specifically, the biochemical and phenotypic consequences of expressing Ras oncoproteins at endogenous levels may differ substantially from what is observed after overexpression in many cell types. Furthermore, although the underlying mechanisms are incompletely understood, there is selective pressure for cancer cell lines and primary tumors to amplify mutant *RAS* genes, delete the corresponding normal allele, or both (Burgess et al. 2017; Bielski et al. 2018).

For decades, oncogenic Ras proteins were widely viewed as "undruggable" due to the combination of high intracellular GTP concentrations, nanomolar affinity of Ras for guanine nucleotides, and the absence of readily accessible "pockets" for drug binding. However, the recent development of K-Ras^{G12C} inhibitors with antitumor activity demonstrated the feasibility of directly targeting Ras oncoproteins and showed that *KRAS*-mutant cancers remain addicted to hyperactive Ras in vivo (Simanshu et al. 2017; Canon et al. 2019; Hallin et al. 2020). Unfortunately, single-agent treatment is invariably followed by disease progression and reactivation of oncogenic Ras signaling through on- and off-target adaptive resistance mechanisms (Awad et al. 2021; Zhao et al. 2021). These observations are consistent with previous studies indicating the need to cotarget mutant Ras and potential possible bypass mechanisms simultaneously (Wang et al. 2017; Lou et al. 2019). While the clinical responses observed in patients treated with K-Ras^{G12C} inhibitors are of exceptional interest, *KRAS*^{G12C} mutations are uncommon in pediatric cancer and it will likely be more difficult to therapeutically target amino acid substitutions that lack a "covalent handle" (i.e., an oncogenic, nucleophilic cysteine residue). We discuss K-Ras^{G12C} inhibitors and ongoing efforts to develop chemical inhibitors of K-Ras^{G12D}, N-Ras^{G12D}, and other common Ras oncoproteins below.

NEUROFIBROMATOSIS TYPE 1 AND OTHER RASOPATHY DISORDERS

The RASopathies are a group of developmental disorders characterized by craniofacial abnormalities and other structural birth defects, short stature, developmental delay, learning disabilities, neurocutaneous abnormalities, cardiovascular disease, and a variably elevated risk of tumorigenesis (Schubbert et al. 2007; Kratz et al. 2015; Castel et al. 2020). The RASopathies include neurofibromatosis type 1 (NF1) and Noonan syndrome (NS) as well as Costello, Cardiofaciocutaneous, Legius, LEOPARD, and Cbl syndromes.

Seminal discoveries showing that NF1 is caused by germline mutations in the *NF1* gene and that the encoded protein (neurofibromin) encodes a GAP for Ras provided the first mechanistic link between a human developmental disorder and hyperactive Ras signaling (Buchberg et al. 1990; Cawthon et al. 1990; Martin et al. 1990; Viskochil et al. 1990; Wallace et al. 1990; Xu et al. 1990a,b). Accordingly, *NF1* functions as a classic tumor suppressor gene. Benign neoplasms and cancers that arise in persons with NF1 are characterized by somatic inactivation of the normal allele due to uniparental disomy with duplication of the mutant copy, somatic mutations in the normal allele, or other genetic mechanisms (Xu et al. 1992; Shannon et al. 1994; Stephens et al. 2006). Over a decade after the discovery of the *NF1* gene, the identification of germline *PTPN11* mutations in ~50% of NS patients by the Gelb laboratory "connected the dots" between hyperactive Ras/MAPK signaling and structural birth defects in NS (Tartaglia et al. 2001). *PTPN11* encodes SHP-2, a protein tyrosine phosphatase (PTPase) that is recruited to activated growth factor receptors where it functions as a "signal relay" protein that increases Ras-GTP levels (Fig. 1; Neel et al. 2003; Tartaglia and Gelb 2005). Research by many groups subsequently identified causative germline mutations in *HRAS*, *KRAS*, *NRAS*, *SOS1*, *RAF1*, *BRAF*, *MEK1*, *RIT1*, and other genes in RASopathy patients and established key genotype/phenotype relationships (Schubbert et al. 2007; Kratz et al. 2015; Tajan et al. 2018; Castel et al. 2020). Deregulated Ras/MAPK signaling is a unifying biochemical feature of these disorders that inspired the term "RASopathies."

The germline *PTPN11* and *KRAS* mutations in RASopathy patients are largely nonoverlapping with the somatic gain-of-function alterations found in cancer (Keilhack et al. 2005; Tartaglia and Gelb 2005; Schubbert et al. 2006). Accordingly, comparative functional and biochemical studies showed that these germline mutations encode proteins with weaker gain-of-function properties than the corresponding SHP-2 or K-Ras oncoproteins (Keilhack et al. 2005; Schubbert et al. 2006, 2007). These data and studies of $Kras^{G12D}$, $Kras^{P34R}$, and $Kras^{T58I}$ mice that harbor conditional mutations found in cancer (G12D) or in RASopathy patients (P34R, T58I) support the idea that mutations that markedly increase K-Ras signal output are incompatible with normal embryonic development (Wong et al. 2020). Interestingly, whereas conditional activation of $Kras^{G12D}$ causes lung tumors and a fatal myeloproliferative neoplasm (MPN), $Kras^{T58I}$ mice exhibit NS-like developmental phenotypes but do not develop benign or malignant tumors (Jackson et al. 2001; Johnson et al. 2001; Braun et al. 2004; Wong et al. 2020). Accordingly, with the notable exception of NF1, for which a somatic "second hit" is essential for tumorigenesis, the incidence of cancer is only modestly elevated in children with RASopathy disorders (Kratz et al. 2015). This clinical observation suggests that most germline RASopathy mutations do not sufficiently increase Ras/MAPK signal output to levels required to initiate or sustain malignant transformation.

SOMATIC AND GERMLINE RAS PATHWAY ALTERATIONS IN PEDIATRIC CANCER

Hematologic Malignancies

Acute Myeloid Leukemia (AML)

AML is an aggressive and molecularly heterogeneous hematopoietic cancer that arises from the clonal expansion of myeloid precursor cells. Molecular analysis of AML samples obtained at diagnosis revealed age-related differences in the respective genomic landscapes of pediatric, adolescent, and young adult (AYA) versus adult leukemias. In older adults, AML is predominantly initiated by somatic mutations in *DNMT3A*, *TET2*, *IDH1/2*, and other epigenetic regulators with secondary mutations in genes encoding signaling proteins cooperating in leukemic transformation (Corces-Zimmerman et al. 2014; Jaiswal et al. 2014). In pediatric and AYA patients, AML is frequently initiated by transcription factor fusions and there is a higher prevalence of mutations in *NRAS*, *KRAS*, and *NF1* (45% of cases compared to <20% in older adults) (Table 1; Bolouri et al. 2018). *RAS* mutations track closely with disease activity in AML, becoming undetectable during remission and reappearing (or being replaced with a different signaling mutation) at relapse (Lindsley and Ebert 2013; Papaemmanuil et al. 2016; McMahon et al. 2019). *RAS* mutations are also a major cause of de novo or adaptive resistance to targeted therapies in AML, including FLT3 (gilteritinib) (McMahon et al. 2019), IDH2 (enasidenib) (Stein et al. 2017), and the Bcl-2 inhibitor venetoclax (DiNardo et al. 2020; Samra et al. 2020; Zhang et al. 2022).

Juvenile Myelomonocytic Leukemia (JMML)

JMML is an aggressive MPN that is almost exclusively diagnosed in early childhood. Hematopoietic stem cell transplantation is the only curative therapy, with few exceptions (Kratz et al. 2005; Matsuda et al. 2007, 2010; Strullu et al. 2014). Ras/MAPK pathway mutations are a hallmark of JMML, as evidenced by their inclusion in the diagnostic criteria for JMML (Chan et al. 2009) and their high prevalence (86% of cases) (Table 1; Stieglitz et al. 2015). These mutations are invariably present at high variant allele frequencies (VAFs) at both diagnosis and relapse. In vivo studies support these human data, with $Nf1$ inactivation and endogenous oncogenic $Kras^{G12D}$, $Nras^{G12D}$, and $Ptpn11^{E76K}$ expression in mice initiating a JMML-like MPN (Braun et al. 2004; Le et al. 2004; Chan et al. 2005; Mohi et al. 2005). Cooperating mutations are found in transcription factor and epigenetic genes in one-third of JMML patients, with a greater number of somatic alterations at diagnosis conferring a worse prognosis (Stieglitz et al. 2015). More recently, global hypermethylation was associated with adverse outcomes (Stieglitz et al. 2017; Schönung et al. 2021). Infants with NS are predisposed to developing a JMML-like MPN, particularly in the context of germline $PTPN11$ and $KRAS$ mutations. Importantly, because the clinical course is frequently characterized by spontaneous resolution, supportive care with close follow-up is the current standard of care (Kratz et al. 2005; Strullu et al. 2014). These clinical observations reinforce the idea that a threshold level of constitutive Ras/MAPK pathway activation is likely required for full malignant transformation that is higher than the signal output of most germline RAS-opathy mutations.

Acute Lymphoblastic Leukemia

B-cell and T-cell acute lymphoblastic leukemias (B-ALL and T-ALL) are the most common child-

hood malignancies and are associated with a lower prevalence of Ras/MAPK pathway mutations (32% for B-ALL) than pediatric AML or JMML. Interestingly, these alterations are highly enriched in some B-ALL subtypes (Brady et al. 2022). For example, in the favorable risk B-ALL category, 61% of hyperdiploid B-ALLs harbor $NRAS$, $KRAS$, or $PTPN11$ mutations (Brady et al. 2022). In the high-risk B-ALL category, alterations in $NF1$ or $NRAS/KRAS$ mutations exist in 64% of near haploid (24–31 chromosomes) B-ALL cases (Table 1). In contrast, low hypodiploid (32–39 chromosomes) B-ALL is characterized by frequent $TP53$ mutations and few Ras/MAPK pathway alterations (Holmfeldt et al. 2013). Overall, pediatric T-ALLs harbor fewer Ras/MAPK mutations than B-ALL and have a higher prevalence of PI3K pathway alterations, particularly $PTEN$ inactivation (Gutierrez et al. 2009; Brady et al. 2022). Early T precursor (ETP) ALL, which is characterized by frequent $KRAS$ and $NRAS$ mutations, is an exception to this general rule (Zhang et al. 2012). Accordingly, retroviral mutagenesis in $Kras^{G12D}$ mice generated T-ALLs that have a similar pattern of molecular alterations as ETP-ALL (Dail et al. 2010; Huang et al. 2019).

Histiocytic Disorders

Langerhans Cell Histiocytosis

Langerhans cell histiocytosis (LCH) is a neoplastic histiocytic disorder that most often arises from bone, skin, bone marrow, liver, spleen, and/or the central nervous system. It can present as a single or multisystem disease and involvement of "risk" organs (bone marrow, liver, or spleen) is associated with a worse prognosis. LCH is characterized by frequent $BRAF^{V600E}$ mutations (Badalian-Very et al. 2010) with other patients harboring mutations in $MAP2K1$, which encodes MEK1, or $ARAF$ (Brown et al. 2014).

Extracranial Solid Tumors

Rhabdomyosarcoma (RMS)

RMS is the most common soft tissue sarcoma of childhood. Broadly speaking, RMS tumors are stratified based on the presence of a *FOXO1-PAX3* or *FOXO1-PAX7* translocation that results in the expression of FOXO1 fusion oncoproteins. Alveolar morphology and older age are often associated with fusion-positive RMS. Conversely, whereas *FOXO1* fusions are very infrequent in RMS tumors with embryonal morphology, 24% of these tumors harbor mutations in *NRAS*, *HRAS*, or *NF1* (Table 1; Shern et al. 2014). Risk stratification for RMS is complex and incorporates fusion status, clinical stage, and postsurgical group, among other criteria. While the presence (or absence) of a *FOXO1* fusion is the most prognostic genetic biomarker, *RAS/NF1* mutations are associated with less favorable outcomes in patients with fusion-negative RMS when corrected for other factors such as stage and group (Chen et al. 2013). Cases of RMS and other sarcomas have been reported in children with RASopathy disorders and germline *HRAS* mutations (Astiazaran-Symonds et al. 2023). Some of these tumors are characterized by somatic uniparental disomy with loss of the normal *HRAS* allele and duplication of the pathogenic germline mutation (Kratz et al. 2007).

Malignant Peripheral Nerve Sheath Tumor (MPNST)

MPNST is an aggressive soft tissue sarcoma that is the major cause of premature death in persons with NF1 (lifetime incidence of 10%–15%). Overall, ~50% of MPNSTs occur in NF1 patients, usually in association with preexisting plexiform neurofibromas. Somatic *NF1* inactivation is common in sporadic MPNSTs and in tumors that develop after radiation therapy (Ducatman et al. 1986; Zadeh et al. 2007). The risk of developing MPNST is greatest in individuals in whom NF1 is caused by germline deletions that acquire additional gene mutations, such as in *SUZ12*. Accordingly, somatic mutations in *SUZ12* and other genes encoding PRC2 complex proteins are highly prevalent in both NF1-associated and sporadic MPNST (De Raedt et al. 2014; Lee et al. 2014). Elegant functional studies showed that the combined mutation of *NF1* and *SUZ12* alters the epigenetic landscape of MPNST cells, rendering them more sensitive to

BET bromodomain inhibitors (De Raedt et al. 2014).

Other Extracranial Solid Tumors

Rarely, Ras/MAPK pathway mutations are found at diagnosis for other extracranial pediatric solid tumors, including osteosarcoma, neuroblastoma, adrenocortical carcinoma, and Wilms tumor (Gröbner et al. 2018; Ma et al. 2018). Though relatively infrequent at diagnosis, *NRAS* mutations are enriched at relapse in neuroblastoma (Eleveld et al. 2015). Malignant ectomesenchymoma is a rare pediatric tumor with a high prevalence of *HRAS* mutations that is characterized by malignant mesenchymal and neuroectodermal components (Huang et al. 2016).

Brain Tumors

Low-Grade Gliomas (LGGs) and Papillary Craniopharyngioma

With the recognition that genetic alterations both have prognostic value and can inform therapeutic decision making, approaches to glioma classification have increasingly incorporated both molecular and histopathological features. The 2021 edition of the World Health Organization (WHO) classification was the first to separate diffuse gliomas based on age and molecular alterations, which differ significantly between pediatric and adult patients diagnosed with brain tumors. Within pediatric cohorts, pilocytic astrocytoma remains the most common LGG based on morphology. Notably, *KIAA1549-BRAF* translocations occur in 83% of pilocytic astrocytoma tumors with $BRAF^{V600E}$ mutations occurring in most other patients (Ryall et al. 2020). Children with NF1 are at elevated risk of LGG, which most often manifests as optic pathway gliomas early in life. While clinically problematic because they frequently cause vision loss, optic pathway gliomas rarely progress to higher-grade tumors in the absence of radiation therapy, which is contraindicated for this reason. $BRAF^{V600E}$ mutations are also highly prevalent in papillary craniopharyngioma, a rare tumor that arises near the stalk of the pituitary gland in children and adults (Müller et al. 2019).

 Cite this article as *Cold Spring Harb Perspect Med* doi: 10.1101/cshperspect.a041572

High-Grade Gliomas (HGGs)

Ras/MAPK pathway mutations are rare in pediatric HGGs, which are predominantly driven by H3-K27 and H3-G34 alterations. This biology differs from the mutational landscape of adult grade 4 astrocytomas (i.e., glioblastoma multiforme), wherein one-third of cases harbor somatic *NF1* mutations that are associated with a distinct mesenchymal histology (Brennan et al. 2013).

MODELS OF PEDIATRIC CANCER FOR PRECLINICAL TESTING

Developing more effective and less toxic treatments for cancers with mutations in *RAS* and other key driver genes is a fundamental challenge in pediatric oncology that has historically been hampered by substantial barriers. As pediatric cancers are orphan diseases, there are few economic incentives for developing new therapies. Recognizing this problem, Congress passed the Childhood Cancer Survivorship, Treatment, Access, Research (STAR) and Research Acceleration for Cure and Equity (RACE) for Children Acts, which require the pharmaceutical industry to evaluate promising drugs developed for adult cancers in pediatric patients. As summarized above, many pediatric cancers harbor germline or somatic mutations in genes encoding protein components of Ras signaling networks such as *KRAS, NRAS, HRAS, NF1, BRAF,* and *FLT3* that are frequently altered in common adult cancers. This fact provides an opportunity for "repurposing" drugs developed to treat adult tumors for testing in selected pediatric cancers. Recent and ongoing trials of Ras pathway inhibitors in specific pediatric cancers are discussed below.

Due to the relatively low incidence of pediatric malignancies and the availability of effective front-line treatment protocols for many childhood cancers, there are relatively few children and adolescents with relapsed/refractory disease to enroll in early phase clinical trials. In addition, some key driver mutations found in pediatric cancers are not directly "druggable" due to the biophysical properties of the encoded mutant proteins or because mutations in genes such as *PTEN, RB1,* and *NF1* result in loss of protein expression. Given these challenges, the development of predictive preclinical models is of central importance for prioritizing the most promising agents and drug combinations for clinical evaluation in pediatric patients with high-risk, relapsed, and refractory cancers.

Cancer cell lines remain a cornerstone of drug discovery efforts for adult and pediatric cancers. Cell line xenografts implanted into immunodeficient mice are widely used to confirm that anticancer drugs that inhibit growth in vitro have similar activity in vivo. Advantages of these models include relatively low cost, extensive preclinical data for previous drug testing for many cell lines, and the ability to investigate drug synergy using Bliss Independence analysis and other approaches (Remsberg et al. 2021; Popescu et al. 2023). Deep genomic characterization, chemical and functional dependency screens, and the advent of efficient gene editing technologies have provided a wealth of new information about many cancer cell lines that have enhanced their value for posing focused questions about drug response and resistance. Recent studies have demonstrated genotype/phenotype relationships between the presence of a specific driver mutation, such as $BRAF^{V600E}$, and the responses of individual cell lines to targeted inhibitors of the respective mutant oncoprotein that have been translated clinically (Solit et al. 2006). With this said, only a small percentage of conventional and targeted anticancer agents that show promising activity in cancer cell lines are proven effective in the clinic. The use of panels of multiple independent cell lines for drug screening has been proposed as a way to overcome this fundamental problem. Based on recent precedents, panels of cancer cell lines will likely prove most informative for evaluating chemical inhibitors that directly target mutant Ras proteins such as the K-Ras^{G12C} inhibitors sotorasib and adagrasib, which were recently approved by the Food and Drug Administration (FDA). For cancers with *NRAS* or *KRAS* mutations, cell line models will serve as valuable tools to characterize genotype/phenotype relationships between the presence/absence of a *RAS* mutation and drug response, as well as to assess selectivity for an individual isoform (e.g.,

K-Ras vs. N-Ras) or amino acid substitution (e.g., G12D, G12V, or Q61K). Investigators who use cancer cell lines to test chemical inhibitors of proteins that deregulate Ras signaling due to upstream or downstream mechanisms of action should remain mindful that selective pressure to amplify mutant *RAS* alleles or to delete the corresponding normal copy occurs frequently and can modulate drug responses (Modrek et al. 2009; Burgess et al. 2017; Bielski et al. 2018). Thus, utilization of multiple, well-characterized cell lines generated from individual pediatric cancers is important for robustly testing mechanism-based inhibitors of oncogenic Ras proteins.

Genetically engineered mouse (GEM) models are appealing preclinical research tools due to the relatively low mutational burden of most pediatric cancers and the identification of key initiating mutations such as *FOXO1-PAX3*, *EWS-FLI1*, *KIAA1549-BRAF*, and *KMT2A* translocations. Indeed, strains of "knockin" mice harboring conditional mutations of many pediatric cancer drivers have been generated and characterized. However, only a subset of these mice exhibits the expected tumor type(s) with high penetrance. While the reasons for this finding are unclear, potential explanations include species-specific developmental vulnerabilities, uncertainty regarding the correct cellular context for expressing relevant driver genes, and/or the requirement for cooperating mutations. Even in pediatric cancers for which GEM models do exist, factors such as prolonged latency and high cost can preclude widespread use. With this said, GEM models of pediatric cancer have important advantages for drug testing because they arise in situ in the context of a normal tissue microenvironment and an intact immune system (Lauchle et al. 2009; Burgess et al. 2014, 2017; Dail et al. 2014; Wandler et al. 2020). We describe some examples of how preclinical testing in GEM models of *RAS*- and *NF1*-mutant pediatric neoplasms has informed clinical translation later in this article.

Patient-derived xenograft (PDX) models reflect the underlying biology of their respective human cancer as well as the genetic diversity across different patients. However, the use of immunocompromised mice for engraftment pre-

cludes examination of any potential effects of the tumor microenvironment, and some investigators have reported reduced genetic complexity in PDX models due to the selective barrier of xenotransplantation, which may skew the results of studies of tumor evolution and relapse (Klco et al. 2014). PDX trials are challenging and expensive to perform at scale, which has led many investigators to use surrogate end points of efficacy, such as the percentage of circulating human leukemia cells in treated versus untreated mice at a specific time point. Despite these limitations, PDX models are increasingly being used to extend and validate promising evidence of antitumor activity from cell line models and to verify the genotype-selective preclinical activity of targeted agents (Townsend et al. 2016; Krivtsov et al. 2019).

TARGETING HYPERACTIVE RAS SIGNALING IN PEDIATRIC CANCER

Allosteric MEK Inhibitors

The allosteric MEK inhibitors trametinib, binimetinib, selumetinib, and cobimetinib are FDA-approved for melanomas, lung cancers, and other tumors harboring oncogenic *BRAF* mutations. These compounds are derived from a single chemotype (i.e., they were all developed by chemical modifications of the same scaffold). Single-agent MEK inhibition in advanced cancers results in the rapid development of adaptive resistance. For this reason, the current standard of care for advanced cancers with *BRAF* mutations is combination treatment with an allosteric MEK inhibitor and a "first generation" Raf kinase inhibitor such as vemurafenib or dabrafenib (Flaherty et al. 2012). A recent phase 2 trial of vemurafenib and cobimetinib that enrolled 16 patients with papillary craniopharyngioma reported a response rate of >90% with most of these patients showing a reduction in tumor volume of ~90% (Brastianos et al. 2023).

Allosteric MEK inhibitors are largely ineffective in advanced cancers with *NRAS*, *KRAS*, or *NF1* mutations. As a representative example, relapse and refractory myeloid malignancies in adults have only shown modest response rates to MEK inhibitors, such as trametinib and binimetinib (Borthakur et al. 2016; Maiti et al. 2019).

In contrast to the disappointing single-agent efficacy data in advanced cancers, selumetinib induced significant regression in NF1-associated plexiform neurofibromas and is FDA approved for this indication (Dombi et al. 2016). These responses are invariably partial, and tumors regrow when selumetinib is discontinued (due to adverse side effects). A recent trial of trametinib in relapsed or refractory JMMLs harboring driver mutations in *KRAS*, *NRAS*, *PTPN11*, or *CBL* showed dramatic regression or long-term disease control in most patients (NCT03190915). This promising efficacy signal informed a risk-stratified clinical trial of trametinib in combination with other agents in newly diagnosed JMML patients as part of an upcoming clinical trial (NCT05849662). Data from controlled preclinical trials in GEM models of plexiform neurofibroma (Jessen et al. 2013) and JMML (Lyubynska et al. 2011; Chang et al. 2013) provided a strong scientific rationale for testing MEK inhibitors in these patients. Correlative molecular analysis in mouse JMML models showed that treatment reduced the proliferation and promoted the differentiation of mutant hematopoietic cells but did not eradicate them (Lyubynska et al. 2011; Chang et al. 2013). A similar mechanism of action was subsequently observed in the recent JMML clinical trial (NCT03190915).

First-Generation Raf Kinase Inhibitors

As immediate downstream effectors of Ras-GTP in the MAPK pathway, the Raf-1 (also known as C-Raf), B-Raf, and A-Raf kinases are compelling therapeutic targets for anticancer drug discovery (Fig. 1). Raf activation by Ras is complex and involves recruitment to the PM followed by the formation of an active signaling complex comprised of Raf homo- and heterodimers (Park et al. 2019; Tran et al. 2021). *BRAF* is mutated in 5%–10% of human cancers (Marranci et al. 2017). The predilection of cancer for *BRAF* mutations is likely explained by the fact that B-Raf has a higher basal kinase activity than Raf-1 or A-Raf due to a phosphomimetic aspartic acid residue at position 448 within the N region, and constitu-

tive phosphorylation at S445 (Lito et al. 2013). Accordingly, B-Raf kinase can be constitutively activated by single amino acid substitutions such as V600E, while this requires multiple individual missense mutations in Raf-1 or A-Raf.

The first successful Raf inhibitors, vemurafenib, dabrafenib, and encorafenib, were developed about a decade ago and potently inhibit B-Raf^{V600E} as well as some other B-Raf oncoproteins that can signal as monomers (Solit et al. 2006; Chapman et al. 2011; Hauschild et al. 2012; Li et al. 2016). While these drugs produced unprecedented response rates in metastatic melanoma, clinical responses were mostly short-lived due to the rapid development of adaptive resistance. Squamous cell carcinomas harboring *RAS* gene mutations also emerged in some patients during treatment with vemurafenib due to the paradoxical activation of MAPK signaling (Modrek et al. 2009; Su et al. 2012). Another important limitation of first-generation Raf inhibitors is that they are ineffective in cancers driven by *KRAS*, *NRAS*, or *NF1* mutations. This finding is consistent with the selective inhibition of B-Raf^{V600E} and other mutant proteins that signal as monomers by this class of compounds. As noted above, combination treatment with Raf and allosteric MEK inhibitors improves antitumor activity by suppressing paradoxical MAPK pathway activation. Thus, while the development of "first generation" Raf inhibitors demonstrated the therapeutic potential of targeting the MAPK pathway, this remains a significant unmet need, particularly in advanced cancers with *KRAS*, *NRAS*, or *NF1* mutations. Although combinations such as Raf/MEK inhibitor therapy are required in most tumors, vemurafenib is highly effective as a single agent in LCH that is refractory to upfront chemotherapy (Mohapatra et al. 2023).

Pan-RAF Inhibitors

Raf dimer inhibitors (also known as "pan-Raf" inhibitors) were developed to overcome some of the liabilities of current FDA-approved Raf kinase inhibitors. A key advantage of this approach is that it reduces paradoxical activation of the MAPK pathway due to dimerization-dependent mechanisms. However, cotargeting the MAPK pathway downstream from Raf may be required to enhance efficacy and to abrogate the development of paradoxical activation due to dimerization-independent mechanisms such as downregulation of negative feedback mediated by DUSP, SPRED, and SPRY proteins.

Raf dimer inhibition is expected to suppress hyperactive MAPK pathway signaling in cancers with *RAS* or *NF1* driver mutations as well as in tumors expressing B-Raf^{V600E} and other B-Raf mutant proteins that signal as monomers. Indeed, objective responses to the Raf dimer inhibitor belvarafenib were reported in some patients with relapsed or refractory *NRAS*-mutant melanoma (Yen et al. 2021). Molecular analysis of tumor biopsies obtained at the time of disease progression identified *ARAF* mutations as a cause of adaptive resistance in this trial. Consistent with these data, belvarafenib potently inhibits B-Raf and Raf-1 homo- and heterodimer formation but is less potent against A-Raf. Belvarafenib and another Raf dimer inhibitor (tovorafenib) are currently being evaluated as single agents and with MEK inhibition in melanoma and adult solid cancers as well as in an ongoing trial for children with recurrent or progressive LGGs with *KIAA1549-BRAF* fusions or *BRAF^{V600E}* mutations (NCT04775485).

K-Ras^{G12C} and Other Switch II Pocket Inhibitors

Oncogenic Ras is a difficult target for rational drug discovery due to its defective enzymatic (GTP hydrolysis) activity, structural properties of the phosphate binding loop, and its high affinity for GTP. Ostrem et al. discovered a novel class of K-Ras^{G12C}-specific small molecule inhibitors that form a covalent bond with the nucleophilic thiol side chain of the mutant cysteine. These compounds irreversibly bind to this cysteine and extend into an adjacent switch II pocket, altering the preference to GDP (over GTP) and impairing binding to Raf (Ostrem et al. 2013). The K-Ras^{G12C} inhibitors sotorasib and adagrasib induced impressive, though transient, regression of lung cancers harboring *KRAS^{G12C}* mutations, and have been approved by the FDA for this indi-

cation (Canon et al. 2019; Hong et al. 2020; Skoulidis et al. 2021; Strickler et al. 2023). Subsequent efforts to develop noncovalent switch II pocket inhibitors for other Ras mutants have thus far resulted in the development of a potent and specific K-Ras^{G12D} inhibitor (MRTX1133) (Wang et al. 2022). Preliminary data suggest that several other G12, G13, and Q61 K-Ras mutants are also amenable to similar switch II pocket-targeting strategies (Vasta et al. 2022). As with other targeted inhibitors, adaptive resistance emerges rapidly in patients treated with sotorasib and adagrasib. Molecular analysis uncovered non-G12C *KRAS*, *NRAS*, *BRAF*, *NF1*, and *MAP2K1* mutations at relapse that reactivate oncogenic MAPK signaling in these tumors (Awad et al. 2021). Whereas *KRAS^{G12C}* mutations are uncommon in pediatric cancers, the development of potent and selective K-Ras^{G12D} and other Ras mutant inhibitors could have a major therapeutic impact. With this said, the propensity for advanced cancers to rapidly develop adaptive resistance to targeted agents will require the development of combination regimens.

SHP2 Inhibitors

SHP2 is a nonreceptor PTPase that is positioned as a common node signaling downstream from receptor tyrosine kinases and upstream of Ras (Fig. 1). SHP2 is activated by binding to phosphorylated tyrosine residues on growth factor receptors by interacting with adaptor proteins. SHP2 then functions to activate Ras through various mechanisms, which include dephosphorylation of SPRY (a negative regulator of Ras), dephosphorylation of Ras at Y32 relieving an autoinhibitory Ras–Raf binding interaction, and interacting with the GAB1–GRB2–SOS1 complex to facilitate the exchange of Ras-GDP to Ras-GTP (Yuan et al. 2020). The somatic *PTPN11* mutations found in JMML, AML, ALL, and other cancers constitutively activate SHP2 PTPase activity by relieving normal autoinhibition by the SH2 domain. Like Ras, SHP2 was previously considered "undruggable" because enzymatic active-site inhibitors generally showed off-target inhibition of other proteins. More recently, allosteric SHP2 inhibitors have shown improved selectivity and potency, effective blockade of signal transduction between receptor tyrosine kinases and Ras/MAPK signaling, and promising efficacy in preclinical cancer models (Fedele et al. 2018; Nichols et al. 2018; Popescu et al. 2023). Several SHP2 inhibitors are undergoing early phase clinical trial evaluation in adults. However, these compounds target the normal "closed" conformation of the protein and are, therefore, ineffective in cancers with oncogenic *PTPN11* mutations. The development of SHP2 inhibitors with activity against mutant proteins with elevated PTPase activity would open exciting opportunities for treating JMML and other cancers with somatic *PTPN11* mutations.

Inhibitors of Ras Posttranslational Modification

Carboxy-terminal posttranslational modifications are required to localize K-Ras, N-Ras, and H-Ras proteins to the PM, where activated growth factor receptors and other external stimuli that increase GEF activity increase guanine nucleotide exchange. The irreversible covalent attachment of a farnesyl group to the carboxy-terminal cysteine in the HVR of Ras proteins is the first posttranslational modification step for all four Ras isoforms (Fig. 2B). Based on data showing that mutating this cysteine abrogated transformation by oncogenic Ras proteins, potent and selective FTase inhibitors were developed and tested over two decades ago (Downward 2003). Unfortunately, these compounds were largely ineffective in *RAS*-mutant cancers in the clinic because K-Ras and N-Ras are geranylgeranylated when FTase is inhibited. This alternative lipid modification is sufficient for membrane localization and oncogenic signaling. Interestingly, because H-Ras is not a substrate for geranylgeranyl transferase, there are several ongoing efforts to reevaluate FTase inhibition in *HRAS*-mutant cancers (NCT04284774).

The failure of FTase inhibitors reinforced the incorrect assumption that oncogenic Ras proteins are "undruggable" and led many academic investigators and pharmaceutical companies to abandon efforts to therapeutically target Ras processing and trafficking. However, the potential for inhibiting Ras posttranslational modifi-

cations recently gained renewed momentum with the recognition that N-Ras palmitoylation is required for transformation in vitro and in vivo (Cuiffo and Ren 2010; Xu et al. 2012; Zambetti et al. 2020). Importantly, because K-Ras4b does not undergo palmitoylation, targeting the dynamic cycle of N-Ras palmitoylation and depalmitoylation might selectively inhibit the growth of *NRAS*-mutant cancers with limited toxicity to normal tissues (Fig. 2B). This idea is consistent with recent studies in which chemical inhibitors of ABHD17 family of SHs, which depalmitoylate N-Ras, selectively reduced the growth of *NRAS*-mutant AML cell lines and inhibited Ras effector activation in vitro (Remsberg et al. 2021).

CONCLUDING REMARKS

Pioneering clinical trials of pediatric ALL conducted in the 1960s and 1970s established the fundamental principle that combining chemically distinct compounds with nonoverlapping mechanisms of action was essential for curing advanced cancers. It is remarkable that these studies were executed decades before the underlying molecular basis of cancer and of clonal evolution in response to the selective pressures imposed by drug treatment were understood. Although the concept had not yet emerged in cancer biology, the exquisite sensitivity of lymphoblasts to glucocorticoids was an early and powerful example of synthetic lethality. The development of imatinib and other targeted inhibitors that block the aberrant biochemical output of mutant oncoproteins illuminated the principle of oncogene addiction and disappointingly highlighted the ability of cancer cells to develop resistance through genetic and nongenetic adaptive mechanisms. If anything, we have perhaps recently placed too much faith in the concept of molecular "magic bullets" and need to recall some of the lessons learned from treating children with ALL six decades ago about the need for rapidly testing drug combinations. The remarkable improvement in outcomes observed when conventional agents were combined with Abl kinase inhibitors to treat *BCR-ABL* positive ALL (Schultz et al. 2009) illustrates an approach that may be applicable to other cancers.

For reasons outlined above, cancers harboring *RAS* and *NF1* driver mutations are likely to pose formidable therapeutic challenges for the foreseeable future. Although relatively rare, pediatric cancer models may be particularly informative for identifying actionable therapeutic vulnerabilities in Ras-regulated signaling networks, due to their relatively simple genetic landscapes. Furthermore, because Ras and other protein components of Ras signaling networks are frequently mutated in common adult cancers, many new agents that are being developed by biotechnology and pharmaceutical companies for common adult cancers should be tested in pediatric patients. However, the relatively small numbers of children and adolescents with relapsed and refractory cancer infer an essential role for comprehensive preclinical testing pipelines for identifying the most promising individual agents and drug combinations. This is particularly true for most *RAS*- and *NF1*-mutant cancers where directly targeting the underlying molecular lesion remains challenging. While preclinical testing pipelines should ideally incorporate a multipronged approach to assess compounds in the conventional and genetically modified isogenic cancer cell, GEM, and PDX models, few laboratories have the expertise or resources to study drugs (and especially drug combinations) across multiple platforms.

The linear signaling pathways shown in this review and elsewhere mask the immense complexity of Ras signaling networks and the cross talk and compensatory feedback that have evolved over eons. Inhibiting distinct components of the MAPK pathway and of other kinase cascades will undoubtedly have variable consequences due to the chemical properties of individual compounds, unique aspects of the target protein, or both. The unexpected insights from comparative studies of Raf and MEK inhibitors are likely to be extended to other targets and compounds in the future. Accordingly, determining the biochemical mechanism of action in parallel with performing phenotypic proliferation and cell death assays is valuable for informing rational preclinical, and ultimately clinical, trial design. We and others have also found that Bliss Independence analysis (Remsberg et al.

2021; Popescu et al. 2023) and other methods for rigorously assessing drug synergy can uncover promising combinations for in vivo testing. Furthermore, it is worth noting that tissue context modulates the relative dependency of different cancer types on specific Ras effectors and, likely, their vulnerability to inhibitors of those proteins. For example, while the MAPK pathway appears to be dominant in myeloid malignancies, genetic and functional studies suggest a central role for abnormal PI3 kinase/Akt signal output in ALL and other lymphoid cancers. Finally, we suggest that pediatric oncologists and their biopharmaceutical and basic sciences collaborators move rapidly to test inhibitors of Ras and Ras-regulated proteins in combination with conventional and other targeted anticancer agents.

REFERENCES

Astiazaran-Symonds E, Ney GM, Higgs C, Oba L, Srivastava R, Livinski AA, Rosenberg PS, Stewart DR. 2023. Cancer in Costello syndrome: a systematic review and meta-analysis. *Br J Cancer* **128:** 2089–2096. doi:10.1038/s41416-023-02229-7

Awad MM, Liu S, Rybkin II, Arbour KC, Dilly J, Zhu VW, Johnson ML, Heist RS, Patil T, Riely GJ, et al. 2021. Acquired resistance to KRAS^{G12C} inhibition in cancer. *N Engl J Med* **384:** 2382–2393. doi:10.1056/NEJMoa2105281

Badalian-Very G, Vergilio JA, Degar BA, MacConaill LE, Brandner B, Calicchio ML, Kuo FC, Ligon AH, Stevenson KE, Kehoe SM, et al. 2010. Recurrent BRAF mutations in Langerhans cell histiocytosis. *Blood* **116:** 1919–1923. doi:10.1182/blood-2010-04-279083

Bielski CM, Donoghue MTA, Gadiya M, Hanrahan AJ, Won HH, Chang MT, Jonsson P, Penson AV, Gorelick A, Harris C, et al. 2018. Widespread selection for oncogenic mutant allele imbalance in cancer. *Cancer Cell* **34:** 852–862.e4. doi:10.1016/j.ccell.2018.10.003

Boguski MS, McCormick F. 1993. Proteins regulating Ras and its relatives. *Nature* **366:** 643–654. doi:10.1038/366643a0

Bolouri H, Farrar JE, Triche T Jr, Ries RE, Lim EL, Alonzo TA, Ma Y, Moore R, Mungall AJ, Marra MA, et al. 2018. The molecular landscape of pediatric acute myeloid leukemia reveals recurrent structural alterations and age-specific mutational interactions. *Nat Med* **24:** 103–112. doi:10.1038/nm.4439

Borthakur G, Popplewell L, Boyiadzis M, Foran J, Platzbecker U, Vey N, Walter RB, Olin R, Raza A, Giagounidis A, et al. 2016. Activity of the oral mitogen-activated protein kinase kinase inhibitor trametinib in *RAS*-mutant relapsed or refractory myeloid malignancies. *Cancer* **122:** 1871–1879. doi:10.1002/cncr.29986

Cite this article as *Cold Spring Harb Perspect Med* doi: 10.1101/cshperspect.a041572

Bos JL, Rehmann H, Wittinghofer A. 2007. GEFs and GAPs: critical elements in the control of small G proteins. *Cell* **129:** 865–877. doi:10.1016/j.cell.2007.05.018

Brady SW, Roberts KG, Gu Z, Shi L, Pounds S, Pei D, Cheng C, Dai Y, Devidas M, Qu C, et al. 2022. The genomic landscape of pediatric acute lymphoblastic leukemia. *Nat Genet* **54:** 1376–1389. doi:10.1038/s41588-022-01159-z

Brastianos PK, Twohy E, Geyer S, Gerstner ER, Kaufmann TJ, Tabrizi S, Kabat B, Thierauf J, Ruff MW, Bota DA, et al. 2023. BRAF-MEK inhibition in newly diagnosed papillary craniopharyngiomas. *N Engl J Med* **389:** 118–126. doi:10.1056/NEJMoa2213329

Braun BS, Tuveson DA, Kong N, Le DT, Kogan SC, Rozmus J, Le Beau MM, Jacks TE, Shannon KM. 2004. Somatic activation of oncogenic *Kras* in hematopoietic cells initiates a rapidly fatal myeloproliferative disorder. *Proc Natl Acad Sci* **101:** 597–602. doi:10.1073/pnas.0307203101

Brennan CW, Verhaak RG, McKenna A, Campos B, Noushmehr H, Salama SR, Zheng S, Chakravarty D, Sanborn JZ, Berman SH, et al. 2013. The somatic genomic landscape of glioblastoma. *Cell* **155:** 462–477. doi:10.1016/j.cell.2013.09.034

Brohl AS, Kahen E, Yoder SJ, Teer JK, Reed DR. 2017. The genomic landscape of malignant peripheral nerve sheath tumors: diverse drivers of Ras pathway activation. *Sci Rep* **7:** 14992. doi:10.1038/s41598-017-15183-1

Brown NA, Furtado LV, Betz BL, Kiel MJ, Weigelin HC, Lim MS, Elenitoba-Johnson KS. 2014. High prevalence of somatic MAP2K1 mutations in BRAF V600E-negative Langerhans cell histiocytosis. *Blood* **124:** 1655–1658. doi:10.1182/blood-2014-05-577361

Buchberg AM, Cleveland LS, Jenkins NA, Copeland NG. 1990. Sequence homology shared by neurofibromatosis type-1 gene and IRA-1 and IRA-2 negative regulators of the RAS cyclic AMP pathway. *Nature* **347:** 291–294. doi:10.1038/347291a0

Burgess MR, Hwang E, Firestone AJ, Huang T, Xu J, Zuber J, Bohin N, Wen T, Kogan SC, Haigis KM, et al. 2014. Preclinical efficacy of MEK inhibition in Nras-mutant AML. *Blood* **124:** 3947–3955. doi:10.1182/blood-2014-05-574582

Burgess MR, Hwang E, Mroue R, Bielski CM, Wandler AM, Huang BJ, Firestone AJ, Young A, Lacap JA, Crocker L, et al. 2017. KRAS allelic imbalance enhances fitness and modulates MAP kinase dependence in cancer. *Cell* **168:** 817–829.e15. doi:10.1016/j.cell.2017.01.020

Canon J, Rex K, Saiki AY, Mohr C, Cooke K, Bagal D, Gaida K, Holt T, Knutson CG, Koppada N, et al. 2019. The clinical KRAS^{G12C} inhibitor AMG 510 drives anti-tumour immunity. *Nature* **575:** 217–223. doi:10.1038/s41586-019-1694-1

Castel P, Rauen KA, McCormick F. 2020. The duality of human oncoproteins: drivers of cancer and congenital disorders. *Nat Rev Cancer* **20:** 383–397. doi:10.1038/s41568-020-0256-z

Cawthon RM, Weiss R, Xu GF, Viskochil D, Culver M, Stevens J, Robertson M, Dunn D, Gesteland R, O'Connell P, et al. 1990. A major segment of the neurofibromatosis type 1 gene: cDNA sequence, genomic structure, and point mutations. *Cell* **62:** 193–201. doi:10.1016/0092-8674(90)90253-b

Chan RJ, Leedy MB, Munugalavadla V, Voorhorst CS, Li Y, Yu M, Kapur R. 2005. Human somatic PTPN11 mutations induce hematopoietic-cell hypersensitivity to granulocyte-macrophage colony-stimulating factor. *Blood* **105:** 3737–3742. doi:10.1182/blood-2004-10-4002

Chan RJ, Cooper T, Kratz CP, Weiss B, Loh ML. 2009. Juvenile myelomonocytic leukemia: a report from the 2nd International JMML Symposium. *Leuk Res* **33:** 355–362. doi:10.1016/j.leukres.2008.08.022

Chang T, Krisman K, Theobald EH, Xu J, Akutagawa J, Lauchle JO, Kogan S, Braun BS, Shannon K. 2013. Sustained MEK inhibition abrogates myeloproliferative disease in Nf1 mutant mice. *J Clin Invest* **123:** 335–339. doi:10.1172/JCI63193

Chapman PB, Hauschild A, Robert C, Haanen JB, Ascierto P, Larkin J, Dummer R, Garbe C, Testori A, Maio M, et al. 2011. Improved survival with vemurafenib in melanoma with BRAF V600E mutation. *N Engl J Med* **364:** 2507–2516. doi:10.1056/NEJMoa1103782

Chen X, Stewart E, Shelat AA, Qu C, Bahrami A, Hatley M, Wu G, Bradley C, McEvoy J, Pappo A, et al. 2013. Targeting oxidative stress in embryonal rhabdomyosarcoma. *Cancer Cell* **24:** 710–724. doi:10.1016/j.ccr.2013.11.002

Chen PY, Huang BJ, Harris M, Boone C, Wang W, Carias H, Mesiona B, Mavrici D, Kohler AC, Bollag G, et al. 2023. Structural and functional analyses of a germline KRAS T50I mutation provide insights into Raf activation. *JCI Insight* **8:** e168445. doi:10.1172/jci.insight.168445

Corces-Zimmerman MR, Hong WJ, Weissman IL, Medeiros BC, Majeti R. 2014. Preleukemic mutations in human acute myeloid leukemia affect epigenetic regulators and persist in remission. *Proc Natl Acad Sci* **111:** 2548–2553. doi:10.1073/pnas.1324297111

Cuiffo B, Ren R. 2010. Palmitoylation of oncogenic NRAS is essential for leukemogenesis. *Blood* **115:** 3598–3605. doi:10.1182/blood-2009-03-213876

Dail M, Li Q, McDaniel A, Wong J, Akagi K, Huang B, Kang HC, Kogan SC, Shokat K, Wolff L, et al. 2010. Mutant *Ikzf1*, *KrasG12D*, and *Notch1* cooperate in T lineage leukemogenesis and modulate responses to targeted agents. *Proc Natl Acad Sci* **107:** 5106–5111. doi:10.1073/pnas.1001064107

Dail M, Wong J, Lawrence J, O'Connor D, Nakitandwe J, Chen SC, Xu J, Lee LB, Akagi K, Li Q, et al. 2014. Loss of oncogenic Notch1 with resistance to a PI3K inhibitor in T-cell leukaemia. *Nature* **513:** 512–516. doi:10.1038/nature13495

De Raedt T, Beert E, Pasmant E, Luscan A, Brems H, Ortonne N, Helin K, Hornick JL, Mautner V, Kehrer-Sawatzki H, et al. 2014. PRC2 loss amplifies Ras-driven transcription and confers sensitivity to BRD4-based therapies. *Nature* **514:** 247–251. doi:10.1038/nature13561

DiNardo CD, Maiti A, Rausch CR, Pemmaraju N, Naqvi K, Daver NG, Kadia TM, Borthakur G, Ohanian M, Alvarado Y, et al. 2020. Ten-day decitabine with venetoclax for newly diagnosed intensive chemotherapy ineligible, and relapsed or refractory acute myeloid leukaemia: a single-centre, phase 2 trial. *Lancet Haematol* **7:** e724–e736. doi:10.1016/S2352-3026(20)30210-6

Dombi E, Baldwin A, Marcus LJ, Fisher MJ, Weiss B, Kim A, Whitcomb P, Martin S, Aschbacher-Smith LE, Rizvi TA, et al. 2016. Activity of selumetinib in neurofibromatosis type

1-related plexiform neurofibromas. *N Engl J Med* **375:** 2550–2560. doi:10.1056/NEJMoa1605943

Donovan S, Shannon KM, Bollag G. 2002. GTPase activating proteins: critical regulators of intracellular signaling. *Biochim Biophys Acta* **1602:** 23–45. doi:10.1016/s0304-419x(01)00041-5

Downward J. 2003. Targeting RAS signalling pathways in cancer therapy. *Nat Rev Cancer* **3:** 11–22. doi:10.1038/nrc969

Ducatman BS, Scheithauer BW, Piepgras DG, Reiman HM, Ilstrup DM. 1986. Malignant peripheral nerve sheath tumors. A clinicopathologic study of 120 cases. *Cancer* **57:** 2006–2021. doi:10.1002/1097-0142(19860515)57:10<2006::aid-cncr2820571022>3.0.co;2-6

Eleveld TF, Oldridge DA, Bernard V, Koster J, Colmet Daage L, Diskin SJ, Schild L, Bentahar NB, Bellini A, Chicard M, et al. 2015. Relapsed neuroblastomas show frequent RAS-MAPK pathway mutations. *Nat Genet* **47:** 864–871. doi:10.1038/ng.3333

Fedele C, Ran H, Diskin B, Wei W, Jen J, Geer MJ, Araki K, Ozerdem U, Simeone DM, Miller G, et al. 2018. SHP2 inhibition prevents adaptive resistance to MEK inhibitors in multiple cancer models. *Cancer Discov* **8:** 1237–1249. doi:10.1158/2159-8290.CD-18-0444

Flaherty KT, Robert C, Hersey P, Nathan P, Garbe C, Milhem M, Demidov LV, Hassel JC, Rutkowski P, Mohr P, et al. 2012. Improved survival with MEK inhibition in BRAF-mutated melanoma. *N Engl J Med* **367:** 107–114. doi:10.1056/NEJMoa1203421

Gröbner SN, Worst BC, Weischenfeldt J, Buchhalter I, Kleinheinz K, Rudneva VA, Johann PD, Balasubramanian GP, Segura-Wang M, Brabetz S, et al. 2018. The landscape of genomic alterations across childhood cancers. *Nature* **555:** 321–327. doi:10.1038/nature25480

Gutierrez A, Sanda T, Grebliunaite R, Carracedo A, Salmena L, Ahn Y, Dahlberg S, Neuberg D, Moreau LA, Winter SS, et al. 2009. High frequency of PTEN, PI3K, and AKT abnormalities in T-cell acute lymphoblastic leukemia. *Blood* **114:** 647–650. doi:10.1182/blood-2009-02-206722

Hallin J, Engstrom LD, Hargis L, Calinisan A, Aranda R, Briere DM, Sudhakar N, Bowcut V, Baer BR, Ballard JA, et al. 2020. The KRAS^G12C inhibitor MRTX849 provides insight toward therapeutic susceptibility of KRAS-mutant cancers in mouse models and patients. *Cancer Discov* **10:** 54–71. doi:10.1158/2159-8290.CD-19-1167

Hancock JF, Paterson H, Marshall CJ. 1990. A polybasic domain or palmitoylation is required in addition to the CAAX motif to localize p21ras to the plasma membrane. *Cell* **63:** 133–139. doi:10.1016/0092-8674(90)90294-o

Hauschild A, Grob JJ, Demidov LV, Jouary T, Gutzmer R, Millward M, Rutkowski P, Blank CU, Miller WH Jr, Kaempgen E, et al. 2012. Dabrafenib in BRAF-mutated metastatic melanoma: a multicentre, open-label, phase 3 randomised controlled trial. *Lancet* **380:** 358–365. doi:10.1016/S0140-6736(12)60868-X

Holmfeldt L, Wei L, Diaz-Flores E, Walsh M, Zhang J, Ding L, Payne-Turner D, Churchman M, Andersson A, Chen SC, et al. 2013. The genomic landscape of hypodiploid acute lymphoblastic leukemia. *Nat Genet* **45:** 242–252. doi:10.1038/ng.2532

Hong DS, Fakih MG, Strickler JH, Desai J, Durm GA, Shapiro GI, Falchook GS, Price TJ, Sacher A, Denlinger CS, et al.

2020. KRAS^g12c inhibition with sotorasib in advanced solid tumors. *N Engl J Med* **383:** 1207–1217. doi:10.1056/NEJMoa1917239

Huang SC, Alaggio R, Sung YS, Chen CL, Zhang L, Kao YC, Agaram NP, Wexler LH, Antonescu CR. 2016. Frequent HRAS mutations in malignant ectomesenchymoma: overlapping genetic abnormalities with embryonal rhabdomyosarcoma. *Am J Surg Pathol* **40:** 876–885. doi:10.1097/PAS.0000000000000612

Huang BJ, Wandler AM, Meyer LK, Dail M, Daemen A, Sampath D, Li Q, Wang X, Wong JC, Nakitandwe J, et al. 2019. Convergent genetic aberrations in murine and human T lineage acute lymphoblastic leukemias. *PLoS Genet* **15:** e1008168. doi:10.1371/journal.pgen.1008168

Jackson EL, Willis N, Mercer K, Bronson RT, Crowley D, Montoya R, Jacks T, Tuveson DA. 2001. Analysis of lung tumor initiation and progression using conditional expression of oncogenic *K-ras*. *Genes Dev* **15:** 3243–3248. doi:10.1101/gad.943001

Jacob K, Albrecht S, Sollier C, Faury D, Sader E, Montpetit A, Serre D, Hauser P, Garami M, Bognar L, et al. 2009. Duplication of 7q34 is specific to juvenile pilocytic astrocytomas and a hallmark of cerebellar and optic pathway tumours. *Br J Cancer* **101:** 722–733. doi:10.1038/sj.bjc.6605179

Jaiswal S, Fontanillas P, Flannick J, Manning A, Grauman PV, Mar BG, Lindsley RC, Mermel CH, Burtt N, Chavez A, et al. 2014. Age-related clonal hematopoiesis associated with adverse outcomes. *N Engl J Med* **371:** 2488–2498. doi:10.1056/NEJMoa1408617

Jessen WJ, Miller SJ, Jousma E, Wu J, Rizvi TA, Brundage ME, Eaves D, Widemann B, Kim MO, Dombi E, et al. 2013. MEK inhibition exhibits efficacy in human and mouse neurofibromatosis tumors. *J Clin Invest* **123:** 340–347. doi:10.1172/JCI60578

Johnson L, Mercer K, Greenbaum D, Bronson RT, Crowley D, Tuveson DA, Jacks T. 2001. Somatic activation of the K-ras oncogene causes early onset lung cancer in mice. *Nature* **410:** 1111–1116. doi:10.1038/35074129

Keilhack H, David FS, McGregor M, Cantley LC, Neel BG. 2005. Diverse biochemical properties of Shp2 mutants. Implications for disease phenotypes. *J Biol Chem* **280:** 30984–30993. doi:10.1074/jbc.M504699200

Klco JM, Spencer DH, Miller CA, Griffith M, Lamprecht TL, O'Laughlin M, Fronick C, Magrini V, Demeter RT, Fulton RS, et al. 2014. Functional heterogeneity of genetically defined subclones in acute myeloid leukemia. *Cancer Cell* **25:** 379–392. doi:10.1016/j.ccr.2014.01.031

Kratz CP, Niemeyer CM, Castleberry RP, Cetin M, Bergsträsser E, Emanuel PD, Hasle H, Kardos G, Klein C, Kojima S, et al. 2005. The mutational spectrum of PTPN11 in juvenile myelomonocytic leukemia and Noonan syndrome/myeloproliferative disease. *Blood* **106:** 2183–2185. doi:10.1182/blood-2005-02-0531

Kratz CP, Steinemann D, Niemeyer CM, Schlegelberger B, Koscielniak E, Kontny U, Zenker M. 2007. Uniparental disomy at chromosome 11p15.5 followed by HRAS mutations in embryonal rhabdomyosarcoma: lessons from Costello syndrome. *Hum Mol Genet* **16:** 374–379. doi:10.1093/hmg/ddl458

Kratz CP, Franke L, Peters H, Kohlschmidt N, Kazmierczak B, Finckh U, Bier A, Eichhorn B, Blank C, Kraus C, et al. 2015.

Cite this article as *Cold Spring Harb Perspect Med* doi: 10.1101/cshperspect.a041572

Cancer spectrum and frequency among children with Noonan, Costello, and cardio-facio-cutaneous syndromes. *Br J Cancer* **112:** 1392–1397. doi:10.1038/bjc .2015.75

Krivtsov AV, Evans K, Gadrey JY, Eschle BK, Hatton C, Uckelmann HJ, Ross KN, Perner F, Olsen SN, Pritchard T, et al. 2019. A menin-MLL inhibitor induces specific chromatin changes and eradicates disease in models of MLL-rearranged leukemia. *Cancer Cell* **36:** 660–673.e11. doi:10.1016/j.ccell.2019.11.001

Lau N, Feldkamp MM, Roncari L, Loehr AH, Shannon P, Gutmann DH, Guha A. 2000. Loss of neurofibromin is associated with activation of RAS/MAPK and PI3-K/ AKT signaling in a neurofibromatosis 1 astrocytoma. *J Neuropathol Exp Neurol* **59:** 759–767. doi:10.1093/jnen/ 59.9.759

Lauchle JO, Kim D, Le DT, Akagi K, Crone M, Krisman K, Warner K, Bonifas JM, Li Q, Coakley KM, et al. 2009. Response and resistance to MEK inhibition in leukaemias initiated by hyperactive Ras. *Nature* **461:** 411–414. doi:10 .1038/nature08279

Le DT, Kong N, Zhu Y, Lauchle JO, Aiyigari A, Braun BS, Wang E, Kogan SC, Le Beau MM, Parada L, et al. 2004. Somatic inactivation of Nf1 in hematopoietic cells results in a progressive myeloproliferative disorder. *Blood* **103:** 4243–4250. doi:10.1182/blood-2003-08-2650

Lee W, Teckie S, Wiesner T, Ran L, Prieto Granada CN, Lin M, Zhu S, Cao Z, Liang Y, Sboner A, et al. 2014. PRC2 is recurrently inactivated through EED or SUZ12 loss in malignant peripheral nerve sheath tumors. *Nat Genet* **46:** 1227–1232. doi:10.1038/ng.3095

Li Z, Jiang K, Zhu X, Lin G, Song F, Zhao Y, Piao Y, Liu J, Cheng W, Bi X, et al. 2016. Encorafenib (LGX818), a potent BRAF inhibitor, induces senescence accompanied by autophagy in BRAFV600E melanoma cells. *Cancer Lett* **370:** 332–344. doi:10.1016/j.canlet.2015.11.015

Lindsley RC, Ebert BL. 2013. The biology and clinical impact of genetic lesions in myeloid malignancies. *Blood* **122:** 3741–3748. doi:10.1182/blood-2013-06-460295

Lito P, Rosen N, Solit DB. 2013. Tumor adaptation and resistance to RAF inhibitors. *Nat Med* **19:** 1401–1409. doi:10 .1038/nm.3392

Lou K, Steri V, Ge AY, Hwang YC, Yogodzinski CH, Shkedi AR, Choi ALM, Mitchell DC, Swaney DL, Hann B, et al. 2019. KRAS^{g12c} inhibition produces a driver-limited state revealing collateral dependencies. *Sci Signal* **12:** eaaw9450. doi:10.1126/scisignal.aaw9450

Lyubynska N, Gorman MF, Lauchle JO, Hong WX, Akutagawa JK, Shannon K, Braun BS. 2011. A MEK inhibitor abrogates myeloproliferative disease in *Kras* mutant mice. *Sci Transl Med* **3:** 76ra27. doi:10.1126/scitranslmed.300 1069

Ma X, Liu Y, Liu Y, Alexandrov LB, Edmonson MN, Gawad C, Zhou X, Li Y, Rusch MC, Easton J, et al. 2018. Pan-cancer genome and transcriptome analyses of 1,699 paediatric leukaemias and solid tumours. *Nature* **555:** 371–376. doi:10.1038/nature25795

Maiti A, Naqvi K, Kadia TM, Borthakur G, Takahashi K, Bose P, Daver NG, Patel A, Alvarado Y, Ohanian M, et al. 2019. Phase II trial of MEK inhibitor binimetinib (MEK162) in RAS-mutant acute myeloid leukemia. *Clin Lymphoma* *Myeloma Leuk* **19:** 142–148.e1. doi:10.1016/j.clml.2018 .12.009

Malumbres M, Barbacid M. 2003. RAS oncogenes: the first 30 years. *Nat Rev Cancer* **3:** 459–465. doi:10.1038/nrc1097

Marranci A, Jiang Z, Vitiello M, Guzzolino E, Comelli L, Sarti S, Lubrano S, Franchin C, Echevarria-Vargas I, Tuccoli A, et al. 2017. The landscape of BRAF transcript and protein variants in human cancer. *Mol Cancer* **16:** 85. doi:10.1186/ s12943-017-0645-4

Martin GA, Viskochil D, Bollag G, McCabe PC, Crosier WJ, Haubruck H, Conroy L, Clark R, O'Connell P, Cawthon RM, et al. 1990. The GAP-related domain of the neurofibromatosis type 1 gene product interacts with ras p21. *Cell* **63:** 843–849. doi:10.1016/0092-8674(90)90150-d

Matsuda K, Shimada A, Yoshida N, Ogawa A, Watanabe A, Yajima S, Iizuka S, Koike K, Yanai F, Kawasaki K, et al. 2007. Spontaneous improvement of hematologic abnormalities in patients having juvenile myelomonocytic leukemia with specific RAS mutations. *Blood* **109:** 5477– 5480. doi:10.1182/blood-2006-09-046649

Matsuda K, Taira C, Sakashita K, Saito S, Tanaka-Yanagisawa M, Yanagisawa R, Nakazawa Y, Shiohara M, Fukushima K, Oda M, et al. 2010. Long-term survival after nonintensive chemotherapy in some juvenile myelomonocytic leukemia patients with CBL mutations, and the possible presence of healthy persons with the mutations. *Blood* **115:** 5429–5431. doi:10.1182/blood-2009-12-260729

McMahon CM, Ferng T, Canaani J, Wang ES, Morrissette JJD, Eastburn DJ, Pellegrino M, Durruthy-Durruthy R, Watt CD, Asthana S, et al. 2019. Clonal Selection with RAS pathway activation mediates secondary clinical resistance to selective FLT3 inhibition in acute myeloid leukemia. *Cancer Discov* **9:** 1050–1063. doi:10.1158/2159-8290 .CD-18-1453

Modrek B, Ge L, Pandita A, Lin E, Mohan S, Yue P, Guerrero S, Lin WM, Pham T, Modrusan Z, et al. 2009. Oncogenic activating mutations are associated with local copy gain. *Mol Cancer Res* **7:** 1244–1252. doi:10.1158/1541-7786 .MCR-08-0532

Mohapatra D, Gupta AK, Haldar P, Meena JP, Tanwar P, Seth R. 2023. Efficacy and safety of vemurafenib in Langerhans cell histiocytosis (LCH): a systematic review and meta-analysis. *Pediatr Hematol Oncol* **40:** 86–97. doi:10.1080/ 08880018.2022.2072986

Mohi MG, Williams IR, Dearolf CR, Chan G, Kutok JL, Cohen S, Morgan K, Boulton C, Shigematsu H, Keilhack H, et al. 2005. Prognostic, therapeutic, and mechanistic implications of a mouse model of leukemia evoked by Shp2 (PTPN11) mutations. *Cancer Cell* **7:** 179–191. doi:10 .1016/j.ccr.2005.01.010

Mor A, Philips MR. 2006. Compartmentalized Ras/MAPK signaling. *Annu Rev Immunol* **24:** 771–800. doi:10.1146/ annurev.immunol.24.021605.090723

Müller HL, Merchant TE, Warmuth-Metz M, Martinez-Barbera JP, Puget S. 2019. Craniopharyngioma. *Nat Rev Dis Primers* **5:** 75. doi:10.1038/s41572-019-0125-9

Neel BG, Gu H, Pao L. 2003. The 'Shp'ing news: SH2 domain-containing tyrosine phosphatases in cell signaling. *Trends Biochem Sci* **28:** 284–293. doi:10.1016/S0968-0004(03) 00091-4

Nichols RJ, Haderk F, Stahlhut C, Schulze CJ, Hemmati G, Wildes D, Tzitzilonis C, Mordec K, Marquez A, Romero J, et al. 2018. RAS nucleotide cycling underlies the SHP2 phosphatase dependence of mutant BRAF-, NF1- and RAS-driven cancers. *Nat Cell Biol* 20: 1064–1073. doi:10.1038/s41556-018-0169-1

Omerovic J, Laude AJ, Prior IA. 2007. Ras proteins: paradigms for compartmentalised and isoform-specific signalling. *Cell Mol Life Sci* 64: 2575–2589. doi:10.1007/s00018-007-7133-8

Ostrem JM, Peters U, Sos ML, Wells JA, Shokat KM. 2013. K-Ras^{G12C} inhibitors allosterically control GTP affinity and effector interactions. *Nature* 503: 548–551. doi:10.1038/nature12796

Papaemmanuil E, Gerstung M, Bullinger L, Gaidzik VI, Paschka P, Roberts ND, Potter NE, Heuser M, Thol F, Bolli N, et al. 2016. Genomic classification and prognosis in acute myeloid leukemia. *N Engl J Med* 374: 2209–2221. doi:10.1056/NEJMoa1516192

Park E, Rawson S, Li K, Kim BW, Ficarro SB, Pino GG, Sharif H, Marto JA, Jeon H, Eck MJ. 2019. Architecture of autoinhibited and active BRAF-MEK1-14-3-3 complexes. *Nature* 575: 545–550. doi:10.1038/s41586-019-1660-y

Popescu B, Stahlhut C, Tarver TC, Wishner S, Lee BJ, Peretz CAC, Luck C, Phojanakong P, Camara Serrano JA, Hongo H, et al. 2023. Allosteric SHP2 inhibition increases apoptotic dependency on BCL2 and synergizes with venetoclax in FLT3- and KIT-mutant AML. *Cell Rep Med* 4: 101290. doi:10.1016/j.xcrm.2023.101290

Prior IA, Lewis PD, Mattos C. 2012. A comprehensive survey of Ras mutations in cancer. *Cancer Res* 72: 2457–2467. doi:10.1158/0008-5472.CAN-11-2612

Remsberg JR, Suciu RM, Zambetti NA, Hanigan TW, Firestone AJ, Inguva A, Long A, Ngo N, Lum KM, Henry CL, et al. 2021. ABHD17 regulation of plasma membrane palmitoylation and N-Ras-dependent cancer growth. *Nat Chem Biol* 17: 856–864. doi:10.1038/s41589-021-00785-8

Ryall S, Zapotocky M, Fukuoka K, Nobre L, Guerreiro Stucklin A, Bennett J, Siddaway R, Li C, Pajovic S, Arnoldo A, et al. 2020. Integrated molecular and clinical analysis of 1,000 pediatric low-grade gliomas. *Cancer Cell* 37: 569–583.e5. doi:10.1016/j.ccell.2020.03.011

Samra B, Konopleva M, Isidori A, Daver N, DiNardo C. 2020. Venetoclax-based combinations in acute myeloid leukemia: current evidence and future directions. *Front Oncol* 10: 562558. doi:10.3389/fonc.2020.562558

Schönung M, Meyer J, Nöllke P, Olshen AB, Hartmann M, Murakami N, Wakamatsu M, Okuno Y, Plass C, Loh ML, et al. 2021. International consensus definition of DNA methylation subgroups in juvenile myelomonocytic leukemia. *Clin Cancer Res* 27: 158–168. doi:10.1158/1078-0432.CCR-20-3184

Schubbert S, Zenker M, Rowe SL, Böll S, Klein C, Bollag G, van der Burgt I, Musante L, Kalscheuer V, Wehner LE, et al. 2006. Germline KRAS mutations cause Noonan syndrome. *Nat Genet* 38: 331–336. doi:10.1038/ng1748

Schubbert S, Shannon K, Bollag G. 2007. Hyperactive Ras in developmental disorders and cancer. *Nat Rev Cancer* 7: 295–308. doi:10.1038/nrc2109

Schultz KR, Bowman WP, Aledo A, Slayton WB, Sather H, Devidas M, Wang C, Davies SM, Gaynon PS, Trigg M, et al. 2009. Improved early event-free survival with imatinib in Philadelphia chromosome–positive acute lymphoblastic

leukemia: a children's oncology group study. *J Clin Oncol* 27: 5175–5181. doi:10.1200/JCO.2008.21.2514

Shannon KM, O'Connell P, Martin GA, Paderanga D, Olson K, Dinndorf P, McCormick F. 1994. Loss of the normal NF1 allele from the bone marrow of children with type 1 neurofibromatosis and malignant myeloid disorders. *N Engl J Med* 330: 597–601. doi:10.1056/NEJM199403033300903

Shern JF, Chen L, Chmielecki J, Wei JS, Patidar R, Rosenberg M, Ambrogio L, Auclair D, Wang J, Song YK, et al. 2014. Comprehensive genomic analysis of rhabdomyosarcoma reveals a landscape of alterations affecting a common genetic axis in fusion-positive and fusion-negative tumors. *Cancer Discov* 4: 216–231. doi:10.1158/2159-8290.CD-13-0639

Simanshu DK, Nissley DV, McCormick F. 2017. RAS proteins and their regulators in human disease. *Cell* 170: 17–33. doi:10.1016/j.cell.2017.06.009

Simanshu DK, Philips MR, Hancock JF. 2023. Consensus on the RAS dimerization hypothesis: strong evidence for lipid-mediated clustering but not for G-domain-mediated interactions. *Mol Cell* 83: 1210–1215. doi:10.1016/j.molcel.2023.03.008

Skoulidis F, Li BT, Dy GK, Price TJ, Falchook GS, Wolf J, Italiano A, Schuler M, Borghaei H, Barlesi F, et al. 2021. Sotorasib for lung cancers with KRAS p.G12C mutation. *N Engl J Med* 384: 2371–2381. doi:10.1056/NEJMoa2103695

Solit DB, Garraway LA, Pratilas CA, Sawai A, Getz G, Basso A, Ye Q, Lobo JM, She Y, Osman I, et al. 2006. BRAF mutation predicts sensitivity to MEK inhibition. *Nature* 439: 358–362. doi:10.1038/nature04304

Stein EM, DiNardo CD, Pollyea DA, Fathi AT, Roboz GJ, Altman JK, Stone RM, DeAngelo DJ, Levine RL, Flinn IW, et al. 2017. Enasidenib in mutant IDH2 relapsed or refractory acute myeloid leukemia. *Blood* 130: 722–731. doi:10.1182/blood-2017-04-779405

Stephen AG, Esposito D, Bagni RK, McCormick F. 2014. Dragging ras back in the ring. *Cancer Cell* 25: 272–281. doi:10.1016/j.ccr.2014.02.017

Stephens K, Weaver M, Leppig KA, Maruyama K, Emanuel PD, Le Beau MM, Shannon KM. 2006. Interstitial uniparental isodisomy at clustered breakpoint intervals is a frequent mechanism of NF1 inactivation in myeloid malignancies. *Blood* 108: 1684–1689. doi:10.1182/blood-2005-11-011486

Stieglitz E, Taylor-Weiner AN, Chang TY, Gelston LC, Wang YD, Mazor T, Esquivel E, Yu A, Seepo S, Olsen S, et al. 2015. The genomic landscape of juvenile myelomonocytic leukemia. *Nat Genet* 47: 1326–1333. doi:10.1038/ng.3400

Stieglitz E, Mazor T, Olshen AB, Geng H, Gelston LC, Akutagawa J, Lipka DB, Plass C, Flotho C, Chehab FF, et al. 2017. Genome-wide DNA methylation is predictive of outcome in juvenile myelomonocytic leukemia. *Nat Commun* 8: 2127. doi:10.1038/s41467-017-02178-9

Strickler JH, Satake H, George TJ, Yaeger R, Hollebecque A, Garrido-Laguna I, Schuler M, Burns TF, Coveler AL, Falchook GS, et al. 2023. Sotorasib in KRAS p.G12C-mutated advanced pancreatic cancer. *N Engl J Med* 388: 33–43. doi:10.1056/NEJMoa2208470

Strullu M, Caye A, Lachenaud J, Cassinat B, Gazal S, Fenneteau O, Pouvreau N, Pereira S, Baumann C, Contet A, et al. 2014. Juvenile myelomonocytic leukaemia and Noonan

Cite this article as *Cold Spring Harb Perspect Med* doi: 10.1101/cshperspect.a041572

syndrome. *J Med Genet* **51:** 689–697. doi:10.1136/jmedge net-2014-102611

Su F, Viros A, Milagre C, Trunzer K, Bollag G, Spleiss O, Reis-Filho JS, Kong X, Koya RC, Flaherty KT, et al. 2012. *RAS* mutations in cutaneous squamous-cell carcinomas in patients treated with BRAF inhibitors. *N Engl J Med* **366:** 207–215. doi:10.1056/NEJMoa1105358

Tajan M, Paccoud R, Branka S, Edouard T, Yart A. 2018. The RASopathy family: consequences of germline activation of the RAS/MAPK pathway. *Endocr Rev* **39:** 676–700. doi:10 .1210/er.2017-00232

Tartaglia M, Gelb BD. 2005. Noonan syndrome and related disorders: genetics and pathogenesis. *Annu Rev Genomics Hum Genet* **6:** 45–68. doi:10.1146/annurev.genom.6.080 604.162305

Tartaglia M, Mehler EL, Goldberg R, Zampino G, Brunner HG, Kremer H, van der Burgt I, Crosby AH, Ion A, Jeffery S, et al. 2001. Mutations in PTPN11, encoding the protein tyrosine phosphatase SHP-2, cause Noonan syndrome. *Nat Genet* **29:** 465–468. doi:10.1038/ng772

Townsend EC, Murakami MA, Christodoulou A, Christie AL, Köster J, DeSouza TA, Morgan EA, Kallgren SP, Liu H, Wu SC, et al. 2016. The public repository of xenografts enables discovery and randomized phase II-like trials in mice. *Cancer Cell* **29:** 574–586. doi:10.1016/j.ccell.2016.06 .008

Tran TH, Chan AH, Young LC, Bindu L, Neale C, Messing S, Dharmaiah S, Taylor T, Denson JP, Esposito D, et al. 2021. KRAS interaction with RAF1 RAS-binding domain and cysteine-rich domain provides insights into RAS-mediated RAF activation. *Nat Commun* **12:** 1176. doi:10.1038/ s41467-021-21422-x

Vasta JD, Peacock DM, Zheng Q, Walker JA, Zhang Z, Zimprich CA, Thomas MR, Beck MT, Binkowski BF, Corona CR, et al. 2022. KRAS is vulnerable to reversible switch-II pocket engagement in cells. *Nat Chem Biol* **18:** 596–604. doi:10.1038/s41589-022-00985-w

Vetter IR, Wittinghofer A. 2001. The guanine nucleotide-binding switch in three dimensions. *Science* **294:** 1299–1304. doi:10.1126/science.1062023

Viskochil D, Buchberg AM, Xu G, Cawthon RM, Stevens J, Wolff RK, Culver M, Carey JC, Copeland NG, Jenkins NA, et al. 1990. Deletions and a translocation interrupt a cloned gene at the neurofibromatosis type 1 locus. *Cell* **62:** 187–192. doi:10.1016/0092-8674(90)90252-a

Wallace MR, Marchuk DA, Andersen LB, Letcher R, Odeh HM, Saulino AM, Fountain JW, Brereton A, Nicholson J, Mitchell AL, et al. 1990. Type 1 neurofibromatosis gene: identification of a large transcript disrupted in three NF1 patients. *Science* **249:** 181–186. doi:10.1126/science.213 4734

Wandler AM, Huang BJ, Craig JW, Hayes K, Yan H, Meyer LK, Scacchetti A, Monsalve G, Dail M, Li Q, et al. 2020. Loss of glucocorticoid receptor expression mediates in vivo dexamethasone resistance in T-cell acute lymphoblastic leukemia. *Leukemia* **34:** 2025–2037. doi:10.1038/ s41375-020-0748-6

Wang T, Yu H, Hughes NW, Liu B, Kendirli A, Klein K, Chen WW, Lander ES, Sabatini DM. 2017. Gene essentiality profiling reveals gene networks and synthetic lethal interactions with oncogenic ras. *Cell* **168:** 890–903.e15. doi:10 .1016/j.cell.2017.01.013

Wang X, Allen S, Blake JF, Bowcut V, Briere DM, Calinisan A, Dahlke JR, Fell JB, Fischer JP, Gunn RJ, et al. 2022. Identification of MRTX1133, a noncovalent, potent, and selective KRAS^{G12D} inhibitor. *J Med Chem* **65:** 3123–3133. doi:10.1021/acs.jmedchem.1c01688

Ward AF, Braun BS, Shannon KM. 2012. Targeting oncogenic Ras signaling in hematologic malignancies. *Blood* **120:** 3397–3406. doi:10.1182/blood-2012-05-378596

Wong JC, Perez-Mancera PA, Huang TQ, Kim J, Grego-Bessa J, Del Pilar Alzamora M, Kogan SC, Sharir A, Keefe SH, Morales CE, et al. 2020. *Krasp34r* and *KrasT58I* mutations induce distinct RASopathy phenotypes in mice. *JCI Insight* **5:** e140495. doi:10.1172/jci.insight.140495

Xu GF, Lin B, Tanaka K, Dunn D, Wood D, Gesteland R, White R, Weiss R, Tamanoi F. 1990a. The catalytic domain of the neurofibromatosis type 1 gene product stimulates ras GTPase and complements ira mutants of *S. cerevisiae*. *Cell* **63:** 835–841. doi:10.1016/0092-8674(90)90149-9

Xu GF, O'Connell P, Viskochil D, Cawthon R, Robertson M, Culver M, Dunn D, Stevens J, Gesteland R, White R, et al. 1990b. The neurofibromatosis type 1 gene encodes a protein related to GAP. *Cell* **62:** 599–608. doi:10.1016/0092-8674(90)90024-9

Xu W, Mulligan LM, Ponder MA, Liu L, Smith BA, Mathew CG, Ponder BA. 1992. Loss of *NF1* alleles in phaeochromocytomas from patients with type I neurofibromatosis. *Genes Chromosomes Cancer* **4:** 337–342. doi:10.1002/gcc .2870040411

Xu J, Hedberg C, Dekker FJ, Li Q, Haigis KM, Hwang E, Waldmann H, Shannon K. 2012. Inhibiting the palmitoylation/depalmitoylation cycle selectively reduces the growth of hematopoietic cells expressing oncogenic Nras. *Blood* **119:** 1032–1035. doi:10.1182/blood-2011-06-358960

Yen I, Shanahan F, Lee J, Hong YS, Shin SJ, Moore AR, Sudhamsu J, Chang MT, Bae I, Dela Cruz D, et al. 2021. ARAF mutations confer resistance to the RAF inhibitor belvarafenib in melanoma. *Nature* **594:** 418–423. doi:10.1038/ s41586-021-03515-1

Yuan X, Bu H, Zhou J, Yang CY, Zhang H. 2020. Recent advances of SHP2 inhibitors in cancer therapy: current development and clinical application. *J Med Chem* **63:** 11368–11396. doi:10.1021/acs.jmedchem.0c00249

Zadeh G, Buckle C, Shannon P, Massicotte EM, Wong S, Guha A. 2007. Radiation induced peripheral nerve tumors: case series and review of the literature. *J Neurooncol* **83:** 205–212. doi:10.1007/s11060-006-9315-x

Zambetti NA, Firestone AJ, Remsberg JR, Huang BJ, Wong JC, Long AM, Predovic M, Suciu RM, Inguva A, Kogan SC, et al. 2020. Genetic disruption of N-Ras^{G12D} palmitoylation perturbs hematopoiesis and prevents myeloid transformation in mice. *Blood* **135:** 1772–1782. doi:10.1182/ blood.2019003530

Zhang J, Ding L, Holmfeldt L, Wu G, Heatley SL, Payne-Turner D, Easton J, Chen X, Wang J, Rusch M, et al. 2012. The genetic basis of early T-cell precursor acute lymphoblastic leukaemia. *Nature* **481:** 157–163. doi:10 .1038/nature10725

Zhang Q, Riley-Gillis B, Han L, Jia Y, Lodi A, Zhang H, Ganesan S, Pan R, Konoplev SN, Sweeney SR, et al. 2022. Activation of RAS/MAPK pathway confers MCL-1 mediated acquired resistance to BCL-2 inhibitor venetoclax in acute myeloid leukemia. *Signal Transduct Target Ther* **7**: 51. doi:10.1038/s41392-021-00870-3

Zhao Y, Murciano-Goroff YR, Xue JY, Ang A, Lucas J, Mai TT, Da Cruz Paula AF, Saiki AY, Mohn D, Achanta P, et al. 2021. Diverse alterations associated with resistance to KRAS^{G12C} inhibition. *Nature* **599**: 679–683. doi:10.1038/s41586-021-04065-2

Zhou M, Philips MR. 2019. Where no Ras has gone before: VPS35 steers N-Ras through the cytosol. *Small GTPases* **10**: 20–25. doi:10.1080/21541248.2016.1263380

Zhou M, Wiener H, Su W, Zhou Y, Liot C, Ahearn I, Hancock JF, Philips MR. 2016. VPS35 binds farnesylated N-Ras in the cytosol to regulate N-Ras trafficking. *J Cell Biol* **214**: 445–458. doi:10.1083/jcb.201604061

Cite this article as *Cold Spring Harb Perspect Med* doi: 10.1101/cshperspect.a041572

Parallels in Canonical Developmental Signaling Pathways between Normal Development and the Tumor Microenvironment

Julia Segal, James Cronk, Brendan Ball, Greta Forbes, Kailey Jackett, Kathy Li, Alondra Martinez Osorno, Emily San Andres Montalvan, Alice Browne, Jessica Lake, and Rosandra N. Kaplan

Pediatric Oncology Branch, Center for Cancer Research, National Cancer Institute, National Institutes of Health, Bethesda, Maryland 20892, USA

Correspondence: rosie.kaplan@nih.gov

The tumor microenvironment (TME) is comprised of both cellular and stromal elements and plays an essential role in the growth, survival, and dissemination of malignancies. The TME is an organized program that develops with a growing tumor, using many processes involved in normal tissue development. In multiple solid tumors, developmental pathways are used to recruit immunosuppressive cells, including immunosuppressive monocytes and neutrophils, tumor-associated macrophages, and regulatory T cells to block the antitumor immune response. In addition, stromal cells sustain tumor growth via trophic support, angiogenesis, repair mechanisms, and associated immunosuppression, driven, at least in part, by canonical developmental signaling pathways. The microenvironmental ecosystem shapes tumor progression from its earliest inception by modulating important programs that dictate tumor behavior, necessitating further consideration when studying the developmental origins of malignancy. Here, we review the role of developmental pathways in the formation and modulation of the TME in pediatric and adult solid tumors, including Wnt, Notch, Hippo, Hedgehog, TGF-β, BMP, SOX, and OCT.

INTRODUCTION TO THE TUMOR MICROENVIRONMENT

The tumor microenvironment (TME) is a complex and dynamic immunosuppressive niche comprising a diverse array of cell types and matrix, characterized by altered signaling pathways supporting cancer development and progression. Across multiple tumor types, programs that regulate TME development are similar to those that regulate normal embryologic organ development, tissue homeostasis and repair, and stem cell niches (Quail and Joyce 2013; Murgai et al. 2015; Dempke et al. 2017).

The highly coordinated and specialized processes in normal tissue settings are altered as malignant tumor tissue adapts for survival. These microenvironmental alterations can shield the tumor tissue from normalizing cues that may otherwise halt tumor progression, culminating in a TME that results in altered homeostasis and an intensely immunosuppressive environment tailored to promote immune escape for the continued growth and survival of malignant cells (Quail and Joyce 2013; Anderson and Simon 2020). Disruption of several developmental pathways, which are key mediators of complex cellular processes, including proliferation, cell patterning, apoptosis, quiescence, migration, angiogenesis, inflammation, and plasticity, plays a prominent role in TME establishment and persistence. Microenvironmental regulation is integral to cancer initiation, progression, and spread, known as metastasis. The pre-metastatic niche (PreMN) is formed in distant tissue sites under the influence of tumor-derived factors, leading to the dysregulation of multiple pathways that promote the establishment and survival of metastatic cells (Peinado et al. 2017). The PreMN is integral to the metastatic process and is marked by the coordinated establishment of a unique microenvironment featuring extracellular matrix (ECM) remodeling and stromal cell activation, as well as the recruitment of bone marrow–derived myeloid cells that are important mediators of immune suppression. The PreMN results from alterations in the activation of tissue-resident niches distant from the primary tumor site, which is reminiscent of a tissue repair program and highlights the systemic nature of cancer. Even with localized tumors, developmental processes in multiple organ systems are impacted, including the bone marrow, lymphatics, vasculature, and peripheral nervous system. The components of the TME that originate at the time of tumorigenesis and evolve during cancer progression ultimately link back to highly preserved developmental pathways. The TME is highly diverse and tumor-specific; however, there are commonalities across tumor types. In this review, we discuss the key components of the TME in their developmental context and highlight their essential role in cancer development and progression across different tumor types.

FEATURES OF THE TUMOR MICROENVIRONMENT

The TME includes immune cells, neovasculature, activated stromal cells, and ECM, in addition to clonal populations of tumor cells with variable progenitor capacity (Wessel and Kaplan 2022). Below, and in Figure 1, we review important elements of the TME, and their roles in cancer based on their functions in developmental biology.

Phenotypic States of the TME Cells

The increased plasticity of TME cells contributes to the protumorigenic state. Epithelial-to-mesenchymal transition (EMT) is a dynamic cellular process fundamental to both organ development and cancer biology. During EMT, epithelial cells undergo a reversible transformation, shedding epithelial markers and acquiring mesenchymal traits, functioning like mesenchymal stromal cells (MSCs). However, it has now been shown that mesenchymal cells including pericytes and endothelial cells can undergo a similar process to acquire an MSC phenotype known as EndoMT (endothelial-to-mesenchymal transition). EMT and EndoMT prevent cell apoptosis and senescence and maintain stemness and plasticity, allowing for migration and subsequent differentiation into different cell phenotypes. In development, EMT is seen in gastrulation, neural crest formation, organogenesis, and tissue damage repair (Paolillo and Schinelli 2019; Scott et al. 2019; Frey et al. 2020).

The features of EMT/EndoMT allow for the tumor cell and TME cell plasticity, migration, invasion, and immune evasion pivotal to the cancer state. EMT/EndoMT results in the up-regulation of mesenchymal and cancer-associated fibroblast (CAF) markers, enhancing the tumor cells' survival capabilities and invasive potential. Disseminated tumor cells (DTCs) metastasize via the vasculature or lymphatic systems, all while exhibiting resistance to apoptosis and phenotypic flexibility for immune evasion and adaptation to a new environment (Quail and Joyce 2013; Paolillo and Schinelli 2019). The reverse counterpart of EMT, mesenchymal-to-epithelial transition

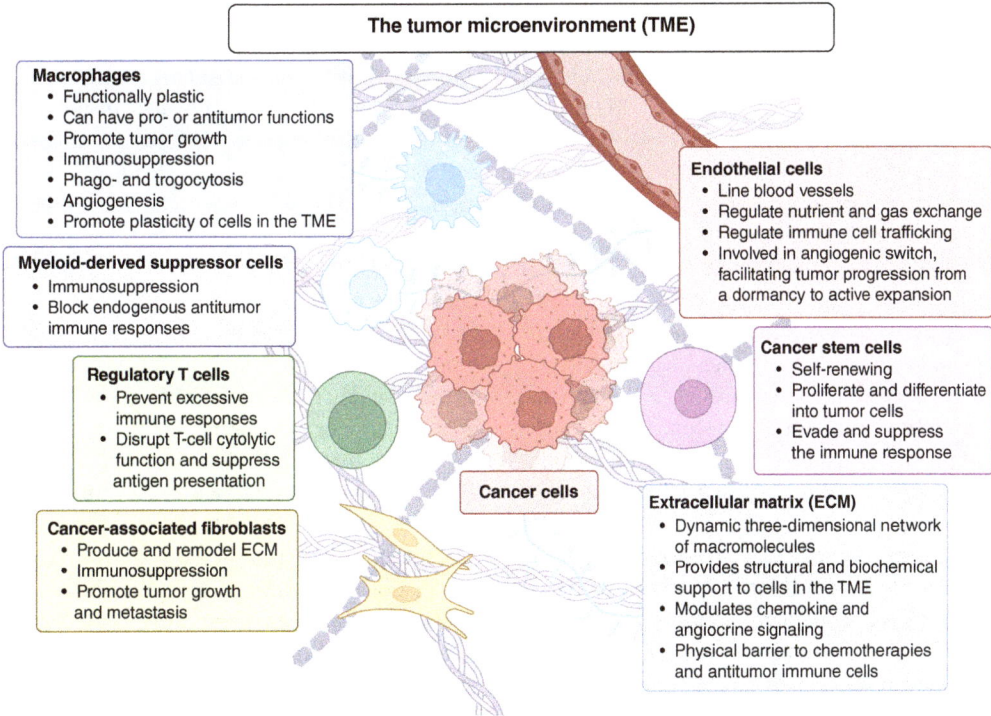

Figure 1. The tumor microenvironment (TME). The TME is comprised of myriad components, each with an important role in cancer initiation and progression. The altered homeostasis in the TME created by the interplay of immune cells, neovasculature, stromal cells, extracellular matrix, as well as clonal cancer cell populations facilitates the protumorigenic state. (Figure created with BioRender.com.)

(MET), also facilitates metastasis by allowing these circulating tumor cells to regain characteristics necessary to effectively colonize the PreMN (Paolillo and Schinelli 2019). Although the precise mechanisms and implications of EMT and MET in both TME/PreMN formation and cancer metastasis are subjects of ongoing research, their significant impact on disease progression is increasingly recognized.

Cell Types and Cell–Cell Interactions of the TME

Macrophages are functionally plastic white blood cells with a multitude of functions including tissue repair, angiogenesis, regulation of adaptive immune cells, and phagocytosis of organisms or debris. They sense and respond to environmental cues along the oversimplified spectrum of "classically activated M1" macro-phages, which are proinflammatory and antitumorigenic, and "alternatively activated M2" macrophages, which are immunosuppressive and protumorigenic (Quail and Joyce 2013). This classification is derived from in vitro skewing paradigms that do not accurately represent the complexity of macrophages within the in vivo TME (Cendrowicz et al. 2021). Tumor-associated macrophages (TAMs) regulate the TME, accumulate under hypoxic conditions, facilitate tumor cell invasion, and secrete proteases to support tumor progression and therapeutic resistance (Quail and Joyce 2013). Macrophages are highly abundant in many pediatric cancers including osteosarcoma, neuroblastoma, and brain tumors such as high-grade gliomas and medulloblastoma (Vakkila et al. 2006; Lin et al. 2018; Koo et al. 2020; Terry et al. 2020; Wessel and Kaplan 2022). Several protumorigenic and immunosuppressive populations of

TAMs have been observed in disparate solid tumors including cervical, colorectal cancer, sarcomas, carcinomas, and gliomas, characterized by markers such as osteopontin (*SPP1*), complement component 1q (*C1Q*), and triggering receptor expressed on myeloid cells 2 (*TREM2*), among others (Katzenelenbogen et al. 2020; Molgora et al. 2020; Li et al. 2021; Qi et al. 2022).

TAMs facilitate cancer initiation and progression by regulating immune response, modulating angiogenesis, and promoting the development and plasticity of other TME-associated cells (Sica et al. 2012; Shen et al. 2021). TAMs limit antitumor immune responses through multiple mechanisms, notably their expression of programmed death-ligand 1 (PD-L1) to signal on programmed death cell protein 1 (PD-1) receptors on cytotoxic T cells, depleting nutrients like arginine, inhibition of dendritic cell (DC) antigen presentation activity, and secretion of direct immunosuppressive ligands such as transforming growth factor β (TGF-β) and interleukin 10 (IL-10) (Mantovani et al. 2022). In addition, TAMs promote tumor growth via the production of growth factors, such as insulin-like growth factor 1 (IGF-1) and vascular endothelial growth factors (VEGFs), and ECM remodeling through matrix metalloproteases (MMPs) (Anderson and Simon 2020; Pan et al. 2020; Mantovani et al. 2022).

Macrophages in Development

A functional parallel exists between macrophages during normal development and TAMs during tumor growth. Animal models lacking colony-stimulating factor 1 receptor (CSF1R) or its ligand, colony-stimulating factor 1 (CSF1), both have a severe deficit of macrophages (Yoshida et al. 1990; Dai et al. 2002; Li et al. 2006). In the absence of macrophages, normal tissue development is disrupted via multiple failed developmental processes, including the lack of macrophage-derived trophic factors and failure to clear apoptotic cells, leading to organ malformations, bone marrow failure, osteopetrosis, global growth deficit, and premature death (Yoshida et al. 1990; Dai et al. 2002; Li et al. 2006). Similarly, TAMs promote the growth of the "develop-ing" tumor by secreting angiogenic factors, growth factors, and MMPs, which restructure the extracellular environment and protect the developing tumor from inflammatory insult via direct immunosuppressive functions (Anderson and Simon 2020; Pan et al. 2020; Mantovani et al. 2022). These homeostatic functions are also seen postdevelopmentally in tissue-resident macrophages throughout the body, which maintain homeostasis through many of these same processes critical to both development and tumorigenesis (Mantovani et al. 2022; Mass et al. 2023). The functions of macrophages as stewards of tissue growth, development, and homeostasis appear to be a strongly conserved property. TAMs execute normal macrophage functions, such as wound healing and immune suppression, promoting growth and preventing immunologic elimination of the tumor. Given the fundamental role of macrophages in tumorigenesis, there is active interest in the development of therapeutic strategies designed to inhibit macrophages from performing their homeostatic, protumorigenic functions, and to redirect them to harness their proinflammatory/antitumorigenic functions.

Myeloid-derived suppressor cells (MDSCs) are a heterogenous population of immature monocytes and/or granulocytes that further support the tumorigenic phenotype of the TME by suppressing antitumor T-cell function. These cells are produced during emergency hematopoiesis as well as in response to chronic stress and inflammation (Hegde et al. 2021). MDSCs are frequently identified in cancer due to their role in modulation of the immune response, but have also been well characterized in the setting of pregnancy and during resolution of injury/inflammation, such as postinfectious arthropathy (Pawelec et al. 2019; Veglia et al. 2021; Shibata et al. 2022). In relation to development, MDSCs can be considered an offshoot of normal myelopoiesis, thought to be triggered by a "two-signal" process. An immature, developing myeloid cell exposed to a canonical myeloid growth factor such as CSF1, granulocyte colony-stimulating factor (GCSF), or granulocyte-macrophage colony-stimulating factor (GMCSF), followed by a second signal (usually an inflammatory cytokine) activating the nuclear factor-κB (NF-κB) pathway, leads to the

Cite this article as *Cold Spring Harb Perspect Med* doi: 10.1101/cshperspect.a041609

development of the immunosuppressive monocytes or granulocytes (Hegde et al. 2021). In the TME, MDSCs contribute significantly to immune evasion by disrupting normal antigen presentation and effector T-cell functions through nutrient depletion, production of reactive oxygen species, altered macrophage polarization, promotion of tumor angiogenesis, and ECM remodeling (Gabrilovich et al. 2012; Quail and Joyce 2013). MDSCs have been demonstrated as prognostic circulating biomarkers of cancer progression and overall survival (Kalathil and Thanavala 2021).

T regulatory (Treg) cells are a heterogeneous group of CD4$^+$ T cells that support an immune evasion phenotype. Under normal conditions, these cells maintain complex regulation of lymphocyte homeostasis by preventing excessive and inappropriate immune responses. In development, Treg cells prevent immune activation during organ development and help in limiting the proliferation of postthymic self-reactive T cells, as several autoimmune diseases are associated with the loss of this regulatory population (Leventhal et al. 2016; Goswami et al. 2022). In contrast, cancer-associated Treg cells disrupt T-cell cytolytic function against tumor cells and suppress antigen presentation, which would otherwise activate an antitumor immune response (Quail and Joyce 2013).

Cancer stem cells (CSCs) are a subpopulation of cells within the TME possessing stem cell–like characteristics, including the ability to self-renew and differentiate into various tumor cell types (Wu et al. 2023). While CSCs can evade and suppress the immune response, the immune system can also select for CSCs more adept at immune evasion, leading to a more aggressive cancer phenotype (Ferguson et al. 2021). These differences in cell behavior are brought about by alterations in developmental signaling pathways regulating cell renewal, differentiation, and survival, including the Notch, Wingless-related integration site (WNT), Hedgehog (HH), and Hippo pathways, leading to further immune evasion and a skewed inflammatory milieu (Clara et al. 2020). Understanding these relationships is critical for the development of effective cancer immunotherapies, which aim to both target CSCs and reverse TME immunosuppression.

CAFs are a heterogeneous cell population functionally distinct from normal fibroblasts and abundant in the TME. CAFs originate from resident and recruited stromal cells, such as MSCs, fibroblasts, pericytes, vascular smooth cells, Schwann cells, endothelial cells, epithelial cells, and adipocytes. In some cancers, tumor cells that have undergone EMT can generate their own population of CAFs that share the same genomic aberrations as the tumor (Sahai et al. 2020). CAFs are activated by several tumor-derived factors and serve as the chief architects of the TME, contributing to ECM remodeling and supporting tumor growth and metastasis through secretion of growth factors, as well as VEGF, supporting angiogenesis (Quail and Joyce 2013). Further, CAFs can effectively cross talk and interact with TAMs in the bone marrow and tumor site, promoting cancer progression (Hashimoto et al. 2016).

Fibroblasts in Development

The role of CAFs in the TME echoes the role of fibroblasts in normal physiological processes during the development of connective tissues and organs during embryogenesis. During development, fibroblasts produce ECM and release signaling molecules guiding the migration, growth, and differentiation of various cell types, paralleling the actions of CAFs in tumors (Quail and Joyce 2013). The regulatory pathways directing organ formation, such as those mediated by TGF-β, Wnt, and Hedgehog signals, are similarly exploited by CAFs to bolster tumor progression. In the process of wound healing, fibroblasts assume an activated state akin to that of CAFs to mend the injured tissue. These cells proliferate and migrate to the injury site, synthesize ECM proteins, and emit growth factors to repair the wound. Myofibroblasts, a specialized but typically transient form of activated fibroblasts prominent in wound repair, share several markers with CAFs, such as α-smooth muscle actin (α-SMA) expression (Yang et al. 2023). CAFs can originate from various cell types, including MSCs, pericytes, or other stromal cells, through the influence of tumor-derived factors. This transformation alters both their function and their impact on neighboring cells. CAFs exhibit significant plasticity and do not differentiate uniformly (Sahai et al. 2020). They can adopt various phenotypes, including reverting to less differentiated states resembling MSCs or pericytes. This phenotypic diversity modulates their function and influence on surrounding cells (Cords et al. 2024). The phenotypic plasticity of fibroblasts during development and CAF activation involves enhanced migratory and secretory functions. Just as CAFs affect the behavior of cancer cells and the immune landscape within the TME, activated fibroblasts in normal development direct the organization of surrounding cells to create structured tissues.

ECM remodeling releases growth factors, cytokines, and chemokines that regulate immune cell movement and effector function. Specific populations of CAFs activated on the leading edges of the tumor and in metastases communicate most closely with immune populations and can take on an inflammatory CAF (iCAF) phenotype (Yang et al. 2023). These populations support MDSCs and TAMs to promote immune suppression and dysfunctional immune activation with tolerogenic antigen presentation to induce tolerance of T cells and alter DC function. In addition, CAFs have been shown to induce a specific type of lipid-associated TAM, STAB1 + TREM2^{high}, which supports tumor growth in triple-negative breast cancer (TNBC) models (Timperi et al. 2022). This example highlights the protumorigenic stromal-immune cross talk.

The physiological activation of fibroblasts during development and tissue repair is regulated and temporary, unlike the pathological activation of CAFs in cancer. This contributes to a persistently deregulated environment fostering tumor growth and metastasis. Studying the parallels between the developmental biology of fibroblasts and the pathobiology of CAFs can shed light on disease mechanisms and identify new therapeutic targets, further bridging the gap between developmental processes and cancer biology.

Endothelial cells form the lining of blood vessels and regulate nutrient and gas exchange to tissues (Lugano et al. 2020). These cells are key regulators of the TME and angiogenesis. This process is regulated by many canonical developmental pathways such as Wnt, Notch, Hedgehog, TGF-β, and bone morphogenetic protein (BMP), which signal through proangiogenic factors such as VEGF, FGF, PDGF, and others (Olsen et al. 2017; Akil et al. 2021; Del Gaudio et al. 2022).

Endothelial Cells during Development

The pattern of blood vessels in the developing embryo starts from a subset of endothelial cells known as tip and stalk cells. These cells work in concert to generate a complex network of functional blood vessels. Surrounding support cells, such as vascular smooth muscle cells and pericytes, regulate vessel tone, and permeability. Endothelial cells also communicate with Schwann cells that surround the nerves that run parallel to blood vessels. These processes are co-opted in cancer; the TME is rich in growth factors such as VEGF that can result in the reappearance of tip and stalk cells in tumors, mimicking the angiogenic processes that normally occur during organogenesis. These parallels have long been understood from older studies that transplanted tumor cells into a developing embryo. The tumor cells subsequently developed into functional components of the organism, and no longer demonstrated tumor features despite persistent genetic alterations. Notably, the tumor cell–derived tissue formed organ-appropriate vasculature, in contrast to the abnormal vasculature of the parent tumor from which the tumor cells were derived (Mintz and Illmensee 1975; Dolberg and Bissell 1984; Stoker et al. 1990).

In the developing TME, developmental angiogenic signaling pathways are often reactivated to support the formation of tumor-associated blood vessels in a process known as the angiogenic switch, which can disrupt tumor dormancy and stimulate growth. The angiogenic switch activates the formation of new blood vessels, fueling the tumor with nutrients and oxygen and facilitating its progression from dormant to growing lesion (Shaked et al. 2014). Tumor-associated endothelial cells also regulate the flux of circulating immune populations into

Cite this article as *Cold Spring Harb Perspect Med* doi: 10.1101/cshperspect.a041609

the TME (Nagl et al. 2020). In addition, EndoMT fosters an important source of CAFs within the TME (Quail and Joyce 2013). Taken together, endothelial cell populations, under the influence of hallmark developmental pathways, are instrumental in developing the TME.

Noncellular Components of the TME

The extracellular matrix is a dynamic three-dimensional network of macromolecules that provides structural support and communicates environmental cues to cells. Each tissue has a unique ECM composition primarily composed of fibrous proteins such as collagen, elastin, and fibronectin; proteoglycans such as versican; and glycosaminoglycan such as hyaluronic acid (HA). Far from being a passive scaffold, the ECM plays crucial roles in maintaining tissue architecture, regulating cell adhesion and migration, orchestrating differentiation, sequestering and presenting growth factors, facilitating mechanotransduction, and mediating cell signaling. This versatile structure is integral to tissue homeostasis and undergoes constant remodeling in response to physiological and pathological cues (Frantz et al. 2010; Yue 2014).

Dysregulation and remodeling of the ECM can influence tumor initiation and progression through dynamic changes in abundance, composition, and architecture. These alterations result in shifts in both chemical and mechanical cues, profoundly impacting cell–ECM interactions and cellular and tissue functions. The tumor ECM is denser and stiffer than normal tissue ECM, primarily due to increased deposition and cross-linking of collagen, fibronectin, and HA, supporting a protumorigenic environment (Sleeboom et al. 2024).

The establishment of the tumor-derived ECM is intricately linked to developmental pathways. Tumors co-opt developmental pathways normally active during tissue growth and wound healing to facilitate the development of tumor-derived ECM. For instance, signaling molecules like TGF-β, Wnt, and Hedgehog, which are crucial in embryogenesis, also play significant roles in modulating the ECM within the TME. These pathways can influence the dep-

osition of ECM components, alter their composition, and change the behavior of stromal cells, such as fibroblasts, which become activated and contribute to ECM remodeling in the TME (Bonnans et al. 2014).

Alterations of ECM promote conditions essential for tumor establishment and growth, such as angiogenesis and immune evasion. ECM deposition and remodeling can create a dense fibrotic rim encapsulating a tumor or metastatic lesion that can contribute to hypoxic conditions and high interstitial pressure, as well as acting as a physical barrier to therapies and activated immune cells (Quail and Joyce 2013; Wessel and Kaplan 2022). This remodeling of the tumor-derived ECM impacts the dynamic interactions between tumor cells and neighboring stromal cells, including fibroblasts, immune cells, and endothelial cells, thus influencing the signaling pathways involved in tumor progression and metastasis.

The distribution of the various cell types present throughout the TME is not homogenous. Immunosuppressive cells like TAMs, CAFs, MDSCs, and myeloid progenitor cells are often concentrated at the leading edge of primary tumors, thus blocking infiltration of the host immune system and therapies to the TME (Quail and Joyce 2013; Ligon et al. 2021; Wessel and Kaplan 2022). The ECM provides structural and functional support to cells within the TME, and its components are key to tissue architecture and cellular communication.

ECM during Development

The ECM actively contributes to tissue development and homeostasis, safeguarding normal physiological function. Furthermore, the ECM regulates essential physiological activities by binding growth factors, engaging with cell-surface receptors, triggering signal transduction, and modulating gene expression through mechanoreceptor signaling. The ECM serves as a reservoir for bioactive molecules and undergoes constant remodeling and turnover (Frantz et al. 2010).

In response to increased mechanical load, such as in aging or injury, ECM remodeling contributes to tissue integrity and repair by replacing damaged ECM components with newly synthesized ones (Hettinger et al. 2022). Additionally, ECM remodeling can expose or conceal signaling molecules, influencing cell signaling pathways and allowing adaptation to different environmental changes. This process enables cells to adapt in response to extracellular signals, impacting cell adhesion, migration, proliferation, and differentiation (Kular et al. 2014). Finally, ECM remodeling is instrumental for effective oxygen delivery by facilitating angiogenesis and modifying the tissue matrix to enhance diffusion processes, thereby supporting overall function (Gilkes et al. 2014).

Pre-Metastatic Niche

The PreMN is a microenvironment distinct and distant from the primary TME, which shapes metastatic outgrowth. It comprises a series of temporal events, including stromal activation, ECM remodeling, vascular permeability, and myeloid cell infiltration (Murgai et al. 2015). The bone marrow microenvironment responds to tumor-derived factors by shifting to protumorigenic myelopoiesis, which ultimately leads to systemic changes to the host immune system (Giles et al. 2016; LaMarche et al. 2024). The PreMN supports DTCs in their invasion, survival, and ultimate metastatic outgrowth (Peinado et al. 2017; Kaczanowska and Kaplan 2020).

Introduction to Developmental Pathways

In addition to the developmental origins of TME, the tumor and the TME are regulated by developmental signaling programs. Developmental signaling pathways such as Wnt, Notch, Hippo, Hedgehog, BMP, TGF-β, SRY-related high mobility group box (Sox), and octamer-binding transcription factor (Oct) are highly preserved gene networks present in almost all multicellular organisms. These act in tandem for embryonic development, regulating cell migration, differentiation, proliferation, and homeostasis. Mutations in one pathway can have downstream effects in other related pathways, resulting in dysmorphia or malignancy

(Dempke et al. 2017). Many features of malignancy, such as uncontrolled proliferation, tissue remodeling, metastasis, and neoangiogenesis, have their roots in normal developmental processes regulated by this complex network of signaling pathways.

Much focus has been placed on the role of these dysregulated pathways in the cell of origin in pediatric cancers—examples include muscle and mesenchymal progenitor cells in rhabdomyosarcoma (Wei et al. 2022), neural crest progenitor cells in neuroblastoma, and neural progenitors in medulloblastoma (Marshall et al. 2014). In development, multiple heterogenous cell types often relay autocrine and paracrine signals that complete a developmental pathway. The role of these developmental signaling pathways in cancer has introduced the possibility of new therapies. However, caution must be taken when targeting pathways that have such far-reaching biological activity (Dempke et al. 2017). Below, we discuss the canonical developmental signaling pathways (Fig. 2; Table 1) that communicate between the tumor and the TME to facilitate cancer initiation and progression. For each pathway, we have introduced its normal role in development as well as the common features of dysregulation broadly in the setting of cancer and in specific cancer types, which are highlighted.

Wnt

Wnt is a secreted growth factor that regulates proliferation, differentiation, and cell function across a spectrum of cell types (Steinhart and Angers 2018; Choudhary et al. 2024). The Wnt signaling axis is indispensable for development, including body axis patterning and cell migration, proliferation, and tissue homeostasis, and can be co-opted in pediatric cancers to facilitate a protumorigenic microenvironment. Activation of canonical Wnt signaling centers around the posttranslational regulation of β-catenin. In the absence of Wnt activation, β-catenin is degraded by glycogen synthase kinase 3 (GSK3). The Wnt ligand binds its receptor, LRP5/6, and coreceptor, Frizzled, to elicit downstream signaling. Wnt receptor binding leads to the re-

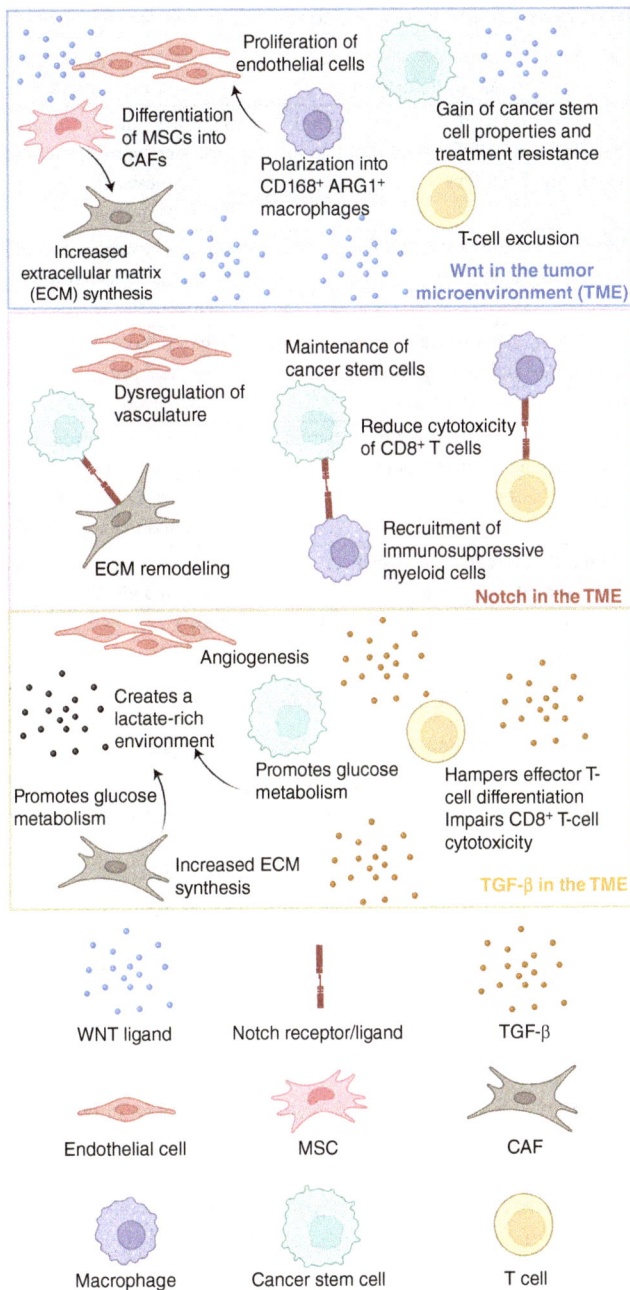

Figure 2. Key developmental signaling pathways and their protumorigenic roles in the tumor microenvironment (TME). Wnt, Notch, and TGFB play critical roles in the TME that can be tumor promoting and tumor suppressing. Here, we specifically highlight the protumorigenic roles of these pathways in the TME across a broad range of tumor types. Wnt signaling in the TME contributes to macrophage polarization, endothelial cell proliferation, differentiation of mesenchymal stromal cells into cancer-associated fibroblasts, extracellular matrix (ECM) synthesis, T-cell exclusion, and tumor cell stemness and resistance to treatment. Notch signaling has been implicated in the recruitment of immunosuppressive myeloid cells, vascular dysregulation, altered metabolism, ECM remodeling, reduced T-cell cytotoxicity, and maintenance of cancer stem cells. TGFB contributes to metabolic rewiring of the TME, angiogenesis, ECM synthesis, and T-cell hampering. (Figure created with BioRender.com.)

Table 1. Developmental pathways during normal development and cancer

Developmental pathways	Normal development	Cancer initiation and progression
Wnt	Regulates proliferation, differentiation, stemness, and cell function across a spectrum of cell types	Supports tumor proliferation and metastasis, promotes immunosuppression and angiogenesis, and induces cancer-associated fibroblast formation
Notch	Involved in cellular differentiation, development, and stemness	Supports cancer stem cells (CSCs), modulates stemness, immunosuppression, and angiogenesis
Hippo	Regulates organ size, cell proliferation, and responds to cell density and tissue regeneration requirements	Drives uncontrolled proliferation, invasiveness, immunosuppression, and extracellular matrix (ECM) remodeling
Hedgehog	Regulates cell migration, differentiation, proliferation, stemness apoptosis, and organ development	Supports CSCs, promotes immunosuppression by suppressing $CD8^+$ T-cell recruitment, promotes Treg differentiation, and tumor-associated macrophage (TAM) polarization
Transforming growth factor β	Stem cell differentiation, T-cell development, wound healing, angiogenesis, and other homeostatic programs	Modulation of immune cells, ECM, angiogenesis, and metabolism in the tumor microenvironment (TME)
Bone morphogenetic protein	Controls osteogenesis, cell growth, apoptosis, differentiation, tissue homeostasis, and immune cell development	Increases motility, stemness, and resistance to apoptosis of cancer cells and supports angiogenesis
Sox/Oct	Regulate pluripotency and maintains stemness	Dedifferentiation and induction of CSCs under hypoxic conditions, epithelial-to-mesenchymal transition (EMT)

Wnt, Notch, Hippo, Hedgehog, BMP, TGF-β, Sox, and Oct are highly preserved developmental signaling pathways that are instrumental in normal developmental processes such as stemness, cellular proliferation, differentiation, and migration, as well as immune system and organ development and angiogenesis. In the setting of cancer, these pathways are co-opted to support the development of malignancy. (Table created with BioRender.com.)

cruitment of the scaffold protein, Disheveled, and the Axin–GSK3 complex. This recruitment inhibits the degradation of β-catenin, enabling its translocation to the nucleus. Once in the nucleus, β-catenin serves as a transcriptional coactivator for Lef/Tcf transcription factors, encouraging the transcription of context-dependent Wnt (Niehrs 2012). Noncanonical Wnt signaling is independent of β-catenin. It modulates Ca^{2+} levels and calmodulin to activate the transcription of nuclear factor of activated T-cell (NFAT)-driven Wnt target genes through the planar cell polarity pathway, which enables the transcription of c-jun and jun-driven Wnt target genes (Niehrs 2012; Ackers and Malgor 2018). Once in the nucleus, β-catenin serves as a transcriptional coactivator for Lef/Tcf transcription factors, encouraging the transcription of context-dependent Wnt responsive genes such as myc, cyclin D, MMPs, CD44, VEGFs, and ID in stem cells and tumor cells, Gremlin in fibroblasts, and IL-8 in endothelial cells (Niehrs 2012).

In development, Wnt signaling promotes cell proliferation and maintains stemness (Steinhart and Angers 2018). In the developing tumor niche, TME components facilitate the formation of a Wnt-rich ecosystem, which supports the proliferation, invasion, and metastasis of tumor cells across a broad range of cancers (Rogers et al. 2012; Patel et al. 2019; Nobre et al. 2020; Xu et al. 2023). For example, in the progression of colorectal adenoma to carcinoma, infiltrating macrophages express increasingly high levels of Wnt. In addition, stromal populations, such as CAFs, secrete Wnt ligands, promoting the Wnt signaling cascade in colorectal cancer cells (Aizawa et al. 2019). Wnt signaling can contribute to increased proliferation, the development of CSCs, and resistance to chemotherapy in medulloblastoma, neuroblastoma, rhabdomyosarcoma, and colorectal cancer (Yuan et al. 2020; Choudhary et al. 2024). Further, the tumor cells of pediatric cancers such as medulloblastoma, acute lymphoblastic leukemia, and retinoblastoma, which harbor multiple genetic aberrations in the Wnt signaling pathway, can be a source of Wnt ligand to their surrounding environment (Bao et al. 2013).

Wnt signaling has also been implicated in dampening antitumor immune response (Patel et al. 2019). The secretion of Wnt by tumor cells may stimulate the polarization of macrophages into CD68[+] ARG1[+] macrophages through the canonical Wnt signaling pathway (Tigue et al. 2023). In addition, the activation of the Wnt pathway correlates with immune exclusion across human cancers and has been implicated in resistance to checkpoint inhibitors (Luke et al. 2019; Muto et al. 2023). These findings suggest that the Wnt pathway is a potent promoter of immunosuppression in the TME. Beyond its role in immune response, in colorectal cancer Wnt signaling involving tumor-secreted factors such as Wnt2 contributes to the activation and functional reprogramming of fibroblasts into CAFs. Similarly, in cervical cancer, tumor-secreted Wnt2B plays a role in activating fibroblasts to promote tumor progression (Aizawa et al. 2019; Liang et al. 2021). These cells produce increased ECM deposition and take on characteristics of regenerative MSCs that, together with CAFs, are fundamental components of the TME and essential to supporting cancer progression (Borriello et al. 2017).

Moreover, Wnt signaling supports angiogenesis in the developing TME by enabling nutrient trafficking to the tumor across tumor types (Olsen et al. 2017). Specifically, the Wnt ligand produced by macrophages within the TME activates canonical Wnt signaling in vascular endothelial cells. As a result, vascular endothelial cells proliferate, promoting angiogenesis (Lobov et al. 2005). In summary, the Wnt signaling axis can support the development of pediatric TME by facilitating tumor cell proliferation and stemness and creating a favorable environment in which the tumor cells can reside.

Notch

The Notch signaling pathway is critical to cellular differentiation and development. The TME of pediatric tumors can exploit these behaviors (Bray 2006; Rota et al. 2012; Uluçkan et al. 2015; LaFoya et al. 2016). Notch signaling occurs through juxtacrine signaling between adjacent cells (Bray 2006). The Notch ligand and its receptor sit on the membrane of distinct adjacent cells. When the Notch receptor binds its ligand on a neighboring cell, it cleaves and enters the nucleus. In the nucleus, Notch binds the transcriptional repressor CSL, converts CSL to an activator of transcription, and promotes the transcription of Notch target genes, which regulate cellular differentiation and stemness (Kopan and Ilagan 2009; Gozlan and Sprinzak 2023).

Evidence suggests that this developmental pathway is dysregulated in pediatric cancers such as T-cell acute lymphoblastic leukemia, rhabdomyosarcoma, and synovial sarcoma (Zweidler-McKay 2008; Rota et al. 2012). Depending on cellular context, notch dysregulation in tumor cells can have both protumorigenic and tumor-suppressing abilities. However, mounting evidence shows that Notch signaling supports the maintenance of CSCs, which are linked to therapeutic resistance (Venkatesh et al. 2018; de Almeida Magalhães et al. 2020; Karthikeyan et al. 2024).

In addition, the Notch signaling pathway is a critical communication track within the developing TME (D'Assoro et al. 2022). Notch signaling encourages myeloid recruitment, induces MDSCs, and facilitates the polarization of macrophages into TAMs (Hossain et al. 2018). Notch signaling in TAMs reduces CD8[+] T-cell cytotoxicity, suggesting that Notch signaling may support an immunosuppressive TME. However, evidence in glioblastoma also suggests that Notch signaling can boost antitumor immune responses by promoting antigen presentation in DCs and the secretion of IFN-γ from natural killer (NK) cells (Li et al. 2023). Thus, Notch has dual roles in modulating immune response in the TME as dictated by the environment.

Beyond modulating the tumor immune compartment, Notch signaling also influences the development of the TME-associated stroma. Notch signaling has been shown to facilitate communication between CAFs and tumor cells in melanoma, but a similar process may be conserved in pediatric cancers (Du et al. 2019). Tumor-associated stromal cells across a range of cancers have heightened levels of Notch, suggesting stromal cell notch signaling may be important for the TME. In addition, Notch signaling in glioblastoma modulates the tumor stroma through ECM remodeling via MMP2 and MMP4 (Bao et al. 2013; Xia et al. 2022).

Notch signaling is critical to normal angiogenesis in a developing embryo, highlighting yet another mechanism through which the TME co-opts developmental processes. Notch signaling plays a major role in endothelial cell development and function. Tip and stalk cells that form new blood vessels and specify their orientation result from Notch pathway activation. Further, Notch target genes in endothelial cells promote vessel stabilization and maturation, including arterial–venous differentiation. However, persistent activation can lead to the dysfunctional vasculature observed in the TME (Akil et al. 2021; Del Gaudio et al. 2022). In summary, Notch signaling modulates hallmark features of the developing TME, such as stemness, immunosuppression, and angiogenesis, suggesting it is a critical modulator of the pediatric TME.

Hippo

The Hippo pathway, which plays critical roles in both normal and cancer development, shares common features in its functions and mechanisms between these two spheres. During development, Hippo regulates organ size and cell proliferation and responds to cell density and tissue regeneration requirements. This signaling pathway primarily involves the key components MST1/2, LATS1/2, YAP, and TAZ (Zhu et al. 2015; Han 2019). Activation of the Hippo pathway consists of a series of phosphorylation events, starting with the phosphorylation of MST1/2 and activation of LATS1/2, which, in turn, inhibits the activity of YAP/TAZ (Oka et al. 2010; Zhu et al. 2015; Han 2019).

Multiple developmentally relevant signaling pathways converge to regulate the Hippo pathway. For example, the Wnt and AMPK pathways inhibit the YAP protein. In addition, the MAPK/ERK and KRAS pathways, which play a role in the cell cycle and tumor initiation, promote the YAP protein (Nussinov et al. 2016; Han 2019). The TAZ protein is also a promotor of the TGF-β pathway, which plays a role in cell proliferation, apoptosis, differentiation, and remodeling of the ECM (Saito et al. 2013; Han 2019). Thus, the Hippo signaling pathway also directs other developmentally relevant signaling pathways that may be co-opted in the setting of the TME (Liu et al. 2024).

In cancer, the Hippo pathway is often altered to drive uncontrolled cell growth, inhibit apoptosis, enhance invasiveness, and facilitate metastasis (Han 2019). Research has shown that increased YAP/TAZ activity is associated with many solid tumors (including multiple carcinomas, sarcomas, and medulloblastoma, among others) and tumorigenesis (Ahmed et al. 2017; Han 2019). YAP and TAZ interact with DNA-binding partners such as the TEAD 1–4 proteins in mammals (Zanconato et al. 2015; Han 2019). They both contain a carboxy-terminal transcription activation domain, and their WW domains interact with PPXY motifs of proteins (Macias et al. 1996; Han 2019).

Interaction of YAP with TEAD can form an oncoprotein that promotes transcription of downstream target genes, such as c-Myc and survivin (Pei et al. 2015). Studies have shown that inhibition of YAP has an antitumor effect in chronic myeloid leukemia (Li et al. 2016; Han 2019). In addition, YAP1 is a key modulator of proliferation in angiosarcomas, rhabdomyosarcomas, and osteosarcomas, and high expression of YAP1 predicts a poor prognosis in both sarcomas and leukemias (Cottini et al. 2014;

Tremblay et al. 2014; Chai et al. 2017; Han 2019).

The Hippo pathway can promote and maintain an immunosuppressive TME. Tregs are maintained by activating MST1/2 via amplification of IL-2R and STAT5 signaling. YAP signaling in CD4$^+$ and CD8$^+$ effector T cells inhibits their activation and cytotoxic function, thus dampening the antitumor adaptive immune response. YAP also directly promotes Treg differentiation via TGF-β signaling to promote forkhead box P3 (FOXP3) expression, a key Treg transcription factor. YAP/TAZ activation in tumor cells promotes immunosuppression by induction of PD-L1 expression and production of cytokines and chemokines, leading to the recruitment of TAMs and MDSCs. In both normal development and cancer, there is significant cross talk and paracrine signaling between epithelial and stromal cells. The Hippo pathway is a key mediator of this signaling, regulating the production of various growth factors, cytokines, and ECM components (Zhou et al. 2018; Liu et al. 2023). Dysregulation of the Hippo pathway in either stromal or epithelial cells can disrupt this signaling balance, contributing to cancer progression (Yoshida 2020).

YAP and TAZ can also directly affect the composition and organization of the ECM. When activated, they can promote the expression of genes involved in ECM remodeling and fibrosis, resulting in increased tissue stiffness and influencing mechanosignaling (Dupont et al. 2011; Noguchi et al. 2018; Huang et al. 2022). Overall, the Hippo pathway promotes an immunosuppressive TME via intrinsic suppression of an antitumor, adaptive immune response through MST1/2 and YAP signaling in T cells, and promotion of the myeloid-based immunosuppressive milieu by tumor cells (Zhou et al. 2018; Liu et al. 2023).

Hedgehog

Hedgehog (HH) is a crucial pathway for stem and progenitor cell homeostasis and function during development. HH signaling is complex. The HH family of secreted proteins are formed through HH lipidation and involves Patched 1 (Ptch1), Smoothened (SMO), GLI proteins, and primary cilia. The HH pathway regulates vital processes such as cellular migration, differentiation, proliferation, apoptosis, and organ development. The HH signaling pathway is responsible for the regulation and morphogenesis of various organs during embryogenesis (Dempke et al. 2017). It also plays a role in cell fate determination, EMT, and cell motility and adhesion (Espinosa-Sánchez et al. 2020).

In mammals, three spatiotemporally confined ligands have been identified with HH, including Sonic hedgehog (Shh), Indian hedgehog (Ihh), and Desert hedgehog (Dhh) (Pak and Segal 2016). The primary receptors of the HH cascade are Ptch1, a 12-pass transmembrane protein; SMO, a seven-pass transmembrane G-protein-coupled receptor (GPCR); and three zinc finger GLI transcription factors (GLI1, GLI2, GLI3) as the final mediators of the transcriptional response of HH signaling.

Dysregulation of HH can result in the promotion of tumor growth and maintenance as well as therapeutic resistance (Pak and Segal 2016; Espinosa-Sánchez et al. 2020). The HH pathway is dysregulated in pancreatic, gastric, prostate, and esophageal cancers (Espinosa-Sánchez et al. 2020). Without a ligand, Ptch1 will block HH by inhibiting SMO (Pak and Segal 2016). This allows low levels of Kif7, Suppressor of Fused (SUFU), and full-length glioma-associated oncogene (GliFL) to enter the primary cilium, resulting in the conversion of GliFL turning into a repressor form (GliR) after phosphorylation by PKA, GSK3, and CK1. GliR inhibits transcription of Hedgehog target genes such as Gli1 and Ptch (Pak and Segal 2016; Dempke et al. 2017). Further, there is also noncanonical HH activation via ligand-dependent SMO activation independent of GLI activation and results in TGF-β, EGFR Ras, and PI3K-Akt-mTOR pathway signaling, which supports the crucial role of HH signaling in cancer cell formation (Javelaud et al. 2012; Bonilla et al. 2016).

With regard to the TME, the HH pathway is commonly altered to support the survival of CSCs (Clara et al. 2020). The dysregulated HH pathway also supports other TME cell types, such as tumor cells, stromal cells, endothelial cells, TAMs, and myofibroblasts. Tumor-derived Shh can promote TAM polarization, upregulating M2-associated genes. Disruption of Hedgehog signaling reduces M2 skew and tumor growth, concurrently suppressing antitumor CD8[+] T-cell recruitment by down-regulating CXCL9 and CXCL10 (Petty et al. 2019). HH inhibition in orthotopic mouse models of mammary tumors reduced the immunosuppressive skew of TAMs (Hinshaw et al. 2021). Tumor-derived Shh has also been shown to induce PD-L1 expression in TAMs directly (Petty et al. 2021). Analysis of tissues from oral squamous cell carcinoma demonstrated expression of HH ligands Shh and Ihh in endothelial cells and Ihh in TAMs (Valverde Lde et al. 2016). This finding suggests that endothelial–macrophage cross talk plays a role in promoting tumor angiogenesis. HH signaling has also been shown to promote Treg differentiation, and inhibition of HH promotes Treg transdifferentiation (Hinshaw et al. 2023). There is growing interest in targeting this pathway to limit CSCs, myeloid-mediated immune suppression, and tumor-associated endothelial cell proliferation, which facilitate cancer progression and therapeutic resistance (Clara et al. 2020; Espinosa-Sánchez et al. 2020).

Transforming Growth Factor β (TGF-β)

TGF-β, part of the TGF-β superfamily, is a versatile cytokine with roles in normal development, homeostasis, and immune response. There are three known mammalian isoforms of the TGF-β ligand: TGF-β1, -β2, and -β3 (Li and Flavell 2008). The interaction of ligand and receptor results in the regulation of gene transcription in either a SMAD protein-dependent manner (canonical pathway) or through other noncanonical pathways, including MAPK, PI3K, and Rho (Derynck and Zhang 2003). In early normal development, TGF-β signaling directs embryonic stem cell differentiation, T-cell development and function, wound healing, angiogenesis, and a host of homeostatic programs, including paracrine signaling, cell proliferation, migration, and apoptosis (Massagué and Xi 2012; Chen 2023).

TGF-β also has roles in tumor and TME development as well as metastasis. In different contexts and stages of cancer development, TGF-β can lead to either antitumor or protumorigenic effects (Batlle and Massagué 2019; Shi et al. 2022). For example, the growth-limiting effects of TGF-β signaling have been extensively studied in many different cell types, particularly epithelial cells. Early on in the process of tumorigenesis, TGF-β signaling results in down-regulation of c-Myc and induction of cyclin-dependent kinase inhibitors to halt cell cycle progression, effectively suppressing tumor development (Chaudhury and Howe 2009). However, later in the tumorigenic process, altered TGF-β signaling has been demonstrated to promote the growth and migration of normal human epithelial and breast cancer cell lines and contribute to metastasis formation in colorectal cancer models (Tang et al. 2003; Calon et al. 2012; Principe et al. 2014; Morikawa et al. 2016). Once a solid tumor is established, TGF-β generally acts protumorigenic and is often overexpressed by the tumor cells.

Within the TME, TGF-β signaling reprograms immune cells and modulates CAFs, ECM, and metabolic pathways. TGF-β signaling hampers the differentiation of naive T cells into effector CD4[+] helper and CD8[+] cytotoxic T cells, which are necessary to mount an effective antitumor response (Chen et al. 2003). The cytotoxic capabilities of CD8[+] T cells are dampened by TGF-β signaling through the down-regulation of granzyme A and B, perforin, and secretion of IFN-γ and IL-2, as demonstrated in the context of TGF-β1 secreting thymoma (Thomas and Massagué 2005). NK cells are epigenetically regulated by TGF-β to express fewer activating receptors and to have impaired cytotoxicity (Castriconi et al. 2003; Jun et al. 2019; Regis et al. 2020). TGF-β also inhibits the maturation of DCs, negatively impacting their antigen-presenting capacity and ability to secrete inflammatory cytokines such as IL-12 (Lyakh et al. 2005; Lin et al. 2013). Other important cytokines in the TME, such as IL-2, which supports the proliferation of T and NK cells, and IFN-γ, which helps activate and recruit immune cells to the area, are also down-regulated in an

 Cite this article as *Cold Spring Harb Perspect Med* doi: 10.1101/cshperspect.a041609

SMAD-dependent manner (Thomas and Massagué 2005; Zhao et al. 2020). Conversely, epigenetic regulation by TGF-β enhances the differentiation of Tregs, which normally function to maintain homeostasis by secreting immunosuppressive cytokines (including TGF-β), ultimately regulating the differentiation and limiting the activation of helper and cytotoxic T cells (Sun et al. 2019; Gu et al. 2022). Altered TGF-β signaling in the TME also increases infiltration of suppressive myeloid cells, with this myeloid-specific TGF-β signaling directly contributing to increased metastasis in breast cancer, melanoma, and colon cancer models (Yang et al. 2008; Pang et al. 2013).

Beyond the realm of immune cells, TGF-β also influences the stroma of the TME. TGF-β differentiates fibroblasts into CAFs and induces EMT (Calon et al. 2012; Hassona et al. 2013). TGF-β signaling also leads to dysregulation of the ECM in the TME across many different solid tumor types including ovarian, head and neck, and colorectal cancers (Noble et al. 1992; Chakravarthy et al. 2018). This increased stromal density hinders the physical penetration of immune cells into the TME, protecting the tumor from immune recognition (Chan et al. 2022).

TGF-β is a main driver of metabolic reprogramming of the TME, termed the tumor metabolic microenvironment (TMME). TGF-β enhances glucose uptake and metabolism in CAFs and cancer cells, including pancreatic, breast, glioma, and gastric, promoting cell proliferation and metastasis (Shi et al. 2022). Increased glycolysis leads to a lactate-rich environment, which is important for maintaining the acidity of the TME. A lactate-rich environment has been found to promote Treg production and enhance their function (Watson et al. 2021; Gu et al. 2022). In contrast, TGF-β favors immunosuppression via glycolysis restriction in T cells and NK cells (Shi et al. 2022). TGF-β also modulates the metabolism of lipids, cholesterol, and amino acids in the TME and increases angiogenesis via up-regulated production of VEGF, providing cancer cells with necessary building blocks and blood supply for rapid proliferation (Pertovaara et al. 1994; Fang et al. 2020; Shi et al. 2022).

The effects of TGF-β have also been seen clinically. Several studies comparing gene expression profiles of colorectal tumors have shown an association between increased TGF-β expression and poor prognosis, including disease relapse (Gulubova et al. 2010; Calon et al. 2012, 2015). The expression of TGF-β, particularly in the TME stromal cells, seems critical for cancer progression. Current studies in translational and clinical research have found the blockade of TGF-β to be a promising target for cancer treatment in breast and colorectal cancer models (Li et al. 2020).

Bone Morphogenetic Protein (BMP)

BMPs are a group of cytokines belonging to the TGF-β family that carry out a wide array of functions, most crucial during embryonic development. BMPs drive bone formation and maintenance and are required for osteoblast and osteoclast formation, thereby controlling osteogenesis (Abe et al. 2000; Lademann et al. 2020). However, BMP signaling also mediates systemic effects, with essential roles in cell proliferation, tissue patterning, migration, differentiation, apoptosis, and tissue homeostasis (Wang et al. 2014; Bach et al. 2018). BMP signaling is activated canonically by receptor-regulated SMAD proteins, while inhibitory SMADs regulate the pathway by feedback inhibition (Heldin et al. 1997; Derynck and Zhang 2003; Bach et al. 2018). BMP signaling can also be activated noncanonically by intracellular kinases or extracellular factors (Corradini et al. 2009; Wang et al. 2014).

Several different BMPs exist with different primary functions. BMP-2, -4, and -7 allow for the proliferation of hematopoietic stem cells (HSCs). BMP signaling plays a role in immune cell differentiation and functionality, specifically DC, NK, and thymic T-cell maturation (Migliorini et al. 2020). In monocytes, BMP-2 induces migration through PI3K and MAPK pathways and potentiates differentiation to M2 macrophages (Katsuno et al. 2008; Polyak and Weinberg 2009; Xu et al. 2009; Scott et al. 2019; Frey et al. 2020; Yuan et al. 2023). Furthermore, the BMP pathway is involved in CD4$^+$ T-cell activation after TCR stimulation, and aids CD4$^+$ T-cell proliferation in the presence of antigen-presenting cells (Martínez et al. 2015). BMP signaling has a role in ECM assembly and remodeling, and, in turn, the ECM can regulate the availability and activity of BMP signaling molecules (Migliorini et al. 2020). BMPs are also involved in angiogenesis. BMP-2, -4, -6, -7 and GDF5 induce VEGF secretion through activation of endothelial cells and phosphorylation of p70, p38, and Erk-1/2 (Kozawa et al. 2001; Langenfeld and Langenfeld 2004; David et al. 2009; Benn et al. 2017). Conversely, BMP-9 can inhibit endothelial cell proliferation by blocking basic fibroblast factor (bFGF) and by inhibiting VEGF in vessel formation, suggesting a tumor-limiting effect on the TME (Scharpfenecker et al. 2007).

Dysregulation of the BMP pathway and downstream target genes can promote tumorigenesis and accelerate disease progression through modulation of the TME, although this is often context dependent (Kiesslich et al. 2012). The activation of the BMP pathway induces EMT, imperative for both embryonic development and solid tumor metastasis. Along with ECM remodeling, EMT promotes metastasis by increasing motility, proliferation, stemness, and resistance to apoptosis (Katsuno et al. 2008; Polyak and Weinberg 2009; Xu et al. 2009; Scott et al. 2019; Frey et al. 2020; Yuan et al. 2023).

BMP's role in angiogenesis for embryo development benefits tumor cell survival through the overexpression of VEGF. BMP-2 stimulates neoangiogenesis in solid tumors, while BMP-4 potentiates proangiogenic factor Drm/gremlin (Langenfeld and Langenfeld 2004; Stabile et al. 2007).

Cross talk mechanisms also exist with other developmental pathways such as HH, Wnt, and Notch, manipulating their physiologic functions in the setting of cancer development (Kiesslich et al. 2012). BMP involvement in development is extensive and the multifaceted roles of BMP are exploited in the setting of cancer.

Sox/Oct

The characterization of SRY-related high mobility group box (SOX) proteins and octamer binding (OCT) transcription factors in the context of the TME remains an open area of study. SOX transcription factors consist of 20 vertebrate members that play a role in various features of healthy development, such as sex determination, chondrocyte differentiation, central nervous system, and cardiovascular system development (Grimm et al. 2020). OCT proteins are termed after a subset of POU domain family transcription factors with an affinity for an octamer motif. OCT proteins vary in their role within development, with roles in tissue homeostasis and developmental patterning, with Sox2 and OCT4 being well-characterized factors to regulate pluripotency (Nichols et al. 1998; Tantin 2013). Both transcription factor families have important roles in self-renewal and fine control over stem cell pluripotency (Rizzino 2009; Tantin 2013).

The important and far-reaching developmental processes of SOX and OCT can be co-opted in the setting of cancer. Several members of these protein families are expressed in multiple cancer types (Bahl et al. 2012; Ling et al. 2012; Matsuoka et al. 2012; Yang et al. 2012; Thu et al. 2014; Kim et al. 2015; Fu et al. 2016; Takeda et al. 2018; Venugopal et al. 2023). Interestingly, depending on the cellular context and through control of epigenetic mechanisms, SOX transcription factors can also act as tumor suppressor genes, as seen in a pancreatic cancer model (Grimm et al. 2020). The promotion of SOX2 and OCT4 has been implicated in the dedifferentiation and induction of CSCs under hypoxic conditions in glioma, hepatoma, and lung cancer cells. Dedifferentiation and CSCs are hallmark key regulatory features of the TME (Wang et al. 2017). Furthermore, an OCT4–SOX4 complex activates an enhancer region of the SOX2 gene to maintain the stemness of glioma-initiating cells, an important factor for the communication of tumor cells and the TME in the context of metastasis and therapeutics (Ikushima et al. 2011). SOX2 has also been implicated in recruiting TAMs to the TME in breast cancer models through STAT3 and NF-κB to promote metastasis and tumorigenicity (Mou et al. 2015). The SOX4 transcription factor is of particular interest, as the expression has been found to interact with several cancer-associated pathways, including Wnt, TGF-β, Hedgehog, Notch, and PI3K signaling (Scharer et al. 2009; Bilir et al. 2016). Additionally, in a meta-analysis of 3700 genes, SOX4 was found to be expressed at a high level in multiple cancer types such as ovarian, bladder, and TNBCs, naming the gene as part of a general cancer signature (Rhodes et al. 2004; Moreno 2020). SOX9 has also been indicated to play an oncogenic role, as it is involved in the promotion of CSCs and EMTs in various cancer types, such as lung carcinoma, prostate cancer, and basal cell carcinoma (Luanpitpong et al. 2016; Huang et al. 2019; Panda et al. 2021). Taken together, these features can support the development of a favorable TME and facilitate cancer growth and progression.

CONCLUDING REMARKS

Although it is tempting to imagine the TME as an entirely unique biological response to malignancy, a closer examination suggests that the TME is an amalgam of co-opted essential physiologic processes that support the tumor's growth and "repair" and, ultimately, its progression. Across multiple tumor types, the TME can be considered a combination of the stem cell niche, the wound healing response, the chronic inflammatory response, and the normal developmental processes associated with a developing tissue (Murgai et al. 2015). In this highly regulated and protected microenvironmental context, inflammatory responses that may otherwise eliminate the tumor are suppressed in favor of pathways promoting tissue regeneration and development, many of which represent canonical pathways essential to the development and maintenance of tissue homeostasis. As described in this article, there are significant parallels between development and tumorigenesis regarding their common molecular pathways, which promote the growth and survival of the forming tissue. The formation of a solid malignancy represents a scenario in which aberrantly proliferating cells have co-opted normal growth and development pathways to evade immunologic elimination and recruit active support for the continued growth of the "developing" tumor tissue. Modulation of the TME is an increasing area of interest for future therapeutics, and therefore, pharmacologic strategies to alter the developmental pathways that promote the immunosuppressive and protumorigenic TME represent a promising approach.

ACKNOWLEDGMENTS

We would like to acknowledge members of the Kaplan laboratory for their review of this manuscript.

REFERENCES

Abe E, Yamamoto M, Taguchi Y, Lecka-Czernik B, O'Brien CA, Economides AN, Stahl N, Jilka RL, Manolagas SC. 2000. Essential requirement of BMPs-2/4 for both osteoblast and osteoclast formation in murine bone marrow cultures from adult mice: antagonism by noggin. *J Bone Miner Res* **15**: 663–673. doi:10.1359/jbmr.2000.15.4.663

Ackers I, Malgor R. 2018. Interrelationship of canonical and non-canonical Wnt signalling pathways in chronic metabolic diseases. *Diab Vasc Dis Res* **15**: 3–13. doi:10.1177/1479164117738442

Ahmed AA, Mohamed AD, Gener M, Li W, Taboada E. 2017. YAP and the Hippo pathway in pediatric cancer. *Mol Cell Oncol* **4**: e1295127. doi:10.1080/23723556.2017.1295127

Aizawa T, Karasawa H, Funayama R, Shirota M, Suzuki T, Maeda S, Suzuki H, Yamamura A, Naitoh T, Nakayama K, et al. 2019. Cancer-associated fibroblasts secrete Wnt2 to promote cancer progression in colorectal cancer. *Cancer Med* **8**: 6370–6382. doi:10.1002/cam4.2523

Akil A, Gutiérrez-García AK, Guenter R, Rose JB, Beck AW, Chen H, Ren B. 2021. Notch signaling in vascular endothelial cells, angiogenesis, and tumor progression: an update and prospective. *Front Cell Dev Biol* **9**: 642352. doi:10.3389/fcell.2021.642352

Anderson NM, Simon MC. 2020. The tumor microenvironment. *Curr Biol* **30**: R921–R925. doi:10.1016/j.cub.2020.06.081

Bach DH, Park HJ, Lee SK. 2018. The dual role of bone morphogenetic proteins in cancer. *Mol Ther Oncolytics* **8**: 1–13. doi:10.1016/j.omto.2017.10.002

Bahl K, Saraya A, Sharma R. 2012. Increased levels of circulating and tissue mRNAs of Oct-4, Sox-2, Bmi-1 and Nanog is ESCC patients: potential tool for minimally invasive cancer diagnosis. *Biomark Insights* **7**: 27–37. doi:10.4137/BMI.S8452

Bao J, Lee HJ, Zheng JJ. 2013. Genome-wide network analysis of Wnt signaling in three pediatric cancers. *Sci Rep* **3**: 2969. doi:10.1038/srep02969

Batlle E, Massagué J. 2019. Transforming growth factor-β signaling in immunity and cancer. *Immunity* **50**: 924–940. doi:10.1016/j.immuni.2019.03.024

Benn A, Hiepen C, Osterland M, Schütte C, Zwijsen A, Knaus P. 2017. Role of bone morphogenetic proteins in sprouting angiogenesis: differential BMP receptor-dependent signaling pathways balance stalk vs. tip cell competence. *FASEB J* **31:** 4720–4733. doi:10.1096/fj.201700193RR

Bilir B, Osunkoya AO, Wiles WG IV, Sannigrahi S, Lefebvre V, Metzger D, Spyropoulos DD, Martin WD, Moreno CS. 2016. SOX4 is essential for prostate tumorigenesis initiated by PTEN ablation. *Cancer Res* **76:** 1112–1121. doi:10.1158/0008-5472.CAN-15-1868

Bonilla X, Parmentier L, King B, Bezrukov F, Kaya G, Zoete V, Seplyarskiy VB, Sharpe HJ, McKee T, Letourneau A, et al. 2016. Genomic analysis identifies new drivers and progression pathways in skin basal cell carcinoma. *Nat Genet* **48:** 398–406. doi:10.1038/ng.3525

Bonnans C, Chou J, Werb Z. 2014. Remodelling the extracellular matrix in development and disease. *Nat Rev Mol Cell Biol* **15:** 786–801. doi:10.1038/nrm3904

Borriello L, Nakata R, Sheard MA, Fernandez GE, Sposto R, Malvar J, Blavier L, Shimada H, Asgharzadeh S, Seeger RC, et al. 2017. Cancer-associated fibroblasts share characteristics and protumorigenic activity with mesenchymal stromal cells. *Cancer Res* **77:** 5142–5157. doi:10.1158/0008-5472.CAN-16-2586

Bray SJ. 2006. Notch signalling: a simple pathway becomes complex. *Nat Rev Mol Cell Biol* **7:** 678–689. doi:10.1038/nrm2009

Calon A, Espinet E, Palomo-Ponce S, Tauriello DV, Iglesias M, Céspedes MV, Sevillano M, Nadal C, Jung P, Zhang XH, et al. 2012. Dependency of colorectal cancer on a TGF-β-driven program in stromal cells for metastasis initiation. *Cancer Cell* **22:** 571–584. doi:10.1016/j.ccr.2012.08.013

Calon A, Lonardo E, Berenguer-Llergo A, Espinet E, Hernando-Momblona X, Iglesias M, Sevillano M, Palomo-Ponce S, Tauriello DVF, Byrom D, et al. 2015. Stromal gene expression defines poor-prognosis subtypes in colorectal cancer. *Nat Genet* **47:** 320–329. doi:10.1038/ng.3225

Castriconi R, Cantoni C, Della Chiesa M, Vitale M, Marcenaro E, Conte R, Biassoni R, Bottino C, Moretta L, Moretta A. 2003. Transforming growth factor β1 inhibits expression of NKp30 and NKG2D receptors: consequences for the NK-mediated killing of dendritic cells. *Proc Natl Acad Sci* **100:** 4120–4125. doi:10.1073/pnas.0730640100

Cendrowicz E, Sas Z, Bremer E, Rygiel TP. 2021. The role of macrophages in cancer development and therapy. *Cancers (Basel)* **13:** 1946. doi:10.3390/cancers13081946

Chai J, Xu S, Guo F. 2017. TEAD1 mediates the oncogenic activities of Hippo-YAP1 signaling in osteosarcoma. *Biochem Biophys Res Commun* **488:** 297–302. doi:10.1016/j.bbrc.2017.05.032

Chakravarthy A, Khan L, Bensler NP, Bose P, De Carvalho DD. 2018. TGF-β-associated extracellular matrix genes link cancer-associated fibroblasts to immune evasion and immunotherapy failure. *Nat Commun* **9:** 4692. doi:10.1038/s41467-018-06654-8

Chan MK, Chung JY, Tang PC, Chan AS, Ho JY, Lin TP, Chen J, Leung KT, To KF, Lan HY, et al. 2022. TGF-β signaling networks in the tumor microenvironment. *Cancer Lett* **550:** 215925. doi:10.1016/j.canlet.2022.215925

Chaudhury A, Howe PH. 2009. The tale of transforming growth factor β (TGF-β) signaling: a soigné enigma. *IUBMB Life* **61:** 929–939. doi:10.1002/iub.239

Chen W. 2023. TGF-β regulation of T cells. *Annu Rev Immunol* **41:** 483–512. doi:10.1146/annurev-immunol-101921-045939

Chen C-H, Seguin-Devaux C, Burke NA, Oriss TB, Watkins SC, Clipstone N, Ray A. 2003. Transforming growth factor β blocks Tec kinase phosphorylation, Ca^{2+} influx, and NFATc translocation causing inhibition of T cell differentiation. *J Exp Med* **197:** 1689–1699. doi:10.1084/jem.20021170

Choudhary S, Singh MK, Kashyap S, Seth R, Singh L. 2024. Wnt/β-catenin signaling pathway in pediatric tumors: implications for diagnosis and treatment. *Children (Basel)* **11:** 700. doi:10.3390/children11060700

Clara JA, Monge C, Yang Y, Takebe N. 2020. Targeting signalling pathways and the immune microenvironment of cancer stem cells—a clinical update. *Nat Rev Clin Oncol* **17:** 204–232. doi:10.1038/s41571-019-0293-2

Cords L, de Souza N, Bodenmiller B. 2024. Classifying cancer-associated fibroblasts—the good, the bad, and the target. *Cancer Cell* **42:** 1480–1485. doi:10.1016/j.ccell.2024.08.011

Corradini E, Babitt JL, Lin HY. 2009. The RGM/DRAGON family of BMP co-receptors. *Cytokine Growth Factor Rev* **20:** 389–398. doi:10.1016/j.cytogfr.2009.10.008

Cottini F, Hideshima T, Xu C, Sattler M, Dori M, Agnelli L, ten Hacken E, Bertilaccio MT, Antonini E, Neri A, et al. 2014. Rescue of Hippo coactivator YAP1 triggers DNA damage-induced apoptosis in hematological cancers. *Nat Med* **20:** 599–606. doi:10.1038/nm.3562

Dai XM, Ryan GR, Hapel AJ, Dominguez MG, Russell RG, Kapp S, Sylvestre V, Stanley ER. 2002. Targeted disruption of the mouse colony-stimulating factor 1 receptor gene results in osteopetrosis, mononuclear phagocyte deficiency, increased primitive progenitor cell frequencies, and reproductive defects. *Blood* **99:** 111–120. doi:10.1182/blood.V99.1.111

D'Assoro AB, Leon-Ferre R, Braune EB, Lendahl U. 2022. Roles of Notch signaling in the tumor microenvironment. *Int J Mol Sci* **23:** 6241. doi:10.3390/ijms23116241

David L, Feige JJ, Bailly S. 2009. Emerging role of bone morphogenetic proteins in angiogenesis. *Cytokine Growth Factor Rev* **20:** 203–212. doi:10.1016/j.cytogfr.2009.05.001

de Almeida Magalhães T, Cruzeiro GAV, de Sousa GR, da Silva KR, Lira RCP, Scrideli CA, Tone LG, Valera ET, Borges KS. 2020. Notch pathway in ependymoma RELA-fused subgroup: upregulation and association with cancer stem cells markers expression. *Cancer Gene Ther* **27:** 509–512. doi:10.1038/s41417-019-0122-x

Del Gaudio F, Liu D, Lendahl U. 2022. Notch signalling in healthy and diseased vasculature. *Open Biol* **12:** 220004. doi:10.1098/rsob.220004

Dempke WCM, Fenchel K, Uciechowski P, Chevassut T. 2017. Targeting developmental pathways: the Achilles heel of cancer? *Oncology* **93:** 213–223. doi:10.1159/000478703

Derynck R, Zhang YE. 2003. Smad-dependent and Smad-independent pathways in TGF-β family signalling. *Nature* **425:** 577–584. doi:10.1038/nature02006

Dolberg DS, Bissell MJ. 1984. Inability of Rous sarcoma virus to cause sarcomas in the avian embryo. *Nature* **309:** 552–556. doi:10.1038/309552a0

Du Y, Shao H, Moller M, Prokupets R, Tse YT, Liu ZJ. 2019. Intracellular Notch1 signaling in cancer-associated fibroblasts dictates the plasticity and stemness of melanoma stem/initiating cells. *Stem Cells* **37:** 865–875. doi:10.1002/stem.3013

Dupont S, Morsut L, Aragona M, Enzo E, Giulitti S, Cordenonsi M, Zanconato F, Le Digabel J, Forcato M, Bicciato S, et al. 2011. Role of YAP/TAZ in mechanotransduction. *Nature* **474:** 179–183. doi:10.1038/nature10137

Espinosa-Sánchez A, Suárez-Martínez E, Sánchez-Díaz L, Carnero A. 2020. Therapeutic targeting of signaling pathways related to cancer stemness. *Front Oncol* **10:** 1533. doi:10.3389/fonc.2020.01533

Fang L, Li Y, Wang S, Li Y, Chang H-M, Yi Y, Yan Y, Thakur A, Leung PCK, Cheng J-C, et al. 2020. TGF-β1 induces VEGF expression in human granulosa-lutein cells: a potential mechanism for the pathogenesis of ovarian hyperstimulation syndrome. *Exp Mol Med* **52:** 450–460. doi:10.1038/s12276-020-0396-y

Ferguson LP, Diaz E, Reya T. 2021. The role of the microenvironment and immune system in regulating stem cell fate in cancer. *Trends Cancer* **7:** 624–634. doi:10.1016/j.trecan.2020.12.014

Frantz C, Stewart KM, Weaver VM. 2010. The extracellular matrix at a glance. *J Cell Sci* **123:** 4195–4200. doi:10.1242/jcs.023820

Frey P, Devisme A, Schrempp M, Andrieux G, Boerries M, Hecht A. 2020. Canonical BMP signaling executes epithelial–mesenchymal transition downstream of SNAIL1. *Cancers (Basel)* **12:** 1019. doi:10.3390/cancers12041019

Fu TY, Hsieh IC, Cheng JT, Tsai MH, Hou YY, Lee JH, Liou HH, Huang SF, Chen HC, Yen LM, et al. 2016. Association of OCT4, SOX2, and NANOG expression with oral squamous cell carcinoma progression. *J Oral Pathol Med* **45:** 89–95. doi:10.1111/jop.12335

Gabrilovich DI, Ostrand-Rosenberg S, Bronte V. 2012. Coordinated regulation of myeloid cells by tumours. *Nat Rev Immunol* **12:** 253–268. doi:10.1038/nri3175

Giles AJ, Reid CM, Evans JD, Murgai M, Vicioso Y, Highfill SL, Kasai M, Vahdat L, Mackall CL, Lyden D, et al. 2016. Activation of hematopoietic stem/progenitor cells promotes immunosuppression within the pre-metastatic niche. *Cancer Res* **76:** 1335–1347. doi:10.1158/0008-5472.CAN-15-0204

Gilkes DM, Semenza GL, Wirtz D. 2014. Hypoxia and the extracellular matrix: drivers of tumour metastasis. *Nat Rev Cancer* **14:** 430–439. doi:10.1038/nrc3726

Goswami TK, Singh M, Dhawan M, Mitra S, Emran TB, Rabaan AA, Mutair AA, Alawi ZA, Alhumaid S, Dhama K. 2022. Regulatory T cells (Tregs) and their therapeutic potential against autoimmune disorders—advances and challenges. *Hum Vaccin Immunother* **18:** 2035117. doi:10.1080/21645515.2022.2035117

Gozlan O, Sprinzak D. 2023. Notch signaling in development and homeostasis. *Development* **150:** dev201138. doi:10.1242/dev.201138

Grimm D, Bauer J, Wise P, Krüger M, Simonsen U, Wehland M, Infanger M, Corydon TJ. 2020. The role of SOX family members in solid tumours and metastasis. *Semin Cancer Biol* **67:** 122–153. doi:10.1016/j.semcancer.2019.03.004

Gu J, Zhou J, Chen Q, Xu X, Gao J, Li X, Shao Q, Zhou B, Zhou H, Wei S, et al. 2022. Tumor metabolite lactate promotes tumorigenesis by modulating MOESIN lactylation and enhancing TGF-β signaling in regulatory T cells. *Cell Rep* **39:** 110986. doi:10.1016/j.celrep.2022.110986

Gulubova M, Manolova I, Ananiev J, Julianov A, Yovchev Y, Peeva K. 2010. Role of TGF-β1, its receptor TGFβRII, and Smad proteins in the progression of colorectal cancer. *Int J Colorect Dis* **25:** 591–599. doi:10.1007/s00384-010-0906-9

Han Y. 2019. Analysis of the role of the Hippo pathway in cancer. *J Transl Med* **17:** 116. doi:10.1186/s12967-019-1869-4

Hashimoto O, Yoshida M, Koma Y, Yanai T, Hasegawa D, Kosaka Y, Nishimura N, Yokozaki H. 2016. Collaboration of cancer-associated fibroblasts and tumour-associated macrophages for neuroblastoma development. *J Pathol* **240:** 211–223. doi:10.1002/path.4769

Hassona Y, Cirillo N, Lim KP, Herman A, Mellone M, Thomas GJ, Pitiyage GN, Parkinson EK, Prime SS. 2013. Progression of genotype-specific oral cancer leads to senescence of cancer-associated fibroblasts and is mediated by oxidative stress and TGF-β. *Carcinogenesis* **34:** 1286–1295. doi:10.1093/carcin/bgt035

Hegde S, Leader AM, Merad M. 2021. MDSC: markers, development, states, and unaddressed complexity. *Immunity* **54:** 875–884. doi:10.1016/j.immuni.2021.04.004

Heldin CH, Miyazono K, ten Dijke P. 1997. TGF-β signalling from cell membrane to nucleus through SMAD proteins. *Nature* **390:** 465–471. doi:10.1038/37284

Hettinger ZR, Wen Y, Peck BD, Hamagata K, Confides AL, Van Pelt DW, Harrison DA, Miller BF, Butterfield TA, Dupont-Versteegden EE. 2022. Mechanotherapy reprograms aged muscle stromal cells to remodel the extracellular matrix during recovery from disuse. *Function (Oxf)* **3:** zqac015. doi:10.1093/function/zqac015

Hinshaw DC, Hanna A, Lama-Sherpa T, Metge B, Kammerud SC, Benavides GA, Kumar A, Alsheikh HA, Mota M, Chen D, et al. 2021. Hedgehog signaling regulates metabolism and polarization of mammary tumor-associated macrophages. *Cancer Res* **81:** 5425–5437. doi:10.1158/0008-5472.CAN-20-1723

Hinshaw DC, Benavides GA, Metge BJ, Swain CA, Kammerud SC, Alsheikh HA, Elhamamsy A, Chen D, Darley-Usmar V, Rathmell JC, et al. 2023. Hedgehog signaling regulates Treg to Th17 conversion through metabolic rewiring in breast cancer. *Cancer Immunol Res* **11:** 687–702. doi:10.1158/2326-6066.CIR-22-0426

Hossain F, Majumder S, Ucar DA, Rodriguez PC, Golde TE, Minter LM, Osborne BA, Miele L. 2018. Notch signaling in myeloid cells as a regulator of tumor immune responses. *Front Immunol* **9:** 1288. doi:10.3389/fimmu.2018.01288

Huang JQ, Wei FK, Xu XL, Ye SX, Song JW, Ding PK, Zhu J, Li HF, Luo XP, Gong H, et al. 2019. SOX9 drives the epithelial-mesenchymal transition in non-small-cell lung cancer through the Wnt/β-catenin pathway. *J Transl Med* **17:** 143. doi:10.1186/s12967-019-1895-2

Huang S, Liu Z, Qian X, Li L, Zhang H, Li S, Liu Z. 2022. YAP/TAZ promote fibrotic activity in human trabecular meshwork cells by sensing cytoskeleton structure alternation. *Chemosensors* **10**: 235. doi:10.3390/chemosensors 10070235

Ikushima H, Todo T, Ino Y, Takahashi M, Saito N, Miyazawa K, Miyazono K. 2011. Glioma-initiating cells retain their tumorigenicity through integration of the Sox axis and Oct4 protein. *J Biol Chem* **286**: 41434–41441. doi:10 .1074/jbc.M111.300863

Javelaud D, Pierrat MJ, Mauviel A. 2012. Crosstalk between TGF-β and hedgehog signaling in cancer. *FEBS Lett* **586**: 2016–2025. doi:10.1016/j.febslet.2012.05.011

Jun E, Song AY, Choi JW, Lee HH, Kim MY, Ko DH, Kang HJ, Kim SW, Bryceson Y, Kim SC, et al. 2019. Progressive impairment of NK cell cytotoxic degranulation is associated with TGF-β1 deregulation and disease progression in pancreatic cancer. *Front Immunol* **10**: 1354. doi:10.3389/ fimmu.2019.01354

Kaczanowska S, Kaplan RN. 2020. Mapping the switch that drives the pre-metastatic niche. *Nat Cancer* **1**: 577–579. doi:10.1038/s43018-020-0076-9

Kalathil SG, Thanavala Y. 2021. Importance of myeloid derived suppressor cells in cancer from a biomarker perspective. *Cell Immunol* **361**: 104280. doi:10.1016/j.cell imm.2020.104280

Karthikeyan S, Casey PJ, Wang M. 2024. RAB4A is a master regulator of cancer cell stemness upstream of NUMB-NOTCH signaling. *Cell Death Dis* **15**: 778. doi:10.1038/ s41419-024-07172-w

Katsuno Y, Hanyu A, Kanda H, Ishikawa Y, Akiyama F, Iwase T, Ogata E, Ehata S, Miyazono K, Imamura T. 2008. Bone morphogenetic protein signaling enhances invasion and bone metastasis of breast cancer cells through Smad pathway. *Oncogene* **27**: 6322–6333. doi:10.1038/onc.2008.232

Katzenelenbogen Y, Sheban F, Yalin A, Yofe I, Svetlichnyy D, Jaitin DA, Bornstein C, Moshe A, Keren-Shaul H, Cohen M, et al. 2020. Coupled scRNA-seq and intracellular protein activity reveal an immunosuppressive role of TREM2 in cancer. *Cell* **182**: 872–885.e19. doi:10.1016/j.cell.2020 .06.032

Kiesslich T, Berr F, Alinger B, Kemmerling R, Pichler M, Ocker M, Neureiter D. 2012. Current status of therapeutic targeting of developmental signalling pathways in oncology. *Curr Pharm Biotechnol* **13**: 2184–2220. doi:10.2174/ 138920112802502114

Kim BW, Cho H, Choi CH, Ylaya K, Chung JY, Kim JH, Hewitt SM. 2015. Clinical significance of OCT4 and SOX2 protein expression in cervical cancer. *BMC Cancer* **15**: 1015. doi:10.1186/s12885-015-2015-1

Koo J, Hayashi M, Verneris MR, Lee-Sherick AB. 2020. Targeting tumor-associated macrophages in the pediatric sarcoma tumor microenvironment. *Front Oncol* **10**: 581107. doi:10.3389/fonc.2020.581107

Kopan R, Ilagan MX. 2009. The canonical Notch signaling pathway: unfolding the activation mechanism. *Cell* **137**: 216–233. doi:10.1016/j.cell.2009.03.045

Kozawa O, Matsuno H, Uematsu T. 2001. Involvement of p70 S6 kinase in bone morphogenetic protein signaling: vascular endothelial growth factor synthesis by bone morphogenetic protein-4 in osteoblasts. *J Cell Biochem*

81: 430–436. doi:10.1002/1097-4644(20010601)81:3 <430::AID-JCB1056>3.0.CO;2-G

Kular JK, Basu S, Sharma RI. 2014. The extracellular matrix: structure, composition, age-related differences, tools for analysis and applications for tissue engineering. *J Tissue Eng* **5**: 2041731414557112. doi:10.1177/204173141455 7112

Lademann F, Hofbauer LC, Rauner M. 2020. The bone morphogenetic protein pathway: the osteoclastic perspective. *Front Cell Dev Biol* **8**: 586031. doi:10.3389/fcell.2020 .586031

LaFoya B, Munroe JA, Mia MM, Detweiler MA, Crow JJ, Wood T, Roth S, Sharma B, Albig AR. 2016. Notch: a multi-functional integrating system of microenvironmental signals. *Dev Biol* **418**: 227–241. doi:10.1016/j .ydbio.2016.08.023

LaMarche NM, Hegde S, Park MD, Maier BB, Troncoso L, Le Berichel J, Hamon P, Belabed M, Mattiuz R, Hennequin C, et al. 2024. An IL-4 signalling axis in bone marrow drives pro-tumorigenic myelopoiesis. *Nature* **625**: 166–174. doi:10.1038/s41586-023-06797-9

Langenfeld EM, Langenfeld J. 2004. Bone morphogenetic protein-2 stimulates angiogenesis in developing tumors. *Mol Cancer Res* **2**: 141–149. doi:10.1158/1541-7786.141.2.3

Leventhal DS, Gilmore DC, Berger JM, Nishi S, Lee V, Malchow S, Kline DE, Kline J, Vander Griend DJ, Huang H, et al. 2016. Dendritic cells coordinate the development and homeostasis of organ-specific regulatory T cells. *Immunity* **44**: 847–859. doi:10.1016/j.immuni.2016.01.025

Li MO, Flavell RA. 2008. TGF-β: a master of all T cell trades. *Cell* **134**: 392–404. doi:10.1016/j.cell.2008.07.025

Li J, Chen K, Zhu L, Pollard JW. 2006. Conditional deletion of the colony stimulating factor-1 receptor (c-fms proto-oncogene) in mice. *Genesis* **44**: 328–335. doi:10.1002/dvg .20219

Li H, Huang Z, Gao M, Huang N, Luo Z, Shen H, Wang X, Wang T, Hu J, Feng W. 2016. Inhibition of YAP suppresses CML cell proliferation and enhances efficacy of imatinib in vitro and in vivo. *J Exp Clin Cancer Res* **35**: 134. doi:10.1186/s13046-016-0414-z

Li S, Liu M, Do MH, Chou C, Stamatiades EG, Nixon BG, Shi W, Zhang X, Li P, Gao S, et al. 2020. Cancer immunotherapy via targeted TGF-β signalling blockade in T_H cells. *Nature* **587**: 121–125. doi:10.1038/s41586-020-2850-3

Li X, Zhang Q, Chen G, Luo D. 2021. Multi-omics analysis showed the clinical value of gene signatures of C1QC[+] and SPP1[+] TAMs in cervical cancer. *Front Immunol* **12**: 694801. doi:10.3389/fimmu.2021.694801

Li X, Yan X, Wang Y, Kaur B, Han H, Yu J. 2023. The Notch signaling pathway: a potential target for cancer immunotherapy. *J Hematol Oncol* **16**: 45. doi:10.1186/s13045-023-01439-z

Liang LJ, Yang Y, Wei WF, Wu XG, Yan RM, Zhou CF, Chen XJ, Wu S, Wang W, Fan LS. 2021. Tumor-secreted exosomal Wnt2B activates fibroblasts to promote cervical cancer progression. *Oncogenesis* **10**: 30. doi:10.1038/ s41389-021-00319-w

Ligon JA, Choi W, Cojocaru G, Fu W, Hsiue EH, Oke TF, Siegel N, Fong MH, Ladle B, Pratilas CA, et al. 2021. Pathways of immune exclusion in metastatic osteosarcoma are associated with inferior patient outcomes. *J Immunother Cancer* **9**: e001772. doi:10.1136/jitc-2020-001772

Lin C-S, Chen M-F, Wang Y-S, Chuang T-F, Chiang Y-L, Chu R-M. 2013. IL-6 restores dendritic cell maturation inhibited by tumor-derived TGF-β through interfering Smad 2/3 nuclear translocation. *Cytokine* **62**: 352–359. doi:10.1016/j.cyto.2013.03.005

Lin GL, Nagaraja S, Filbin MG, Suvà ML, Vogel H, Monje M. 2018. Non-inflammatory tumor microenvironment of diffuse intrinsic pontine glioma. *Acta Neuropathol Commun* **6**: 51. doi:10.1186/s40478-018-0553-x

Ling GQ, Chen DB, Wang BQ, Zhang LS. 2012. Expression of the pluripotency markers Oct3/4, Nanog and Sox2 in human breast cancer cell lines. *Oncol Lett* **4**: 1264–1268. doi:10.3892/ol.2012.916

Liu C, Song Y, Li D, Wang B. 2023. Regulation of the tumor immune microenvironment by the Hippo pathway: implications for cancer immunotherapy. *Int Immunopharmacol* **122**: 110586. doi:10.1016/j.intimp.2023.110586

Liu K, Wehling L, Wan S, Weiler SME, Tóth M, Ibberson D, Marhenke S, Ali A, Lam M, Guo T, et al. 2024. Dynamic YAP expression in the non-parenchymal liver cell compartment controls heterologous cell communication. *Cell Mol Life Sci* **81**: 115. doi:10.1007/s00018-024-05126-1

Lobov IB, Rao S, Carroll TJ, Vallance JE, Ito M, Ondr JK, Kurup S, Glass DA, Patel MS, Shu W, et al. 2005. WNT7b mediates macrophage-induced programmed cell death in patterning of the vasculature. *Nature* **437**: 417–421. doi:10.1038/nature03928

Luanpitpong S, Li J, Manke A, Brundage K, Ellis E, McLaughlin SL, Angsutararux P, Chanthra N, Voronkova M, Chen YC, et al. 2016. SLUG is required for SOX9 stabilization and functions to promote cancer stem cells and metastasis in human lung carcinoma. *Oncogene* **35**: 2824–2833. doi:10.1038/onc.2015.351

Lugano R, Ramachandran M, Dimberg A. 2020. Tumor angiogenesis: causes, consequences, challenges and opportunities. *Cell Mol Life Sci* **77**: 1745–1770. doi:10.1007/s00018-019-03351-7

Luke JJ, Bao R, Sweis RF, Spranger S, Gajewski TF. 2019. WNT/β-catenin pathway activation correlates with immune exclusion across human cancers. *Clin Cancer Res* **25**: 3074–3083. doi:10.1158/1078-0432.CCR-18-1942

Lyakh LA, Sanford M, Chekol S, Young HA, Roberts AB. 2005. TGF-β and vitamin D3 utilize distinct pathways to suppress IL-12 production and modulate rapid differentiation of human monocytes into CD83+ dendritic cells. *J Immunol* **174**: 2061–2070. doi:10.4049/jimmunol.174.4.2061

Macias MJ, Hyvönen M, Baraldi E, Schultz J, Sudol M, Saraste M, Oschkinat H. 1996. Structure of the WW domain of a kinase-associated protein complexed with a proline-rich peptide. *Nature* **382**: 646–649. doi:10.1038/382646a0

Mantovani A, Allavena P, Marchesi F, Garlanda C. 2022. Macrophages as tools and targets in cancer therapy. *Nat Rev Drug Discov* **21**: 799–820. doi:10.1038/s41573-022-00520-5

Marshall GM, Carter DR, Cheung BB, Liu T, Mateos MK, Meyerowitz JG, Weiss WA. 2014. The prenatal origins of cancer. *Nat Rev Cancer* **14**: 277–289. doi:10.1038/nrc3679

Martínez VG, Sacedón R, Hidalgo L, Valencia J, Fernández-Sevilla LM, Hernández-López C, Vicente A, Varas A. 2015. The BMP pathway participates in human naive CD4+ T cell activation and homeostasis. *PLoS ONE* **10**: e0131453. doi:10.1371/journal.pone.0131453

Mass E, Nimmerjahn F, Kierdorf K, Schlitzer A. 2023. Tissue-specific macrophages: how they develop and choreograph tissue biology. *Nat Rev Immunol* **23**: 563–579. doi:10.1038/s41577-023-00848-y

Massagué J, Xi Q. 2012. TGF-β control of stem cell differentiation genes. *FEBS Lett* **586**: 1953–1958. doi:10.1016/j.febslet.2012.03.023

Matsuoka J, Yashiro M, Sakurai K, Kubo N, Tanaka H, Muguruma K, Sawada T, Ohira M, Hirakawa K. 2012. Role of the stemness factors sox2, oct3/4, and nanog in gastric carcinoma. *J Surg Res* **174**: 130–135. doi:10.1016/j.jss.2010.11.903

Migliorini E, Guevara-Garcia A, Albiges-Rizo C, Picart C. 2020. Learning from BMPs and their biophysical extracellular matrix microenvironment for biomaterial design. *Bone* **141**: 115540. doi:10.1016/j.bone.2020.115540

Mintz B, Illmensee K. 1975. Normal genetically mosaic mice produced from malignant teratocarcinoma cells. *Proc Natl Acad Sci* **72**: 3585–3589. doi:10.1073/pnas.72.9.3585

Molgora M, Esaulova E, Vermi W, Hou J, Chen Y, Luo J, Brioschi S, Bugatti M, Omodei AS, Ricci B, et al. 2020. TREM2 modulation remodels the tumor myeloid landscape enhancing anti-PD-1 immunotherapy. *Cell* **182**: 886–900.e17. doi:10.1016/j.cell.2020.07.013

Moreno CS. 2020. SOX4: the unappreciated oncogene. *Semin Cancer Biol* **67**: 57–64. doi:10.1016/j.semcancer.2019.08.027

Morikawa M, Derynck R, Miyazono K. 2016. TGF-β and the TGF-β family: context-dependent roles in cell and tissue physiology. *Cold Spring Harb Perspect Biol* **8**: a021873. doi:10.1101/cshperspect.a021873

Mou W, Xu Y, Ye Y, Chen S, Li X, Gong K, Liu Y, Chen Y, Li X, Tian Y, et al. 2015. Expression of Sox2 in breast cancer cells promotes the recruitment of M2 macrophages to tumor microenvironment. *Cancer Lett* **358**: 115–123. doi:10.1016/j.canlet.2014.11.004

Murgai M, Giles A, Kaplan R. 2015. Physiological, tumor, and metastatic niches: opportunities and challenges for targeting the tumor microenvironment. *Crit Rev Oncog* **20**: 301–314. doi:10.1615/CritRevOncog.2015013668

Muto S, Enta A, Maruya Y, Inomata S, Yamaguchi H, Mine H, Takagi H, Ozaki Y, Watanabe M, Inoue T, et al. 2023. Wnt/β-catenin signaling and resistance to immune checkpoint inhibitors: from non-small-cell lung cancer to other cancers. *Biomedicines* **11**: 190. doi:10.3390/biomedicines11010190

Nagl L, Horvath L, Pircher A, Wolf D. 2020. Tumor endothelial cells (TECs) as potential immune directors of the tumor microenvironment—new findings and future perspectives. *Front Cell Dev Biol* **8**: 766. doi:10.3389/fcell.2020.00766

Nichols J, Zevnik B, Anastassiadis K, Niwa H, Klewe-Nebenius D, Chambers I, Schöler H, Smith A. 1998. Formation of pluripotent stem cells in the mammalian embryo depends on the POU transcription factor Oct4. *Cell* **95**: 379–391. doi:10.1016/S0092-8674(00)81769-9

Niehrs C. 2012. The complex world of WNT receptor signalling. *Nat Rev Mol Cell Biol* **13**: 767–779. doi:10.1038/nrm3470

Noble NA, Harper JR, Border WA. 1992. In vivo interactions of TGF-β and extracellular matrix. *Prog Growth Factor Res* **4**: 369–382. doi:10.1016/0955-2235(92)90017-C

Nobre L, Zapotocky M, Khan S, Fukuoka K, Fonseca A, McKeown T, Sumerauer D, Vicha A, Grajkowska WA, Trubicka J, et al. 2020. Pattern of relapse and treatment response in WNT-activated medulloblastoma. *Cell Rep Med* **1**: 100038. doi:10.1016/j.xcrm.2020.100038

Noguchi S, Saito A, Nagase T. 2018. YAP/TAZ signaling as a molecular link between fibrosis and cancer. *Int J Mol Sci* **19**: 3674. doi:10.3390/ijms19113674

Nussinov R, Tsai CJ, Jang H, Korcsmáros T, Csermely P. 2016. Oncogenic KRAS signaling and YAP1/β-catenin: similar cell cycle control in tumor initiation. *Semin Cell Dev Biol* **58**: 79–85. doi:10.1016/j.semcdb.2016.04.001

Oka T, Remue E, Meerschaert K, Vanloo B, Boucherie C, Gfeller D, Bader GD, Sidhu SS, Vandekerckhove J, Gettemans J, et al. 2010. Functional complexes between YAP2 and ZO-2 are PDZ domain-dependent, and regulate YAP2 nuclear localization and signalling. *Biochem J* **432**: 461–472. doi:10.1042/BJ20100870

Olsen JJ, Pohl S, Deshmukh A, Visweswaran M, Ward NC, Arfuso F, Agostino M, Dharmarajan A. 2017. The role of Wnt signalling in angiogenesis. *Clin Biochem Rev* **38**: 131–142.

Pak E, Segal RA. 2016. Hedgehog signal transduction: key players, oncogenic drivers, and cancer therapy. *Dev Cell* **38**: 333–344. doi:10.1016/j.devcel.2016.07.026

Pan Y, Yu Y, Wang X, Zhang T. 2020. Tumor-associated macrophages in tumor immunity. *Front Immunol* **11**: 583084. doi:10.3389/fimmu.2020.583084

Panda M, Tripathi SK, Biswal BK. 2021. SOX9: an emerging driving factor from cancer progression to drug resistance. *Biochim Biophys Acta Rev Cancer* **1875**: 188517. doi:10.1016/j.bbcan.2021.188517

Pang Y, Gara SK, Achyut BR, Li Z, Yan HH, Day CP, Weiss JM, Trinchieri G, Morris JC, Yang L. 2013. TGF-β signaling in myeloid cells is required for tumor metastasis. *Cancer Discov* **3**: 936–951. doi:10.1158/2159-8290.CD-12-0527

Paolillo M, Schinelli S. 2019. Extracellular matrix alterations in metastatic processes. *Int J Mol Sci* **20**: 4947. doi:10.3390/ijms20194947

Patel S, Alam A, Pant R, Chattopadhyay S. 2019. Wnt signaling and its significance within the tumor microenvironment: novel therapeutic insights. *Front Immunol* **10**: 2872. doi:10.3389/fimmu.2019.02872

Pawelec G, Verschoor CP, Ostrand-Rosenberg S. 2019. Myeloid-derived suppressor cells: not only in tumor immunity. *Front Immunol* **10**: 1099. doi:10.3389/fimmu.2019.01099

Pei T, Li Y, Wang J, Wang H, Liang Y, Shi H, Sun B, Yin D, Sun J, Song R, et al. 2015. YAP is a critical oncogene in human cholangiocarcinoma. *Oncotarget* **6**: 17206–17220. doi:10.18632/oncotarget.4043

Peinado H, Zhang H, Matei IR, Costa-Silva B, Hoshino A, Rodrigues G, Psaila B, Kaplan RN, Bromberg JF, Kang Y, et al. 2017. Pre-metastatic niches: organ-specific homes for metastases. *Nat Rev Cancer* **17**: 302–317. doi:10.1038/nrc.2017.6

Pertovaara L, Kaipainen A, Mustonen T, Orpana A, Ferrara N, Saksela O, Alitalo K. 1994. Vascular endothelial growth factor is induced in response to transforming growth factor-β in fibroblastic and epithelial cells. *J Biol Chem* **269**: 6271–6274. doi:10.1016/S0021-9258(17)37365-9

Petty AJ, Li A, Wang X, Dai R, Heyman B, Hsu D, Huang X, Yang Y. 2019. Hedgehog signaling promotes tumor-associated macrophage polarization to suppress intratumoral CD8[+] T cell recruitment. *J Clin Invest* **129**: 5151–5162. doi:10.1172/JCI128644

Petty AJ, Dai R, Lapalombella R, Baiocchi RA, Benson DM, Li Z, Huang X, Yang Y. 2021. Hedgehog-induced PD-L1 on tumor-associated macrophages is critical for suppression of tumor-infiltrating CD8[+] T cell function. *JCI Insight* **6**: e146707. doi:10.1172/jci.insight.146707

Polyak K, Weinberg RA. 2009. Transitions between epithelial and mesenchymal states: acquisition of malignant and stem cell traits. *Nat Rev Cancer* **9**: 265–273. doi:10.1038/nrc2620

Principe DR, Doll JA, Bauer J, Jung B, Munshi HG, Bartholin L, Pasche B, Lee C, Grippo PJ. 2014. TGF-β: duality of function between tumor prevention and carcinogenesis. *J Natl Cancer Inst* **106**: djt369. doi:10.1093/jnci/djt369

Qi J, Sun H, Zhang Y, Wang Z, Xun Z, Li Z, Ding X, Bao R, Hong L, Jia W, et al. 2022. Single-cell and spatial analysis reveal interaction of FAP[+] fibroblasts and SPP1[+] macrophages in colorectal cancer. *Nat Commun* **13**: 1742. doi:10.1038/s41467-022-29366-6

Quail DF, Joyce JA. 2013. Microenvironmental regulation of tumor progression and metastasis. *Nat Med* **19**: 1423–1437. doi:10.1038/nm.3394

Regis S, Dondero A, Caliendo F, Bottino C, Castriconi R. 2020. NK cell function regulation by TGF-β-induced epigenetic mechanisms. *Front Immunol* **11**: 311. doi:10.3389/fimmu.2020.00311

Rhodes DR, Yu J, Shanker K, Deshpande N, Varambally R, Ghosh D, Barrette T, Pandey A, Chinnaiyan AM. 2004. Large-scale meta-analysis of cancer microarray data identifies common transcriptional profiles of neoplastic transformation and progression. *Proc Natl Acad Sci* **101**: 9309–9314. doi:10.1073/pnas.0401994101

Rizzino A. 2009. Sox2 and Oct-3/4: a versatile pair of master regulators that orchestrate the self-renewal and pluripotency of embryonic stem cells. *Wiley Interdisc Rev Syst Biol Med* **1**: 228–236. doi:10.1002/wsbm.12

Rogers HA, Sousa S, Salto C, Arenas E, Coyle B, Grundy RG. 2012. WNT/β-catenin pathway activation in Myc immortalised cerebellar progenitor cells inhibits neuronal differentiation and generates tumours resembling medulloblastoma. *Br J Cancer* **107**: 1144–1152. doi:10.1038/bjc.2012.377

Rota R, Ciarapica R, Miele L, Locatelli F. 2012. Notch signaling in pediatric soft tissue sarcomas. *BMC Med* **10**: 141. doi:10.1186/1741-7015-10-141

Sahai E, Astsaturov I, Cukierman E, DeNardo DG, Egeblad M, Evans RM, Fearon D, Greten FR, Hingorani SR, Hunter T, et al. 2020. A framework for advancing our understanding of cancer-associated fibroblasts. *Nat Rev Cancer* **20**: 174–186. doi:10.1038/s41568-019-0238-1

 Cite this article as *Cold Spring Harb Perspect Med* doi: 10.1101/cshperspect.a041609

Saito A, Suzuki HI, Horie M, Ohshima M, Morishita Y, Abiko Y, Nagase T. 2013. An integrated expression profiling reveals target genes of TGF-β and TNF-α possibly mediated by microRNAs in lung cancer cells. *PLoS ONE* **8:** e56587. doi:10.1371/journal.pone.0056587

Scharer CD, McCabe CD, Ali-Seyed M, Berger MF, Bulyk ML, Moreno CS. 2009. Genome-wide promoter analysis of the *SOX4* transcriptional network in prostate cancer cells. *Cancer Res* **69:** 709–717. doi:10.1158/0008-5472.CAN-08-3415

Scharpfenecker M, van Dinther M, Liu Z, van Bezooijen RL, Zhao Q, Pukac L, Löwik CW, ten Dijke P. 2007. BMP-9 signals via ALK1 and inhibits bFGF-induced endothelial cell proliferation and VEGF-stimulated angiogenesis. *J Cell Sci* **120:** 964–972. doi:10.1242/jcs.002949

Scott LE, Weinberg SH, Lemmon CA. 2019. Mechanochemical signaling of the extracellular matrix in epithelial–mesenchymal transition. *Front Cell Dev Biol* **7:** 135. doi:10.3389/fcell.2019.00135

Shaked Y, McAllister S, Fainaru O, Almog N. 2014. Tumor dormancy and the angiogenic switch: possible implications of bone marrow–derived cells. *Curr Pharm Des* **20:** 4920–4933. doi:10.2174/1381612819666131125153536

Shen M, Du Y, Ye Y. 2021. Tumor-associated macrophages, dendritic cells, and neutrophils: biological roles, crosstalk, and therapeutic relevance. *Med Rev* **1:** 222–243. doi:10.1515/mr-2021-0014

Shi X, Yang J, Deng S, Xu H, Wu D, Zeng Q, Wang S, Hu T, Wu F, Zhou H. 2022. TGF-β signaling in the tumor metabolic microenvironment and targeted therapies. *J Hematol Oncol* **15:** 135. doi:10.1186/s13045-022-01349-6

Shibata M, Nanno K, Yoshimori D, Nakajima T, Takada M, Yazawa T, Mimura K, Inoue N, Watanabe T, Tachibana K, et al. 2022. Myeloid-derived suppressor cells: cancer, autoimmune diseases, and more. *Oncotarget* **13:** 1273–1285. doi:10.18632/oncotarget.28303

Sica A, Porta C, Morlacchi S, Banfi S, Strauss L, Rimoldi M, Totaro MG, Riboldi E. 2012. Origin and functions of tumor-associated myeloid cells (TAMCs). *Cancer Microenviron* **5:** 133–149. doi:10.1007/s12307-011-0091-6

Sleeboom JJF, van Tienderen GS, Schenke-Layland K, van der Laan LJW, Khalil AA, Verstegen MMA. 2024. The extracellular matrix as hallmark of cancer and metastasis: from biomechanics to therapeutic targets. *Sci Transl Med* **16:** eadg3840. doi:10.1126/scitranslmed.adg3840

Stabile H, Mitola S, Moroni E, Belleri M, Nicoli S, Coltrini D, Peri F, Pessi A, Orsatti L, Talamo F, et al. 2007. Bone morphogenic protein antagonist Drm/gremlin is a novel proangiogenic factor. *Blood* **109:** 1834–1840. doi:10.1182/blood-2006-06-032276

Steinhart Z, Angers S. 2018. Wnt signaling in development and tissue homeostasis. *Development* **145:** dev146589. doi:10.1242/dev.146589

Stoker AW, Hatier C, Bissell MJ. 1990. The embryonic environment strongly attenuates v-src oncogenesis in mesenchymal and epithelial tissues, but not in endothelia. *J Cell Biol* **111:** 217–228. doi:10.1083/jcb.111.1.217

Sun X, Cui Y, Feng H, Liu H, Liu X. 2019. TGF-β signaling controls *Foxp3* methylation and T reg cell differentiation by modulating Uhrf1 activity. *J Exp Med* **216:** 2819–2837. doi:10.1084/jem.20190550

Takeda K, Mizushima T, Yokoyama Y, Hirose H, Wu X, Qian Y, Ikehata K, Miyoshi N, Takahashi H, Haraguchi N, et al. 2018. Sox2 is associated with cancer stem-like properties in colorectal cancer. *Sci Rep* **8:** 17639. doi:10.1038/s41598-018-36251-0

Tang B, Vu M, Booker T, Santner SJ, Miller FR, Anver MR, Wakefield LM. 2003. TGF-β switches from tumor suppressor to prometastatic factor in a model of breast cancer progression. *J Clin Invest* **112:** 1116–1124. doi:10.1172/JCI200318899

Tantin D. 2013. Oct transcription factors in development and stem cells: insights and mechanisms. *Development* **140:** 2857–2866. doi:10.1242/dev.095927

Terry RL, Meyran D, Ziegler DS, Haber M, Ekert PG, Trapani JA, Neeson PJ. 2020. Immune profiling of pediatric solid tumors. *J Clin Invest* **130:** 3391–3402. doi:10.1172/JCI137181

Thomas DA, Massagué J. 2005. TGF-β directly targets cytotoxic T cell functions during tumor evasion of immune surveillance. *Cancer Cell* **8:** 369–380. doi:10.1016/j.ccr.2005.10.012

Thu KL, Becker-Santos DD, Radulovich N, Pikor LA, Lam WL, Tsao MS. 2014. SOX15 and other SOX family members are important mediators of tumorigenesis in multiple cancer types. *Oncoscience* **1:** 326–335. doi:10.18632/oncoscience.46

Tigue ML, Loberg MA, Goettel JA, Weiss WA, Lee E, Weiss VL. 2023. Wnt signaling in the phenotype and function of tumor-associated macrophages. *Cancer Res* **83:** 3–11. doi:10.1158/0008-5472.CAN-22-1403

Timperi E, Gueguen P, Molgora M, Magagna I, Kieffer Y, Lopez-Lastra S, Sirven P, Baudrin LG, Baulande S, Nicolas A, et al. 2022. Lipid-associated macrophages are induced by cancer-associated fibroblasts and mediate immune suppression in breast cancer. *Cancer Res* **82:** 3291–3306. doi:10.1158/0008-5472.CAN-22-1427

Tremblay AM, Missiaglia E, Galli GG, Hettmer S, Urcia R, Carrara M, Judson RN, Thway K, Nadal G, Selfe JL, et al. 2014. The Hippo transducer YAP1 transforms activated satellite cells and is a potent effector of embryonal rhabdomyosarcoma formation. *Cancer Cell* **26:** 273–287. doi:10.1016/j.ccr.2014.05.029

Uluçkan O, Segaliny A, Botter S, Santiago JM, Mutsaers AJ. 2015. Preclinical mouse models of osteosarcoma. *Bonekey Rep* **4:** 670. doi:10.1038/bonekey.2015.37

Vakkila J, Jaffe R, Michelow M, Lotze MT. 2006. Pediatric cancers are infiltrated predominantly by macrophages and contain a paucity of dendritic cells: a major nosologic difference with adult tumors. *Clin Cancer Res* **12:** 2049–2054. doi:10.1158/1078-0432.CCR-05-1824

Valverde Lde F, Pereira Tde A, Dias RB, Guimarães VS, Ramos EA, Santos JN, Gurgel Rocha CA. 2016. Macrophages and endothelial cells orchestrate tumor-associated angiogenesis in oral cancer via hedgehog pathway activation. *Tumour Biol* **37:** 9233–9241. doi:10.1007/s13277-015-4763-6

Veglia F, Sanseviero E, Gabrilovich DI. 2021. Myeloid-derived suppressor cells in the era of increasing myeloid cell diversity. *Nat Rev Immunol* **21:** 485–498. doi:10.1038/s41577-020-00490-y

Venkatesh V, Nataraj R, Thangaraj GS, Karthikeyan M, Gnanasekaran A, Kaginelli SB, Kuppanna G, Kallappa CG,

Basalingappa KM. 2018. Targeting Notch signalling pathway of cancer stem cells. *Stem Cell Investig* **5**: 5. doi:10.21037/sci.2018.02.02

Venugopal DC, Caleb CL, Kirupakaran NP, Shyamsundar V, Ravindran S, Yasasve M, Krishnamurthy A, Harikrishnan T, Sankarapandian S, Ramshankar V. 2023. Clinicopathological significance of cancer stem cell markers (OCT-3/4 and SOX-2) in oral submucous fibrosis and oral squamous cell carcinoma. *Biomedicines* **11**: 1040. doi:10.3390/biomedicines11041040

Wang RN, Green J, Wang Z, Deng Y, Qiao M, Peabody M, Zhang Q, Ye J, Yan Z, Denduluri S, et al. 2014. Bone morphogenetic protein (BMP) signaling in development and human diseases. *Genes Dis* **1**: 87–105. doi:10.1016/j.gendis.2014.07.005

Wang P, Wan WW, Xiong SL, Feng H, Wu N. 2017. Cancer stem-like cells can be induced through dedifferentiation under hypoxic conditions in glioma, hepatoma and lung cancer. *Cell Death Discov* **3**: 16105. doi:10.1038/cddiscovery.2016.105

Watson MJ, Vignali PDA, Mullett SJ, Overacre-Delgoffe AE, Peralta RM, Grebinoski S, Menk AV, Rittenhouse NL, DePeaux K, Whetstone RD, et al. 2021. Metabolic support of tumour-infiltrating regulatory T cells by lactic acid. *Nature* **591**: 645–651. doi:10.1038/s41586-020-03045-2

Wei Y, Qin Q, Yan C, Hayes MN, Garcia SP, Xi H, Do D, Jin AH, Eng TC, McCarthy KM, et al. 2022. Single-cell analysis and functional characterization uncover the stem cell hierarchies and developmental origins of rhabdomyosarcoma. *Nat Cancer* **3**: 961–975. doi:10.1038/s43018-022-00414-w

Wessel KM, Kaplan RN. 2022. Targeting tumor microenvironment and metastasis in children with solid tumors. *Curr Opin Pediatr* **34**: 53–60. doi:10.1097/MOP.0000000000001082

Wu B, Shi X, Jiang M, Liu H. 2023. Cross-talk between cancer stem cells and immune cells: potential therapeutic targets in the tumor immune microenvironment. *Mol Cancer* **22**: 38. doi:10.1186/s12943-023-01748-4

Xia R, Xu M, Yang J, Ma X. 2022. The role of Hedgehog and Notch signaling pathway in cancer. *Mol Biomed* **3**: 44. doi:10.1186/s43556-022-00099-8

Xu J, Lamouille S, Derynck R. 2009. TGF-β-induced epithelial to mesenchymal transition. *Cell Res* **19**: 156–172. doi:10.1038/cr.2009.5

Xu L, Pierce JL, Sanchez A, Chen KS, Shukla AA, Fustino NJ, Stuart SH, Bagrodia A, Xiao X, Guo L, et al. 2023. Integrated genomic analysis reveals aberrations in WNT signaling in germ cell tumors of childhood and adolescence. *Nat Commun* **14**: 2636. doi:10.1038/s41467-023-38378-9

Yang L, Huang J, Ren X, Gorska AE, Chytil A, Aakre M, Carbone DP, Matrisian LM, Richmond A, Lin PC, et al. 2008. Abrogation of TGFβ signaling in mammary carcinomas recruits Gr⁻1⁺CD11b⁺ myeloid cells that promote metastasis. *Cancer Cell* **13**: 23–35. doi:10.1016/j.ccr.2007.12.004

Yang S, Zheng J, Ma Y, Zhu H, Xu T, Dong K, Xiao X. 2012. Oct4 and Sox2 are overexpressed in human neuroblastoma and inhibited by chemotherapy. *Oncol Rep* **28**: 186–192.

Yang D, Liu J, Qian H, Zhuang Q. 2023. Cancer-associated fibroblasts: from basic science to anticancer therapy. *Exp Mol Med* **55**: 1322–1332. doi:10.1038/s12276-023-01013-0

Yoshida GJ. 2020. Regulation of heterogeneous cancer-associated fibroblasts: the molecular pathology of activated signaling pathways. *J Exp Clin Cancer Res* **39**: 112. doi:10.1186/s13046-020-01611-0

Yoshida H, Hayashi S, Kunisada T, Ogawa M, Nishikawa S, Okamura H, Sudo T, Shultz LD, Nishikawa S. 1990. The murine mutation osteopetrosis is in the coding region of the macrophage colony stimulating factor gene. *Nature* **345**: 442–444. doi:10.1038/345442a0

Yuan S, Tao F, Zhang X, Zhang Y, Sun X, Wu D. 2020. Role of Wnt/β-catenin signaling in the chemoresistance modulation of colorectal cancer. *Biomed Res Int* **2020**: 9390878. doi:10.1155/2020/9390878

Yuan Z, Li Y, Zhang S, Wang X, Dou H, Yu X, Zhang Z, Yang S, Xiao M. 2023. Extracellular matrix remodeling in tumor progression and immune escape: from mechanisms to treatments. *Mol Cancer* **22**: 48. doi:10.1186/s12943-023-01744-8

Yue B. 2014. Biology of the extracellular matrix: an overview. *J Glaucoma* **23**: S20–S23. doi:10.1097/IJG.0000000000000108

Zanconato F, Forcato M, Battilana G, Azzolin L, Quaranta E, Bodega B, Rosato A, Bicciato S, Cordenonsi M, Piccolo S. 2015. Genome-wide association between YAP/TAZ/TEAD and AP-1 at enhancers drives oncogenic growth. *Nat Cell Biol* **17**: 1218–1227. doi:10.1038/ncb3216

Zhao H, Wei J, Sun J. 2020. Roles of TGF-β signaling pathway in tumor microenvironment and cancer therapy. *Int Immunopharmacol* **89**: 107101. doi:10.1016/j.intimp.2020.107101

Zhou Y, Huang T, Zhang J, Cheng ASL, Yu J, Kang W, To KF. 2018. Emerging roles of Hippo signaling in inflammation and YAP-driven tumor immunity. *Cancer Lett* **426**: 73–79. doi:10.1016/j.canlet.2018.04.004

Zhu C, Li L, Zhao B. 2015. The regulation and function of YAP transcription co-activator. *Acta Biochim Biophys Sin (Shanghai)* **47**: 16–28. doi:10.1093/abbs/gmu110

Zweidler-McKay PA. 2008. Notch signaling in pediatric malignancies. *Curr Oncol Rep* **10**: 459–468. doi:10.1007/s11912-008-0071-2

Osteosarcoma through the Lens of Bone Development, Signaling, and Microenvironment

Elizabeth P. Young,[1] Amanda E. Marinoff,[1] Eunice Lopez-Fuentes, and E. Alejandro Sweet-Cordero

Division of Pediatric Oncology, Department of Pediatrics, University of California San Francisco, San Francisco, California 94158, USA

Correspondence: alejandro.sweet-Cordero@ucsf.edu

In this work, we review the multifaceted connections between osteosarcoma (OS) biology and normal bone development. We summarize and critically analyze existing research, highlighting key areas that merit further exploration. The review addresses several topics in OS biology and their interplay with normal bone development processes, including OS cell of origin, genomics, tumor microenvironment, and metastasis. We examine the potential cellular origins of OS and how their roles in normal bone growth may contribute to OS pathogenesis. We survey the genomic landscape of OS, highlighting the developmental roles of genes frequently altered in OS. We then discuss the OS microenvironment, emphasizing the transformation of the bone niche in OS to facilitate tumor growth and metastasis. The role of stromal and immune cells is examined, including their impact on tumor progression and therapeutic response. We further provide insights into potential development-informed opportunities for novel therapeutic strategies.

Osteosarcoma (OS), the most prevalent primary bone cancer in adolescents and young adults, presents a unique intersection between oncology and developmental biology. The process of bone growth, intricately influenced by environmental and genetic factors, plays a pivotal role in the development of OS, although the precise mechanisms underlying this relationship are not yet fully understood. The peak incidence of OS during adolescence, a period marked by rapid bone growth and hormonal changes, strongly supports an important role for growth hormones and bone development processes in its pathogenesis. Intriguingly, OS peak incidence occurs at a younger age in girls compared to boys, hinting at underlying hormonal influences (Glass and Fraumeni 1970). Moreover, the anatomical predilection of OS, primarily affecting the metaphysis of long bones with the knee being the most prevalent site, raises questions about the interaction between biological and mechanical factors. In contrast, <20% of OSs occur in the pelvis or skull, areas with differing bone composition and growth dynamics. Despite advancements in our understanding OS biology, unraveling the connection between bone development

---

[1]These authors contributed equally to this work.

Cite this article as *Cold Spring Harb Perspect Med* doi: 10.1101/cshperspect.a041635

and osteosarcomagenesis is a nascent area of study. Through a developmental lens, this paper aims to explore the complex interplay between OS biology and bone growth, hormonal regulation, and bone microenvironment, shedding light on the current state of knowledge and identifying avenues for future research.

BONE DEVELOPMENT AND OSTEOSARCOMA CELL OF ORIGIN

Osteocytes comprise 95% of the total cell population of normal bone and have a half-life of 25 yr (Pathak et al. 2020). Because they are terminally differentiated cells, their potential role as either a cell of origin or a contributing element to osteosarcomagenesis has been overlooked. However, osteocytes are well known to be responsive to mechanotransduction, and they regulate the formation and function of osteoblasts and osteoclasts (OCs) to maintain bone homeostasis, so may contribute to OS pathogenesis in ways that remain to be understood. The precursor to osteocytes are osteoblasts, which are thought to have a half-life of only a few days or weeks and have been considered to be the most likely cell of origin given that OS produces malignant osteoid. Because OS can have a chondroblastic, fibroblastic, osteoblastic, or rarely telangiectatic appearance, it is assumed that the skeletal precursors that give rise to OS still maintain some pluripotency.

Signaling Pathways Involved in Osteoblastic Lineage

During bone formation, mesenchymal stem cells (MSCs) differentiate at the epiphyseal growth plate into resting chondrocytes. Chondrogenesis is regulated by the Indian Hedgehog (IHH)/ parathyroid hormone-related protein (PTHrP) pathway. IHH produced by proliferating chondrocytes induces chondrocyte cell division as well as the secretion of PTHrP, which inhibits chondrocyte differentiation and maintains chondrocytes in a proliferative state. PTHrP also negatively regulates this process to allow the chondrocytes to differentiate in a controlled manner. Proliferative chondrocytes leave the epiphyseal plate and enter into a hypertrophic phase, after which they undergo apoptosis, leaving behind a hypocellular hyaline cartilage matrix into which osteoblasts migrate and initiate ossification (Chung et al. 2001). Signaling pathways important for bone development could be coopted to promote oncogenesis. For example, knockdown of the PTHrP receptor reduces activation through cyclic AMP–dependent protein kinase A (PKA) and decreases tumor differentiation, invasion, and proliferation in vivo (Walkley et al. 2014).

Osteoblasts secrete several extracellular matrix (ECM) proteins including type 1 collagen, osteocalcin, osteopontin, and alkaline phosphatase. Calcium is deposited as hydroxyapatite, which, together with type I collagen, forms the structural support of the skeleton. The differentiation of osteoblasts is thought to occur through at least three distinct stages: osteoprogenitors, preosteoblasts, and mature osteoblasts (Fig. 1). Expression of the transcription factor SOX9 is the first step in the commitment to an osteoprogenitor cell. RUNX2 then commits these osteoprogenitors toward to the preosteoblast stage. RUNX2 forms a heterodimer with core binding factor B and regulates expression of osteoblast genes including osteopontin (SPP1), bone sialoprotein 2 (IBSP), and osteocalcin 2 (GLAB2). RUNX2 is strongly expressed in immature osteoblasts but expression decreases in mature osteoblasts (Komori 2017; 2020). Further maturation proceeds through WNT/β-catenin-mediated expression of osterix (OSX or SP7). Once cells are embedded in the bone matrix, they become mature osteocytes and up-regulate expression of genes such as SOST and DKK1 (Zhou et al. 2010).

The process of bone maintenance is sensitive to mechanical forces; during mechanical unloading, osteocytes express receptor activator of nuclear factor κB ligand (RANKL), which promotes bone resorption through the activation of OCs. Conversely, in response to mechanical loading, osteocytes decrease the expression of Dickkopf-related protein 1 (DKK1) and sclerostin, leading to increased bone formation through activation of WNT/β-catenin signaling in osteoblasts. Osteocytes respond to hormonal and mechanical signals to tightly control bone

Cite this article as Cold Spring Harb Perspect Med doi: 10.1101/cshperspect.a041635

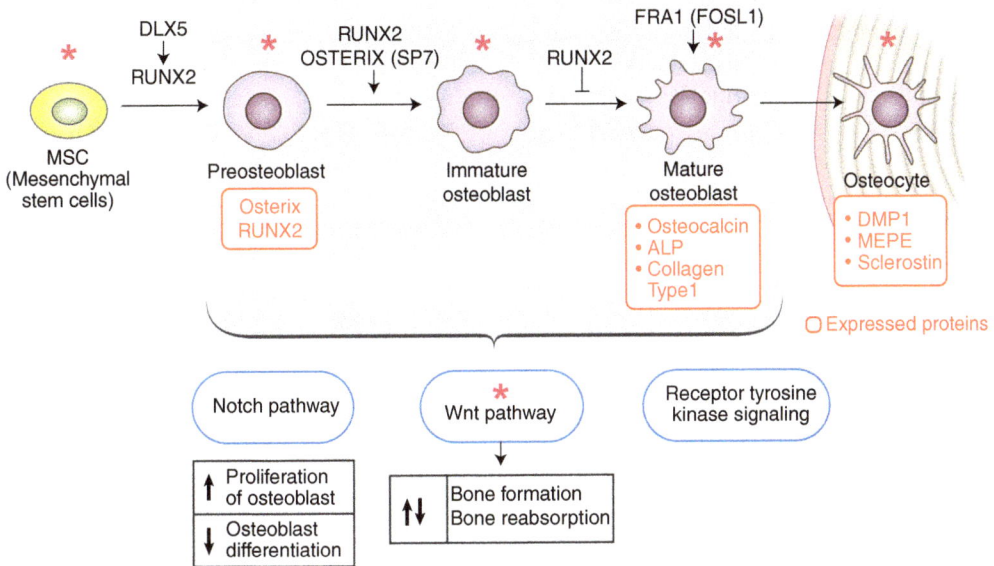

Figure 1. Transcription factors (TFs) and pathways involved in bone development. (*Top*) Key TFs involved in osteosarcomagenesis. Osteoblasts originate from mesenchymal progenitors driven by RUNX2, DLX5, and SP7. In the final stages of osteoblast differentiation, another set of TFs (including FOSL1) plays a central role. The cell of origin of osteosarcoma is thought to be a malignant transformation of the osteoblastic lineage. (*Bottom*) Signaling pathways involved in the initiation and proliferation of osteosarcoma cells. Canonical and noncanonical Wnt pathway is involved in most of the differentiation stages of the osteoblastic lineage.

remodeling through signaling pathways discussed below. Other transcriptional circuits are also important in bone development. For example, the transcription factors Fos, Jun, and ATF4 are members of the AP1 transcription factor complex and have a critical role in bone development (Wagner 2010; Bozec et al. 2013). In addition, Fra2 regulates bone formation by regulating expression of Col1a2 and osteocalcin in both mouse and human osteoblasts (Fig. 1; Bozec et al. 2010).

Other signaling pathways are also well known to be critical for osteoblast differentiation and may also be involved in OS. For example, osteoblasts require GLI1 or GLI2 activation via IHH. In the absence of IHH, osteoblast differentiation cannot proceed, and bone formation is incomplete (Jemtland et al. 2003). IHH signaling is important for endochondral bone formation. IHH stimulates GL2/3 via PTCH1, which also stimulates the transition from osteoblast to osteocyte via RUNX2 expression. In addition to HH, WNT and BMP signaling also cooperate in the formation of bone.

Furthermore, Notch enhances proliferation of osteoblasts and suppresses osteoblast differentiation. This may be in part through the repression of RUNX2 function (Tu et al. 2012b). Overall, the balance between Notch and WNT signaling is critical for the commitment of immature osteoblasts to mature osteoblasts. BMP signaling in turn complements and amplifies the effect of WNT on osteoblast differentiation.

Role of Osteoclasts in Bone Development

Normal bone homeostasis is maintained through the balance of osteoblast and OC function, as coordinated by osteocytes. OCs are produced by the hematopoietic lineage, have resorptive capacity, and form the bone marrow cavity itself during development (Yahara et al. 2022). The early wave of OC development occurs from yolk-sac derived early erythromyeloid progenitors. Later in development, bone marrow–derived hematopoietic stem cells also generate OCs. A recent single-cell RNA-seq study provides further detailed insight into the origin of OCs, demonstrating that OCs are derived from CD11c-expressing precursors and that the receptor activator of nuclear factor κB expression is required for OC development (Tsukasaki et al. 2020).

Key Mouse Models of OS

Mouse models have also been informative regarding the developmental origins of OS. For example, the genetic cooperativity between p53/Rb loss at accelerating OS is apparent only within the committed osteoblast populations (as marked by Osx1-Cre expression). As noted above, in mouse bone development, RUNX2 expression decreases and OSX increases as osteoblasts become mature osteocytes (Nakashima et al. 2002). In contrast, when p53/Rb is deleted in more immature populations, marked by Prx1-Cre, undifferentiated sarcomas predominate (Lin et al. 2009). Other recent work has demonstrated that OS can arise across a spectrum of cells from the osteoblastic lineage. Notably, even committed osteoblasts were able to initiate OS in vivo (Mutsaers et al. 2013; Mutsaers and Walkley 2014; Quist et al. 2015).

OS GENOMICS THROUGH A DEVELOPMENTAL LENS

OS is characterized by significant somatic copy-number alteration (SCNA) and structural variation (SV) with few recurrent point mutations in protein-coding genes. This characteristically chaotic genomic landscape suggests that copy number (CN)-amplified genes within SCNAs may be critical drivers of disease progression and maintenance. The most frequently occurring SCNAs involve loss of *RB1*, *CDNK2A/B*, *PTEN*, and amplification in *MYC*, *CDK4*, *CCNE1*, *CCND3*, *ATRX* (Fig. 2; Chen et al. 2014; Perry et al. 2014; Behjati et al. 2017; Sayles et al. 2019; Marinoff et al. 2023). The single most frequently altered gene is the tumor suppressor *TP53*, which is lost in >90% of OS tumors, with the majority lost through intron 1 rearrangements or deletions, rather than through point mutations. Genes commonly implicated as OS drivers also have critical roles in development and bone homeostasis. For example, p53 plays

an essential role in controlling the differentiation of MSCs by regulating genes involved in cell cycle and early steps of the differentiation process. Crucially, p53 functions as a negative regulator of osteogenesis, osteoblast differentiation, and bone remodeling, in part through its role in negatively regulating the expression of MSC-derived osteoprotegerin (OPG) (Wang et al. 2006; Velletri et al. 2021). Similarly, Rb1 deficiency in MSCs promotes osteogenesis, increases bone mass, and inhibits osteoclastogenesis through up-regulation of OPG via activation of yes-associated protein (YAP) signaling (Chinnam and Goodrich 2011; Li et al. 2022).

Myc family transcription factors, *c-Myc* and *N-Myc*, strongly influence the development and maintenance of pluripotent and multipotent cells in a number of tissues, including bone. *N-Myc* and *c-Myc* are sequentially expressed in developing limbs and play complementary roles in limb formation, with *N-Myc* expressed primarily in undifferentiated limb bud mesenchyme and *c-Myc* expressed in proliferating chondrocytes of the growth plate and osteoblasts in the perichondrium. While the normal expression of *c-MYC* is vital for coordinating the intricate processes of limb formation and postnatal bone development, its overexpression via *MYC* amplification, observed in 8%–15% of OS, has been associated with poor outcomes (Ladanyi et al. 1993; Smida et al. 2010; Marinoff et al. 2023).

Cell cycle regulators, including cyclin E1 (CCNE1), cyclin D3 (CCND3), cyclin-dependent kinase 4 (CDK4), and CDK6, are key regulators of the G_1-S phase transition and thus are essential for cell cycle progression and DNA replication across multiple developmental contexts. In the context of bone development, these and other cyclins and CDKs are essential for proliferation of osteoblasts and chondroblasts (Ogasawara et al. 2011; Shaikh et al. 2023). Conversely, cyclin-dependent kinase inhibitor 2A and 2B (CDKN2A/B) are tumor suppressors that encode cell cycle inhibitors p16INK4a and p15INK4b. These critical regulators of cell cycle progression in normal development may serve as ripe substrates for oncogenesis and osteosarcomagenesis, specifically, when dysregulated in the bone. Other frequently altered genes in OS also

Cite this article as *Cold Spring Harb Perspect Med* doi: 10.1101/cshperspect.a041635

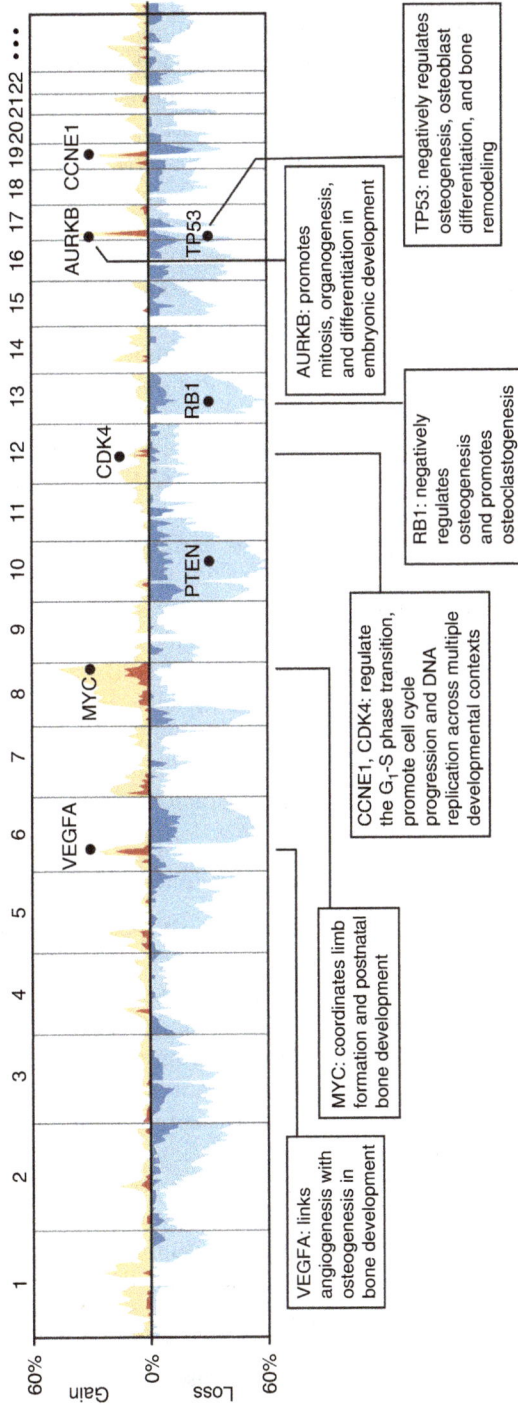

Figure 2. Representative plots of somatic copy number alterations in osteosarcoma (modified from Sayles et al. 2019). Frequently altered genes and their key roles in development are highlighted. Chromosomes are numbered across the *top*. Copy number gains are shown in yellow/red and losses are shown in blue. Yellow-CN gain >4 copies. Red-CN gain >8 copies. Light blue <1.2 copies (minor loss), dark blue <0.8 copies (major loss).

have critical roles in bone development and underscore OS as normal development gone awry. For example, bone formation relies on the coupling of angiogenesis and osteogenesis, and *VEGF*, amplified in a subset of OS, plays an essential role in mediating this link. PDGFRA/B play important roles in recruitment and differentiation of osteoblast precursors, as well as bone remodeling and fracture healing.

How do the complex genomic landscapes of OS develop? Genome-wide sequencing studies suggest several different mechanisms, including chromothripsis, chromoplexy, and kataegis. Chromothripsis is characterized by massive genomic rearrangements that are often generated in a single catastrophic event and localized to isolated chromosomal regions. In contrast to the traditional view of oncogenesis as the gradual process of the accumulation of mutations, chromothripsis provides a mechanism for the rapid accrual of hundreds of rearrangements in a few cell divisions. A comprehensive study of chromothripsis across over 2000 human cancers using WGS found that 77% of OS exhibit high confidence chromothripsis, and at least five chromosomes were affected in 66% of OSs (Cortés-Ciriano et al. 2020). Chromothripsis is uncommonly observed across pediatric cancers relative to adult cancers, but it has been reported in a subset of pediatric cancers, including OS, Wilms tumor, B-ALL, AML, and neuroblastoma (Ma et al. 2018), as well as in the Sonic hedgehog subtype of medulloblastomas in the context of *TP53* mutations (Rausch et al. 2012).

Two models have been proposed to explain the evolution of chromosomal structure and CNs in cancer genomes. One model suggests that underlying genomic instability gives rise to populations of cells with diverse phenotypic variations and that ongoing selection of advantageous phenotypes drives evolution and adaptation (Fearon and Vogelstein 1990; Höglund et al. 2001). A second model posits that discrete periods of genomic instability, isolated in tumor developmental time, give rise to extreme chromosomal complexity driven by a small number of impactful catastrophic events (Umbreit et al. 2020). A recent report using single-cell DNA sequencing in osteosarcoma supports the emerging hypothesis

that early catastrophic events, rather than sustained genomic instability, give rise to structural complexity, which is then preserved over long periods of tumor developmental time (Rajan et al. 2023). Ongoing research will continue to inform our understanding of the contributions of initial catastrophic events and ongoing mechanisms of genomic evolution and how they influence clinical outcomes.

Although the stepwise genomic mechanisms underlying osteosarcomagenesis are incompletely understood, it has long been understood that *TP53* is required for genome stability, and *TP53* alterations in OS correlate with high levels of genomic instability (Overholtzer et al. 2003). Therefore, it is highly plausible that osteosarcomagenesis involves loss of *TP53* via loss of heterozygosity (LOH) as an initiating or early event preceding a catastrophic event leading to massive genomic rearrangements. Alternatively, or in addition, *TP53* loss (via LOH and intron 1 rearrangements) may arise as a consequence of the massive chromosomal rearrangements. One can deduce that cells with gene CN patterns that confer a growth advantage form a dominant clone and subsequently acquire additional alterations that are evolutionarily favorable for survival, growth, proliferation, and metastasis.

Germline Predisposition Syndromes

Germline predispositions to OS provide a window into links to developmental processes. While most cases of OS are sporadic, a subset of cases occur in the context of known germline cancer predisposition syndromes (Table 1) nominating connections between OS pathogenesis and molecular mechanisms of bone maturation early in development. The reported frequency of germline alterations in OS patients ranges from 18% to 28% (Mirabello et al. 2015; Zhang et al. 2015). The substantial proportion of OS cases that occur within the context of a cancer-predisposing germline mutation suggests these genes are early initiating events in tumorigenesis.

Most cases of Li–Fraumeni syndrome (LFS) are caused by germline *TP53* alternations, underscoring the critical role of *TP53* in bone homeostasis and the differentiation of osteoblasts. Simi-

Cite this article as *Cold Spring Harb Perspect Med* doi: 10.1101/cshperspect.a041635

Table 1. Germline predisposition syndromes linked to osteosarcoma and development

Syndrome	Gene	Inheritance pattern	Relevance to development	References
Li–Fraumeni syndrome	TP53	Autosomal dominant	Bone homeostasis and the differentiation of osteoblasts	Velletri et al. 2016, 2021
Retinoblastoma	RB1	Autosomal dominant	Plays a critical role early in commitment to the osteoblast lineage and osteoblast differentiation	Gutierrez et al. 2008
Paget's disease	SQSTM1 (most common)	Autosomal dominant	Osteoclast dysregulation	Kansara and Thomas 2007; Gianferante et al. 2017; Cundy 2018; Beird et al. 2022
Bloom syndrome	RECQL3 (BLM)	Autosomal recessive	Maintenance of genome stability	Lu et al. 2020
Werner syndrome	RECQL2 (WRN)	Autosomal recessive	Maintenance of genome stability	Lu et al. 2020
Rothmund–Thomson syndrome (RTS)	RECQL4	Autosomal recessive	Bone differentiation and/or maintenance of genomic stability in osteoblasts; skeletal development by modulating p53 activity	Lu et al. 2015, 2020; Ng et al. 2015
Diamond Blackfan anemia (DBA)	Ribosomal proteins	Autosomal dominant	Potential dysregulation in cell differentiation	Costa et al. 2020

larly, hereditary retinoblastoma, linked to *RB1* mutations, highlights the dual role of Rb in retinal development and bone tissue regulation. Paget disease of bone is a benign metabolic bone disorder associated with OC dysregulation (Beird et al. 2022). It affects bone remodeling by increasing bone breakdown and bone formation. The RecQ family of DNA helicases includes proteins encoded by genes that are disrupted in Bloom syndrome, Werner syndrome, and Rothmund–Thomson syndrome (RTS). DNA helicases are integral to DNA repair and thus interruption of their function may herald genomic instability and thus contribute to tumorigenesis. Interestingly, patients with RTS patients often have skeletal defects, suggesting a potential role for RECQL4 in bone differentiation and/or maintenance of genomic stability in osteoblasts (Ramaswamy et al. 2016). Diamond Blackfan anemia (DBA) is a rare congenital bone marrow failure syndrome caused by mutations in ribosomal proteins. From a developmental perspective, both bone and bone marrow arise from MSCs, which possess the capability to differentiate into osteoblasts and hematopoietic

stem cells. A disruption in the finely tuned genetic programs that regulate these differentiation processes can result in both bone marrow failure syndromes and bone malignancies. In addition, given the role of protein synthesis in governing cell growth, differentiation, and function, it is conceivable that the same genetic aberrations causing marrow failure in DBA might also disturb the osteoblastic lineage's normal maturation. Over time, these disturbances may predispose to both congenital bone abnormality, as seen in DBA, and predisposition to OS. Thus, the onset of OS in the context of these germline predisposition syndromes can be viewed as a derailment of the developmental trajectory, leading to uncontrolled proliferation, and subsequently malignancy.

Genomics at the Intersection of Puberty and Growth

OS most often develops in the appendicular skeleton during puberty and the developmental growth spurt. This observation has spurred interest in investigating the potential role of hormonal signaling pathways activated during puberty in OS pathogenesis. This association first arose in the context of dog OS, in which OS is more common in large or giant breeds (Tjalma 1966; Withrow et al. 1991). In humans, a similar relationship between body size and the risk of developing OS has been suggested by case reports of OS occurring in individuals with acromegaly (Lima et al. 2006) and a case series demonstrating an association between height and OS risk (Longhi et al. 2005).

The biological basis for the association with growth is incompletely understood. However, insulin-like growth factor I (IGF-I) is a potent mitogen for human OS cells and plays a key role in longitudinal bone growth with its receptor in response to growth hormone receptor (Pollak et al. 1990). IGF-1 receptor (IGF1R) activation induces the recruitment and stimulation of signaling adaptor proteins such as IRS-1/2 and SHC that trigger the PI3K/AKT and the RAS/MAP kinase signaling cascades (Ameline et al. 2021). In the context of deregulated IGF-1 signaling, the PI3K/AKT pathway appears to be overactivated during early tumor development and pulmonary spread, whereas RAS/MAPK pathway activation could rather play a role at later stages of pulmonary dissemination (Wang et al. 2019). Interestingly, overexpression of IGF1R also has been shown in canine OS to strongly correlate with tumor stage and adverse outcome (Sergi et al. 2019). Recurrent alterations of the IGF family of genes have been detected in 7% of OS cases, and although clinical trials of IGF1R inhibitors have not shown benefit to date across all OS cases, it is possible a subset of those with underlying aberrations in IGF signaling may benefit.

Puberty is associated with profound physical changes, and there is an association between sex hormones and OS pathogenesis. OS is slightly more common in males than females, with an average male-to-female ratio of 1.4:1 (Brown et al. 2018). Estrogen levels rise during puberty and estrogens regulate osteoblast differentiation and bone mineral density. However, there is little understanding the role of this estrogen peak in OS development during puberty. Estrogens induce the transcription of osteoblast differentiation genes, such as alkaline phosphatase (AP)

and BMP2 (Krum 2011). OS tumors and cell lines models do not express estrogen receptor α (ERα), whereas normal osteoblasts do express ERα (Dohi et al. 2008; Lillo Osuna et al. 2019). Lillo Osuna et al. showed that ERs can be reexpressed with exposure to decitabine (Lillo Osuna et al. 2019), an FDA-approved DNA methyltransferase inhibitor. Treatment with decitabine induced differentiation, an effect replicated simply with plasmid-driven expression of *ESR1* and enhanced with the addition of exogenous estrogen. The same treatment decreased proliferation and colony formation for OS cell lines in vitro and inhibited tumor growth and decreased metastasis in a mouse xenograft model of OS. Decitabine treatment additionally had profound effects on the formation of metastatic lesions in the lungs (Lillo Osuna et al. 2019). These results suggest that reversible epigenetic silencing of ERs could be a therapeutically viable strategy for inducing differentiation in OS tumors.

Androgen receptor (AR) is highly expressed in various OS cell lines and patient tumor tissues (Wagle et al. 2015; Liao et al. 2017). Liao et al. found by differential gene expression several genes that might be the AR downstream target genes in OS, such as cyclin G2, CDC20, caspase-8, and JNK (Liao et al. 2017). Most of those genes have been reported to be involved in AR signaling in prostate cancer and are plausibly involved in OS. The higher incidence in males than females suggests an additional factor involved in osteosarcomagenesis during pubertal development. These and other studies underscore the need for further evaluation of the link between OS and endocrine pathways.

Genome- and Development-Informed Therapeutic Opportunities for OS

Insights into the genomic landscape and dysregulated developmental pathways in OS have led to opportunities for exploration of targeted therapies. Here we highlight several potential strategies drawing from advances in knowledge of OS genomics and links to development. Targeting patient-specific SCNAs, including recurrent alterations in CDK4, CCNE1, MYC, and AURKB as well as aberrant signaling in PI3K/AKT/MTOR and

VEGF signaling, is a strategy that has shown promise in patient-derived xenograft (PDX) models of OS (Sayles et al. 2019). This provides a blueprint for developing rational combination therapies either with multiple targeted agents or targeted agents in combination with cytotoxic chemotherapy and setting the stage for biomarker-driven clinical trial designs, such as umbrella trials evaluating multiple targeted therapies for OS stratified into molecularly defined subgroups. While targeting mutant p53 has to date proven a scientific and clinical challenge, mutant p53-reactivating compounds are being explored as a possible mechanism by which to selectively target mutant *TP53*, an attractive strategy that may prove beneficial in treating OS (Bykov et al. 2018).

There has also been significant interest in incorporating tyrosine kinase inhibitors (TKIs) into combination therapies in clinical trials for OS and other bone tumors (Grignani et al. 2012; Duffaud et al. 2019; Italiano et al. 2020; Gaspar et al. 2021). One such TKI, cabozantinib, which inhibits VEGF and c-MET, is currently being investigated in combination with cytotoxic chemotherapy in a randomized phase II/III study for newly diagnosed OS (NCT05691478). Targeting aberrant signaling in pathways involved in bone development is an additional strategy under investigation in preclinical studies and early phase clinical trials, including targeting the Wnt pathway (NCT04851119), the Hedgehog pathway with SMO inhibitors (vismodegib and sonidegib, which are FDA-approved for basal cell carcinoma and currently being investigated for other indications, including sarcomas) and Notch signaling with γ-secretase inhibitors (such as RO4929097 studied in an adult sarcoma phase I/II study, NCT01154452) (Gounder et al. 2022).

Given emerging evidence highlighting the role of epigenetics in regulating OS pathogenesis, drugs targeting DNA methylation (such as azacitidine and decitabine) (Lillo Osuna et al. 2019) and histone deacetylases (HDAC inhibitors, such as vorinostat and panobinostat) (Nerlakanti et al. 2019) may serve as important components in novel OS combination strategies. The dysregulation of pathways involved in both OS pathogenesis and developmental processes, including development of bone and the immune response, suggest a

potential avenue for differentiation-directed therapy, such as all-*trans* retinoic acid (Zhou et al. 2017) and engineered MSCs. The DNA damage response (DDR) has also emerged as an additional promising therapeutic target. The combination of the PARP inhibitor, olaparib, and ATR inhibitor, ceralasertib, is currently the subject of investigation in an ongoing phase II study in recurrent OS (NCT04417062) and emerging preclinical data suggest targeting the ATR/Wee1/Chk1 pathway may be a fruitful strategy in pediatric sarcomas (unpublished data presented at the Children's Oncology Group fall meeting, 2023, combined sarcoma biology session).

DEVELOPMENTAL BASIS OF OS MICROENVIRONMENT AND METASTASIS

The developmental origin of OS also has important implications for the tumor microenvironment. In this developmental model, bone-specific microenvironmental influences beyond cell-intrinsic oncogenic events may promote OS development and metastasis. The bone niche, containing ECM, stromal, and immune factors, is likely co-opted in OS to create a permissive tumor microenvironment in the bone. In normal bone, a variety of cell types and soluble factors function in concert to provide a scaffold and vascular supply that can also support rapid bone growth arising from the epiphyseal plate during puberty. These productive roles and relationships are transformed in the context of OS tumorigenesis to promote tumor growth and metastasis (Fig. 3). A biomechanical study of the capacity of different cell lines to undergo gap junction intracellular communication has suggested that cell–cell communication networks are established in the bone, including between OS cells in a homotypic manner, as well as between OS cells and osteoblasts (but not OCs) (Tellez-Gabriel et al. 2017).

Stromal Cell and Vascular Contributions

Insight into the complexity of the OS microenvironment is emerging from studies using single-cell technologies. A recent study performed scRNA-seq analysis on 11 OS tumor samples, ob-

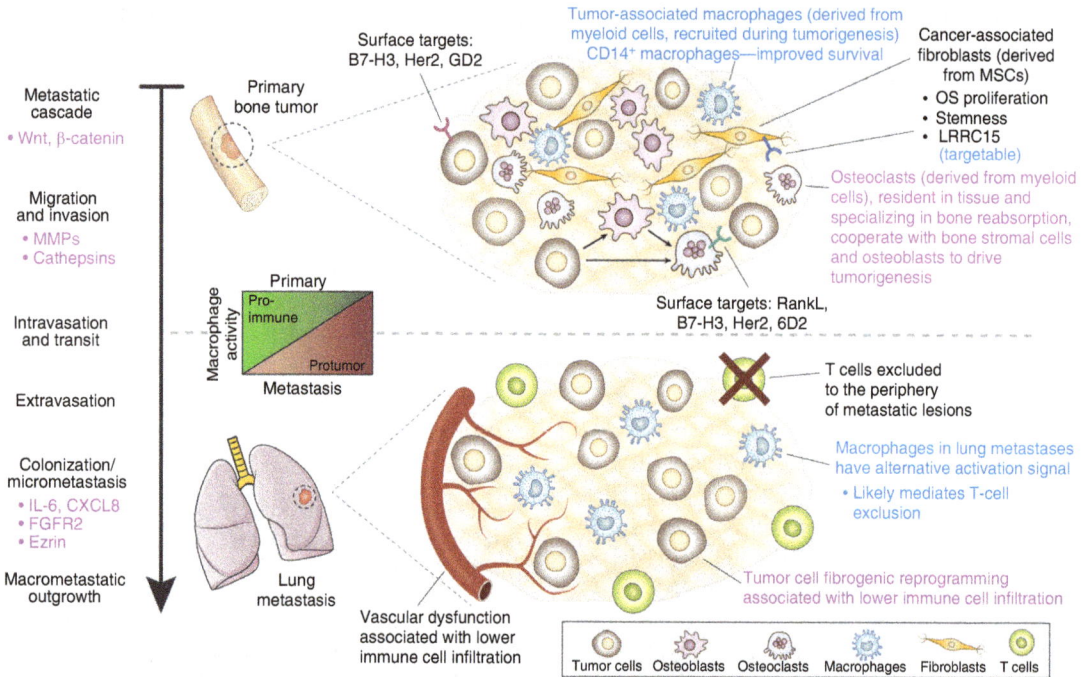

Figure 3. Osteosarcoma primary bone and metastatic niche. (*Left*) Key contributors to steps in metastatic cascade. (*Right, top*) Primary bone niche with cellular and structural contributions from bone microenvironment, including the following key cell types: cancer-associated fibroblasts, osteoclasts, osteoblasts, and tumor-associated macrophages (TAMs), highlighting interaction between tumor cells, osteoclasts, and osteoblasts via RANK-RANKL signaling. Macrophages in primary tumor appear to be more proimmune in contrast to TAMs in lung metastases. Low T-cell infiltration in primary bone tumor is likely due to some extent to bone structural factors; important surface targets for engineered T-cell therapies are shown. (*Right, bottom*) Lung metastatic niche, showing vascular dysfunction that likely mediates lower immune cell infiltration; macrophages in lung are characterized by an alternative activation signature (protumor) and this may mediate T-cell exclusion to periphery of lesions.

taining single-cell transcriptomes from ~100,000 cells (2/3 primary tumors and 1/3 recurrent or metastatic tumors), identifying 11 major cell type clusters (Zhou et al. 2020). This group identified three stromal MSC subclusters and three fibroblast subclusters. A proposed model suggests that OS cells directly induce differentiation of MSCs into cancer-associated fibroblasts (CAFs) (Cortini et al. 2017). Fibroblasts expressing COL14A1, MYL9, and LUM were enriched in primary and recurrent bone tumors, in contrast to fibroblasts expressing low COL14A1 and high DES, which were enriched in lung metastases. The precise cellular functions of these CAFs in OS tumors has heretofore not been established.

MSCs, however, have been more extensively studied in OS in vitro models. MSCs induce proliferation of OS and promote OS stemness, and acidosis derived from tumor growth and osteolysis in turn promotes stemness of MSCs (Cortini et al. 2017). OS cells educate MSCs via tumor-secreted extracellular vesicles (EVs) (Baglio et al. 2017). TGF-β from tumor EVs induces production of IL-6 in MSCs. In the Zhou et al. (2020) single-cell data set, three cellular subclusters of MSCs were defined, based on the following markers: NT5E, WISP2, and CLEC11A, suggesting differing roles for these MSCs in the context of the OS TME. Primary tumors appear to have a higher enrichment for WISP2$^+$ MSCs, which are presumed to be undergoing a MSC proliferation program, in contrast to recurrent tumors that enrich much more highly for NT5E$^+$ MSCs. This shift supports transition to a more immunosuppressive role for MSCs in recurrent disease as NT5E (CD73) produces adenosine and these NT5E$^+$ MSCs also coexpress higher levels of TGFB1, as well as VEGFA. A preclinical study targeting a highly expressed protein on CAFs (*LRRC15*) with an antibody-drug conjugate supports the protumor role of the OS stroma (Hingorani et al. 2021).

Vascular proliferation and dysfunction have also been linked to OS tumorigenesis and metastasis primarily through the contributions of VEGF and VEGF receptors, which are associated with development of pulmonary metastasis and decreased survival (Sorenson et al. 2019). This data has nominated vascular normalization therapy with multityrosine kinase inhibitor (MTKI) as a possible avenue for therapeutic benefit in OS, and indeed some benefit has been shown in relapsed disease, prompting creating of a cooperative group trial studying addition of MTKIs to standard of care chemotherapy, as discussed above (Sayles et al. 2019; Italiano et al. 2020).

Conflicting Roles of Osteoclasts in OS Pathogenesis

OCs are a crucial component of the bone OS niche (Fig. 3). Our understanding of the role of OCs in OS derives in part from OS-specific studies correlating OC abundance with prognosis or other clinical features, as well as from the established role of OCs in promoting bone metastasis of tumors of nonskeletal origin (e.g., breast cancer). Based on pathophysiology of bone metastatic epithelial cancers, there is a proposed model wherein OCs cooperate with bone stromal cells and osteoblasts to drive progression, supported by findings that OS tumor growth increases OC activity resulting in OC-mediated bone resorption and release of protumor factors from the bone matrix (Endo-Munoz et al. 2012). Radiologically, evidence of osteolysis and bone remodeling supports the presence of OC activity in OS primary tumor development. In the Zhou et al. (2020) scRNA-seq data set, three subclusters of OC are observed: (1) high myeloid markers (CD74 and CD14), with dim OC markers (CTSK and ACP5), (2) coexpression of myeloid and OC markers, and (3) high CTSK and ACP5 with low myeloid markers. The proportion of mature OC was found to be much lower in recurrent and lung metastatic lesions, supporting a role for interplay between osteoblastic lesions and OC activity as a unique feature of the OS primary tumor bone niche (Zhou et al. 2020).

Literature on the role of OCs in OS metastasis is conflicting, as OCs seem to have a tumor promoting role in primary tumor development but an opposite role in advanced disease including metastasis (Akiyama et al. 2008; Endo-Munoz et al. 2012). Moreover, a significant inverse relationship has been identified between OC markers and time to development of pulmonary metastasis, suggesting a protective effect of OCs against metastasis; this study also identified a role of tumor cells in secreting factors that directly inhibit OC proliferation (Endo-Munoz et al. 2010). These contradictory reports have been reconciled by a proposed model wherein OCs contribute to the foundational primary tumor niche as described above, but subsequent tumor evolution begets expansion of tumor clones that inhibit this OC activity and therefore promote tumor invasion and metastasis away from this niche (Endo-Munoz et al. 2012). In preclinical models, bisphosphonates, which block bone resorption/OC activity, inhibit tumor growth, but this approach is not recommended clinically due to negative results in a phase III clinical trial combining conventional chemotherapy with zoledronate (Piperno-Neumann et al. 2016; Brown et al. 2018).

Innate Immunity: Myeloid Cells and Tumor-Associated Macrophages

Both OCs and macrophages derive from a common myeloid precursor, but OCs subsequently become tissue-resident cells that specialize in bone resorption, whereas tumor-associated macrophages are recruited to the TME and retain capacity for phagocytosis and alternative immunosuppressive functions. Mature OCs release proteolytic enzymes that degrade bone matrix proteins and decalcify bone and are supported by cooperative signaling between RANK (expressed on OCs) and RANKL, among others. In contrast, macrophages are activated into a range of functional states depending on the balance of secreted factors and are also observed to shift between these phenotypes as well as *trans*-differentiate into OCs.

In the Zhou et al. (2020) data set, 10 subclusters of myeloid cells were identified, with monocytes and macrophages accounting for 70%–80% of this total. Among macrophages, three clusters were classified as follows: M1_TAM (proinflammatory), M2_TAM (anti-inflammatory), and FABP4⁺_TAM, which was observed only in lung metastases and likely reflected alveolar macrophages expressing *FABP4*. Another recent single-cell analysis of treatment-naive OS patient samples reinforced the abundance of myeloid cells in the OS TME, primarily monocytes and macrophages (Cillo et al. 2022). In line with the emerging shift away from dichotomous M1 versus M2 terminology, this group identified multiple distinct subsets of $CD14^+ CD16^+$ macrophages and found that higher proportions of $CD14^+ CD16^+$ macrophages were associated with improved overall survival (Cillo et al. 2022). The association between myeloid infiltration and improved prognosis is also supported by immunostaining of patient samples for myeloid markers, demonstrating significantly better overall survival in samples with higher infiltration of TAMs, denoted by $CD163^+$ score by immunohistochemistry (IHC) (Gomez-Brouchet et al. 2017). What remains unclear from these studies is the cellular pathways that are involved in promoting antitumor immunity in these specific macrophages, and conversely the "checkpoints" or features that restrain infiltration of this cell population in other tumors, which appear to represent the higher-risk disease subset.

These studies highlight the antitumor function of certain macrophage populations in the OS TME and support therapeutic reprogramming of myeloid cells toward this antitumor phenotype. Canine studies have demonstrated a response in lung metastatic lesions to MTP, but trials of this immune activating agent have yielded conflicting results in patients (Kleinerman et al. 1992; Kurzman et al. 1995; Meyers et al. 2008). Administration of *Listeria* antigen to reprogram macrophages also has shown response in early phase canine OS trials, and there is preclinical rationale from animal models that targeting CSF1/CSF1R may have therapeutic benefit in multiple sarcomas including OS (Mason et al. 2016; Fujiwara et al. 2021a,b).

T-Cell Activation and Targeting the OS Surfaceome

Bulk and single-cell analyses suggest low antitumor T-cell activity in the OS microenvironment, with terminally exhausted $CD8^+$ T cells expressing coinhibitory receptors (Cillo et al. 2022). Thus, $CD8^+$ T-cell exhaustion is likely multifactorial in OS and may explain the disappointing clinical experience with immune checkpoint blockade (ICB) (Tawbi et al. 2017; Le Cesne et al. 2019). Another potential factor to explain the lack of T-cell activation, or priming to respond to ICB, is the relatively lower mutational burden in OS, in addition to low T-cell receptor clonality and neoantigen expression, compared to adult tumors that have been shown to be responsive to these therapies (Gröbner et al. 2018). In addition, older adult patients who develop OS tend to have higher mutational burden and have a greater likelihood of ICB response, possibly reflecting a different developmental origin of this malignant process with an increased component of somatic mutagenesis in the aged population (Wu et al. 2020). Nonmetastatic OS specimens have been shown to demonstrate increased expression of various immunotherapeutic targets in comparison metastatic specimens (Sorenson et al. 2019). Lymphocyte exclusion and vascular dysfunction in metastatic specimens further render ICB therapy challenging.

Efforts to overcome these challenges have focused on identifying novel cell surface antigens as well as proof-of-concept studies reinforcing a need to target multiple activating/inhibitory mechanisms simultaneously, such as cotargeting of the surface antigen GD2 and the inhibitory signal CD47 (Theruvath et al. 2022). Another inhibitory signal that may mediate immunosuppression in OS is B7-H3, and this surface molecule has been found to be expressed at high levels in the majority of OS samples tested; an antibody-drug conjugate targeting B7-H3 and a B7-H3 CART product have both demonstrated significant antitumor activity in patient-derived animal models, and this target is currently under investigation in clinical trials (Kendsersky et al. 2021; Zhang et al. 2022). A subset of OS tumors express HER2 and there has been some preclinical efficacy of the anti-HER2 antibody-drug conjugate trastuzumab deruxtecan

(Hingorani et al. 2022) and HER2 CART (Rainusso et al. 2012); however, clinical translation has been met with mixed results with some studies ongoing.

TME characterization by IHC suggests a significantly higher concentration of tumor-infiltrating leukocytes (TILs) in pulmonary metastases compared to primary tumors (Ligon et al. 2021). These TILs express multiple immunoregulatory molecules and are excluded from the metastases at the tumor–normal interface, possibly due to contributions from protumor myeloid cells infiltrating the metastatic tumors (Ligon et al. 2021). Confinement of T cells to the periphery of lesions was also observed in an independent cohort of OS metastatic lesions, which suggested that immune exclusion may be due to vascular dysfunction in the context of metastasis given the observed inverse relationship between lymphocyte abundance and markers of vascular dysfunction (increased expression of VEGFA, ANGPT2: ANGPT1 ratio, and decreased expression of SELE, gene encoding the adhesion molecule E-selectin) (Sorenson et al. 2019). Similarly, a study employing multiplex immunofluorescence (IF) observed increased immune infiltration in relapsed versus primary disease, perhaps due to increased immune infiltration in lung and some soft tissue primaries compared to bone (Cillo et al. 2022). A mechanistic understanding of the features of the primary bone niche that prevent immune infiltration is lacking, and furthering this understanding will likely improve efficacy of immunotherapeutic approaches especially if tested in the upfront setting.

Mechanisms of Metastasis and Lung Microenvironment

Efforts to characterize OS through a developmental lens must also address the major clinical problem of lung metastasis, which is the most common cause of mortality in patients with OS (Khanna et al. 2014). Preclinical investigations have nominated many cellular and secreted factors arising from the OS primary tumor as key effectors of the lung metastatic niche, in line with Paget's "seed and soil" hypothesis (Fig. 3; Brown et al. 2018; Heymann et al. 2019). For one, the role of MSCs

in priming the lung niche has been well described. MSCs are influenced by EVs containing TGF-β secreted by OS cells, and then release IL-6 and STAT3, factors that drive lung metastasis (Tu et al. 2012a; Baglio et al. 2017). CXCL8, in addition to IL-6, are key factors that promote lung colonization and are targetable (Gross et al. 2018). The role of OS-derived EVs has also been shown to promote myeloid cell infiltration in premetastatic lungs (Mazumdar et al. 2020). Other literature suggests a role for myofibroblastic reprogramming in OS stem cells (OSCs) as one factor that confers a survival advantage in the establishment of lung macro-metastases (Zhang et al. 2018). FGFR2 appears to be a key molecule involved in the initiation of fibrosis in OSCs, and is targetable via inhibition with nintedanib, which disrupts lung metastasis in animal models.

The Wnt/β-catenin pathway has been implicated in the pathophysiology of OS metastasis, which may shed light on the bone-specific origins of the OS metastatic program (see above) (Chen et al. 2008). ECM proteins including the cytoskeletal linker ezrin have been shown to facilitate OS tumor survival during early seeding of the lung (Khanna et al. 2004; Bulut et al. 2012). It has been described that OS cells release proteolytic enzymes such as matrix metalloproteinases (MMPs) and cathepsins, which induce local tissue ECM and basement membrane disintegration. Modification of TIMP3, MMP1, MMP3, and MMP11 has been found to alter in vitro invasiveness of OS cells and increase tumorigenicity in vivo (Han et al. 2016; Waresijiang et al. 2016; Chen et al. 2019). To some extent, conclusions to be drawn about the influences of the primary bone niche on metastasis are limited by prior models used that study metastasis in an experimental setting (i.e., by tail vein injection of tumor cells). A focus on spontaneous metastasis via orthotopic (Schott et al. 2023) or genetically engineered mouse models (Walkley et al. 2008) is of critical importance in the field to elucidate the bone niche-specific factors involved.

CONCLUDING REMARKS

We survey the complex intersection between OS biology and normal bone development and how this has not received substantial attention in OS research. We highlight here a few key aspects of the role of development in osteosarcomagenesis that may help guide further research on this topic.

ACKNOWLEDGMENTS

Artwork was done by Sarah Pyle. We thank all members of the Sweet-Cordero laboratory for helpful discussions that have shaped our understanding of osteosarcoma. We also thank the patients and families who inspire this work, including those have generously donated biospecimens to advance our understanding of this disease. The authors declare there are no potential conflicts of interest.

REFERENCES

Akiyama T, Dass CR, Choong PF. 2008. Novel therapeutic strategy for osteosarcoma targeting osteoclast differentiation, bone-resorbing activity, and apoptosis pathway. *Mol Cancer Ther* 7: 3461–3469. doi:10.1158/1535-7163.MCT-08-0530

Ameline B, Kovac M, Nathrath M, Barenboim M, Witt O, Krieg AH, Baumhoer D. 2021. Overactivation of the IGF signalling pathway in osteosarcoma: a potential therapeutic target? *J Pathol Clin Res* 7: 165–172. doi:10.1002/cjp2.191

Baglio SR, Lagerweij T, Pérez-Lanzón M, Ho XD, Léveillé N, Melo SA, Cleton-Jansen AM, Jordanova ES, Roncuzzi L, Greco M, et al. 2017. Blocking tumor-educated MSC paracrine activity halts osteosarcoma progression. *Clin Cancer Res* 23: 3721–3733. doi:10.1158/1078-0432.CCR-16-2726

Behjati S, Tarpey PS, Haase K, Ye H, Young MD, Alexandrov LB, Farndon SJ, Collord G, Wedge DC, Martincorena I, et al. 2017. Recurrent mutation of IGF signalling genes and distinct patterns of genomic rearrangement in osteosarcoma. *Nat Commun* 8: 15936. doi:10.1038/ncomms15936

Beird HC, Bielack SS, Flanagan AM, Gill J, Heymann D, Janeway KA, Livingston JA, Roberts RD, Strauss SJ, Gorlick R. 2022. Osteosarcoma. *Nat Rev Dis Primers* 8: 77. doi:10.1038/s41572-022-00409-y

Bozec A, Bakiri L, Jimenez M, Schinke T, Amling M, Wagner EF. 2010. Fra-2/AP-1 controls bone formation by regulating osteoblast differentiation and collagen production. *J Cell Biol* 190: 1093–1106. doi:10.1083/jcb.201002111

Bozec A, Bakiri L, Jimenez M, Rosen ED, Catala-Lehnen P, Schinke T, Schett G, Amling M, Wagner EF. 2013. Osteoblast-specific expression of Fra-2/AP-1 controls adiponectin and osteocalcin expression and affects metabolism. *J Cell Sci* 126: 5432–5440. doi:10.1242/jcs.134510

Brown HK, Schiavone K, Gouin F, Heymann MF, Heymann D. 2018. Biology of bone sarcomas and new therapeutic developments. *Calcif Tissue Int* 102: 174–195. doi:10.1007/s00223-017-0372-2

Bulut G, Hong SH, Chen K, Beauchamp EM, Rahim S, Kosturko GW, Glasgow E, Dakshanamurthy S, Lee HS, Daar I, et al. 2012. Small molecule inhibitors of ezrin inhibit the invasive phenotype of osteosarcoma cells. *Oncogene* 31: 269–281. doi:10.1038/onc.2011.245

Bykov VJN, Eriksson SE, Bianchi J, Wiman KG. 2018. Targeting mutant p53 for efficient cancer therapy. *Nat Rev Cancer* 18: 89–102. doi:10.1038/nrc.2017.109

Chen K, Fallen S, Abaan HO, Hayran M, Gonzalez C, Wodajo F, MacDonald T, Toretsky JA, Üren A. 2008. Wnt10b induces chemotaxis of osteosarcoma and correlates with reduced survival. *Pediatr Blood Cancer* 51: 349–355. doi:10.1002/pbc.21595

Chen X, Bahrami A, Pappo A, Easton J, Dalton J, Hedlund E, Ellison D, Shurtleff S, Wu G, Wei L, et al. 2014. Recurrent somatic structural variations contribute to tumorigenesis in pediatric osteosarcoma. *Cell Rep* 7: 104–112. doi:10.1016/j.celrep.2014.03.003

Chen CL, Zhang L, Jiao YR, Zhou Y, Ge QF, Li PC, Sun XJ, Lv Z. 2019. miR-134 inhibits osteosarcoma cell invasion and metastasis through targeting MMP1 and MMP3 in vitro and *in vivo*. *FEBS Lett* 593: 1089–1101. doi:10.1002/1873-3468.13387

Chinnam M, Goodrich DW. 2011. RB1, development, and cancer. *Curr Top Dev Biol* 94: 129–169. doi:10.1016/B978-0-12-380916-2.00005-X

Chung UI, Schipani E, McMahon AP, Kronenberg HM. 2001. Indian hedgehog couples chondrogenesis to osteogenesis in endochondral bone development. *J Clin Invest* 107: 295–304. doi:10.1172/JCI11706

Cillo AR, Mukherjee E, Bailey NG, Onkar S, Daley J, Salgado C, Li X, Liu D, Ranganathan S, Burgess M, et al. 2022. Ewing sarcoma and osteosarcoma have distinct immune signatures and intercellular communication networks. *Clin Cancer Res* 28: 4968–4982. doi:10.1158/1078-0432.CCR-22-1471

Cortés-Ciriano I, Lee JJK, Xi R, Jain D, Jung YL, Yang L, Gordenin D, Klimczak LJ, Zhang CZ, Pellman DS, et al. 2020. Comprehensive analysis of chromothripsis in 2,658 human cancers using whole-genome sequencing. *Nat Genet* 52: 331–341. doi:10.1038/s41588-019-0576-7

Cortini M, Avnet S, Baldini N. 2017. Mesenchymal stroma: role in osteosarcoma progression. *Cancer Lett* 405: 90–99. doi:10.1016/j.canlet.2017.07.024

Costa LD, Leblanc T, Mohandas N. 2020. Diamond–Blackfan anemia. *Blood* 136: 1262–1273.

Cundy T. 2018. Paget's disease of bone. *Metabolism* 80: 5–14.

Dohi O, Hatori M, Suzuki T, Ono K, Hosaka M, Akahira J, Miki Y, Nagasaki S, Itoi E, Sasano H. 2008. Sex steroid receptors expression and hormone-induced cell proliferation in human osteosarcoma. *Cancer Sci* 99: 518–523. doi:10.1111/j.1349-7006.2007.00673.x

Duffaud F, Mir O, Boudou-Rouquette P, Piperno-Neumann S, Penel N, Bompas E, Delcambre C, Kalbacher E, Italiano A, Collard O, et al. 2019. Efficacy and safety of regorafenib in adult patients with metastatic osteosarcoma: a noncomparative, randomised, double-blind, placebo-controlled, phase 2 study. *Lancet Oncol* 20: 120–133. doi:10.1016/S1470-2045(18)30742-3

Endo-Munoz L, Cumming A, Rickwood D, Wilson D, Cueva C, Ng C, Strutton G, Cassady AI, Evdokiou A, Sommerville S, et al. 2010. Loss of osteoclasts contributes to development of osteosarcoma pulmonary metastases. *Cancer Res* **70:** 7063–7072. doi:10.1158/0008-5472.CAN-09-4291

Endo-Munoz L, Evdokiou A, Saunders NA. 2012. The role of osteoclasts and tumour-associated macrophages in osteosarcoma metastasis. *Biochim Biophys Acta* **1826:** 434–442. doi:10.1016/j.bbcan.2012.07.00

Fearon ER, Vogelstein B. 1990. A genetic model for colorectal tumorigenesis. *Cell* **61:** 759–767. doi:10.1016/0092-8674 (90)90186-i

Fujiwara T, Healey J, Ogura K, Yoshida A, Kondo H, Hata T, Kure M, Tazawa H, Nakata E, Kunisada T, et al. 2021a. Role of tumor-associated macrophages in sarcomas. *Cancers (Basel)* **13:** 1086. doi:10.3390/cancers13051086

Fujiwara T, Yakoub MA, Chandler A, Christ AB, Yang G, Ouerfelli O, Rajasekhar VK, Yoshida A, Kondo H, Hata T, et al. 2021b. CSF1/CSF1R signaling inhibitor pexidartinib (PLX3397) reprograms tumor-associated macrophages and stimulates T-cell infiltration in the sarcoma microenvironment. *Mol Cancer Ther* **20:** 1388–1399. doi:10.1158/1535-7163.MCT-20-0591

Gaspar N, Venkatramani R, Hecker-Nolting S, Melcon SG, Locatelli F, Bautista F, Longhi A, Lervat C, Entz-Werle N, Casanova M, et al. 2021. Lenvatinib with etoposide plus ifosfamide in patients with refractory or relapsed osteosarcoma (ITCC-050): a multicentre, open-label, multicohort, phase 1/2 study. *Lancet Oncol* **22:** 1312–1321. doi:10.1016/S1470-2045(21)00387-9

Gianferante DM, Mirabello L, Savage SA. 2017. Germline and somatic genetics of osteosarcoma—connecting aetiology, biology and therapy. *Nat Rev Endocrinol* **13:** 480–491.

Glass AG, Fraumeni JF Jr. 1970. Epidemiology of bone cancer in children. *J Natl Cancer Inst* **44:** 187–199.

Gomez-Brouchet A, Illac C, Gilhodes J, Bouvier C, Aubert S, Guinebretiere JM, Marie B, Larousserie F, Entz-Werlé N, de Pinieux G, et al. 2017. CD163-positive tumor-associated macrophages and CD8-positive cytotoxic lymphocytes are powerful diagnostic markers for the therapeutic stratification of osteosarcoma patients: an immunohistochemical analysis of the biopsies from the French OS2006 phase 3 trial. *Oncoimmunology* **6:** e1331193. doi:10.1080/2162402X.2017.1331193

Gounder MM, Rosenbaum E, Wu N, Dickson MA, Sheikh TN, D'Angelo SP, Chi P, Keohan ML, Erinjeri JP, Antonescu CR, et al. 2022. A phase Ib/II randomized study of RO4929097, a γ-secretase or notch inhibitor with or without vismodegib, a hedgehog inhibitor, in advanced sarcoma. *Clin Cancer Res* **28:** 1586–1594. doi:10.1158/1078-0432.CCR-21-3874

Grignani G, Palmerini E, Dileo P, Asaftei SD, D'Ambrosio L, Pignochino Y, Mercuri M, Picci P, Fagioli F, Casali PG, et al. 2012. A phase II trial of sorafenib in relapsed and unresectable high-grade osteosarcoma after failure of standard multimodal therapy: an Italian sarcoma group study. *Ann Oncol* **23:** 508–516. doi:10.1093/annonc/mdr151

Gröbner SN, Worst BC, Weischenfeldt J, Buchhalter I, Kleinheinz K, Rudneva VA, Johann PD, Balasubramanian GP, Segura-Wang M, Brabetz S, et al. 2018. The landscape of genomic alterations across childhood cancers. *Nature* **555:** 321–327. doi:10.1038/nature25480

Gross AC, Cam H, Phelps DA, Saraf AJ, Bid HK, Cam M, London CA, Winget SA, Arnold MA, Brandolini L, et al. 2018. IL-6 and CXCL8 mediate osteosarcoma-lung interactions critical to metastasis. *JCI Insight* **3:** e99791. doi:10.1172/jci.insight.99791

Gutierrez GM, Kong E, Sabbagh Y, Brown NE, Lee J-S, Demay MB, Thomas DM, Hinds PW. 2008. Impaired bone development and increased mesenchymal progenitor cells in calvaria of *RB1*$^{-/-}$ mice. *Proc Natl Acad Sci* **105:** 18402–18407.

Han XG, Li Y, Mo HM, Li K, Lin D, Zhao CQ, Zhao J, Tang TT. 2016. TIMP3 regulates osteosarcoma cell migration, invasion, and chemotherapeutic resistances. *Tumour Biol* **37:** 8857–8867. doi:10.1007/s13277-015-4757-4

Heymann MF, Lézot F, Heymann D. 2019. The contribution of immune infiltrates and the local microenvironment in the pathogenesis of osteosarcoma. *Cell Immunol* **343:** 103711. doi:10.1016/j.cellimm.2017.10.011

Hingorani P, Roth ME, Wang Y, Zhang W, Gill JB, Harrison DJ, Teicher B, Erickson S, Gatto G, Smith MA, et al. 2021. ABBV-085, antibody-drug conjugate targeting LRRC15, is effective in osteosarcoma: a report by the pediatric preclinical testing consortium. *Mol Cancer Ther* **20:** 535–540. doi:10.1158/1535-7163.MCT-20-0406

Hingorani P, Zhang W, Zhang Z, Xu Z, Wang WL, Roth ME, Wang Y, Gill JB, Harrison DJ, Teicher BA, et al. 2022. Trastuzumab deruxtecan, antibody-drug conjugate targeting HER2, is effective in pediatric malignancies: a report by the pediatric preclinical testing consortium. *Mol Cancer Ther* **21:** 1318–1325. doi:10.1158/1535-7163.MCT-21-0758

Höglund M, Gisselsson D, Mandahl N, Johansson B, Mertens F, Mitelman F, Säll T. 2001. Multivariate analyses of genomic imbalances in solid tumors reveal distinct and converging pathways of karyotypic evolution. *Genes Chromosomes Cancer* **31:** 156–171. doi:10.1002/gcc.1129

Italiano A, Mir O, Mathoulin-Pelissier S, Penel N, Piperno-Neumann S, Bompas E, Chevreau C, Duffaud F, Entz-Werlé N, Saada E, et al. 2020. Cabozantinib in patients with advanced Ewing sarcoma or osteosarcoma (CABONE): a multicentre, single-arm, phase 2 trial. *Lancet Oncol* **21:** 446–455. doi:10.1016/S1470-2045(19)30825-3

Jemtland R, Divieti P, Lee K, Segre GV. 2003. Hedgehog promotes primary osteoblast differentiation and increases PTHrP mRNA expression and iPTHrP secretion. *Bone* **32:** 611–620. doi:10.1016/S8756-3282(03)00092-9

Kansara M, Thomas DM. 2007. Molecular pathogenesis of osteosarcoma. *DNA Cell Biol* **26:** 1–18.

Kendsersky NM, Lindsay J, Kolb EA, Smith MA, Teicher BA, Erickson SW, Earley EJ, Mosse YP, Martinez D, Pogoriler J, et al. 2021. The B7-H3-targeting antibody-drug conjugate m276-SL-PBD is potently effective against pediatric cancer preclinical solid tumor models. *Clin Cancer Res* **27:** 2938–2946. doi:10.1158/1078-0432.CCR-20-4221

Khanna C, Wan X, Bose S, Cassaday R, Olomu O, Mendoza A, Yeung C, Gorlick R, Hewitt SM, Helman LJ. 2004. The membrane-cytoskeleton linker ezrin is necessary for osteosarcoma metastasis. *Nat Med* **10:** 182–186. doi:10.1038/nm982

Khanna C, Fan TM, Gorlick R, Helman LJ, Kleinerman ES, Adamson PC, Houghton PJ, Tap WD, Welch DR, Steeg PS, et al. 2014. Toward a drug development path that targets metastatic progression in osteosarcoma. *Clin Cancer Res* **20**: 4200–4209. doi:10.1158/1078-0432.CCR-13-2574

Komori T. 2017. Roles of Runx2 in skeletal development. *Adv Exp Med Biol* **962**: 83–93. doi:10.1007/978-981-10-3233-2_6

Komori T. 2020. Molecular mechanism of Runx2-dependent bone development. *Mol Cells* **43**: 168–175. doi:10.14348/molcells.2019.0244

Krum SA. 2011. Direct transcriptional targets of sex steroid hormones in bone. *J Cell Biochem* **112**: 401–408. doi:10.1002/jcb.22970

Kurzman ID, MacEwen EG, Rosenthal RC, Fox LE, Keller ET, Helfand SC, Vail DM, Dubielzig RR, Madewell BR, Rodriguez CO Jr, et al. 1995. Adjuvant therapy for osteosarcoma in dogs: results of randomized clinical trials using combined liposome-encapsulated muramyl tripeptide and cisplatin. *Clin Cancer Res* **1**: 1595–1601.

Ladanyi M, Park CK, Lewis R, Jhanwar SC, Healey JH, Huvos AG. 1993. Sporadic amplification of the MYC gene in human osteosarcomas. *Diagn Mol Pathol* **2**: 163–167. doi:10.1097/00019606-199309000-00004

Le Cesne A, Marec-Berard P, Blay JY, Gaspar N, Bertucci F, Penel N, Bompas E, Cousin S, Toulmonde M, Bessede A, et al. 2019. Programmed cell death 1 (PD-1) targeting in patients with advanced osteosarcomas: results from the PEMBROSARC study. *Eur J Cancer* **119**: 151–157. doi:10.1016/j.ejca.2019.07.018

Li Y, Yang S, Yang S. 2022. Rb1 negatively regulates bone formation and remodeling through inhibiting transcriptional regulation of YAP in Glut1 and OPG expression and glucose metabolism in male mice. *Mol Metab* **66**: 101630. doi:10.1016/j.molmet.2022.101630

Liao Y, Sassi S, Halvorsen S, Feng Y, Shen J, Gao Y, Cote G, Choy E, Harmon D, Mankin H, et al. 2017. Androgen receptor is a potential novel prognostic marker and oncogenic target in osteosarcoma with dependence on CDK11. *Sci Rep* **7**: 43941. doi:10.1038/srep43941

Ligon JA, Choi W, Cojocaru G, Fu W, Hsiue EH, Oke TF, Siegel N, Fong MH, Ladle B, Pratilas CA, et al. 2021. Pathways of immune exclusion in metastatic osteosarcoma are associated with inferior patient outcomes. *J Immunother Cancer* **9**: e001772. doi:10.1136/jitc-2020-001772

Lillo Osuna MA, Garcia-Lopez J, El Ayachi I, Fatima I, Khalid AB, Kumpati J, Slayden AV, Seagroves TN, Miranda-Carboni GA, Krum SA. 2019. Activation of estrogen receptor α by decitabine inhibits osteosarcoma growth and metastasis. *Cancer Res* **79**: 1054–1068. doi:10.1158/0008-5472.CAN-18-1255

Lima GA, Gomes EM, Nunes RC, Vieira Neto L, Sieiro AP, Brabo EP, Gadelha MR. 2006. Osteosarcoma and acromegaly: a case report and review of the literature. *J Endocrinol Invest* **29**: 1006–1011. doi:10.1007/BF03349215

Lin PP, Pandey MK, Jin F, Raymond AK, Akiyama H, Lozano G. 2009. Targeted mutation of p53 and Rb in mesenchymal cells of the limb bud produces sarcomas in mice. *Carcinogenesis* **30**: 1789–1795. doi:10.1093/carcin/bgp180

Longhi A, Pasini A, Cicognani A, Baronio F, Pellacani A, Baldini N, Bacci G. 2005. Height as a risk factor for osteosarcoma. *J Pediatr Hematol Oncol* **27**: 314–318. doi:10.1097/01.mph.0000169251.57611.8e

Lu L, Harutyunyan K, Jin W, Wu J, Yang T, Chen Y, Joeng KS, Bae Y, Tao J, Dawson BC, et al. 2015. RECQL4 regulates p53 function in vivo during skeletogenesis. *J Bone Miner Res* **30**: 1077–1089.

Lu L, Jin W, Wang LL. 2020. RECQ DNA helicases and osteosarcoma. *Adv Exp Med Biol* **1258**: 37–54.

Ma X, Liu Y, Liu Y, Alexandrov LB, Edmonson MN, Gawad C, Zhou X, Li Y, Rusch MC, Easton J, et al. 2018. Pan-cancer genome and transcriptome analyses of 1,699 paediatric leukaemias and solid tumours. *Nature* **555**: 371–376. doi:10.1038/nature25795

Marinoff AE, Spurr LF, Fong C, Li YY, Forrest SJ, Ward A, Doan D, Corson L, Mauguen A, Pinto N, et al. 2023. Clinical targeted next-generation panel sequencing reveals *MYC* amplification is a poor prognostic factor in osteosarcoma. *JCO Precis Oncol* **7**: e2200334. doi:10.1200/PO.22.00334

Mason NJ, Gnanandarajah JS, Engiles JB, Gray F, Laughlin D, Gaurnier-Hausser A, Wallecha A, Huebner M, Paterson Y. 2016. Immunotherapy with a HER2-targeting *Listeria* induces HER2-specific immunity and demonstrates potential therapeutic effects in a phase I trial in canine osteosarcoma. *Clin Cancer Res* **22**: 4380–4390. doi:10.1158/1078-0432.CCR-16-0088

Mazumdar A, Urdinez J, Boro A, Arlt MJE, Egli FE, Niederöst B, Jaeger PK, Moschini G, Muff R, Fuchs B, et al. 2020. Exploring the role of osteosarcoma-derived extracellular vesicles in pre-metastatic niche formation and metastasis in the 143-B xenograft mouse osteosarcoma model. *Cancers (Basel)* **12**: 3457. doi:10.3390/cancers12113457

Meyers PA, Schwartz CL, Krailo MD, Healey JH, Bernstein ML, Betcher D, Ferguson WS, Gebhardt MC, Goorin AM, Harris M, et al. 2008. Osteosarcoma: the addition of muramyl tripeptide to chemotherapy improves overall survival —a report from the Children's Oncology Group. *J Clin Oncol* **26**: 633–638. doi:10.1200/JCO.2008.14.0095

Mirabello L, Yeager M, Mai PL, Gastier-Foster JM, Gorlick R, Khanna C, Patiño-Garcia A, Sierrasesúmaga L, Lecanda F, Andrulis IL, et al. 2015. Germline TP53 variants and susceptibility to osteosarcoma. *J Natl Cancer Inst* **107**: djv101. doi:10.1093/jnci/djv101

Mutsaers AJ, Walkley CR. 2014. Cells of origin in osteosarcoma: mesenchymal stem cells or osteoblast committed cells? *Bone* **62**: 56–63. doi:10.1016/j.bone.2014.02.003

Mutsaers AJ, Ng AJ, Baker EK, Russell MR, Chalk AM, Wall M, Liddicoat BJ, Ho PW, Slavin JL, Goradia A, et al. 2013. Modeling distinct osteosarcoma subtypes in vivo using Cre:lox and lineage-restricted transgenic shRNA. *Bone* **55**: 166–178. doi:10.1016/j.bone.2013.02.016

Nakashima K, Zhou X, Kunkel G, Zhang Z, Deng JM, Behringer RR, de Crombrugghe B. 2002. The novel zinc finger-containing transcription factor osterix is required for osteoblast differentiation and bone formation. *Cell* **108**: 17–29. doi:10.1016/S0092-8674(01)00622-5

Nerlakanti N, McGuire J, Yu D, Reed DR, Lynch CC. 2019. Abstract 2016: HDAC inhibition significantly reduces primary and lung metastatic osteosarcoma progression. *Cancer Res* **79**: 2016. doi:10.1158/1538-7445.AM2019-2016

Ng AJ, Walia MK, Smeets MF, Mutsaers AJ, Sims NA, Purton LE, Walsh NC, Martin TJ, Walkley CR. 2015. The DNA helicase Recql4 is required for normal osteoblast expansion and osteosarcoma formation. *PLoS Genet* **11**: e1005160.

Ogasawara T, Mori Y, Abe M, Suenaga H, Kawase-Koga Y, Saijo H, Takato T. 2011. Role of cyclin-dependent kinase (Cdk)6 in osteoblast, osteoclast, and chondrocyte differentiation and its potential as a target of bone regenerative medicine. *Oral Sci Int* **8:** 2–6. doi:10.1016/S1348-8643(11)00007-3

Overholtzer M, Rao PH, Favis R, Lu XY, Elowitz MB, Barany F, Ladanyi M, Gorlick R, Levine AJ. 2003. The presence of p53 mutations in human osteosarcomas correlates with high levels of genomic instability. *Proc Natl Acad Sci* **100:** 11547–11552. doi:10.1073/pnas.1934852100

Pathak JL, Bravenboer N, Klein-Nulend J. 2020. The osteocyte as the new discovery of therapeutic options in rare bone diseases. *Front Endocrinol (Lausanne)* **11:** 405. doi:10.3389/fendo.2020.00405

Perry JA, Kiezun A, Tonzi P, Van Allen EM, Carter SL, Baca SC, Cowley GS, Bhatt AS, Rheinbay E, Pedamallu CS, et al. 2014. Complementary genomic approaches highlight the PI3K/mTOR pathway as a common vulnerability in osteosarcoma. *Proc Natl Acad Sci* **111:** E5564–E5573. doi:10.1073/pnas.1419260111

Piperno-Neumann S, Le Deley MC, Rédini F, Pacquement H, Marec-Bérard P, Petit P, Brisse H, Lervat C, Gentet JC, Entz-Werlé N, et al. 2016. Zoledronate in combination with chemotherapy and surgery to treat osteosarcoma (OS2006): a randomised, multicentre, open-label, phase 3 trial. *Lancet Oncol* **17:** 1070–1080. doi:10.1016/S1470-2045(16)30096-1

Pollak MN, Polychronakos C, Richard M. 1990. Insulinlike growth factor I: a potent mitogen for human osteogenic sarcoma. *J Natl Cancer Inst* **82:** 301–305. doi:10.1093/jnci/82.4.301

Quist T, Jin H, Zhu JF, Smith-Fry K, Capecchi MR, Jones KB. 2015. The impact of osteoblastic differentiation on osteosarcomagenesis in the mouse. *Oncogene* **34:** 4278–4284. doi:10.1038/onc.2014.354

Rainusso N, Brawley VS, Ghazi A, Hicks MJ, Gottschalk S, Rosen JM, Ahmed N. 2012. Immunotherapy targeting HER2 with genetically modified T cells eliminates tumor-initiating cells in osteosarcoma. *Cancer Gene Ther* **19:** 212–217. doi:10.1038/cgt.2011.83

Rajan S, Zaccaria S, Cannon MV, Cam M, Gross AC, Raphael BJ, Roberts RD. 2023. Structurally complex osteosarcoma genomes exhibit limited heterogeneity within individual tumors and across evolutionary time. *Cancer Res Commun* **3:** 564–575. doi:10.1158/2767-9764.CRC-22-0348

Ramaswamy V, Hielscher T, Mack SC, Lassaletta A, Lin T, Pajtler KW, Jones DT, Luu B, Cavalli FM, Aldape K, et al. 2016. Therapeutic impact of cytoreductive surgery and irradiation of posterior fossa ependymoma in the molecular era: a retrospective multicohort analysis. *J Clin Oncol* **34:** 2468–2477. doi:10.1200/JCO.2015.65.7825

Rausch T, Jones DT, Zapatka M, Stütz AM, Zichner T, Weischenfeldt J, Jäger N, Remke M, Shih D, Northcott PA, et al. 2012. Genome sequencing of pediatric medulloblastoma links catastrophic DNA rearrangements with TP53 mutations. *Cell* **148:** 59–71. doi:10.1016/j.cell.2011.12.013

Sayles LC, Breese MR, Koehne AL, Leung SG, Lee AG, Liu HY, Spillinger A, Shah AT, Tanasa B, Straessler K, et al. 2019. Genome-informed targeted therapy for osteosarco-

ma. *Cancer Discov* **9:** 46–63. doi:10.1158/2159-8290.CD-17-1152

Schott CR, Koehne AL, Sayles LC, Young EP, Luck C, Yu K, Lee AG, Breese MR, Leung SG, Xu H, et al. 2023. Osteosarcoma PDX-derived cell line models for preclinical drug evaluation demonstrate metastasis inhibition by dinaciclib through a genome-targeted approach. *Clin Cancer Res* **30:** 849–864.

Sergi C, Shen F, Liu SM. 2019. Insulin/IGF-1R, SIRT1, and FOXOs pathways—an intriguing interaction platform for bone and osteosarcoma. *Front Endocrinol (Lausanne)* **10:** 93. doi:10.3389/fendo.2019.00093

Shaikh A, Wesner AA, Abuhattab M, Kutty RG, Premnath P. 2023. Cell cycle regulators and bone: development and regeneration. *Cell Biosci* **13:** 35. doi:10.1186/s13578-023-00988-7

Smida J, Baumhoer D, Rosemann M, Walch A, Bielack S, Poremba C, Remberger K, Korsching E, Scheurlen W, Dierkes C, et al. 2010. Genomic alterations and allelic imbalances are strong prognostic predictors in osteosarcoma. *Clin Cancer Res* **16:** 4256–4267. doi:10.1158/1078-0432.CCR-10-0284

Sorenson L, Fu Y, Hood T, Warren S, McEachron TA. 2019. Targeted transcriptional profiling of the tumor microenvironment reveals lymphocyte exclusion and vascular dysfunction in metastatic osteosarcoma. *Oncoimmunology* **8:** e1629779. doi:10.1080/2162402X.2019.1629779

Tawbi HA, Burgess M, Bolejack V, Van Tine BA, Schuetze SM, Hu J, D'Angelo S, Attia S, Riedel RF, Priebat DA, et al. 2017. Pembrolizumab in advanced soft-tissue sarcoma and bone sarcoma (SARC028): a multicentre, two-cohort, single-arm, open-label, phase 2 trial. *Lancet Oncol* **18:** 1493–1501. doi:10.1016/S1470-2045(17)30624-1

Tellez-Gabriel M, Charrier C, Brounais-Le Royer B, Mullard M, Brown HK, Verrecchia F, Heymann D. 2017. Analysis of gap junctional intercellular communications using a dielectrophoresis-based microchip. *Eur J Cell Biol* **96:** 110–118. doi:10.1016/j.ejcb.2017.01.003

Theruvath J, Menard M, Smith BAH, Linde MH, Coles GL, Dalton GN, Wu W, Kiru L, Delaidelli A, Sotillo E, et al. 2022. Anti-GD2 synergizes with CD47 blockade to mediate tumor eradication. *Nat Med* **28:** 333–344. doi:10.1038/s41591-021-01625-x

Tjalma RA. 1966. Canine bone sarcoma: estimation of relative risk as a function of body size. *J Natl Cancer Inst* **36:** 1137–1150.

Tsukasaki M, Huynh NCN, Okamoto K, Muro R, Terashima A, Kurikawa Y, Komatsu N, Pluemsakunthai W, Nitta T, Abe T, et al. 2020. Stepwise cell fate decision pathways during osteoclastogenesis at single-cell resolution. *Nat Metab* **2:** 1382–1390. doi:10.1038/s42255-020-00318-y

Tu B, Du L, Fan QM, Tang Z, Tang TT. 2012a. STAT3 activation by IL-6 from mesenchymal stem cells promotes the proliferation and metastasis of osteosarcoma. *Cancer Lett* **325:** 80–88. doi:10.1016/j.canlet.2012.06.006

Tu X, Chen J, Lim J, Karner CM, Lee SY, Heisig J, Wiese C, Surendran K, Kopan R, Gessler M, et al. 2012b. Physiological notch signaling maintains bone homeostasis via RBPjk and Hey upstream of NFATc1. *PLoS Genet* **8:** e1002577. doi:10.1371/journal.pgen.1002577

Umbreit NT, Zhang CZ, Lynch LD, Blaine LJ, Cheng AM, Tourdot R, Sun L, Almubarak HF, Judge K, Mitchell TJ, et al. 2020. Mechanisms generating cancer genome complexity from a single cell division error. *Science* **368:** eaba0712. doi:10.1126/science.aba0712

Velletri T, Xie N, Wang Y, Huang Y, Yang Q, Chen X, Shou P, Gan Y, Cao G, Melino G, et al. 2016. P53 functional abnormality in mesenchymal stem cells promotes osteosarcoma development. *Cell Death Dis* **7:** e2015.

Velletri T, Huang Y, Wang Y, Li Q, Hu M, Xie N, Yang Q, Chen X, Chen Q, Shou P, et al. 2021. Loss of p53 in mesenchymal stem cells promotes alteration of bone remodeling through negative regulation of osteoprotegerin. *Cell Death Differ* **28:** 156–169. doi:10.1038/s41418-020-0590-4

Wagle S, Park SH, Kim KM, Moon YJ, Bae JS, Kwon KS, Park HS, Lee H, Moon WS, Kim JR, et al. 2015. DBC1/CCAR2 is involved in the stabilization of androgen receptor and the progression of osteosarcoma. *Sci Rep* **5:** 13144. doi:10.1038/srep13144

Wagner EF. 2010. Bone development and inflammatory disease is regulated by AP-1 (Fos/Jun). *Ann Rheum Dis* **69:** i86–i88. doi:10.1136/ard.2009.119396

Walkley CR, Qudsi R, Sankaran VG, Perry JA, Gostissa M, Roth SI, Rodda SJ, Snay E, Dunning P, Fahey FH, et al. 2008. Conditional mouse osteosarcoma, dependent on p53 loss and potentiated by loss of Rb, mimics the human disease. *Genes Dev* **22:** 1662–1676.

Walkley CR, Walia MK, Ho PW, Martin TJ. 2014. PTHrp, its receptor, and protein kinase A activation in osteosarcoma. *Mol Cell Oncol* **1:** e965624. doi:10.4161/23723548.2014.965624

Wang X, Kua HY, Hu Y, Guo K, Zeng Q, Wu Q, Ng HH, Karsenty G, de Crombrugghe B, Yeh J, et al. 2006. P53 functions as a negative regulator of osteoblastogenesis, osteoblast-dependent osteoclastogenesis, and bone remodeling. *J Cell Biol* **172:** 115–125. doi:10.1083/jcb.200507106

Wang D, Niu X, Wang Z, Song CL, Huang Z, Chen KN, Duan J, Bai H, Xu J, Zhao J, et al. 2019. Multiregion sequencing reveals the genetic heterogeneity and evolutionary history of osteosarcoma and matched pulmonary metastases. *Cancer Res* **79:** 7–20. doi:10.1158/0008-5472.CAN-18-1086

Waresijiang N, Sun J, Abuduaini R, Jiang T, Zhou W, Yuan H. 2016. The downregulation of miR-125a-5p functions as a tumor suppressor by directly targeting MMP-11 in osteosarcoma. *Mol Med Rep* **13:** 4859–4864. doi:10.3892/mmr.2016.5141

Withrow SJ, Powers BE, Straw RC, Wilkins RM. 1991. Comparative aspects of osteosarcoma. Dog versus man. *Clin Orthop Relat Res* **270:** 159–168.

Wu CC, Beird HC, Andrew Livingston J, Advani S, Mitra A, Cao S, Reuben A, Ingram D, Wang WL, Ju Z, et al. 2020. Immuno-genomic landscape of osteosarcoma. *Nat Commun* **11:** 1008. doi:10.1038/s41467-020-14646-w

Yahara Y, Nguyen T, Ishikawa K, Kamei K, Alman BA. 2022. The origins and roles of osteoclasts in bone development, homeostasis and repair. *Development* **149:** dev199908. doi:10.1242/dev.199908

Zhang J, Walsh MF, Wu G, Edmonson MN, Gruber TA, Easton J, Hedges D, Ma X, Zhou X, Yergeau DA, et al. 2015. Germline mutations in predisposition genes in pediatric cancer. *N Engl J Med* **373:** 2336–2346. doi:10.1056/NEJMoa1508054

Zhang W, Zhao JM, Lin J, Hu CZ, Zhang WB, Yang WL, Zhang J, Zhang JW, Zhu J. 2018. Adaptive fibrogenic reprogramming of osteosarcoma stem cells promotes metastatic growth. *Cell Rep* **24:** 1266–1277.e5. doi:10.1016/j.celrep.2018.06.103

Zhang Q, Zhang Z, Liu G, Li D, Gu Z, Zhang L, Pan Y, Cui X, Wang L, Liu G, et al. 2022. B7-H3 targeted CAR-T cells show highly efficient anti-tumor function against osteosarcoma both in vitro and in vivo. *BMC Cancer* **22:** 1124. doi:10.1186/s12885-022-10229-8

Zhou X, Zhang Z, Feng JQ, Dusevich VM, Sinha K, Zhang H, Darnay BG, de Crombrugghe B. 2010. Multiple functions of osterix are required for bone growth and homeostasis in postnatal mice. *Proc Natl Acad Sci* **107:** 12919–12924. doi:10.1073/pnas.0912855107

Zhou Q, Xian M, Xiang S, Xiang D, Shao X, Wang J, Cao J, Yang X, Yang B, Ying M, et al. 2017. All-*trans* retinoic acid prevents osteosarcoma metastasis by inhibiting M2 polarization of tumor-associated macrophages. *Cancer Immunol Res* **5:** 547–559. doi:10.1158/2326-6066.CIR-16-0259

Zhou Y, Yang D, Yang Q, Lv X, Huang W, Zhou Z, Wang Y, Zhang Z, Yuan T, Ding X, et al. 2020. Single-cell RNA landscape of intratumoral heterogeneity and immunosuppressive microenvironment in advanced osteosarcoma. *Nat Commun* **11:** 6322. doi:10.1038/s41467-020-20059-6

Cite this article as *Cold Spring Harb Perspect Med* doi: 10.1101/cshperspect.a041635

Lineage-Selective Dependencies in Pediatric Cancers

K. Elaine Ritter and Adam D. Durbin

Division of Molecular Oncology, Department of Oncology, St. Jude Children's Research Hospital, Memphis, Tennessee 38015, USA

Correspondence: adam.durbin@stjude.org

The quest for effective cancer therapeutics has traditionally centered on targeting mutated or overexpressed oncogenic proteins. However, challenges arise in cancers with low mutational burden or when the mutated oncogene is not conventionally targetable, which are common situations in childhood cancers. This obstacle has sparked large-scale unbiased screens to identify collateral genetic dependencies crucial for cancer cell growth. These screens have revealed promising targets for therapeutic intervention in the form of lineage-selective dependency genes, which may have an expanded therapeutic window compared to pan-lethal dependencies. Many lineage-selective dependencies regulate gene expression and are closely tied to the developmental origins of pediatric tumors. Placing lineage-selective dependencies in a transcriptional network model is helpful for understanding their roles in driving malignant cell behaviors. Here, we discuss the identification of lineage-selective dependencies and how two transcriptional models, core regulatory circuits and gene regulatory networks, can serve as frameworks for understanding their individual and collective actions, particularly in cancers affecting children and young adults.

Classically, strategies for therapeutic intervention in a myriad of cancers have focused on mutated or overexpressed proteins, such as BCR-ABL in chronic myelogenous leukemia, K-Ras^{G12C} in non-small-cell lung carcinoma, and HER2 in breast carcinoma (Druker et al. 2001; Eiermann and International Herceptin Study Group 2001; Skoulidis et al. 2021). Inhibition of these enzymes has revolutionized the clinical management of cancer by attacking cancer-driving catalytic oncoproteins. This has sparked a search for similarly specific targets in all cancers, which has been complicated by the observation that most cancers arising in childhood typically display a relatively reduced mutational burden compared to adult malignancies (Lee et al. 2012; Gröbner et al. 2018; Ma et al. 2018). Additionally, many of the mutations that have been identified are found in proteins that are challenging to disrupt because of a lack of clear binding pockets for the design of therapeutic inhibitors. These proteins are commonly involved in cellular processes such as DNA repair, cell cycle control, telomere maintenance, mRNA transcription, and splicing (Fig. 1A; Gröbner et al. 2018; Ma et al. 2018). This observation

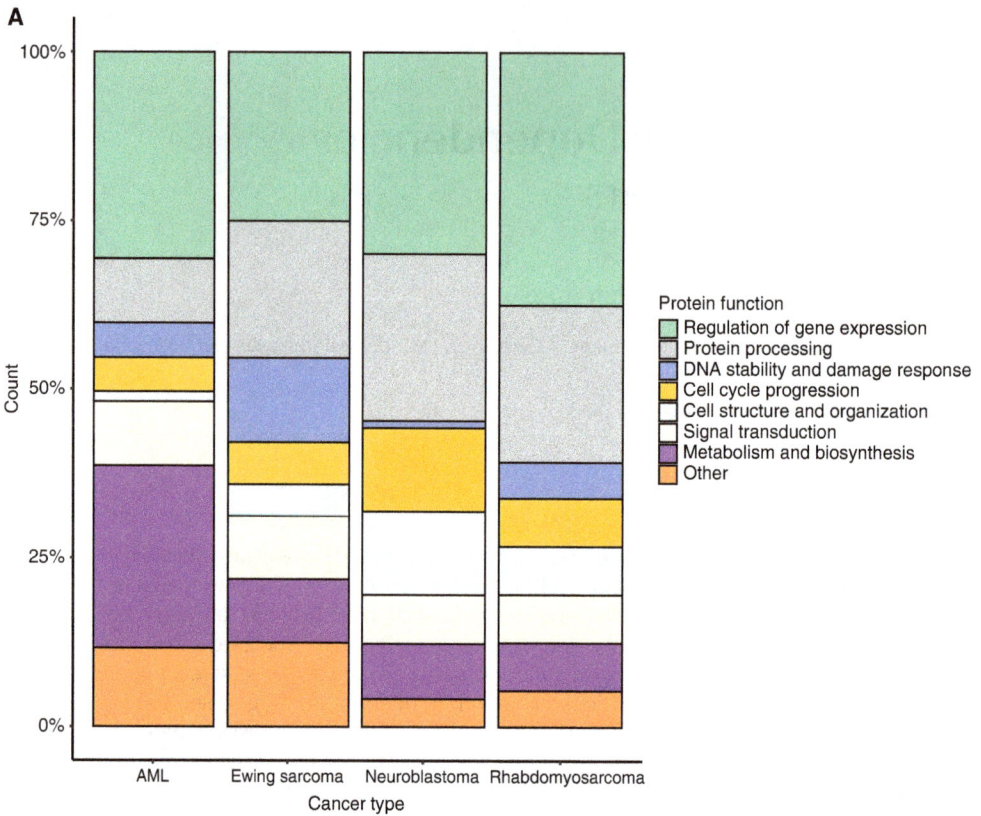

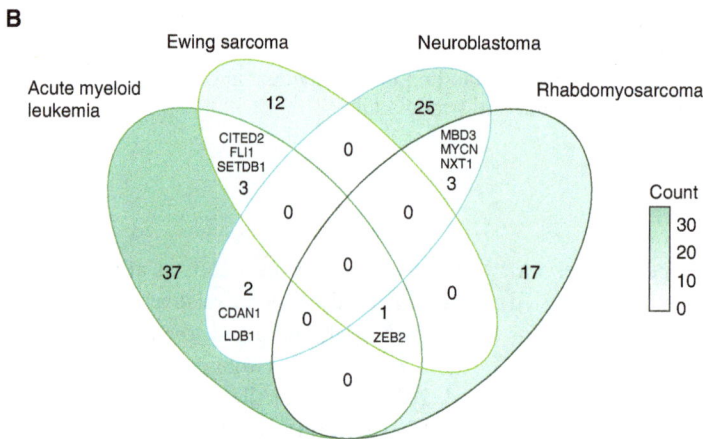

Figure 1. Lineage-selective dependencies are enriched in factors regulating gene expression. Gene dependency scores were extracted from the DepMap repository (23Q4 release) from four pediatric malignancies: acute myeloid leukemia (AML), Ewing sarcoma (ES), neuroblastoma (NB), and rhabdomyosarcoma (RMS). The gene dependency list was then intersected with the top enriched dependencies called by the two-class comparison tool in DepMap. Genes were classified as dependencies if the lower quartile of gene effect score was less than −0.5 across all cell lines of a given cancer. This strict dependency score cutoff mitigated the effects of skewing by small numbers of cell lines that are heavily dependent on a given gene. Note that some genes, such as *TFAP2B* and *PHOX2B*, have been previously reported as lineage-selective dependencies in NB (Durbin et al. 2018) but did not meet our strict dependency criteria. The protein function of each gene dependency was annotated manually. (*A*) Stacked proportional bar plot of protein functions assigned to enriched lineage-selective dependencies in AML, ES, NB, and RMS. (*B*) Venn diagram of lineage-selective dependencies implicated in gene expression (RNA processing, epigenetic regulation, transcription factors). Dependencies shared between two or more malignancies are indicated. Color saturation indicates the number of genes in each ring of the diagram.

Cite this article as *Cold Spring Harb Perspect Med* doi: 10.1101/cshperspect.a041573

has led to a refocused search for new mechanisms to identify potential vulnerabilities in cancers without an obvious enzymatic "Achilles' heel."

To develop more precise and targeted cancer therapeutics with reduced toxicity, research groups have focused on exome-scale, unbiased screening to identify genetic dependencies—genes that are required for the viability or growth of cancer cells (Box 1). The identification of genetic dependencies and determining which represent actionable vulnerabilities across and within distinct cancer subtypes has, therefore, been the subject of intensive research. These herculean efforts, largely using functional genomics technologies such as short-hairpin RNA (shRNA) or CRISPR–Cas9 gene editing, have been performed by the Sanger-Wellcome Trust, The Broad Institute of MIT and Harvard, Project Achilles, and the Childhood Cancer Model Atlas (Meyers et al. 2017; Tsherniak et al. 2017; Behan et al. 2019; Sun et al. 2023). As the depth and scope of large-scale genetic dependency assays have increased, with more cancer cell lines from diverse organs, tissues, and cell types being incorporated, researchers have made key discoveries in identifying cohorts of nonmutationally driven genetic dependencies that are necessary to promote the growth of cells from a particular disease. These "lineage-selective dependencies" may be unique to a specific cancer subtype or may be enriched for that cancer subtype (Box 1). Importantly, these observations have validated prior discoveries of nonmutationally driven genes that control the properties of different tu-

---

BOX 1. KEY DEFINITIONS WITH EXAMPLES

Dependency
An individual component, usually a gene, that a cancer cell requires for growth. This is measured by retention in a competitive assay of cellular fitness, such as a genetic knockout screen.
   Example: c-MYC is required for the majority of cancer cells to grow in genetic knockout screens.

Lineage-selective dependency
A dependency that is related to a cell's lineage or origin. The dependency may not necessarily be unique to a given lineage, although it is typically found in only a few lineages. In contrast, dependencies found in ≥90% of cancer cells are termed "common essential dependencies."
   Example: ISL1 is a lineage-selective dependency in neuroblastoma (NB) (Fig. 1A).

Core regulatory circuit
A model of gene regulation composed of a positive autoregulatory feedback loop of interconnected, super-enhancer-regulated transcription factors and nucleic acid-binding factors. Each member of the core regulatory circuit (CRC) physically binds to its own and each others' super-enhancer elements to cooperatively drive the expression of each component of the circuit. The CRC transcription factors may cobind with other genes, termed the "extended regulatory network," to drive their expression. The CRC model indicates a mechanism by which master transcription factors function to drive and stabilize the cancer cell transcriptome, which thereby defines the fundamental properties of the cell, known as the cancer cell state.
   Example: The general structure of a CRC is outlined in Figure 3A; the CRC characterizing the adrenergic cell state in NB is shown in Figure 4A.

Gene regulatory network
A unidirectional, modular framework of genes defining a given cell fate, typically in the context of development. Gene regulatory networks (GRNs) are usually comprised of multiple subcircuits or modules that perform specific functions in promoting one fate and restricting other cell fates and highlight the connections between these subcircuits.
   Example: Three of the most commonly used GRN subcircuits are illustrated in Figure 3B. The GRN governing migratory neural crest cell state and the specification of autonomic and mesenchymal lineages are depicted in Figure 4B.

---

mor subtypes while also revealing new targets for exploration.

Functional genomic screens have also shed light on two other key observations. The first is the existence of paralog synthetic lethal genetic relationships, in which mutation or loss of one gene in a paralogous pair results in selective reliance on the second gene. Examples of this include synthetic lethality of *VSP4A/B* in rhabdomyosarcoma (RMS) and pancreatic carcinoma, histone deacetylases *HDAC1/2* in pancreatic carcinoma and neuroblastoma (NB), chromatin remodeling enzymes *SMARCA2/4* in non-small-cell lung carcinoma, and histone acetyltransferases *EP300/CBP* in breast carcinomas (Oike et al. 2013; Ogiwara et al. 2016; Neggers et al. 2021; Zhang et al. 2023). In each case, genetic or pharmacologic targeting of the nonmutated paralog causes exceptional responses, suggesting a strategy for inducing synthetic lethality in cancer.

A second important observation is that the genetic dependencies of many different types of cancer may reflect, at least in part, their presumptive cell of origin. This observation is particularly pronounced in pediatric malignancies, in which functional genomics experiments have shown that the most enriched lineage-selective dependencies are implicated in their respective lineage-selective developmental processes, particularly in transcriptional regulation of cell state (Dharia et al. 2021). For example, the master regulator of the myogenic lineage, *MYOD1*, along with a host of other myogenic transcription factors, are enriched in the myogenic childhood solid tumor RMS (Gryder et al. 2017; Tenente et al. 2017; Stewart et al. 2018; Hsu et al. 2022). Along the same lines, other childhood malignancies feature enrichment of master transcription factors with known key roles in lineage-specific development, such as those found in acute myeloid leukemia (AML), Ewing sarcoma (ES), and NB (Fig. 1B; see Supplemental Table S1). Collectively, these findings recapitulate the long-held notion that the neoplastic state reflects disordered development and reframes our interpretation of lineage-selective dependencies as those that are enriched in classes of cancer cells derived from specific cell lineages. Thus, parallel examination of developmental genetics and lineage-specific functional

genomic surveys of diverse cancer cell lineages affords the opportunity to view cancer biology through the lens of normal development and highlights the relationships between them.

There are two commonly used models of transcriptional regulation of cell state, which we will discuss in depth. One model, the core regulatory circuit (CRC), was first described in the context of normal embryonic stem cell pluripotency but has become a common heuristic used for understanding the relationship between lineage, cell state, and genetic dependencies in cancer (Chen et al. 2020). Importantly, although the CRC model has been highly beneficial for contextualizing key transcription factors driving the malignant phenotype in cancer, it is considered a relatively static entity that focuses on an individual cell state, and therefore does not capture the dynamic range of cell states that are found in a given heterogeneous cancer nor illustrate how transitions between cell states are controlled. In contrast to the focused CRC model, the broader gene regulatory network (GRN) model describes the normal development of cell types, tissues, and organs in a spatiotemporal manner. In the GRN model of development, modules or circuits of transcriptional regulators work in hierarchical cascades to specify cell lineages, maintain cell states, carry out cell state transitions, and drive differential gene batteries (Peter and Davidson 2015). The GRN model has been helpful to understand master regulators of individual cell states of differentiation during normal development, but also the drivers and connections between these states. As our understanding of normal embryonic development expands in parallel with our understanding of cancer cell heterogeneity, it is important to consider what insights can be gained from restructuring our view of the CRC model through the lens of developmental GRNs. Therefore, we posit that an examination of GRNs alongside CRCs and lineage-selective dependencies will shed light on new vulnerabilities and the mechanisms by which they are controlled and highlight new strategies to target these for therapeutic benefit.

Here, we discuss the principles of lineage-selective dependencies, CRCs, and GRNs using childhood cancers as example tumors of interest.

Cite this article as *Cold Spring Harb Perspect Med* doi: 10.1101/cshperspect.a041573

To illustrate these points, we focus on high-risk NB, a highly lethal pediatric extracranial solid tumor. NB originates from the neural crest, a cellularly heterogeneous lineage that is characterized by a complex GRN (Martik and Bronner 2017). We use neural crest development as a framework for discussing the overarching principles of GRN logic. Using these principles, we consider NB as a disease of disordered neural crest development and highlight opportunities to improve our understanding of lineage-selective dependencies by viewing them in both CRC and GRN contexts.

THE RELATIONSHIP BETWEEN LINEAGE AND DEPENDENCY

The term "genetic dependency" is defined as a gene required by cancer cells for growth and/or survival (Box 1). This requirement is generally measured by selective depletion of the gene from a specific tumor cell line and measurement of outgrowth compared with normal growth. Many such dependencies are shared among nearly all cells, cancerous and noncancerous, and are termed "common essential dependencies." Common essentiality makes these genes less attractive for therapeutic intervention, because of the likelihood of any targeted therapy against these proteins resulting in collateral damage to untransformed cells. In contrast to the previously mentioned targeted catalytic inhibitors, conventional therapies used in cancer medicine may affect common essential dependencies, such as the *Vinca* alkaloids or taxanes, which disrupt multiple common essential tubulin proteins, or topotecan or irinotecan, which disrupt the common essential DNA topoisomerase 1 enzyme (Liu et al. 2000; Weaver 2014). Despite their potent activity in neoplastic cells, these agents also have toxicity to untransformed tissues. As such, targeting common essential genes, although effective, may have a narrow therapeutic window. In addition to these types of genes, large-scale functional genomic assays across more than 1000 cancer cell lines have also revealed lineage-selective dependencies that are specific to only one or a small handful of cell lineages. This is true in pediatric cancers, despite their relative paucity of oncogenic mutations compared to adult cancers (Gröbner et al. 2018; Ma et al. 2018; Dharia et al. 2021). Lineage-selective dependencies are intriguing candidates for disruption, because the lack of effect in other cancer cells (and, therefore, presumably, untransformed cells) implies a potential therapeutic window that is potentially suitable for clinical translation.

The first genetic knockout surveys of pediatric cancer cell dependencies demonstrated that distinct regulators of mRNA transcription and splicing are heavily and selectively enriched in specific cancer subtypes (Durbin et al. 2018; Dharia et al. 2021; Sun et al. 2023). However, many lineage-selective dependencies are proteins involved in cell signaling pathways, metabolic processes, and immune system interactions that may be therapeutically targeted, either in a single or restricted subset of cancers. As a demonstration of the diversity of cellular functions of lineage-selective dependencies, we analyzed genetic dependency data mined from the Cancer Dependency Map 23Q4 release for four pediatric malignancies derived from distinct cell lineages: hematopoietic cells (AML), neural crest (NB), and myogenic (RMS) lineages. In addition, we analyzed data from Ewing sarcoma (ES), a tumor derived from an unknown lineage that is hypothesized to be mesenchymal (MES). This analysis demonstrates links between these distinct cancer subtypes, in which specific genes are required for the growth of multiple cancer lineages (Fig. 1A; see Supplemental Tables S1–S4). For example, the *ZEB2* gene encodes a transcription factor required to support the growth of AML, ES, and RMS cells, but not NB cells. Similarly, the nuclear export protein-coding gene *NXT1* is required for the growth of NB and RMS, but not AML or ES. These findings are validated in separate investigations of these cancer types (Wiles et al. 2013; Li et al. 2017; Malone et al. 2021). Importantly, these genes are two examples of lineage-selective dependencies that are necessary for the growth of a small number of tumor subtypes, thereby making them viable candidates for therapeutic development because of their expanded presence outside of a single rare malignancy.

Separately, this summative analysis of the dependency landscapes of these four unrelated childhood malignancies reiterates earlier findings that many of the most enriched lineage-selective dependencies are regulators of transcription and gene expression (Durbin et al. 2018; Dharia et al. 2021). These dependencies include transcription factors, epigenetic regulators, and RNA processing factors that are critical to establish and maintain the cancer cell transcriptional state (Fig. 1B). Importantly, a closer examination of lineage-selective transcription factor–encoding dependencies reveals genes with well-characterized roles in cell fate specification and differentiation during embryonic development. For example, the transcription factor–coding gene *ISL1* is one of the most highly enriched dependencies in NB compared to all other cancer cell lines in the DepMap database and is essential for specifying the sympathetic neuronal lineage during neural crest development (see Supplemental Table S3; Huber et al. 2013; Durbin et al. 2018; Zhang et al. 2018; Dharia et al. 2021). Other examples of transcription factors selectively enriched for dependency in NB that are also required for normal sympathetic neuronal specification include *ASCL1, HAND2, PHOX2A/ B, GATA2/3*, and *TFAP2A/B* (Stanke et al. 1999; Tsarovina et al. 2004; Doxakis et al. 2008; Boeva et al. 2017; van Groningen et al. 2017; Durbin et al. 2018, 2022; Dong et al. 2020; Jansky et al. 2021; Kameneva et al. 2021; Kildisiute et al. 2021). These examples indicate that, for a subset of lineage-selective dependencies, these genes are not simply enriched in a given lineage but are critical to the definition and control of lineage itself.

But what of those transcriptional regulators with known critical functions in lineage specification that fail to meet dependency criteria? Lineage-selective dependencies are genetic vulnerabilities to cancer cell viability, in which the loss of a given gene leads to negative selection in a competitive assay of cellular fitness (Box 1). Although large-scale genomic screens for lineage dependencies have led to enormous strides in the identification and interrogation of lineage-selective dependencies in cancer, these assays reflect a relatively simple metric: the effect of loss of one gene on the growth of a small number of incompletely representative cell lines in vitro. Independent of the caveats inherent to in vitro cell culture systems that remove tumor cells from their supportive microenvironment, and in which the abundance of metabolites may mask reliance on metabolic pathways

uniquely present in vivo, additional aspects of malignancy that are controlled specifically by transcriptional regulators may go undetected in traditional cellular fitness assays. For instance, if a cancer cell uses lineage-specific transcription factors to orchestrate cell state switching as a means to evade chemotherapy or acquire metastatic potential, without an effect on proliferation, differentiation, or cell death, then these targetable factors may not be identified in large-scale genetic dependency screens. Similarly, although some transcriptional regulators may have stage-specific critical roles during development, reactivation of paralogous genes in cancer cells may provide a buffering mechanism that would rule out enrichment as a lineage-selective dependency. One example of this is the paralogous function of the *EP300* and *CBP* transcriptional coactivators in many cancer cell lines (Durbin et al. 2022). An additional caveat to consider is the heterogeneity of cancer cell lines included in large-scale functional genomic screens. Not only may cell lines within one lineage vary significantly in phenotypic traits, genetic mutation profiles, or primary versus metastatic site of origin, but even within one cell line there may be substantial variance in gene expression that may mask the identification of lineage-selective dependencies (Ben-David et al. 2018).

Thus, for the generation of more precisely targeted cancer therapies, it is important to consider lineage-selective dependencies not just in the context of cell growth and survival, but also in the context of (1) establishment, and (2) maintenance of the malignant cell state. Further, we predict that evolving technologies using combinations of digenic knockout CRISPR screening will dramatically enhance the detection of synthetic lethal and overlapping gene interactions, although scalability across large numbers of cell lines required to detect functional enrichments in specific cancer lineages remains a challenge (Ito et al. 2021).

NEUROBLASTOMA AS A MODEL FOR UNDERSTANDING LINEAGE-SELECTIVE DEPENDENCIES

NB tumors arise in the peripheral sympathetic nervous system, which is formed during embryonic development from the neural crest (Yntema and

Hammond 1947; Weston 1963; Le Douarin and Teillet 1974). The neural crest is a transient, multipotent cell population arising from the process of neurulation during the fourth week of gestation in humans (Betters et al. 2010). Neural crest cells (NCCs) emigrate from the cranial, vagal, trunk, and sacral regions of the nascent neural tube and give rise to a host of diverse lineages throughout the body (Fig. 2A). Although many NB tumors arise from trunk neural crest–derived sympathetic progenitors in the adrenal glands or sympathetic chain ganglia, primary NB tumors can also rise in regions populated by cranial, vagal, or sacral neural crest derivatives (Fig. 2B; Sung et al. 2009; Vo et al. 2014). In addition to classical prognostic factors in NB, such as patient age, stage, *MYCN* amplification status, and other chromosomal aberrations, some have suggested that the primary tumor site may be prognostically significant in NB, suggesting that the type, and perhaps stage, of NCC from which the tumor develops may impact clinical prognosis (Haase et al. 1995; Leclair et al. 2004; Sung et al. 2009; Vo et al. 2014).

Long-standing in vitro evidence indicates that NB is characterized by interconverting cells with distinct metastable phenotypes, which more recently have been linked to transcriptional and epigenetic profiles (Ross et al. 1983, 2003; Ciccarone et al. 1989; Piskareva et al. 2015; Boeva et al. 2017; van Groningen et al. 2017). Although the majority of NB cells are termed "adrenergic" (ADRN) and

have a well-described and experimentally validated CRC model, there are rarer mesenchymal (MES) cells that are epigenetically and transcriptionally distinct and have been implicated in resistance to therapy and potentially patient relapse (Boeva et al. 2017; van Groningen et al. 2017; Gartlgruber et al. 2021). Integrated transcriptomic and epigenomic analyses of MES cells have revealed close similarity to neural crest progenitor cells, suggesting that there may be a distinct CRC driving the MES cell state, although a limited number of models and apparent heterogeneity in this MES state has limited full resolution of an MES-specific CRC (Boeva et al. 2017; van Groningen et al. 2017, 2019). Further, recent evidence has implicated a role for loss of *LMO1* expression, an ADRN CRC cofactor, in eliciting an MES-like gene expression pattern distinct from high-risk MES cells with similarity to low-risk NB (Weichert-Leahey et al. 2023). Low-risk NB is a disease with striking differences from high-risk NB in terms of expression patterns and clinical characteristics, suggesting key roles for transcriptional circuitries in fate determination (Qiu and Matthay 2022).

MODELS OF TRANSCRIPTIONAL NETWORKS CONTROLLING CELL STATE

Core Regulatory Circuits

The transcriptional circuitries that control cell fate can be conceptualized using different model

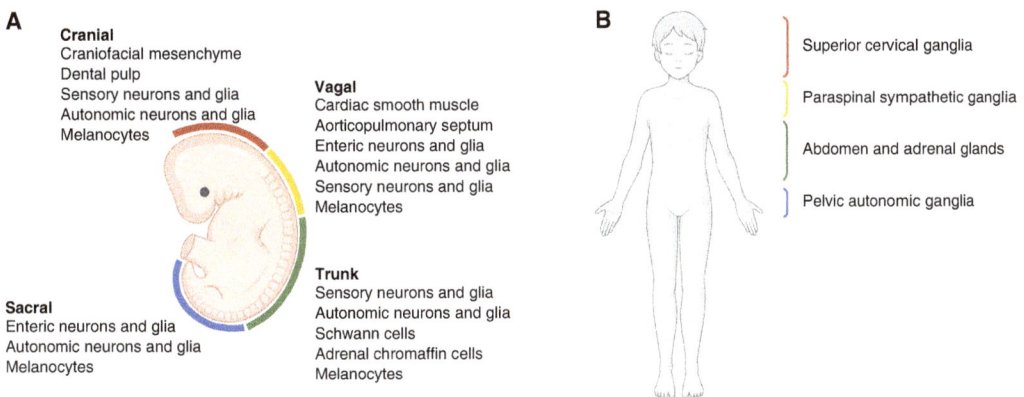

Figure 2. Neuroblastoma (NB) tumors originate from all axial levels of the neural crest. (*A*) Schematic of a human embryo with the four axial levels of the neural crest highlighted. Key derivatives of each axial level of the neural crest are listed. (*B*) Sites of primary NB tumors found in children in relation to neural crest axial levels. (Figure created using biorender.com.)

types. One of these, the CRC, is structured as an interconnected positive feedback loop of super-enhancer-regulated transcription factors that co-operatively regulate the expression of all other components of the circuit, as well as themselves (Fig. 3A; Boyer et al. 2005). By doing so, the CRC model predicts that a small cohort of master transcription factors autoregulate each other and establish and drive the cell type–specific gene expression programs that define a given cell state. The CRC model reflects the observation that cancer cells intrinsically depend on a set of transcription factors that establish the transcriptome to drive the malignant cell state (Weinstein and Joe 2006; Bradner et al. 2017). Indeed, this concept aligns with the observation that many highly expressed lineage-selective dependencies are, in fact, CRC transcription factors (Fig. 1; Supplemental Tables S1–S4). Additionally, as the CRC model was originally derived to explain how embryonic stem cells maintain stemness and growth potential, it also serves this function well in describing cancer cells locked in a growth state resembling stem cell self-renewal.

Many efforts to define initial working models of CRCs in human tumor types focus on the activation of master transcription factors for largely technical reasons and thereby implicitly adopt several assumptions: (1) Each gene component of the CRC plays an equal role in perpetuating the circuit (i.e., there is no hierarchy within the CRC), (2) each gene component plays an activating or stabilizing role in regulating gene expression (i.e., no factors are repressing expression), and (3) the circuit resembles a "house of cards" in which loss of individual components causes the circuit to collapse. Although each of these assumptions is true in some instances, evidence to the contrary indicates that exceptions are found. For example, recent studies indicate that the CRC of AML may be more hierarchically organized (Eagle et al. 2022; Harada et al. 2022). Similarly, initial studies demonstrated that cooperativity between SOX2 and OCT4 was required to induce *NANOG* expression, leading to the formation of the ES cell CRC (Boyer et al. 2005). Thus, the formation and maintenance of the CRC may be distinct processes, requiring different inputs. Additionally, recent studies have identified super-enhancer-regulated transcription factors that function in a repressive manner, such as ETV6 in ES cells that are driven by the strong transcriptional activator EWS–FLI1 fusion oncogene (Gao et al. 2023; Lu et al. 2023). This is reflective of prior evidence that repressive components of core regulatory circuitries exist,

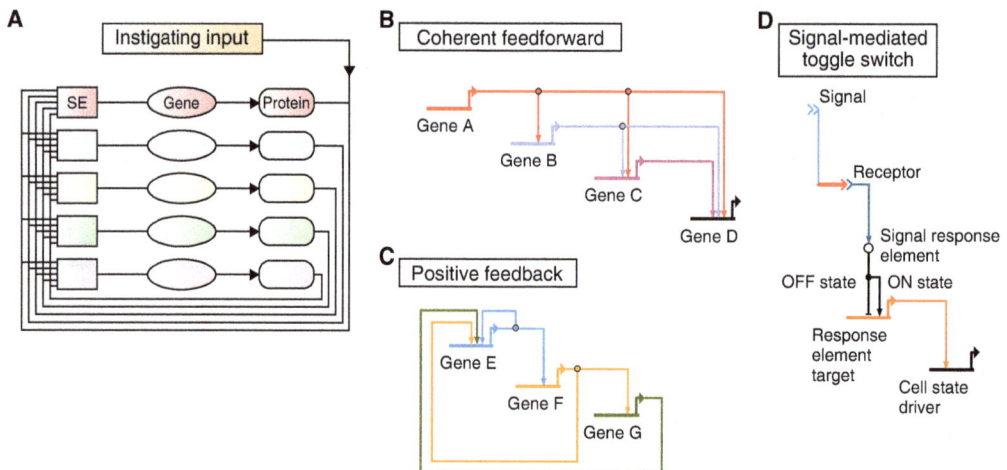

Figure 3. Core regulatory circuits (CRCs) and gene regulatory networks (GRNs) serve as transcriptional models of cancer and development. (*A*) Schematic of CRC structure often used to model the gene regulatory program driving cancer cell state. (SE) Super-enhancer. (*B–D*) Architecture of three subcircuit architectures frequently used in GRNs in development: coherent feedforward (*B*), positive feedback (*C*), and signal-mediated toggle switch (*D*).

such as negative regulatory miRNAs and the function of Polycomb in gene repression (Lee et al. 2006; Marson et al. 2008). Furthermore, chemical agonism of the retinoic acid receptor (RARA) in NB enforces the loss of several members of a prototypical neuroblastoma ADRN CRC (Fig. 4A), resulting not in circuit collapse, but in an altered cell state involving the generation of a second CRC that promotes differentiation (Zimmerman et al. 2021). Thus, a "static" and reductive CRC model of transcriptional regulation of cell state is not truly reflective of cancer cell complexity and may require refinement in the setting of cellular context, microenvironmental signaling, oncogenic drivers, and lineage. In doing so, careful attention to the definition of CRC properties will be needed. Further, and perhaps most notably, the CRC model of transcriptional regulation reflects a single cellular property. In the setting of conventional dependency screenings, this property is cell growth. Thus, interpretation of cancer cell CRC dynamics that integrate cell state shifts in response to extrinsic and intrinsic stimuli, combined with the multifaceted functions of transcription factors in orchestrating the transcriptional landscape, is needed. These complexities highlight the need to refine the CRC model to a more dynamic and flexible model.

Gene Regulatory Networks

In contrast to CRCs, GRNs are comprised of modules or subcircuits of genes that drive gene expression programs in a spatiotemporal-specific manner (Box 1). GRNs have several core features (Peter and Davidson 2015). First, GRNs, like CRCs, are constructed from functional studies of gene regulation and are thus models of causal, not correlative, gene relationships (Rothenberg 2019). To this end, master regulators of individual cell states are typically identified through knockout models in vivo, with subsequent validation in vitro to determine the presumed targets of each individual master regulator. Second, GRNs are modular, with each module or subcircuit representing a discrete cell state, which may be thought of as a CRC. As such, a complete GRN outlines the progression of cell state changes as new modules emerge under the influence of master regulators. Third, to reflect cell state progression over time, GRN modules are hierarchically organized in a unidirectional manner to reflect the stepwise nature of cell fate commitment and differentiation. This aspect of GRNs is particularly challenging to integrate with cancer cell biology, in which flexibility in cell state is an emerging concept related to disruption of cell fate commitment and the capability of cells to dedifferentiate (Hanahan 2022). Overall, however, the GRN model lays out a regulatory logic that underlies the dynamic shifts in cell state that are responsible for development.

The modules and subcircuits comprising GRNs can be topologically complex, with conserved wiring patterns serving specific functions in development. There are a host of GRN subcircuits used throughout development; three subcircuit organizational schemas that are particularly relevant for cancer biology include coherent feedforward, positive feedback, and signal-mediated toggle switch (Fig. 3B–D; Peter and Davidson 2015). These three types of circuitry play critical roles in cell state definition and are variably used by cancer cells to enforce malignant transcriptional programs.

Coherent feedforward circuits are defined by sequential and combinatorial activation of gene expression and are responsible for the stepwise progression of gene expression programs observed in cell fate commitment (Fig. 3B). The combinatorial nature of the genes comprising the circuit enables finer control of cell state and allows individual genes to be used reiteratively over the course of development (Istrail and Davidson 2005). Several lineage-selective dependencies in NB function in coherent feedforward subcircuits in the autonomic neuronal specification GRN, including *ASCL1*, *HAND2*, *PHOX2A*, and *ISL1* (Fig. 4B). Coherent feedforward circuits are similar to the classical definition of CRCs in that gene activation is the direct result of positive inputs from multiple transcription factors, but they are distinct in that the architecture of coherent feedforward circuits is hierarchically organized and not necessarily fully interconnected and directly autoregulatory.

Positive feedback circuits, in which a gene directly or indirectly activates its own expression, are

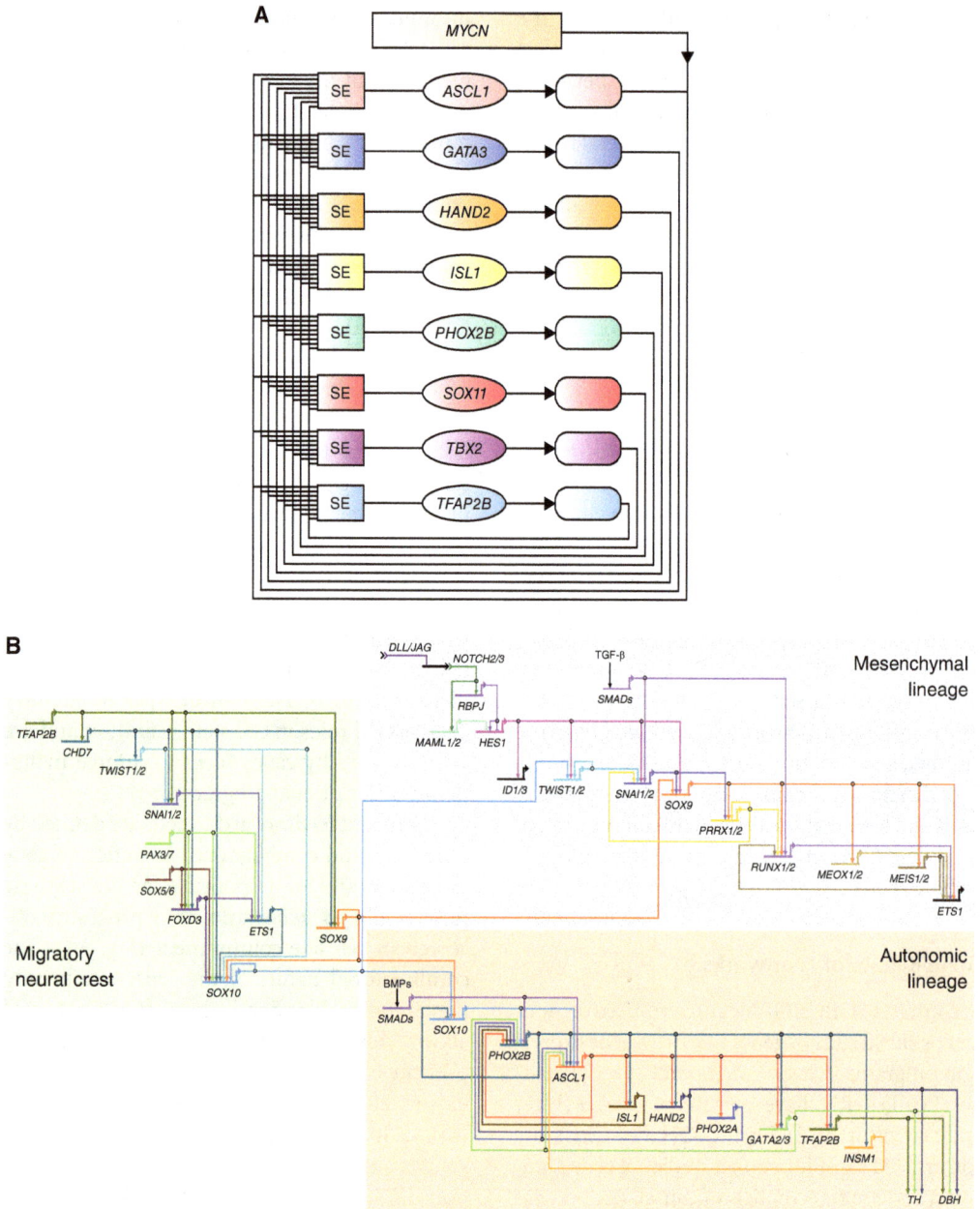

Figure 4. Adrenergic core regulatory circuit (CRC) model versus neural crest gene regulatory network (GRN). (*A*) The CRC model of the adrenergic (ADRN) cell state in neuroblastoma (NB). (*B*) The GRN model of migratory neural crest cell state and specification of autonomic neurons and mesenchymal lineages. Note that some genes, such as *TFAP2B* and *PHOX2B*, have been previously reported as lineage-selective dependencies in NB (Durbin et al. 2018) but did not meet the dependency criteria described in Figure 1.

perhaps the most common GRN subcircuit and are required for the maintenance of cell fate to prevent premature differentiation (Fig. 3C; Peter and Davidson 2015; Simões-Costa and Bronner 2015). The CRC is, in essence, an individual positive feedback circuit in which the master transcription factors promote their own and each other's expression and a stable cell state. Positive feedback circuits help to mitigate variability in signaling inputs that impact cell state by establishing a stable transcriptional program, another tactic that cancer cells must use to maintain malignancy against repressive inputs from neighboring non-malignant cells (Peter and Davidson 2015). The regulation of *SOX10* expression in migratory NCCs is an excellent example of a positive feedback circuit, as sustained, high levels of *SOX10* are required to maintain a multipotent, migratory cell state in the neural crest (Fig. 4B; Kim et al. 2003; Wahlbuhl et al. 2012). In addition to receiving activating input from numerous transcription factors, including *PAX3/7*, *SNAI1/2*, *FOXD3*, and *ETS1*, *SOX10* also regulates itself (Bondurand et al. 2000; Lang et al. 2000; Wahlbuhl et al. 2012; Martik and Bronner 2017). Loss of *SOX10* function leads to cessation of migration and premature neuronal differentiation (Southard-Smith et al. 1998; Kim et al. 2003; Carney et al. 2006). However, as described above in the context of ADRN MES state switching in NB, there is a selective advantage for cancer cells to be able to disengage such state-stabilizing positive feedback circuits in favor of other gene expression programs and corresponding phenotypes, such as enhanced resistance to chemotherapy or acquisition of other cancer hallmarks. Therefore, it is important to consider how lineage-selective dependencies that participate in positive feedback circuits may take on additional functionality to further drive malignancy and disease progression.

Signal-mediated "toggle-switch" circuits are one of the most evolutionarily ancient mechanisms by which spatiotemporal transcriptional programs are encoded (Gazave et al. 2009; Peter and Davidson 2015). In this circuit design, an extracellular signal from one cell is detected by a cognate receptor on a neighboring cell, which triggers a downstream transcriptional regulator to convert from a default repressor state to an activator state (Fig. 3D; Barolo and Posakony 2002; Peter and Davidson 2015). There are several essential design features of signal-mediated toggle-switch circuits that enable them to execute highly specific, fine-tuned transcription programs that control cell fate decisions. First, signal-mediated toggle-

switch circuits incorporate "AND" Boolean logic, which refers to the requirement of both receptor activation by the signal and the presence of cell type–specific transcriptional regulators to mediate downstream changes in gene expression (Barolo and Posakony 2002; Saadatpour and Albert 2013; Peter and Davidson 2015). Second, the cell type–specific transcriptional regulators function as repressors of gene expression in the absence of active signaling. This feature prevents, or at least mitigates, leaky transcription that would dampen the robust gene expression programs driving a given cell state. Third, the "toggle-switch" mechanistic component confers reversibility to cell fate choices resulting from signaling activity, rather than absolute commitment to cell fate (Barolo and Posakony 2002). A classic example of signal-mediated toggle switch circuitry is the Notch pathway, which drives cell fate decisions in numerous tissues and organ systems over the course of development and is implicated in several malignancies, including NB (for review, see Zhou et al. 2022). In developing NCCs, Notch signaling mediates neuronal versus glial cell fate choice in the autonomic and enteric nervous systems (Tsarovina et al. 2008; Ngan et al. 2011). Activation of the *NOTCH1–4* receptor by ligands *DLL1,3,4* or *JAG1/2* results in proteolytic cleavage of the NOTCH receptor intracellular domain, which is then transported to the nucleus and binds to the RBPJ transcription factor. The binding of Notch-ICD to RBPJ also leads to the recruitment of the transcriptional coactivators MAML1–3, with the overall effect of converting RBPJ from a transcriptional repressor to an activator (Tamura et al. 1995; Nam et al. 2006). RBPJ target genes, including members of the HES and HEY transcription factor families, can then drive the expression of proglial specification genes (Lui et al. 2011). The ligand-expressing cells, in the absence of activated Notch, are biased toward a neuronal fate. As a corollary, in NB, activated NOTCH signaling drives cell state switching from the ADRN neuronal state to the MES cell state, primarily via *NOTCH2* and *MAML2* (van Groningen et al. 2019). The activated NOTCH signaling targets *HES1* and *HEYL* in MES cells then participate in a positive feedback circuit (discussed above) to drive the expression of key MES genes, whereas key ADRN genes are down-regulated (van Groningen et al. 2019). As mentioned above, cell state switching is a possible means by which malignant cells may evade immune system attacks or drug treatments that target lineage-selective dependencies. This example of the Notch signal–mediated toggle-switch circuit highlights how this may occur in NB.

LINEAGE-SELECTIVE DEPENDENCIES IN NEUROBLASTOMA FROM A GENE REGULATORY NETWORK PERSPECTIVE

The mechanisms by which lineage-selective dependencies function in core regulatory circuitries may parallel their function in the neural crest GRN are now beginning to be understood. Here, we review two of these transcription factors, *ASCL1* and *SOX11*, as representative examples of links that can be made between these two models.

ASCL1

ASCL1 (previously known as *MASH1*) encodes a basic helix–loop–helix transcription factor with well-established roles in nervous system development, including the autonomic division. Our analysis of lineage-selective dependencies using DepMap revealed enrichment of *ASCL1* as a lineage-selective dependency in many ADRN NB cell lines, consistent with prior reports of *ASCL1* functioning as a core regulatory circuitry member (Supplemental Table S3; Wang et al. 2019). Numerous studies in neural stem cells and cancer cell lines have demonstrated dual roles for ASCL1—first as an essential driver of the proliferative cell state in neural progenitors and then as an instigator of the neuronal differentiation program (Moriguchi et al. 2006; Wildner et al. 2008; Raposo et al. 2015; Park et al. 2017; Parkinson et al. 2022).

The function of ASCL1 in maintaining a proliferative cell state fits well with the canonical view of lineage-selective dependencies as factors that are required for retention in cellular growth assays; factors required for proliferation and promoting cell cycle progression are particularly sensitive to such measures of fitness. In neural stem cells, ASCL1 accomplishes this by promoting gene expression of cell cycle progression genes, including *CDK1*, *CDK2*, *SKP2*, and *BIRC5* (Castro et al. 2011). A similar requirement for *ASCL1* in cell growth was demonstrated in several neural lineage cancers, including NB and glioblastoma (Park et al. 2017; Parkinson et al. 2022). The NB cell lines SH-SY5Y and IMR32 grow more slowly in culture with loss of *ASCL1* expression compared to their respective parental cell lines (Parkinson et al. 2022). In the Kelly NB cell line, *ASCL1* has also been shown to bind, and be bound by, the other members of the ADRN CRC, including *PHOX2B*, *ISL1*, *TBX2*, *GATA3*, *HAND2*, and *MYCN* (Wang et al. 2019). Importantly, high-level expression of *ASCL1* in NB correlates with poor survival, indicating that *ASCL1* may contribute to aggressive malignancy (Wang et al. 2019).

However, there are other facets to lineage-selective dependency that should be considered beyond the metric of cancer cell viability or growth. This is seen clearly with ASCL1, due at least in part to its function as a pioneer factor by binding closed chromatin and facilitating the accessibility of essential enhancers of proneurogenic genes (Wapinski et al. 2013; Raposo et al. 2015; Park et al. 2017). During normal development, *ASCL1* expression in sympathetic progenitors is transient in nature—ASCL1 activates the expression of *INSM1* and *GATA3*, which then go on to inhibit *ASCL1* expression (Fig. 3B; Moriguchi et al. 2006; Wildner et al. 2008). Forced prolonged expression of *ASCL1* prevents the onset of the sympathetic neuronal differentiation program (Moriguchi et al. 2006; Wildner et al. 2008). In addition to having a neurogenic function, ASCL1 also acts to repress alternative cell fate choices in the developing nervous system and in neuroendocrine cancers (Imayoshi et al. 2013; Olsen et al. 2021). In small-cell lung cancer, ASCL1 actively represses a neural crest stem cell–like state marked by coexpression of *SOX9*, *RUNX1*, and *RUNX2* (Olsen et al. 2021). In mouse neural crest development, *Ascl1*, but not other CRC members such as *Phox2b*, presents an impenetrable barrier to mesenchymal cell fate. Lineage tracing experiments in *Ascl1*[CreER/+] mice showed that *Ascl1*[Cre/ER+] lineage-traced NCCs do not contribute to any mesenchymal

Cite this article as *Cold Spring Harb Perspect Med* doi: 10.1101/cshperspect.a041573

cell lineages in the vagal neural crest, whereas scattered $Phox2b^{Cre/+}$-traced cells were found in heart mesenchyme (Soldatov et al. 2019). Moreover, the function of ASCL1 depends on its phosphorylation status. In the developing neural crest, NB, and glioblastoma, ASCL1 dephosphorylation is required for neuronal progenitors to exit the cell cycle and proceed to neuronal differentiation (Wylie et al. 2015; Ali et al. 2020; Azzarelli et al. 2022). ASCL1 also regulates the expression of Notch signaling pathway components, thereby contributing to the maintenance of stem cell identity and inhibiting premature differentiation (Park et al. 2017). Reflecting this, NOTCH signaling drives ADRN to MES cell state switching in NB, with activation of the NOTCH signaling pathway causing transcriptomic and epigenetic reprogramming, including enrichment of key MES genes SOX9, RUNX1, and RUNX2 and repression of ASCL1 and PHOX2B (Boeva et al. 2017; van Groningen et al. 2017, 2019). Opposingly, a recent study demonstrated that induction of ASCL1 instigates and partially drives the ADRN gene regulatory program in MES NB cell lines, which are endogenously deplete of ASCL1 gene expression (Wang et al. 2023). The effect of ASCL1 overexpression on cell state was due at least in part to its activity as a pioneer factor at regions of closed chromatin harboring ADRN genes (Wang et al. 2023). These results strengthen the link between ASCL1 as a lineage-selective dependency and the NOTCH toggle-switch gene regulatory circuit in controlling cell state in NB.

Integration of these studies with the neural crest GRN reveals two key messages. First, with its pioneer function and position at the top of the GRN driving sympathetic neuronal specification and differentiation, ASCL1 may be uniquely positioned as a master regulator of the transcription factors comprising the traditional CRC model of NB, functioning by initiating a MES to ADRN cell state switch. In this role, ASCL1 expression itself is a dependency for other transcription factors that are essential for defining the ADRN cell state. Second, it is possible that the intersection of ASCL1 and NOTCH signaling may present a unique vulnerability that can be targeted in the treatment of NB.

SOX11

SOX11 is a member of the SoxC family of SRY-related HMG box-containing (SOX) transcription factors, which includes SOX4 and SOX12 (Schock and LaBonne 2020). Our analysis of NB lineage-selective dependencies described above revealed SOX11 is strongly enriched among many ADRN NB cell lines (Supplemental Table S3), which is supported by recent studies where SOX11 was identified as a genetic dependency in ADRN NB (Decaesteker et al. 2023). Studies in mouse embryos indicate an important role for SOX11 in the proliferation of sympathetic neuronal progenitors (Potzner et al. 2010). Interestingly, loss of Sox11 in embryonic mice does not appear to impact the specification of sympathetic neuronal fate or differentiation, as sympathetic chain ganglia and heart innervation remain intact in Sox11 knockout animals but with reduced numbers of neurons (Potzner et al. 2010). The function of SOX11 in proliferation is primarily in the early stages of vagal and trunk neural crest development, after the onset of Phox2b expression but before expression of Gata3 and Hand2 (Potzner et al. 2010). These observations place SOX11 near the top of the sympathetic neuronal specification GRN, before postmitotic neuronal differentiation. As with ASCL1, the functions of SOX11 in neural crest and NB appear to be multifaceted and go beyond simply maintaining cell growth.

In NB, SOX11 was found to regulate the expression of numerous factors involved in orchestrating the transcriptional and epigenetic landscape (Decaesteker et al. 2023). In this study, SOX11 was shown to physically interact with SMARCA4, a core catalytic component of the SWI/SNF chromatin remodeling complex, which is also a lineage-selective dependency in NB that controls chromatin accessibility (Supplemental Table S1). Here, SOX11 expression is regulated by multiple super-enhancers that are specific to the ADRN cell state, classifying it as a component of the ADRN CRC (Decaesteker et al. 2023). However, as we describe above, viewing the dependency of SOX11 within a CRC context somewhat masks its unique functions in driving the *formation* of the ADRN cell state. Earlier studies of Sox11 in mouse sympathetic neural development suggest that this transcription factor functions in a non-cell-autonomous manner to regulate the survival of sympathetic neurons undergoing differentiation. Interestingly, *Sox11* expression is not restricted to the autonomic lineage of the vagal neural crest but is also expressed in nonneural crest-derived mesenchyme (Potzner et al. 2010). By comparing the effects of constitutive *Sox11* loss and neural crest-specific *Sox11* knockout, it was determined that, by an unknown signaling mechanism, mesenchymal *Sox11* regulates the expression of sympathetic *Sox4* to promote survival and prevent apoptosis (Potzner et al. 2010). This finding suggests that there are other functions of SOX11 in regulating gene expression outside of the cell-autonomous manner depicted by a CRC model. Reflecting this, enforced treatment of ADRN NB cell lines with retinoic acid causes a rewiring to a prodifferentiation, more neuronal state marked by up-regulation of *SOX4* (Zimmerman et al. 2021).

CONCLUDING REMARKS

Lineage-selective dependencies are highly diverse in cellular and molecular function, but regulators of gene expression comprise a significant proportion of dependencies in many pediatric malignancies (Fig. 1A; Dharia et al. 2021). Placing these dependencies in the context of a CRC model of transcriptional control is helpful for understanding the interconnected nature of gene expression regulators in maintaining a malignant cell state. However, emerging research suggests that more dynamic models are needed to describe complex transcriptional networks to accurately encompass cell state flexibility, acquisition and loss of cancer hallmarks, and variable responses to treatment. By integrating GRN logic into CRC models, lineage-selective dependencies can be contextualized to understand these cellular behaviors. Additionally, new dependencies can be nominated that lie at the GRN subcircuit transition points between distinct cell states. For example, in NB treatment, retinoic acid is used to induce neuronal differentiation and cell cycle exit of tumor cells (Thiele et al. 1985). Retinoic acid accomplishes this by reprogramming the ADRN cell state. Mechanistically, it does this by decommissioning the super-enhanc-

ers driving *GATA3* and *PHOX2B* expression and deploying new enhancers of *MEIS1* and *SOX4* that drive a more differentiated state (Zimmerman et al. 2021). With the advent of new chemical biology techniques, such as transcription factor targeting chimeras (TRAFTACs), additional lineage-selective dependencies regulating gene expression can be targeted alongside state-controlling treatments, such as retinoic acid, to more durably induce differentiation and cell cycle exit (Samarasinghe et al. 2021).

It is important to consider that the study of lineage-selective dependencies has been, up to this point, largely gene-centric. There are likely numerous dependencies that may be found in the noncoding genome, such as enhancers and insulators, but are more challenging to identify. As the screening methods used to discover lineage-selective dependencies evolve, the role of noncoding regulatory elements in driving malignancy will become more apparent.

As chemical approaches to target control of the epigenome and transcriptome evolve, we will realize enormous effects on tumor cell biology and malignant cell behaviors. Harnessing these chemical approaches to target lineage-selective dependencies, therefore, has great potential to improve treatment outcomes while limiting off-target toxicities.

For additional tables, see Supplemental Material.

ACKNOWLEDGMENTS

This work was supported by National Institutes of Health (NIH) grants K08-CA245251, R01-CA286444, P30-CA021765, the V Foundation for Cancer Research, the Hyundai Hope on Wheels Foundation, the Rally Foundation for Childhood Cancer Research and the American Lebanese Syrian Associated Charities. We thank Dr. Brian Abraham (St. Jude Children's Research Hospital) for critical feedback. We apologize to authors of manuscripts not cited herein due to space limitations.

REFERENCES

Ali FR, Marcos D, Chernukhin I, Woods LM, Parkinson LM, Wylie LA, Papkovskaia TD, Davies JD, Carroll JS, Philpott

A. 2020. Dephosphorylation of the proneural transcription factor ASCL1 re-engages a latent post-mitotic differentiation program in neuroblastoma. *Mol Cancer Res* **18:** 1759–1766. doi:10.1158/1541-7786.MCR-20-0693

Azzarelli R, McNally A, Dell'Amico C, Onorati M, Simons B, Philpott A. 2022. ASCL1 phosphorylation and ID2 upregulation are roadblocks to glioblastoma stem cell differentiation. *Sci Rep* **12:** 2341. doi:10.1038/s41598-022-06248-x

Barolo S, Posakony JW. 2002. Three habits of highly effective signaling pathways: principles of transcriptional control by developmental cell signaling. *Genes Dev* **16:** 1167–1181. doi:10.1101/gad.976502

Behan FM, Iorio F, Picco G, Gonçalves E, Beaver CM, Migliardi G, Santos R, Rao Y, Sassi F, Pinnelli M, et al. 2019. Prioritization of cancer therapeutic targets using CRISPR–Cas9 screens. *Nature* **568:** 511–516. doi:10.1038/s41586-019-1103-9

Ben-David U, Siranosian B, Ha G, Tang H, Oren Y, Hinohara K, Strathdee CA, Dempster J, Lyons NJ, Burns R, et al. 2018. Genetic and transcriptional evolution alters cancer cell line drug response. *Nature* **560:** 325–330. doi:10.1038/s41586-018-0409-3

Betters E, Liu Y, Kjaeldgaard A, Sundström E, García-Castro MI. 2010. Analysis of early human neural crest development. *Dev Biol* **344:** 578–592. doi:10.1016/j.ydbio.2010.05.012

Boeva V, Louis-Brennetot C, Peltier A, Durand S, Pierre-Eugène C, Raynal V, Etchevers HC, Thomas S, Lermine A, Daudigeos-Dubus E, et al. 2017. Heterogeneity of neuroblastoma cell identity defined by transcriptional circuitries. *Nat Genet* **49:** 1408–1413. doi:10.1038/ng.3921

Bondurand N, Pingault V, Goerich DE, Lemort N, Sock E, Le Caignec C, Wegner M, Goossens M. 2000. Interaction among *SOX10, PAX3* and *MITF,* three genes altered in Waardenburg syndrome. *Hum Mol Genet* **9:** 1907–1917. doi:10.1093/hmg/9.13.1907

Boyer LA, Lee TI, Cole MF, Johnstone SE, Levine SS, Zucker JP, Guenther MG, Kumar RM, Murray HL, Jenner RG, et al. 2005. Core transcriptional regulatory circuitry in human embryonic stem cells. *Cell* **122:** 947–956. doi:10.1016/j.cell.2005.08.020

Bradner JE, Hnisz D, Young RA. 2017. Transcriptional addiction in cancer. *Cell* **168:** 629–643. doi:10.1016/j.cell.2016.12.013

Carney TJ, Dutton KA, Greenhill E, Delfino-Machín M, Dufourcq P, Blader P, Kelsh RN. 2006. A direct role for Sox10 in specification of neural crest–derived sensory neurons. *Development* **133:** 4619–4630. doi:10.1242/dev.02668

Castro DS, Martynoga B, Parras C, Ramesh V, Pacary E, Johnston C, Drechsel D, Lebel-Potter M, Garcia LG, Hunt C, et al. 2011. A novel function of the proneural factor Ascl1 in progenitor proliferation identified by genome-wide characterization of its targets. *Genes Dev* **25:** 930–945. doi:10.1101/gad.627811

Chen Y, Xu L, Lin RY, Müschen M, Koeffler HP. 2020. Core transcriptional regulatory circuitries in cancer. *Oncogene* **39:** 6633–6646. doi:10.1038/s41388-020-01459-w

Ciccarone V, Spengler BA, Meyers MB, Biedler JL, Ross RA. 1989. Phenotypic diversification in human neuroblastoma cells: expression of distinct neural crest lineages. *Cancer Res* **49:** 219–225.

Decaesteker B, Louwagie A, Loontiens S, De Vloed F, Bekaert SL, Roels J, Vanhauwaert S, De Brouwer S, Sanders E, Berezovskaya A, et al. 2023. SOX11 regulates SWI/SNF complex components as member of the adrenergic neuroblastoma core regulatory circuitry. *Nat Commun* **14:** 1267. doi:10.1038/s41467-023-36735-2

Dharia NV, Kugener G, Guenther LM, Malone CF, Durbin AD, Hong AL, Howard TP, Bandopadhayay P, Wechsler CS, Fung I, et al. 2021. A first-generation pediatric cancer dependency map. *Nat Genet* **53:** 529–538. doi:10.1038/s41588-021-00819-w

Dong R, Yang R, Zhan Y, Lai HD, Ye CJ, Yao XY, Luo WQ, Cheng XM, Miao JJ, Wang JF, et al. 2020. Single-cell characterization of malignant phenotypes and developmental trajectories of adrenal neuroblastoma. *Cancer Cell* **38:** 716–733.e6. doi:10.1016/j.ccell.2020.08.014

Doxakis E, Howard L, Rohrer H, Davies AM. 2008. HAND transcription factors are required for neonatal sympathetic neuron survival. *EMBO Rep* **9:** 1041–1047. doi:10.1038/embor.2008.161

Druker BJ, Talpaz M, Resta DJ, Peng B, Buchdunger E, Ford JM, Lydon NB, Kantarjian H, Capdeville R, Ohno-Jones S, et al. 2001. Efficacy and safety of a specific inhibitor of the BCR-ABL tyrosine kinase in chronic myeloid leukemia. *N Engl J Med* **344:** 1031–1037. doi:10.1056/NEJM200104053441401

Durbin AD, Zimmerman MW, Dharia NV, Abraham BJ, Iniguez AB, Weichert-Leahey N, He S, Krill-Burger JM, Root DE, Vazquez F, et al. 2018. Selective gene dependencies in MYCN-amplified neuroblastoma include the core transcriptional regulatory circuitry. *Nat Genet* **50:** 1240–1246. doi:10.1038/s41588-018-0191-z

Durbin AD, Wang T, Wimalasena VK, Zimmerman MW, Li D, Dharia NV, Mariani L, Shendy NAM, Nance S, Patel AG, et al. 2022. EP300 selectively controls the enhancer landscape of MYCN-amplified neuroblastoma. *Cancer Discov* **12:** 730–751. doi:10.1158/2159-8290.CD-21-0385

Eagle K, Harada T, Kalfon J, Perez MW, Heshmati Y, Ewers J, Koren JV, Dempster JM, Kugener G, Paralkar VR, et al. 2022. Transcriptional plasticity drives leukemia immune escape. *Blood Cancer Discov* **3:** 394–409. doi:10.1158/2643-3230.BCD-21-0207

Eiermann W; International Herceptin Study Group. 2001. Trastuzumab combined with chemotherapy for the treatment of HER2-positive breast cancer: pivotal trial data. *Ann Oncol* **12** (Suppl 1): S57–S62. doi:10.1093/annonc/12.suppl_1.S57

Gao Y, He XY, Wu XS, Huang YH, Toneyan S, Ha T, Ipsaro JJ, Koo PK, Joshua-Tor L, Bailey KM, et al. 2023. ETV6 dependency in Ewing sarcoma by antagonism of EWS-FLI1-mediated enhancer activation. *Nat Cell Biol* **25:** 298–308. doi:10.1038/s41556-022-01060-1

Gartlgruber M, Sharma AK, Quintero A, Dreidax D, Jansky S, Park YG, Kreth S, Meder J, Doncevic D, Saary P, et al. 2021. Super enhancers define regulatory subtypes and cell identity in neuroblastoma. *Nat Cancer* **2:** 114–128. doi:10.1038/s43018-020-00145-w

Gazave E, Lapébie P, Richards GS, Brunet F, Ereskovsky AV, Degnan BM, Borchiellini C, Vervoort M, Renard E. 2009. Origin and evolution of the Notch signalling pathway: an overview from eukaryotic genomes. *BMC Evol Biol* **9:** 249. doi:10.1186/1471-2148-9-249

Gröbner SN, Worst BC, Weischenfeldt J, Buchhalter I, Kleinheinz K, Rudneva VA, Johann PD, Balasubramanian GP, Segura-Wang M, Brabetz S, et al. 2018. The landscape of genomic alterations across childhood cancers. *Nature* **555**: 321–327. doi:10.1038/nature25480

Gryder BE, Yohe ME, Chou HC, Zhang X, Marques J, Wachtel M, Schaefer B, Sen N, Song Y, Gualtieri A, et al. 2017. PAX3-FOXO1 establishes myogenic super enhancers and confers BET bromodomain vulnerability. *Cancer Discov* **7**: 884–899. doi:10.1158/2159-8290.CD-16-1297

Haase GM, O'Leary MC, Stram DO, Lukens JN, Seeger RC, Shimada H, Matthay KK. 1995. Pelvic neuroblastoma—implications for a new favorable subgroup: a children's cancer group experience. *Ann Surg Oncol* **2**: 516–523. doi:10.1007/BF02307085

Hanahan D. 2022. Hallmarks of cancer: new dimensions. *Cancer Discov* **12**: 31–46. doi:10.1158/2159-8290.CD-21-1059

Harada T, Heshmati Y, Kalfon J, Perez MW, Xavier Ferrucio J, Ewers J, Hubbell Engler B, Kossenkov A, Ellegast JM, Yi JS, et al. 2022. A distinct core regulatory module enforces oncogene expression in KMT2A-rearranged leukemia. *Genes Dev* **36**: 368–389. doi:10.1101/gad.349284.121

Hsu JY, Danis EP, Nance S, O'Brien JH, Gustafson AL, Wessells VM, Goodspeed AE, Talbot JC, Amacher SL, Jedlicka P, et al. 2022. SIX1 reprograms myogenic transcription factors to maintain the rhabdomyosarcoma undifferentiated state. *Cell Rep* **38**: 110323. doi:10.1016/j.celrep.2022.110323

Huber K, Narasimhan P, Shtukmaster S, Pfeifer D, Evans SM, Sun Y. 2013. The LIM-Homeodomain transcription factor Islet-1 is required for the development of sympathetic neurons and adrenal chromaffin cells. *Dev Biol* **380**: 286–298. doi:10.1016/j.ydbio.2013.04.027

Imayoshi I, Isomura A, Harima Y, Kawaguchi K, Kori H, Miyachi H, Fujiwara T, Ishidate F, Kageyama R. 2013. Oscillatory control of factors determining multipotency and fate in mouse neural progenitors. *Science* **342**: 1203–1208. doi:10.1126/science.1242366

Istrail S, Davidson EH. 2005. Logic functions of the genomic *cis*-regulatory code. *Proc Natl Acad Sci* **102**: 4954–4959. doi:10.1073/pnas.0409624102

Ito T, Young MJ, Li R, Jain S, Wernitznig A, Krill-Burger JM, Lemke CT, Monducci D, Rodriguez DJ, Chang L, et al. 2021. Paralog knockout profiling identifies DUSP4 and DUSP6 as a digenic dependence in MAPK pathway-driven cancers. *Nat Genet* **53**: 1664–1672. doi:10.1038/s41588-021-00967-z

Jansky S, Sharma AK, Körber V, Quintero A, Toprak UH, Wecht EM, Gartlgruber M, Greco A, Chomsky E, Grünewald TGP, et al. 2021. Single-cell transcriptomic analyses provide insights into the developmental origins of neuroblastoma. *Nat Genet* **53**: 683–693. doi:10.1038/s41588-021-00806-1

Kameneva P, Artemov AV, Kastriti ME, Faure L, Olsen TK, Otte J, Erickson A, Semsch B, Andersson ER, Ratz M, et al. 2021. Single-cell transcriptomics of human embryos identifies multiple sympathoblast lineages with potential implications for neuroblastoma origin. *Nat Genet* **53**: 694–706. doi:10.1038/s41588-021-00818-x

Kildisiute G, Kholosy WM, Young MD, Roberts K, Elmentaite R, van Hooff SR, Pacyna CN, Khabirova E, Piapi A,

Thevanesan C, et al. 2021. Tumor to normal single-cell mRNA comparisons reveal a pan-neuroblastoma cancer cell. *Sci Adv* **7**: eabd3311. doi:10.1126/sciadv.abd3311

Kim J, Lo L, Dormand E, Anderson DJ. 2003. SOX10 maintains multipotency and inhibits neuronal differentiation of neural crest stem cells. *Neuron* **38**: 17–31. doi:10.1016/s0896-6273(03)00163-6

Lang D, Chen F, Milewski R, Li J, Lu MM, Epstein JA. 2000. Pax3 is required for enteric ganglia formation and functions with Sox10 to modulate expression of c-ret. *J Clin Invest* **106**: 963–971. doi:10.1172/JCI10828

Leclair MD, Hartmann O, Heloury Y, Fourcade L, Laprie A, Mechinaud F, Munzer C, Rubie H. 2004. Localized pelvic neuroblastoma: excellent survival and low morbidity with tailored therapy—the 10-year experience of the French Society of Pediatric Oncology. *J Clin Oncol* **22**: 1689–1695. doi:10.1200/JCO.2004.04.069

Le Douarin NM, Teillet MA. 1974. Experimental analysis of the migration and differentiation of neuroblasts of the autonomic nervous system and of neurectodermal mesenchymal derivatives, using a biological cell marking technique. *Dev Biol* **41**: 162–184. doi:10.1016/0012-1606(74)90291-7

Lee TI, Jenner RG, Boyer LA, Guenther MG, Levine SS, Kumar RM, Chevalier B, Johnstone SE, Cole MF, Isono K, et al. 2006. Control of developmental regulators by Polycomb in human embryonic stem cells. *Cell* **125**: 301–313. doi:10.1016/j.cell.2006.02.043

Lee RS, Stewart C, Carter SL, Ambrogio L, Cibulskis K, Sougnez C, Lawrence MS, Auclair D, Mora J, Golub TR, et al. 2012. A remarkably simple genome underlies highly malignant pediatric rhabdoid cancers. *J Clin Invest* **122**: 2983–2988. doi:10.1172/JCI64400

Li H, Mar BG, Zhang H, Puram RV, Vazquez F, Weir BA, Hahn WC, Ebert B, Pellman D. 2017. The EMT regulator ZEB2 is a novel dependency of human and murine acute myeloid leukemia. *Blood* **129**: 497–508. doi:10.1182/blood-2016-05-714493

Liu LF, Desai SD, Li TK, Mao Y, Sun M, Sim SP. 2000. Mechanism of action of camptothecin. *Ann NY Acad Sci* **922**: 1–10. doi:10.1111/j.1749-6632.1996.tb26375.x

Lu DY, Ellegast JM, Ross KN, Malone CF, Lin S, Mabe NW, Dharia NV, Meyer A, Conway A, Su AH, et al. 2023. The ETS transcription factor ETV6 constrains the transcriptional activity of EWS-FLI to promote Ewing sarcoma. *Nat Cell Biol* **25**: 285–297. doi:10.1038/s41556-022-01059-8

Lui JH, Hansen DV, Kriegstein AR. 2011. Development and evolution of the human neocortex. *Cell* **146**: 18–36. doi:10.1016/j.cell.2011.06.030

Ma X, Liu Y, Liu Y, Alexandrov LB, Edmonson MN, Gawad C, Zhou X, Li Y, Rusch MC, Easton J, et al. 2018. Pan-cancer genome and transcriptome analyses of 1,699 paediatric leukaemias and solid tumours. *Nature* **555**: 371–376. doi:10.1038/nature25795

Malone CF, Dharia NV, Kugener G, Forman AB, Rothberg MV, Abdusamad M, Gonzalez A, Kuljanin M, Robichaud AL, Conway AS, et al. 2021. Selective modulation of a pan-essential protein as a therapeutic strategy in cancer. *Cancer Discov* **11**: 2282–2299. doi:10.1158/2159-8290.CD-20-1213

Marson A, Levine SS, Cole MF, Frampton GM, Brambrink T, Johnstone S, Guenther MG, Johnston WK, Wernig M,

Newman J, et al. 2008. Connecting microRNA genes to the core transcriptional regulatory circuitry of embryonic stem cells. *Cell* **134:** 521–533. doi:10.1016/j.cell.2008.07.020

Martik ML, Bronner ME. 2017. Regulatory logic underlying diversification of the neural crest. *Trends Genet* **33:** 715–727. doi:10.1016/j.tig.2017.07.015

Meyers RM, Bryan JG, McFarland JM, Weir BA, Sizemore AE, Xu H, Dharia NV, Montgomery PG, Cowley GS, Pantel S, et al. 2017. Computational correction of copy number effect improves specificity of CRISPR–Cas9 essentiality screens in cancer cells. *Nat Genet* **49:** 1779–1784. doi:10.1038/ng.3984

Moriguchi T, Takako N, Hamada M, Maeda A, Fujioka Y, Kuroha T, Huber RE, Hasegawa SL, Rao A, Yamamoto M, et al. 2006. Gata3 participates in a complex transcriptional feedback network to regulate sympathoadrenal differentiation. *Development* **133:** 3871–3881. doi:10.1242/dev.02553

Nam Y, Sliz P, Song L, Aster JC, Blacklow SC. 2006. Structural basis for cooperativity in recruitment of MAML coactivators to Notch transcription complexes. *Cell* **124:** 973–983. doi:10.1016/j.cell.2005.12.037

Neggers JE, Paolella BR, Asfaw A, Rothberg MV, Skipper TA, Yang A, Kalekar RL, Krill-Burger JM, Dharia NV, Kugener G, et al. 2021. Synthetic lethal interaction between the ESCRT paralog enzymes VPS4A and VPS4B in cancers harboring loss of chromosome 18q or 16q. *Cell Rep* **36:** 109367. doi:10.1016/j.celrep.2021.109367

Ngan ES, Garcia-Barceló MM, Yip BH, Poon HC, Lau ST, Kwok CK, Sat E, Sham MH, Wong KK, Wainwright BJ, et al. 2011. Hedgehog/Notch-induced premature gliogenesis represents a new disease mechanism for Hirschsprung disease in mice and humans. *J Clin Invest* **121:** 3467–3478. doi:10.1172/JCI43737

Ogiwara H, Sasaki M, Mitachi T, Oike T, Higuchi S, Tominaga Y, Kohno T. 2016. Targeting p300 addiction in CBP-deficient cancers causes synthetic lethality by apoptotic cell death due to abrogation of MYC expression. *Cancer Discov* **6:** 430–445. doi:10.1158/2159-8290.CD-15-0754

Oike T, Ogiwara H, Tominaga Y, Ito K, Ando O, Tsuta K, Mizukami T, Shimada Y, Isomura H, Komachi M, et al. 2013. A synthetic lethality-based strategy to treat cancers harboring a genetic deficiency in the chromatin remodeling factor BRG1. *Cancer Res* **73:** 5508–5518. doi:10.1158/0008-5472.CAN-12-4593

Olsen RR, Ireland AS, Kastner DW, Groves SM, Spainhower KB, Pozo K, Kelenis DP, Whitney CP, Guthrie MR, Wait SJ, et al. 2021. ASCL1 represses a SOX9[+] neural crest stem-like state in small cell lung cancer. *Genes Dev* **35:** 847–869. doi:10.1101/gad.348295.121

Park NI, Guilhamon P, Desai K, McAdam RF, Langille E, O'Connor M, Lan X, Whetstone H, Coutinho FJ, Vanner RJ, et al. 2017. ASCL1 reorganizes chromatin to direct neuronal fate and suppress tumorigenicity of glioblastoma stem cells. *Cell Stem Cell* **21:** 209–224.e7. doi:10.1016/j.stem.2017.06.004

Parkinson LM, Gillen SL, Woods LM, Chaytor L, Marcos D, Ali FR, Carroll JS, Philpott A. 2022. The proneural transcription factor ASCL1 regulates cell proliferation and primes for differentiation in neuroblastoma. *Front Cell Dev Biol* **10:** 942579. doi:10.3389/fcell.2022.942579

Peter IS, Davidson EH. 2015. *Genomic control process: development and evolution.* Academic, New York.

Piskareva O, Harvey H, Nolan J, Conlon R, Alcock L, Buckley P, Dowling P, Henry M, O'Sullivan F, Bray I, et al. 2015. The development of cisplatin resistance in neuroblastoma is accompanied by epithelial to mesenchymal transition in vitro. *Cancer Lett* **364:** 142–155. doi:10.1016/j.canlet.2015.05.004

Potzner MR, Tsarovina K, Binder E, Penzo-Méndez A, Lefebvre V, Rohrer H, Wegner M, Sock E. 2010. Sequential requirement of Sox4 and Sox11 during development of the sympathetic nervous system. *Development* **137:** 775–784. doi:10.1242/dev.042101

Qiu B, Matthay KK. 2022. Advancing therapy for neuroblastoma. *Nat Rev Clin Oncol* **19:** 515–533. doi:10.1038/s41571-022-00643-z

Raposo A, Vasconcelos FF, Drechsel D, Marie C, Johnston C, Dolle D, Bithell A, Gillotin S, van den Berg DLC, Ettwiller L, et al. 2015. Ascl1 coordinately regulates gene expression and the chromatin landscape during neurogenesis. *Cell Rep* **10:** 1544–1556. doi:10.1016/j.celrep.2015.02.025

Ross RA, Spengler BA, Biedler JL. 1983. Coordinate morphological and biochemical interconversion of human neuroblastoma cells. *J Natl Cancer Inst* **71:** 741–747.

Ross RA, Biedler JL, Spengler BA. 2003. A role for distinct cell types in determining malignancy in human neuroblastoma cell lines and tumors. *Cancer Lett* **197:** 35–39. doi:10.1016/s0304-3835(03)00079-x

Rothenberg EV. 2019. Causal gene regulatory network modeling and genomics: second-generation challenges. *J Comput Biol* **26:** 703–718. doi:10.1089/cmb.2019.0098

Saadatpour A, Albert R. 2013. Boolean modeling of biological regulatory networks: a methodology tutorial. *Methods* **62:** 3–12. doi:10.1016/j.ymeth.2012.10.012

Samarasinghe KTG, Jaime-Figueroa S, Burgess M, Nalawansha DA, Dai K, Hu Z, Bebenek A, Holley SA, Crews CM. 2021. Targeted degradation of transcription factors by TRAFTACs: TRAnscription Factor TArgeting Chimeras. *Cell Chem Biol* **28:** 648–661.e5. doi:10.1016/j.chembiol.2021.03.011

Schock EN, LaBonne C. 2020. Sorting sox: diverse roles for sox transcription factors during neural crest and craniofacial development. *Front Physiol* **11:** 606889. doi:10.3389/fphys.2020.606889

Simões-Costa M, Bronner ME. 2015. Establishing neural crest identity: a gene regulatory recipe. *Development* **142:** 242–257. doi:10.1242/dev.105445

Skoulidis F, Li BT, Dy GK, Price TJ, Falchook GS, Wolf J, Italiano A, Schuler M, Borghaei H, Barlesi F, et al. 2021. Sotorasib for lung cancers with KRAS p.G12C mutation. *N Engl J Med* **384:** 2371–2381. doi:10.1056/NEJMoa2103695

Soldatov R, Kaucka M, Kastriti ME, Petersen J, Chontorotzea T, Englmaier L, Akkuratova N, Yang Y, Häring M, Dyachuk V, et al. 2019. Spatiotemporal structure of cell fate decisions in murine neural crest. *Science* **364:** eaas9536. doi:10.1126/science.aas9536

Southard-Smith EM, Kos L, Pavan WJ. 1998. Sox10 mutation disrupts neural crest development in Dom Hirschsprung mouse model. *Nat Genet* **18:** 60–64. doi:10.1038/ng0198-60

Stanke M, Junghans D, Geissen M, Goridis C, Ernsberger U, Rohrer H. 1999. The Phox2 homeodomain proteins are sufficient to promote the development of sympathetic neurons. *Development* **126:** 4087–4094. doi:10.1242/dev.126.18.4087

Stewart E, McEvoy J, Wang H, Chen X, Honnell V, Ocarz M, Gordon B, Dapper J, Blankenship K, Yang Y, et al. 2018. Identification of therapeutic targets in rhabdomyosarcoma through integrated genomic, epigenomic, and proteomic analyses. *Cancer Cell* **34:** 411–426. doi:10.1016/j.ccell.2018.07.012

Sun CX, Daniel P, Bradshaw G, Shi H, Loi M, Chew N, Parackal S, Tsui V, Liang Y, Koptyra M, et al. 2023. Generation and multi-dimensional profiling of a childhood cancer cell line atlas defines new therapeutic opportunities. *Cancer Cell* **41:** 660–677.e7. doi:10.1016/j.ccell.2023.03.007

Sung KW, Yoo KH, Koo HH, Kim JY, Cho EJ, Seo YL, Kim J, Lee SK. 2009. Neuroblastoma originating from extra-abdominal sites: association with favorable clinical and biological features. *J Korean Med Sci* **24:** 461–467. doi:10.3346/jkms.2009.24.3.461

Tamura K, Taniguchi Y, Minoguchi S, Sakai T, Tun T, Furukawa T, Honjo T. 1995. Physical interaction between a novel domain of the receptor Notch and the transcription factor RBP-Jκ/Su(H). *Curr Biol* **5:** 1416–1423. doi:10.1016/s0960-9822(95)00279-x

Tenente IM, Hayes MN, Ignatius MS, McCarthy K, Yohe M, Sindiri S, Gryder B, Oliveira ML, Ramakrishnan A, Tang Q, et al. 2017. Myogenic regulatory transcription factors regulate growth in rhabdomyosarcoma. *eLife* **6:** e19214. doi:10.7554/eLife.19214

Thiele CJ, Reynolds CP, Israel MA. 1985. Decreased expression of N-myc precedes retinoic acid-induced morphological differentiation of human neuroblastoma. *Nature* **313:** 404–406. doi:10.1038/313404a0

Tsarovina K, Pattyn A, Stubbusch J, Müller F, van der Wees J, Schneider C, Brunet JF, Rohrer H. 2004. Essential role of Gata transcription factors in sympathetic neuron development. *Development* **131:** 4775–4786. doi:10.1242/dev.01370

Tsarovina K, Schellenberger J, Schneider C, Rohrer H. 2008. Progenitor cell maintenance and neurogenesis in sympathetic ganglia involves Notch signaling. *Mol Cell Neurosci* **37:** 20–31. doi:10.1016/j.mcn.2007.08.010

Tsherniak A, Vazquez F, Montgomery PG, Weir BA, Kryukov G, Cowley GS, Gill S, Harrington WF, Pantel S, Krill-Burger JM, et al. 2017. Defining a cancer dependency map. *Cell* **170:** 564–576.e16. doi:10.1016/j.cell.2017.06.010

van Groningen T, Koster J, Valentijn LJ, Zwijnenburg DA, Akogul N, Hasselt NE, Broekmans M, Haneveld F, Nowakowska NE, Bras J, et al. 2017. Neuroblastoma is composed of two super-enhancer-associated differentiation states. *Nat Genet* **49:** 1261–1266. doi:10.1038/ng.3899

van Groningen T, Akogul N, Westerhout EM, Chan A, Hasselt NE, Zwijnenburg DA, Broekmans M, Stroeken P, Haneveld F, Hooijer GKJ, et al. 2019. A NOTCH feed-forward loop drives reprogramming from adrenergic to mesenchymal state in neuroblastoma. *Nat Commun* **10:** 1530. doi:10.1038/s41467-019-09470-w

Vo KT, Matthay KK, Neuhaus J, London WB, Hero B, Ambros PF, Nakagawara A, Miniati D, Wheeler K, Pearson AD, et al. 2014. Clinical, biologic, and prognostic differences on the basis of primary tumor site in neuroblastoma: a report from the international neuroblastoma risk group project. *J Clin Oncol* **32:** 3169–3176. doi:10.1200/JCO.2014.56.1621

Wahlbuhl M, Reiprich S, Vogl MR, Bosl MR, Wegner M. 2012. Transcription factor Sox10 orchestrates activity of a neural crest-specific enhancer in the vicinity of its gene. *Nucleic Acids Res* **40:** 88–101. doi:10.1093/nar/gkr734

Wang L, Tan TK, Durbin AD, Zimmerman MW, Abraham BJ, Tan SH, Ngoc PCT, Weichert-Leahey N, Akahane K, Lawton LN, et al. 2019. ASCL1 is a MYCN- and LMO1-dependent member of the adrenergic neuroblastoma core regulatory circuitry. *Nat Commun* **10:** 5622. doi:10.1038/s41467-019-13515-5

Wang L, Tan TK, Kim H, Kappei D, Tan SH, Look AT, Sanda T. 2023. ASCL1 characterizes adrenergic neuroblastoma via its pioneer function and cooperation with core regulatory circuit factors. *Cell Rep* **42:** 113541. doi:10.1016/j.celrep.2023.113541

Wapinski OL, Vierbuchen T, Qu K, Lee QY, Chanda S, Fuentes DR, Giresi PG, Ng YH, Marro S, Neff NF, et al. 2013. Hierarchical mechanisms for direct reprogramming of fibroblasts to neurons. *Cell* **155:** 621–635. doi:10.1016/j.cell.2013.09.028

Weaver BA. 2014. How Taxol/paclitaxel kills cancer cells. *Mol Biol Cell* **25:** 2677–2681. doi:10.1091/mbc.E14-04-0916

Weichert-Leahey N, Shi H, Tao T, Oldridge DA, Durbin AD, Abraham BJ, Zimmerman MW, Zhu S, Wood AC, Reyon D, et al. 2023. Genetic predisposition to neuroblastoma results from a regulatory polymorphism that promotes the adrenergic cell state. *J Clin Invest* **133:** e166919. doi:10.1172/JCI166919

Weinstein IB, Joe AK. 2006. Mechanisms of disease: oncogene addiction—a rationale for molecular targeting in cancer therapy. *Nat Clin Pract Oncol* **3:** 448–457. doi:10.1038/ncponc0558

Weston JA. 1963. A radioautographic analysis of the migration and localization of trunk neural crest cells in chick. *Dev Biol* **6:** 279–310. doi:10.1016/0012-1606(63)90016-2

Wildner H, Gierl MS, Strehle M, Pla P, Birchmeier C. 2008. Insm1 (IA-1) is a crucial component of the transcriptional network that controls differentiation of the sympatho-adrenal lineage. *Development* **135:** 473–481. doi:10.1242/dev.011783

Wiles ET, Bell R, Thomas D, Beckerle M, Lessnick SL. 2013. ZEB2 represses the epithelial phenotype and facilitates metastasis in Ewing sarcoma. *Genes Cancer* **4:** 486–500. doi:10.1177/1947601913506115

Wylie LA, Hardwick LJ, Papkovskaia TD, Thiele CJ, Philpott A. 2015. Ascl1 phospho-status regulates neuronal differentiation in a *Xenopus* developmental model of neuroblastoma. *Dis Model Mech* **8:** 429–441. doi:10.1242/dmm.018630

Yntema CL, Hammond WS. 1947. The development of the autonomic nervous system. *Biol Rev Camb Philos Soc* **22:** 344–359. doi:10.1111/j.1469-185x.1947.tb00339.x

Zhang Q, Huang R, Ye Y, Guo X, Lu J, Zhu F, Gong X, Zhang Q, Yan J, Luo L, et al. 2018. Temporal requirements for

Cite this article as *Cold Spring Harb Perspect Med* doi: 10.1101/cshperspect.a041573

ISL1 in sympathetic neuron proliferation, differentiation, and diversification. *Cell Death Dis* **9**: 247. doi:10.1038/s41419-018-0283-9

Zhang Y, Remillard D, Onubogu U, Karakyriakou B, Asiaban JN, Ramos AR, Bowland K, Bishop TR, Barta PA, Nance S, et al. 2023. Collateral lethality between HDAC1 and HDAC2 exploits cancer-specific NuRD complex vulnerabilities. *Nat Struct Mol Biol* **30**: 1160–1171. doi:10.1038/s41594-023-01041-4

Zhou B, Lin W, Long Y, Yang Y, Zhang H, Wu K, Chu Q. 2022. Notch signaling pathway: architecture, disease, and therapeutics. *Signal Transduct Target Ther* **7**: 95. doi:10.1038/s41392-022-00934-y

Zimmerman MW, Durbin AD, He S, Oppel F, Shi H, Tao T, Li Z, Berezovskaya A, Liu Y, Zhang J, et al. 2021. Retinoic acid rewires the adrenergic core regulatory circuitry of childhood neuroblastoma. *Sci Adv* **7**: eabe0834. doi:10.1126/sciadv.abe0834

Immunotherapies for Childhood Cancer

Jeong A. Park[1,3] and Nai-Kong V. Cheung[2,3]

[1]Department of Pediatrics, Inha University Hospital, Inha University College of Medicine, Incheon, Korea
[2]Department of Pediatrics, Memorial Sloan Kettering Cancer Center, New York, New York 10065, USA

Correspondence: cheungn@mskcc.org

Children are surviving cancer in greater numbers than ever. Over the last 50 years, substantial advancements in pediatric cancer treatment have resulted in an 85% 5-year survival rate. Nonetheless, a notable 10%–15% of patients encounter relapse or develop refractory disease, leading to significantly lower survival. Recent attempts to further intensify cytotoxic chemotherapy have failed due to either severe toxicities or ineffectiveness, highlighting the need for new treatment strategies. Immunotherapies are emerging and expanding their clinical application to a wide array of cancers, including those affecting children. In pediatric cancers, monoclonal antibodies targeting GD2 have demonstrated durable radiographic and histologic responses in neuroblastoma (NB), and CD19-targeted bispecific antibodies (BsAbs) and chimeric antigen receptor (CAR) T cells have likewise changed the outlook for refractory acute lymphoblastic leukemia (ALL) in children. This review discusses the clinical development of immunotherapies for pediatric cancers, focusing on pediatric ALL and NB, two major pediatric cancers transformed by immunotherapy, updates on the recent advancements in immunotherapies, and further discusses the future directions of immunotherapy for pediatric cancers.

Multimodal treatment has significantly improved outcomes for the children with cancers over the last decades. However, a considerable proportion of the patients still experience relapse or have refractory disease, and the prognosis for those is still dismal (an interactive website for National Childhood Cancer Registry [NCCR] cancer statistics; nccrexplorer.ccdi.cancer.gov). In addition, many patients who survive often suffer from debilitating treatment-related complications that can permanently impact their lives. Immunotherapies have emerged as a promising avenue for these high-risk patients, which are designed to harness patients' immune cells to eradicate cancers while avoiding the chronic complications associated with conventional treatments. Cancer immunotherapy has shown promising results and improved patient survival in many adult cancers, promoting their application in a variety of pediatric cancers as well. However, while the development of immune checkpoint inhibitors (ICIs) has yielded considerable success in metastatic melanoma, non-small-cell lung cancer (NSCLC), and other adult cancers, the ICIs have not been sufficiently effective against pediatric cancers characterized by low

---

[3]Both authors wrote and edited this manuscript and approved the final manuscript.

Cite this article as *Cold Spring Harb Perspect Med* doi: 10.1101/cshperspect.a041574

tumor mutational burden and "immune-desert" or "cold" tumors (Park and Cheung 2017; Long et al. 2022). In the absence of de novo immune surveillance, synthetic immunity has been the obvious alternative. To develop passive immunity, the identification of targetable tumor antigens is critical. Ideal targets should be highly expressed on cancer cells with limited expression on normal tissues to avoid the potential of on-target, off-tumor toxicities. For that reason, B lymphocyte antigen CD19 and disialogangliosides GD2 have been the focus of major immunotherapeutics against pediatric leukemia and neuroblastoma (NB). CD19 is overexpressed in most human B-cell malignancies, and two major CD19-directed approaches engaging T cells have been proven successful; bispecific antibody (BsAb) (e.g., blinatumomab) and chimeric antigen receptor (CAR) T cells (e.g., tisagenlecleucel) significantly improved outcomes in pediatric patients with relapsed/refractory B-cell acute lymphoblastic leukemia (ALL). GD2 is a membrane-proximal glycolipid antigen overexpressed in a variety of embryonal tumors and bone and soft tissue sarcomas (Dobrenkov and Cheung 2014; Suzuki and Cheung 2015), and NB cells express high levels of the GD2 (estimated at 1×10^6 to 5×10^6 molecules/cell), with relatively rare occurrences of antigen loss (Wu et al. 1986). Monoclonal antibodies (mAbs) targeting GD2 have produced objective responses, leading to durable remission and prolonged survival among high-risk NB patients. In addition to IgG mAbs, various immunotherapeutic modalities targeting GD2 have been actively explored including BsAbs, CAR T cells, radioimmunotherapy (RIT), and vaccines.

This work aims to provide a comprehensive overview of the advancements in immunotherapies for childhood cancers, with a specific focus on their clinical application and efficacy in pediatric ALL and NB, two major pediatric cancers transformed by immunotherapy. In addition, this review updates the recent advancements in immunotherapies for childhood cancers, and further discusses immunotherapy-based combinational strategies improving objective responses and prolonging survival among the pediatric patients.

IMMUNOTHERAPIES TARGETING B-CELL LYMPHOCYTE ANTIGENS

Pediatric ALL represents a significant proportion of childhood cancers. Although ALL occurs in both adults and children, 55.4% of patients are diagnosed at younger than 20 years of age (Brown et al. 2020). Although traditional treatments including chemotherapy and hematopoietic stem cell transplantation (HCT) have made significant improvements in survival rates, relapsed/refractory ALL remains a challenge, necessitating continuous research and innovation. In the past decade, T-cell-based immunotherapies using BsAb and CAR have undergone rapid development and transformed treatment outcomes, especially for high-risk hematologic malignancies. The Food and Drug Administration (FDA) approval of blinatumomab, a bispecific CD19/CD3T cell engager, and tisagenlecleucel, a CD19-CAR T-cell therapy, has accelerated further progress in the treatment of pediatric relapsed/refractory B-cell precursor (BCP) ALL.

CD19 Bispecific Antibodies

BsAbs are engineered to possess specificity for two antigens; one is usually a TAA and the other is an immune cell receptor, usually CD3 on T cells. Blinatumomab is composed of two different single-chain variable fragments (ScFvs) linked via a glycin–serine linker, one scFv engaging CD19 overexpressed on the surface of leukemic cells and the other one activating T cells through their surface CD3 (Löffler et al. 2000). Blinatumomab demonstrated good efficiency and safety in a phase I/II trial for pediatric patients with relapsed/refractory B-cell ALL (NCT01471782) (von Stackelberg et al. 2016). Further subsequent study (RIALTO trial) obtained >80% response rate and 65% of complete remission (CR) accompanied by negative MRD (minimal residual disease) in pediatric patients (NCT02187354), confirming the efficacy of blinatumomab as a single agent (Locatelli et al. 2020, 2022). Further, when incorporated into postreduction consolidation or chemotherapy before allogeneic HCT in relapsed patients, blinatumomab was highly effective in eradicating the MRD and improving disease-free survival (DFS) and overall survival (OS) (NCT02393859 and

NCT02101853) (Brown et al. 2021; Hogan et al. 2023). Blinatumomab became FDA approved for relapsed/refractory disease in 2014 and CD19-positive B-cell ALL in first or second remission with MRD positivity in 2018. Moreover, a phase II study has proven the feasibility of blinatumomab maintenance following HCT in patients at high-risk for relapse (NCT02807883), achieving 71% of 1-year DFS and 0% of nonrelapse mortality (NRM) (Gaballa et al. 2022). In the low-risk group, a clinical trial (NCT02101853) assessing the survival of patients experiencing their first relapse of B-cell ALL also showed that patients with bone marrow (BM) relapse significantly benefited from the addition of blinatumomab to chemotherapy, showing improved disease-free and OS rates, compared to those treated with chemotherapy alone, while no benefit was observed for patients with an isolated extramedullary (EM) relapse (Hogan et al. 2023). Building on these successes, clinical trials to incorporate blinatumomab as a part of upfront therapy for pediatric patients with newly diagnosed B-cell ALL are ongoing (NCT03914625).

To further improve the treatment efficacy of blinatumomab, many combinational strategies are being investigated. Patients responding to blinatumomab were observed to have higher proportions of effector memory CD8[+] T cells, while the nonresponders were T-cell deficient or carried more inhibitory checkpoint molecules on the T-cell surface (Gaballa et al. 2022; Sandhu et al. 2022). The addition of pembrolizumab in patients with relapsed or refractory ALL was tested in a phase I/II clinical trial; this combination was tolerable and showed favorable outcomes in the majority of treated patients (79% of CR with MRD negativity) (Sandhu et al. 2022). On the other hand, a phase II clinical trial of inotuzumab ozogamicin (IO) and blinatumomab combination without chemotherapy is ongoing for adult patients with newly diagnosed, relapsed, or refractory CD22-positive B-cell ALL (NCT03739814). Preliminary results were encouraging with cumulative CR rates reaching 97%, and the 1-year event-free survival (EFS) at 75%, with high tolerability even in elderly patients (Wieduwilt et al. 2023). Likewise, to improve the response rate and overcome CD19-negative relapse following blinatu-momab, trispecific CD3/CD19/CD20 BsAb linking the anti-CD19, anti-CD20, and anti-CD3 scFvs was developed. The trispecific BsAbs (CMG1A46 and A2019) with a potential to reduce the off-target toxicity while increasing the antitumor potency are in active phase I/II clinical trials in patients with CD19 ± CD20-positive B-cell ALL or non-Hodgkin lymphoma (NHL) (NCT05348889) (Tapia-Galisteo et al. 2023).

CD19-CAR T-Cell Therapy

CAR consisting of an antigen-binding domain and costimulatory signaling domain(s) such as CD28 and/or 41BB is another promising immunotherapeutic approach. CAR T cells are designed to recognize specific antigens on the surface of leukemic cells, such as CD19 and CD22, which do not require presentation by the major histocompatibility complex (MHC). Tisagenlecleucel, autologous T cells engineered ex vivo with a CAR containing a 41BB domain, has shown dramatic responses in pediatric patients with relapsed/refractory BCP ALL in a phase I/II study (NCT01626495), achieving a 90% of CR and 68% of a 6-month EFS rate (Maude et al. 2014). Another phase II study of tisagenlecleucel in pediatric and young adult patients with relapsed/refractory B-cell ALL achieved a CR rate of 81% and a 6-month EFS rate of 73% (NCT02435849) (Maude et al. 2018). With the success of tisagenlecleucel, CAR T-cell therapy has emerged as a revolutionary cancer treatment for pediatric ALL, and tisagenlecleucel was FDA approved for the treatment of relapsed/refractory B-cell ALL in patients up to 25 years old.

Besides using 41BB as a costimulatory domain for CAR T cells, CD28 was adopted in brexucabtagene autoleucel (brexucel, KTE-X19), another CD19-CAR T, which became FDA approved (Frey 2022). Brexucel was successfully administered as a single agent in 55 adult patients with relapsed/refractory BCP ALL and achieved 71% of CR and 70% of MRD negativity (Shah et al. 2021a,b). With these successes, a phase I/II study to evaluate the efficacy of brexucel in pediatric and young adult patients was initiated (ZUMA-4, NCT02625480). In this study, patients received conditioning chemotherapy with fludarabine and cyclophosphamide followed by a single infusion of KTE-X19 at a target dose of 1×10^6 cells/kg. There were no dose-limiting toxicities; overall CR rates were 67%; overall MRD negativity rates were 100% among responders; and 88% of responders underwent subsequent allogeneic HCT (Wayne et al. 2023).

Despite these impressive results, those refractory to, or those who relapsed after CD19-CAR T-cell therapy, faced dire outcomes (Lee et al. 2015; Shah et al. 2021c). A repeat infusion of CD19-CAR T cells did not prevent CD19-positive relapse due to limited expansion of second infused CAR T cells, nor prevent CD19-negative relapse caused by mutations or alternate splice variants of CD19 (NCT01593696, NCT02315612, NCT0344839) (Sotillo et al. 2015; Fischer et al. 2017; Holland et al. 2022). A long-term follow-up study of CD19-CAR T-cell therapy (CD19.28ζ-CAR T cells) presented 3.1 months of EFS and 10.5 months of OS in 50 children and young adult patients, despite 62% of patients initially achieving CR with MRD negativity following CAR T-cell infusion. On the other hand, those who proceeded to HCT after remission seemed to fare better: median OS was 70.2 months and the 5-year EFS was 62%, supporting the utility of CAR T cell as a bridge therapy for the subsequent allogeneic HCT (Shah et al. 2021).

CD22-Targeted Antibody-Drug Conjugate

To address these challenges, alternative targets are required. CD22 represents another promising target with a high expression level on most B-cell leukemia cells and is usually retained following CD19 loss, while restricted in tissue expression to only normal B cells. Inotuzumab IO, a CD22-targeted humanized IgG4 antibody covalently linked to N-Ac-γ-calicheamicin dimethylhydrazide, is a potent antibody-drug conjugate (ADC) (DiJoseph et al. 2004). Upon binding to surface CD22 on leukemic cells, IO is rapidly internalized, transported into the acidic lysosomal vesicular compartment, where the drug is released to induce DNA damage, resulting in apoptosis of the leukemic cells (DiJoseph et al. 2004; Shor et al. 2015). IO demonstrated impressive activity against relapsed or refractory disease

(Kantarjian et al. 2012) and was FDA approved in 2017 for adult patients with relapsed B-cell ALL. IO was also tested in pediatric patients. Bhojwani et al. (2019) reported that IO was also well tolerated and effective in children with relapsed/refractory ALL, similar to the outcomes in adult trials. A phase I clinical study of IO (ITCC-059 study) showed a high response rate of 80% and a 40% 1-year OS rate in heavily pretreated children with CD22$^+$ ALL with identical doses as in adults (1.8 mg/m^2/course) (Brivio et al. 2021). The COG AALL1621 trial also demonstrated a high CR rate (58%) with high MRD negativity (67% among the responders) in children with multiple relapsed or refractory ALL treated with IO (NCT02981628). However, compared with blinatumomab, the experience with IO in pediatric patients is still limited, and the risk of hepatic sinusoidal obstruction syndrome (SOS), mostly observed in patients who proceeded to HCT following IO treatment, remains a consideration (McNeer et al. 2020). The risk of SOS following HCT, along with prolonged cytopenia and either loss or down-modulation of CD22, could all pose challenges to IO therapy (O'Brien et al. 2022). A phase III clinical trial of IO with postconsolidation chemotherapy for newly diagnosed high-risk B-cell ALL is accruing patients (COG AALL1732, NCT03959085); preliminary results showed a higher likelihood of SOS post-HCT and severe infections following IO therapy (O'Brien et al. 2023).

CD22-CAR T-Cell Therapy

CD22-CAR T cells have also been proven to be effective in treating patients with B-cell ALL naive or resistant to CD19-CAR T-cell therapy (Fry et al. 2018). While CD22-CAR having a long amino acid linker failed to show effectiveness in relapsed/refractory B-cell leukemia (NCT02588 456), short scFv linker with tonic signaling seemed to enhance antileukemic activity of 41BB-based CD22-CAR T cells (NCT02650414) (Singh et al. 2021). A phase I study of CD22-CAR T cell with 41BB costimulatory domain achieved high CR rates (70%–80%) in pediatric and young adult patients with relapsed/refractory CD22$^+$ B-cell ALL including patients previously treated with CD19-

Cite this article as *Cold Spring Harb Perspect Med* doi: 10.1101/cshperspect.a041574

CAR T cells (Fry et al. 2018; Pan et al. 2019; Shah et al. 2020). Yet, the median remission duration was 6 months, and relapse was associated with diminished CD22 density on leukemic cells, again highlighting the critical role of target antigen density in escaping CAR T cells and the necessity of subsequent HCT to maintain remission (Fry et al. 2018; Shah et al. 2020).

Dual CD19 and CD22 Targeting CAR T-Cell Therapy

The down-regulation or loss of target antigen is one of the escape mechanisms leading to treatment resistance following CAR T-cell therapy. To address this constraint, dual CD19 and CD22 antigen-targeting CAR T-cell strategies such as CD19/CD22 tandem CAR T cells or sequential CD19- and CD22-CAR T-cell therapy have been explored. Liu et al. (2021) administered CD19-CAR T cells followed by CD22-CAR T cells to patients with prior relapse after transplantation. Among the 21 patients receiving sequential CAR T-cell therapy, 19 patients (90%) achieved CR, and 14 (67%) remained in CR at a median follow-up of 20 months, demonstrating the effectiveness of the sequential combination of CD19- and CD22-CAR T cells in patients with relapsed ALL following HCT. Tandem CD19/CD22-CAR T cells also have been evaluated. A phase I/II study tested the safety and efficacy of AUTO3, tandem CD19/CD22-CAR T cells, in pediatric and young adult patients with relapsed/refractory B-cell ALL and attained a promising outcome with a tolerable safety profile: 86% (13 of 15 patients) achieved CR, and the 1-year OS and EFS rates were 60% and 32%, respectively (Cordoba et al. 2021). Consequently, the FDA has designated AUTO3 as an orphan drug for its application in treating relapsed/refractory B-cell ALL. Another tandem CAR T cell, CTA101, a CRISPR-Cas9-CD19/CD22 dual-targeted allogeneic CAR T cell, was developed, where the CAR is composed of two scFvs targeting CD19 and CD22, along with a 4-1BB domain and CRISPR-Cas9-disrupted T-cell receptor α constant (TRAC) region (Hu et al. 2021). The phase I clinical trial of CTA101 in patients with relapsed/refractory B-cell ALL is ongoing (NCT04227015), where preliminary results

showed a high response rate (83% of CR with MRD negativity) with a manageable safety profile and no GVHD (Hu et al. 2020).

To increase the therapeutic potential of CAR T-cell therapy, strategies to prolong the CAR T-cell persistence are actively investigated as well. These include a murine stem cell virus (MSCV)-CD19/CD22-41BB bivalent CAR T cell (CD19.22.BBζ) (Shalabi et al. 2022), IL-15 expressing autologous CD19-CAR T cells following dual CD19/CD22-CAR T cells (Sun et al. 2021), T-cell immunoglobulin mucin-3 (TIM3)–CD28 fusion proteins combined with CD19-CAR T cells (Sun et al. 2021; Zhao et al. 2021), and B-cell-activating factor receptor (BAFF-R) and CD19 dual targeting CAR T cells, all being tested among pediatric patients with relapsed/refractory B-cell ALL (Wang et al. 2022). The positive initial findings of these studies raise expectations for their potential if used earlier before developing resistance.

In spite of the remarkable success of CAR T-cell therapy, life-threatening toxicities, the limited persistence of CAR T cells, and the emergence of antigen-positive or negative relapse have been challenges. PD-L1 and PD-L2 gene expression are among the obstacles for effective CAR T-cell therapy, and checkpoint inhibitors in combination with CAR T-cell therapy is one of the strategies to overcome this issue not only in solid tumors but also in hematologic malignancies (Wang et al. 2019a). In a clinical study among 14 children with relapsed/refractory B-cell ALL treated with CD19-CAR T-cell therapy, PD-1 blockade resulted in improved persistence of CAR T cells and better outcomes (Li et al. 2018). An early combination of these immunotherapeutic modalities with, or substitution of, conventional chemotherapy remains a prominent focus in ongoing research efforts to further improve the cure rates and quality of life among pediatric patients with ALL (Capitini 2018).

IMMUNOTHERAPIES TARGETING DISIALOGANGLIOSIDE GD2

Another pivotal component of immunotherapy in pediatric cancers is the immunotherapy for NB mainly targeting GD2. NB is one of the few cancers transformed by immunotherapy, changing its natural history from a uniformly lethal metastatic cancer in children to >60% of cure (Park and Cheung 2020a). Unlike other malignancies in adults, neoantigens are extremely rare in NB resulting in few targets; thus, GD2 remains the most frequently targeted and representative TAA for NB immunotherapy. Various immunotherapeutic strategies targeting GD2 including mAbs, RIT, immunocytokines, ADCs, T-cell-engaging BsAbs, CAR T cells, and cancer vaccines have been actively explored, and the GD2-targeted cancer immunotherapy is continuously evolving.

GD2 Monoclonal Antibodies

GD2 mAbs bind to cell surface disialogangliosides and induce apoptosis of NB cells, by triggering antibody-dependent cellular cytotoxicity (ADCC) by neutrophils and natural killer (NK) cells, complement-dependent cytotoxicity (CDC), as well as macrophage-dependent phagocytosis (Park and Cheung 2020a). Two anti-GD2 mAbs murine 3F8 and ch14.18 (dinutuximab) are studied the most and incorporated to consolidate remission in patients with high-risk NB following high-dose chemotherapy and autologous HCT (Yu et al. 2010; Cheung et al. 2012a; Moreno et al. 2017; Kushner et al. 2018a; Ladenstein et al. 2018). The 3F8 and ch14.18 therapies have notably improved the survival rates of patients with high-risk disease, both during their initial remission and among those experiencing relapsed/refractory NB (Yu et al. 2010; Cheung et al. 2012a; Mody et al. 2020). Upfront treatment with dinutuximab combined with induction chemotherapy is currently ongoing (NCT03786783), and preliminary results suggested a high overall response rate (87%) with manageable toxicities (Federico et al. 2022). Another phase II study of hu14.18K322A administered during induction chemotherapy has also reported a high response rate of 97% after six cycles of induction therapy (Furman et al. 2022), compared with the end-induction response rates at 75%–80% in the previous COG trials ANBL12P1 and ANBL0532 (Park et al. 2019; Granger et al. 2021). Humanized 3F8 (hu3F8, naxitamab) was developed to reduce antidrug immune response,

while enhancing ADCC through the human IgG1-Fc and retaining CDC potency through maintaining antibody affinity (Cheung et al. 2012b) received FDA breakthrough therapy designation in 2018 (Markham 2021). Naxitamab induced more potent cytotoxicity than 3F8, without worsening pain side effects while avoiding neutralizing antibodies (Cheung et al. 2012b). The phase I/II studies of naxitamab have confirmed the low immunogenicity and favorable pharmacokinetics with improved tumor control (Cheung et al. 2017; Kushner et al. 2018b), and a clinical trial of naxitamab with GM-CSF for relapsed/refractory NB achieved a high response rate of 58% with a favorable safety profile (NCT03363373) (Mora et al. 2022, 2023). Naxitamab obtained final FDA approval for relapsed/refractory NB in 2020. Despite these approvals, access to anti-GD2 IgG antibodies remained a global challenge (Larrosa et al. 2023).

GD2-CAR T-Cell Therapy

Despite the upfront integration of GD2 mAbs into chemotherapy protocols, current treatments remain suboptimal with a long-term EFS of 40%–50%, and many patients who survive still suffer from chronic treatment-related complications (Sait and Modak 2017). Harnessing the adoptive immune system has the potential to sustain disease remission or even to cure patients with resistant disease, and approaches to exploit GD2-specific cytotoxic T cells using BsAbs and CARs also have been thoroughly investigated. Although the risk of on-target off-tumor toxicity against neural cells expressing GD2 was one of the major concerns of GD2-CAR T-cell therapy (Richman et al. 2018), clinical studies of GD2-CAR T cells have shown to be tolerable but yielded a modest antitumor effect with transient CAR T-cell proliferation and a striking expansion of myeloid cells in patients treated (NCT02761915) (Heczey et al. 2017). On the other hand, Del Bufalo et al. reported exceptional results of GD2-CAR T cells in patients with relapsed/refractory NB. Twenty-seven patients were treated with third-generation GD2-CAR T cells, and the overall response was 63%; nine patients had a complete response, eight had a partial response; toxicities were tolerable, and the

Cite this article as *Cold Spring Harb Perspect Med* doi: 10.1101/cshperspect.a041574

inducible caspase 9 suicide gene was needed only in one patient (GD2-CART01, NCT03373097) (Del Bufalo et al. 2023). Another third-generation GD2-CAR T cells combined with a safety switch (GD2-CAR.OX40.28.z.ICD9) are being tested for GD2$^+$ tumors including NB and osteosarcoma in a phase I clinical trial (NCT02107963). In the preliminary result, 15% of patients experienced grade 1 cytokine release syndrome (CRS), and no neurologic toxicity was observed with limited expansion and persistence of GD2-CAR T cells. Although 77% of patients had stable disease at day 28 post-CAR T-cell infusion, all patients eventually progressed (Ramakrishna et al. 2022), implicating that the tumor microenvironment (TME) and the proliferation of myeloid cells represent significant hurdles to overcome to unlock the potential of CAR T cells in pediatric solid tumors.

GD2-BsAb Therapy

With the clinical success of blinatumomab, a number of BsAbs targeting GD2 are currently in clinical development as well. Earlier, a bispecific Fab × Fab anti-GD2/anti-FcγRI (CD64) antibody was developed to engage antigen-presenting cells (APCs), monocytes, and macrophages against NB (Michon et al. 1995). Another early BsAb, chemically conjugated Fab dimer consisting of anti-GD2 IgG2a (ME361) and anti-CD3 IgG2a was developed (Bernhard et al. 1993). Using the ME361, IgG-based heterodimeric trifunctional GD2-BsAb (TRBs07, ektomab) with retaining Fc function was developed and demonstrated preclinical antitumor activity against GD2$^+$ tumor models (Ruf et al. 2004). Another ME361-based BsAbs, ektomun (GD2/human CD3) and surek (GD2/mouse Cd3), presented superior tumor control compared with the ch14.18 in NB mouse models (Zirngibl et al. 2021). However, the low affinity and weak specificity of ME361 for GD2 and high immunogenic potential for human antidrug antibodies limited clinical applications (Kholodenko et al. 2018). On the other hand, another GD2-BsAb chemically conjugated mOKT3 and murine 3F8 demonstrated a significant antitumor effect in preclinical NB models (Yankelevich et al. 2012), and a phase I/II study of hu3F8-BiAb (using humanized 3F8 instead of murine

3F8) armed T cells (BATs) for NB and osteosarcoma is ongoing (NCT02173093). The preliminary results presented that one of the seven NB patients had a complete BM response and remained progression free, one had a minor response on the metaiododobenzylguanidine (MIBG) scan, and four patients were alive longer than 1-year post-BATs injection (Yankelevich et al. 2019). Another GD2-BsAb combining anti-GD2 m5F11-scFv and huOKT3-scFv was developed and showed potent cytotoxicity against NB in preclinical models (Cheng et al. 2015). The substitution of 5F11-scFv with a higher affinity (13-fold) hu3F8-scFv more potentiated the antitumor activity (Cheng et al. 2016), and, further, IgG-based hu3F8-BsAb, where the huOKT3-scFv was linked to the carboxyl end of the light chain of hu3F8 IgG1 (IgG-[L]-scFv), exerted potent cytotoxicity with a prolonged half-life. This hu3F8-BsAb showed high tumor cell killing potency, high safety, and high ability to drive T cells into GD2$^+$ tumors without neurotoxicity in preclinical models (Xu et al. 2015; Wang et al. 2019b; Park and Cheung 2020b). A phase I clinical trial (NCT 03860207) of this IgG-[L]-scFv hu3F8-BsAb (nivatrotamab) in patients with relapsed/refractory NB, osteosarcoma, and other GD2$^+$ solid tumors was started in 2019; a total of 12 patients were enrolled, and no patients experienced fatal adverse events. Although this study was suspended due to company business priorities, CRS- and GD2-directed cytotoxicity were observed.

GD2 Targeting Radioimmunotherapy

^{131}I-MIBG has played a role in the treatment of high-risk NB since the 1980s, with response rates of ∼30% (Wilson et al. 2014). Grade 3–4 dose-dependent hematological toxicities and liver toxicities are common, with late-onset toxicities including hypothyroidism and, rarely, secondary leukemia and myelodysplasia, limiting dose-escalation (Garaventa et al. 1999; Polishchuk et al. 2011; Huibregtse et al. 2016). While ^{131}I-MIBG therapy delivers radiation to NB cells through human norepinephrine transporter (hNET)-mediated uptake (Glowniak et al. 1993), RIT uses antibodies to deliver radionuclides that emit α- or β-particles, or Auger electrons, having the potential to yield more precise and powerful tumoricidal effect while avoiding the toxicities of external beam radiation, which can be debilitating and severe in children (Modak et al. 2014). ^{131}I-labeled GD2 mAb, ^{131}I-3F8, or ^{131}I-14G2.a have been tested for NB and have shown to be more specific and sensitive at detecting NB cells than ^{131}I-MIBG therapy (Cheung et al. 1986). The exceptional tumor selectivity of m3F8 in preclinical models quickly translated into clinical studies with ^{131}I- or ^{124}I-labeled 3F8 for clinical diagnostics in the early 1990s, building a strong clinical rationale for RIT with the same platform (Yeh et al. 1991; Larson et al. 1992). However, patient survival did not improve even with myeloablative doses of 20 mCi/kg of intravenous ^{131}I-3F8, which required stem cell rescue (Cheung et al. 2023b).

To increase the therapeutic index (TI) while reducing the myelotoxicity or on-target off-tumor toxicity, compartmental RIT, intrathecal ^{131}I-3F8 or ^{131}I-omburtamab (8H9, anti-B7H3 mAb) was administered to patients with CNS metastasis of NB. In phase I trials, intra-Ommaya ^{131}I-3F8 for GD2$^+$ CNS diseases achieved high TI (14.9–56 cGy/mCi to the CSF and <2 cGy/mCi to blood and non-CNS organs) with major antitumor responses (NCT00445965) (Kramer et al. 2007), and intra-Ommaya ^{131}I-omburtamab effectively treated CNS relapse, producing long-term survivors past 10 years (NCT00089245) (Yerrabelli et al. 2021). α-Particle-emitting ^{225}Ac has also been conjugated to 3F8 via radiometal chelator DOTA (1,4,7,10-tetra-azacyclododecane)(^{225}Ac-3F8) and demonstrated a significant antitumor effect in a preclinical model of NB meningeal carcinomatosis when administered intrathecally (Miederer et al. 2004).

In parallel, to improve TI for systemic treatment, pretargeted radioimmunotherapy (PRIT) strategies were developed. Anti-GD2 5F11-scFv and streptavidin (SA) fusion protein (5F11-scFv-SA) followed by a clearing agent (CA) and radiolabeled small molecule hapten or peptide achieved high TI compared with ^{131}I-3F8 therapy (Cheung et al. 2004). To avoid SA immunogenicity and its renal trapping/toxicity, a fully humanized PRIT was developed using the 3F8

BsAb platform where the anti-CD3-scFv was replaced by an anti-DOTA (lanthanide) scFv (clone C825). Hu3F8-C825 followed by CA and radionuclides caged in DOTA achieved much higher TI (>100:1) and cured NB tumors with minimal to none toxicities in preclinical models (Cheal et al. 2014).

The next generation of PRIT has since emerged using a novel two-step approach without the necessity of any CA, a platform called the self-assembling and disassembling (SADA) system (Santich et al. 2021). It uses a tetramerization tag derived from p53, p63, or p73 to force tandem scFvs (BiTE format) to form homotetramers. In each BiTE format, one scFv is the fully humanized hu3F8 anti-GD2 and the second scFv is the humanized anti-DOTA (huC825). As tetramers, these SADAs (220 kDa) have long serum half-life and high avidity for the target GD2. But as they clear from circulation, they start monomerizing at a critical serum concentration and are rapidly cleared via renal filtration as monomers (55 kDa). The pharmacokinetics of SADAs are such that they can penetrate any solid tumor, while avoiding the reticuloendothelial system (absence of affinity for FcR, FcR(n), or other serum proteins), normal epithelial tissues, and immunogenicity. The 2-step SADA-PRIT safely delivered massive doses of α-emitting (^{225}Ac, 1.48 MBq/kg) or β-emitting (^{177}Lu, 6660 MBq/kg or 180 mCi/kg) radionuclides to NB xenografts with precision, ablating tumors without systemic or organ toxicities (Santich et al. 2021). As a "plug-and-play" system, SADA could be applied to a wide range of human cancer targets, including GD2 in NB, CD38 in NHL, GPA33 in colon cancer, and B7H3 in solid tumors. Also, using the novel DOTA derivatives (Proteus), β emitters (^{175}Lu, ^{90}Y, ^{67}Cu), α-emitters (^{225}Ac, ^{212}Pb), positron-emitters (^{86}Y, ^{89}Zr, ^{68}Ga, ^{64}Cu), and γ-emitters (^{111}In, ^{203}Pb) could be successfully delivered to tumors for both diagnosis and therapy of a broad spectrum of human solid cancers (Santich et al. 2021; Espinosa-Cotton and Cheung 2022). Recently, SADA-PRIT was used to deliver β$^+$ γ-emitting radioisotope ^{177}Lu as a theranostic, to treat GD2$^+$ solid tumors in a phase I clinical trial, with no toxicities seen so far, pharmacokinetics as expected and tumor localization documented by scintigraphy (NCT05130255) (Owonikoko et al. 2023).

PEDIATRIC CANCER VACCINES

The cancer vaccine is another potential approach in the realm of immunotherapy to induce tumor antigen-specific and persistent long-term immune responses. The cancer vaccine platforms are classified into four categories: cancer cell-based vaccines, peptide-based vaccines, virus-based vaccines, and nucleic acid–based vaccines (Vishweshwaraiah and Dokholyan 2022). Autologous or allogeneic patient-derived tumor cells can be genetically modified by introducing cytokines, chemokines, and costimulated molecule-encoding genes or by silencing immunosuppressive genes, and these tumor antigens loaded with dendritic cells (DCs) activate naive T cells and induce strong immune responses (Igarashi and Sasada 2020; Yu et al. 2022). Virus-based vaccines use replication-defective engineered viral vectors to present tumor antigens to the immune system, and tumor cells infected with oncolytic virus produce reactive oxygen specifies (ROS) and cytokines that can stimulate immune cells to assist in oncolysis (Russell et al. 2012; Russell and Barber 2018). Peptide-based cancer vaccines consist of immunogenic tumor-specific peptide antigens to elicit the desired immune responses. Antigenic peptides are taken up by APCs and presented by the MHC complex on the cell surface, followed by T-cell recognition and cancer cell-specific immune responses (Slingluff 2011; Nelde et al. 2021). Nucleic acid-based cancer vaccines deliver tumor antigens encoded by either DNA or RNA to the immune system and allow multiple antigens to be easily administered with one immunization and to induce strong MHC-I mediated CD8$^+$ T-cell responses (Qin et al. 2021). Recently, mRNA vaccines have emerged as a potential cancer treatment platform with the COVID-19 pandemic and the success of several mRNA vaccines, indicating the possibility to eliminate or reduce tumor volume by proficiently presenting the tumor antigens to APCs, facilitating innate and adoptive antitumor immune responses (Lorentzen et al. 2022). They offer personalization, a favorable safety profile, and the ability to rapidly adapt to different types of cancer, but still having challenges, such as ensuring the stability of RNA molecules, efficient delivery to target cells, and achieving a strong and sustained antitumor immune response (Duan et al. 2022).

In pediatric cancers, the development of a safe and effective cancer vaccine has been hampered by the heterogeneity of tumor pathobiology and down-regulation of MHC and costimulatory molecules, restricting tumor-specific T-cell immune responses (Shilyansky et al. 2007; Verneris and Wagner 2007). Despite these limitations, efforts have been made to develop an effective childhood cancer vaccine, and some have succeeded in producing considerable tumor responses. Autologous tumor cell vaccines genetically modified to secrete IL-2 were proven safe when administered to NB patients in remission and stimulated tumor-specific responses by both Th1 and Th2T cells (Russell et al. 2007). Gene-modified allogeneic NB cell vaccines secreting IL-2 or lymphotoxin also have been tested in phase I/II trials and proved stimulating antitumor immune responses with little toxicity (Bowman et al. 1998; Rousseau et al. 2003). To increase the immunogenicity of tumor cell–based vaccines, DC vaccines targeting cancer testis (CT) antigens, such as melanoma-associated antigen 1 (MAGE-A1), MAGE-A3, and NY-ESO-1, have been tested in combination with decitabine (DAC) for high-risk NB patients. The DAC/DC-CT vaccine was well tolerated and resulted in CR of NB in one patient (Krishnadas et al. 2013, 2015). In addition, a phase I study using DC pulsed with tumor RNA in stage IV NB patients also suggested that the RNA vaccine provoked tumor-specific humoral immune responses without significant toxicity although the efficacy was modest (Caruso et al. 2005).

GD2 Cancer Vaccines

While T-cell-based vaccines for pediatric cancers have failed to show clinical benefit, there is renewed interest in gangliosides-based vaccine strategy as more clinical data accumulates. Although carbohydrates like GD2 or GD3 are poorly immunogenic and do not engage T cells, conjugation of gangliosides to the highly immunogenic foreign protein called keyhole limpet hemocyanin (KLH) when combined with immunologic adjuvant QS-21 did induce weak IgM response against GD2 (Ragupathi et al. 2003), although the clinical benefit was insufficient. With the addition of oral β-glucan as an adjuvant, IgG response against GD2 emerged and persisted in patients with high-risk NB in second remission (NCT00911560, NCT04936529, NCT06057948). The GD2/GD3 vaccine was well tolerated and induced measurable anti-GD2 and anti-GD3 IgG immune responses, with a long-term EFS of 75% (Kushner et al. 2014). The seroconversion was associated with the disappearance of MRD, and the patients with seropositive for anti-GD2 IgG1 (>150 ng/mL) by week 8 showed a significantly better progression-free survival (PFS) and OS than patients who did not (Cheung et al. 2021). In a subsequent phase II trial, starting oral β-glucan in the vaccination priming phase helped push the IgG titer even higher (Cheung et al. 2023a). However, the strong genetic linkage of seroconversion with dectin-1 SNP appeared to create a genetic barrier for a significant subset of patients (Cheung et al. 2023a).

OTHER TARGETS FOR PEDIATRIC CANCERS

Additional NB antigens currently under investigation include glypican 2 (GPC2), neural cell adhesion molecule L1 (CD171), and B7H3 (CD276). B7H3 (CD276) is an immune checkpoint ligand expressed on various human cancers including NB and other pediatric cancers. Hernandez et al. (2020) reported heterodimeric BsAb targeting both B7H3 and GD2 showed the potential to increase the antitumor effect while reducing neuropathic pain or on-target off-tumor toxicity. CAR T cells both targeting GD2 and B7H3 also have demonstrated significant antitumor effect against pediatric solid tumors including NB without significant toxicities (Du et al. 2019; Majzner et al. 2019; Birley et al. 2022). The synthetic Notch (SynNotch) gated GD2-B7H3-CAR T cell was developed and showed high specificity and improved cytotoxicity against NB (Moghimi et al. 2021). Further, multiantigen targeting CAR T cells, bicistronic GPC2-B7H3-CAR T cells were highly effective against NB and showed longer CAR T-cell persistence and higher resistance to exhaus-

tion compared with each single antigen targeting CAR T cells, suggesting the advantage to potentially overcome the heterogeneous antigen expression in NB (Tian et al. 2022). However, although preclinical studies of B7H3 targeting CAR T cells showed antitumor activity and tolerable safety, their clinical efficacy has been limited so far (Li et al. 2022). In addition, specific target organ damage and the potential of cytokine storm still need to be evaluated in the upcoming clinical trials.

STRATEGIES TO OVERCOME TREATMENT RESISTANCE

Despite the innovations in cancer immunotherapy, many hurdles still exist in achieving cures. Breaking through treatment resistance is a key to clinical success, and one strategy is represented by combination approaches targeting parallel or alternative pathways involved in cancer progression. However, given the complexity inherent in the molecular evolution of cancer, both conceptual and empirical screening for effective combinations are formidable challenges.

While targeting TME alone has not been a major focus of cancer therapy, literally and mechanistically, TME appears to be intertwined with almost all immune-based treatment modalities. Hence one strategic roadmap is including TME modifiers in combination strategies (Fig. 1A). Immunomodulatory antibodies antagonizing inhibitory immune checkpoints, antibodies, and small molecules targeting immunosuppressive cytokines or immune effector cells including regulatory T cells (Tregs), myeloid-derived suppressor cells (MDSCs), and tumor-associated macrophages (TAMs) are undergoing evaluation in preclinical and clinical trials (Aghanejad et al. 2022). This attention derives from the observation that tumor cells frequently produce soluble factors that favor myelopoiesis and recruitment of myeloid cells to the TME (Awad et al. 2018). Moreover, as previous studies have suggested, the rapid expansion of MDSCs and TAMs has consistently posed a complex dilemma of CAR T-cell therapy, limiting the clinical efficacy (Heczey et al. 2017; Ramakrishna et al. 2022). Given the crucial role of myeloid cells orchestrating im-

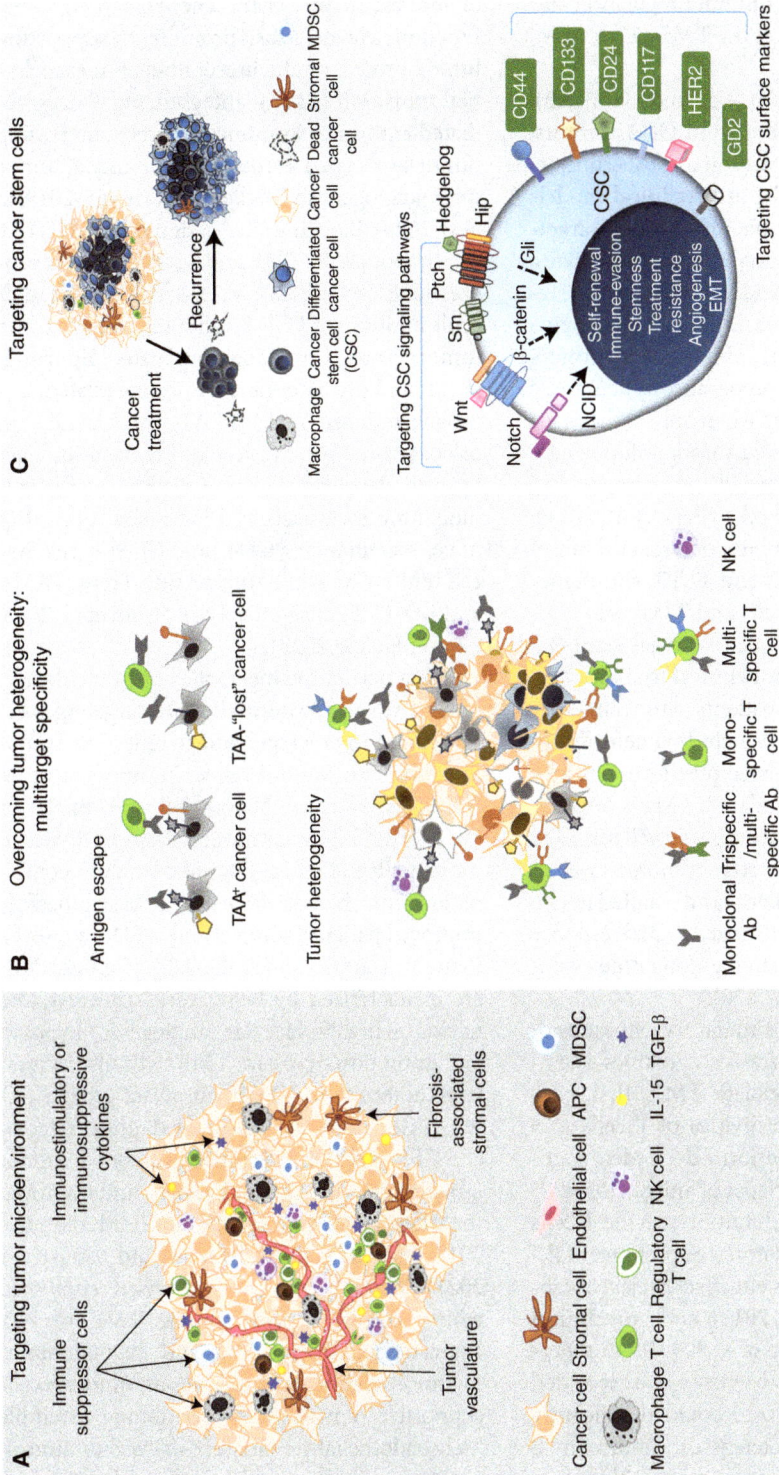

Figure 1. Strategies to overcome treatment resistance. To surmount resistance to cancer immunotherapy, a multifaceted approach is necessary. One approach is including tumor microenvironment (TME) modifiers in combination strategies by targeting suppressive immune cells, modifying immune-stimulatory or suppressive cytokines, targeting fibrosis-associated stromal cells, or targeting abnormal tumor microvasculature (A) (Zhang and Xu 2023). A second strategy is multiple-antigen targeting to overcome tumor antigen escape and tumor heterogeneity (B) (Park and Cheung 2022). A third strategy is targeting cancer stem cells (CSCs) by targeting their abnormal signaling pathways or CSC-specific surface markers to prevent tumor recurrence (C) (Annett and Robson 2018).

mune responses in solid tumors (Mantovani et al. 2021), a T-cell-centric immune regulation has become more inclusive of the TME (Mantovani et al. 2021).

When driven by BsAb (e.g., hu3F8T-BsAb), T cells successfully infiltrated in GD2$^+$ tumors, but accompanied by expansion of TIMs (tumor-infiltrating myeloid cells) in preclinical models (Park et al. 2021). Combination with TIM targeting therapies depleting monocytes, granulocytes, or macrophages could yield favorable results: significantly increasing GD2-EATs (T-cells ex vivo armed with GD2-BsAb) infiltration in tumors and regressing tumor xenografts. When these TIMs were depleted, even when only one specific lineage of cells was depleted, tumor-infiltrating T cells exponentially expanded, proliferated, and effectively killed tumor cells (Park et al. 2021). Pexidartinib, an oral inhibitor of tyrosine kinases including CSF-1R, KIT, and FLT3, simultaneously targeting tumor cells and TIMs was FDA approved in adults with tenosynovial giant cell tumor (TGCT). A phase I clinical trial for pediatric and young adult patients with refractory leukemia or solid tumors including neurofibromatosis type 1 (NF1) related plexiform neurofibromas containing abundant TAMs is ongoing (NCT02390752). Pexidartinib was well tolerated in pediatric patients with solid tumors as well as in heavily pretreated patients and resulted in two stable diseases and one sustained partial response with 67% RECIST reduction among nine evaluable patients (Boal et al. 2020).

Fortifying immunostimulatory cytokines or blocking immunosuppressive cytokines is another approach to modulate TME. IL-15 enhanced the antitumor activities of T cells in a noncognate TCR- and perforin-dependent manner and prevented the relapse of antigen loss variants after treatment with tumor-specific T cells (Liu et al. 2013). Furthermore, the presence of IL-15 enabled NK cells to effectively reject established tumors (Liu et al. 2012), and a preclinical study of the combination of N-803 (IL-15 super-agonist) and dinutuximab with ex vivo expanded NK cells showed that IL-15 could significantly enhance antitumor potency of dinutuximab and improve survival of mice with GD2$^+$ pediatric solid tumors (Chu et al. 2021). Transforming

growth factor β (TGF-β) is another key cytokine of interest in the TME. The TGF-β signaling functions as a metastasis promoter by supporting tumor growth, inducing epithelial–mesenchymal transition (EMT), antagonizing T-cell-mediated antitumor immune responses, increasing tumor-associated fibrosis and enhancing tumor microangiogenesis (Batlle and Massagué 2019; Li et al. 2020). Blocking TGF-β signaling in CD4$^+$ T cells remodels the TME and restrains cancer progression (Li et al. 2020). Vactosertib, a small molecule inhibitor of TGF-β, inhibited osteosarcoma tumor growth and down-regulated Ephrin-2, IL-11, and prostate transmembrane protein androgen-induced-1 (PMEPA1) associated with osteosarcoma progression and metastasis. Vactosertib also reduced osteosarcoma tumor volume, lung metastasis and increased survival of mice, via enhancing CD4$^+$ and CD8$^+$ T cell, NK cell infiltration while suppressing Tregs, TAMs, and PD-1$^+$ T cells in the TME (Choi et al. 2023; Ge and Huang 2023).

As a part of the metabolic approach, the hypoxic TME can be normalized by targeting aberrant tumor microvasculature to renew the antitumor efficacy of various cancer immunotherapies. Although vascular endothelial growth factor (VEGF) inhibitors as a monotherapy have yielded modest clinical efficacy, bevacizumab in combination with chemotherapies or ICIs significantly improved patients' survival and is FDA approved (Saltz et al. 2008; Finn et al. 2020). High-risk NBs are characterized by being very aggressive and having a highly vascular, angiogenic, hypoxic, and immunosuppressive TME. NBs also express large amounts of VEGF, fibroblast growth factor (FGF), and platelet-derived growth factor (PDGF), and bevacizumab inhibited tumor growth, normalized blood vessels, and improved the efficacy of chemotherapy in NB (Modak et al. 2017; Joshi 2020; Ollauri-Ibáñez and Astigarraga 2021). When VEGF inhibitors were combined with GD2-EATs, the hypoxic TME of NB, characterized by tortuous and chaotic tumor microvasculature, reverted into an immune cell-supportive environment, exhibiting multiple high-endothelial venules (HEVs) within tumors, facilitating T-cell migration and proliferation, and resulting in effective tumor eradication (Park et al.

 Cite this article as *Cold Spring Harb Perspect Med* doi: 10.1101/cshperspect.a041574

2023). GD2-CAR T cells when combined with low-dose bevacizumab massively infiltrated tumor mass where they produced IFN-γ and exerted significantly enhanced antitumor activity against NB xenografts (Bocca et al. 2018). Low doses of antiangiogenic therapies lead to vascular normalization, increasing perfusion and facilitating the arrival of chemotherapy, immunotherapy, and oxygen, instead of depriving tumor cells of nutrition or complete inhibition of tumor angiogenesis, suggesting the potential to surmount the metabolically hostile TME (Jain 2005, 2009).

The second roadmap is multiantigen targeting (cocktailing) approaches to overcome tumor heterogeneity and treatment resistance (Fig. 1B). Most of human cancers show heterogeneous antigen expression, and single antigen-targeted therapies (with few exceptions such as GD2 or CD19) are rarely curative. The tumor surface antigens can undergo down-regulation, mutation, or loss under selective immune pressure following preceding immunotherapies, leading to treatment resistance (Sterner and Sterner 2021). Simultaneous or sequential CAR T cells targeting multiple antigens have shown the potential to overcome this tumor escape mechanism (Han et al. 2019). Furthermore, we demonstrated the potential of multiantigen targeting T cells by ex vivo arming of T cells with multiple combinations of BsAbs (multi-EATs). Multi-EATs were more effective than monospecific EATs in ablating tumors with heterogeneous antigen expression and sustaining remission in preclinical models. Contrary to monospecific EATs, which are exclusively effective in target antigen-positive tumors and lead to treatment resistance when retreated due to target antigen loss, multi-EATs have consistently demonstrated effectiveness in heterogeneous tumors with a variety of antigens, even in cases of relapse, presenting prospects to overcome complex tumor heterogeneity and immune escape (Park and Cheung 2022). Multi-EATs expanding the capability to attach specific antibodies to a diverse range of target antigens have the potential to create highly effective precision T cells tailored to individual patients.

The third roadmap is targeting CSCs (Fig. 1C). Most tumors possess a distinct subset of cancer cells that plays a crucial role in metastasis and recurrence, displaying a range of properties, including quiescence, plasticity, self-renewal, heterogeneity, immune escape, and the capacity for treatment resistance (Yu et al. 2012). Dysregulated CSC signaling pathways such as Hedgehog (Hh), Notch, PI3K/Akt, EGFR, JAK/STAT, and Wnt, and CSC's surface markers including CD24, CD44, CD47, CD117, CD123, CD133, epithelial adhesion molecule (EpCAM), GD2, and HER2 have been identified to serve a pivotal function in acquiring and maintaining CSC's property (Batlle and Clevers 2017; Annett and Robson 2018; Yang et al. 2020; Izadpanah et al. 2023). Therefore, targeting CSC-related signaling pathways and surface markers could effectively eradicate tumors and improve the curative potential of pediatric cancers (Izadpanah et al. 2023). While small molecule inhibitors mainly targeting CSC signaling pathways (Takebe et al. 2011), immunotherapies including DC vaccines, mAbs, BsAbs, NK cells, and CAR T cells can target the surface markers, such as CD44 and CD133 known as CSC markers in pediatric solid tumors including rhabdomyosarcoma, Wilms tumor, yolk sac tumor, lymphoma, NB, hepatoblastoma, and pediatric brain tumors including medulloblastoma, pilocytic astrocytoma, ependymoma, and glioblastoma (Bahmad et al. 2019; Singh et al. 2023). Synchronous or sequential targeting of TAA and CSC markers has a prospect to achieve a more definitive eradication of cancers. GD2 was identified as one of the breast CSC markers and promotes tumorigenesis, and anti-GD2-CAR T cells against breast CSCs prevented local tumor growth and lung metastasis (Battula et al. 2012). The possibility of targeting CSCs with multi-EATs, multitarget CAR T cells, or multitarget BsAb highlights the potential of such an approach in achieving a cure (Battula et al. 2012; Korkaya and Wicha 2013; Fathi et al. 2020; Jin et al. 2022).

Last but not least, the final roadmap is finding optimal combinations of immunotherapies or other targeted therapies by exploiting their underlying mode of action, mechanisms of resistance, and toxicity profiles and by optimizing dose, administration schedule, and supportive care. Targeted therapies and systemic immunotherapies have mostly been used individually, with their combinations emerging based on empiric logic or availability. Immunotherapy is typically administered to patients after experiencing disease progression following conventional treatment in the clinical setting. Here, accumulated toxicities and cross-resistance pose another challenge for evaluating novel therapeutic modalities, especially in rare diseases such as pediatric solid tumors (Loria et al. 2022). To address this cross-resistance, molecular reevaluation of recurred tumors either by surgical sampling or by blood test will become increasingly critical for choosing targeted therapies to maximize clinical benefit. Recent advancements in liquid biopsy techniques have shown equivalent clinical efficacy when compared to invasive surgical sampling for patient selection in genomic-driven specific therapies (De Mattos-Arruda and Siravegna 2021). Analysis of circulating tumor DNA (ctDNA) facilitates the rapid identification of emergent resistance mutations to targeted therapies, underscoring its potential in precision oncology (Siravegna et al. 2017). The ctDNA has also shown promise in evaluating molecular resistance to immunotherapies, including pembrolizumab, in lung cancer patients (NCT04093167), facilitating the incorporation of ctDNA-driven intervention into immunotherapy-based clinical trials (Anagnostou et al. 2023). On the other hand, the identification of collateral sensitivities and the evolution of therapeutic resistance, using a temporal collateral sensitivity map and transcriptomic analysis could improve treatment planning (Scarborough et al. 2020). Furthermore, for rational mechanistic-based cancer treatment, mathematical and computational modeling have emerged as pivotal tools in delineating drug resistance mechanisms and optimizing treatment regimens to mitigate toxicity and prevent acquired resistance (Ghaffari Laleh et al. 2022). Recent advancements in computational modeling platforms and software packages should assist in finding optimal dosing in combination therapies for oncogene-driven cancers (Irurzun-Arana et al. 2020). The development of mechanistic-based combination treatments relies on an extensive understanding of the complexity of cancer in space and in time, an evolving dynamic of the genetic, transcriptomic, and epigenetic landscape not known and not druggable until today. There lies the challenge and the opportunity, not to be

missed, and we are forever grateful we have come this far.

CONCLUSION

Immunotherapies have brought forth a paradigm shift in the treatment landscape for high-risk pediatric cancers, offering excitement, hope, and new challenges. They address a fundamental weakness in cytotoxic therapies (i.e., their mode of action does not rely on dividing cells). They can eliminate dormant cancer stem or stem-like cells that manage to escape chemoradiotherapy, even when microscopic and occult. They can tackle another Archilles' heel in conventional drug delivery where the therapeutic windows are generally narrow with dose-limiting toxicities. And perhaps the most impactful is their ability to spread their immune benefit, in the forms of epitope spread, abscopal, and vaccination effects, to ensure durable remissions. Despite these advantages, pediatric cancers are characterized by immunologic cold tumor with low tumor mutational burden, limiting the de novo immune responses that might be unleashed by ICIs. To date, the majority of immunotherapies for these cancers have relied on targeting TAA, etc., CD19 and GD2, using mAbs or BsAbs, CARs, or vaccines. With the recent advances in cancer immunotherapy, these targeted antigen-specific immunotherapies have achieved an impressive milestone in achieving cures for both liquid and solid tumors, with increasingly high precision and potency, resulting in survivals with decent quality of life. Yet beyond these successes lie obstacles, including immune-hostile TME, tumor heterogeneity, antigen down-regulation or loss, T-cell early exhaustion and anergy, and severe although manageable toxicities. Combination approaches to reprogram the immune-suppressive TME, targeting CSC, and multiantigen targeting cocktail therapies to overcome treatment resistance hold promise in realizing the true potential of cancer immunotherapy. While radiation is a proven therapeutic modality in pediatric oncology, the recent advances in RIT, offering laser-like precision and high potency α particle therapy, should be leveraged for their theranostic potential and integration with immuno-oncology. While challenges persist, the outlook on pediatric cancers is brighter than ever.

COMPETING INTEREST STATEMENT

N-K.V.C. and J.A.P. were named as inventors on the patent of EATs filed by MSKCC (Memorial Sloan Kettering Cancer Center) and N-K.V.C. have financial interests in Y-mAbs and Eureka Therapeutics. N-K.V.C. reports receiving past commercial research grants from Y-mabs Therapeutics. N-K.V.C. was named as an inventor on multiple other patents filed by MSKCC, including those licensed to Y-mabs Therapeutics and Biotec Pharma-con. N-K.V.C. is a consultant for Eureka Therapeutics.

ACKNOWLEDGMENTS

This work was supported by funds from Enid A. Haupt Endowed Chair, the Robert Steel Foundation, Kids Walk for Kids with Cancer, NCI Cancer Center Support Grant P30 CA008748 (awarded to N-K.V.C.), and by the Inha University Research grant and the National Research Foundation of Korea (NRF) grant funded by the Korean government (No. 2022R1F1A1076390) (awarded to J.A.P.). The funders were not involved in the study design, collection, analysis, interpretation of data, the writing of this article or the decision to submit it for publication.

REFERENCES

Aghanejad A, Bonab SF, Sepehri M, Haghighi FS, Tarighatnia A, Kreiter C, Nader ND, Tohidkia MR. 2022. A review on targeting tumor microenvironment: the main paradigm shift in the mAb-based immunotherapy of solid tumors. *Int J Biol Macromol* **207:** 592–610. doi:10.1016/j.ijbiomac .2022.03.057

Anagnostou V, Ho C, Nicholas G, Juergens RA, Sacher A, Fung AS, Wheatley-Price P, Laurie SA, Levy B, Brahmer JR, et al. 2023. ctDNA response after pembrolizumab in non-small cell lung cancer: phase 2 adaptive trial results. *Nat Med* **29:** 2559–2569. doi:10.1038/s41591-023-02598-9

Annett S, Robson T. 2018. Targeting cancer stem cells in the clinic: current status and perspectives. *Pharmacol Ther* **187:** 13–30. doi:10.1016/j.pharmthera.2018.02.001

Awad RM, De Vlaeminck Y, Maebe J, Goyvaerts C, Breckpot K. 2018. Turn back the TIMe: targeting tumor infiltrating myeloid cells to revert cancer progression. *Front Immunol* **9:** 1977. doi:10.3389/fimmu.2018.01977

Bahmad HF, Chamaa F, Assi S, Chalhoub RM, Abou-Antoun T, Abou-Kheir W. 2019. Cancer stem cells in neuroblastoma: expanding the therapeutic frontier. *Front Mol Neurosci* **12:** 131. doi:10.3389/fnmol.2019.00131

Batlle E, Clevers H. 2017. Cancer stem cells revisited. *Nat Med* **23:** 1124–1134. doi:10.1038/nm.4409

Batlle E, Massagué J. 2019. Transforming growth factor-β signaling in immunity and cancer. *Immunity* **50:** 924–940. doi:10.1016/j.immuni.2019.03.024

Battula VL, Shi Y, Evans KW, Wang RY, Spaeth EL, Jacamo RO, Guerra R, Sahin AA, Marini FC, Hortobagyi G, et al. 2012. Ganglioside GD2 identifies breast cancer stem cells and promotes tumorigenesis. *J Clin Invest* **122:** 2066–2078. doi:10.1172/JCI59735

Bernhard H, Karbach J, Strittmatter W, Meyer zum Büschenfelde KH, Knuth A. 1993. Induction of tumor-cell lysis by bi-specific antibody recognizing ganglioside GD2 and T-cell antigen CD3. *Int J Cancer* **55:** 465–470. doi:10.1002/ijc .2910550324

Bhojwani D, Sposto R, Shah NN, Rodriguez V, Yuan C, Stetler-Stevenson M, O'Brien MM, McNeer JL, Qureshi A, Cabannes A, et al. 2019. Inotuzumab ozogamicin in pediatric patients with relapsed/refractory acute lymphoblastic leukemia. *Leukemia* **33:** 884–892. doi:10.1038/s41375-018-0265-z

Birley K, Leboreiro-Babe C, Rota EM, Buschhaus M, Gavriil A, Vitali A, Alonso-Ferrero M, Hopwood L, Parienti L, Ferry G, et al. 2022. A novel anti-B7-H3 chimeric antigen receptor from a single-chain antibody library for immunotherapy of solid cancers. *Mol Ther Oncolytics* **26:** 429–443. doi:10.1016/j.omto.2022.08.008

Boal LH, Glod J, Spencer M, Kasai M, Derdak J, Dombi E, Ahlman M, Beury DW, Merchant MS, Persenaire C, et al. 2020. Pediatric PK/PD phase I trial of pexidartinib in relapsed and refractory leukemias and solid tumors including neurofibromatosis type I-related plexiform neurofibromas. *Clin Cancer Res* **26:** 6112–6121. doi:10.1158/1078-0432.CCR-20-1696

Bocca P, Di Carlo E, Caruana I, Emionite L, Cilli M, De Angelis B, Quintarelli C, Pezzolo A, Raffaghello L, Morandi F, et al. 2018. Bevacizumab-mediated tumor vasculature remodelling improves tumor infiltration and antitumor efficacy of GD2-CAR T cells in a human neuroblastoma preclinical model. *Oncoimmunology* **7:** e1378843. doi:10 .1080/2162402X.2017.1378843

Bowman LC, Grossmann M, Rill D, Brown M, Zhong WY, Alexander B, Leimig T, Coustan-Smith E, Campana D, Jenkins J, et al. 1998. Interleukin-2 gene-modified allogeneic tumor cells for treatment of relapsed neuroblastoma. *Hum Gene Ther* **9:** 1303–1311. doi:10.1089/hum.1998.9 .9-1303

Brivio E, Locatelli F, Lopez-Yurda M, Malone A, Díaz-de-Heredia C, Bielorai B, Rossig C, van der Velden VHJ, Ammerlaan ACJ, Thano A, et al. 2021. A phase 1 study of inotuzumab ozogamicin in pediatric relapsed/refractory acute lymphoblastic leukemia (ITCC-059 study). *Blood* **137:** 1582–1590. doi:10.1182/blood.2020007848

Brown P, Inaba H, Annesley C, Beck J, Colace S, Dallas M, DeSantes K, Kelly K, Kitko C, Lacayo N, et al. 2020. Pediatric acute lymphoblastic leukemia, version 2.2020, NCCN clinical practice guidelines in oncology. *J Natl Compr Canc Netw* 18: 81–112. doi:10.6004/jnccn.2020.0001

Brown PA, Ji L, Xu X, Devidas M, Hogan LE, Borowitz MJ, Raetz EA, Zugmaier G, Sharon E, Bernhardt MB, et al. 2021. Effect of postreinduction therapy consolidation with blinatumomab vs chemotherapy on disease-free survival in children, adolescents, and young adults with first relapse of B-cell acute lymphoblastic leukemia: a randomized clinical trial. *J Am Med Assoc* 325: 833–842. doi:10.1001/jama.2021.0669

Capitini CM. 2018. CAR-T immunotherapy: how will it change treatment for acute lymphoblastic leukemia and beyond? *Expert Opin Orphan Drugs* 6: 563–566. doi:10.1080/21678707.2018.1529562

Caruso DA, Orme LM, Amor GM, Neale AM, Radcliff FJ, Downie P, Tang ML, Ashley DM. 2005. Results of a phase I study utilizing monocyte-derived dendritic cells pulsed with tumor RNA in children with stage 4 neuroblastoma. *Cancer* 103: 1280–1291. doi:10.1002/cncr.20911

Cheal SM, Xu H, Guo HF, Zanzonico PB, Larson SM, Cheung NK. 2014. Preclinical evaluation of multistep targeting of diasialoganglioside GD2 using an IgG-scFv bispecific antibody with high affinity for GD2 and DOTA metal complex. *Mol Cancer Ther* 13: 1803–1812. doi:10.1158/1535-7163.MCT-13-0933

Cheng M, Ahmed M, Xu H, Cheung NK. 2015. Structural design of disialoganglioside GD2 and CD3-bispecific antibodies to redirect T cells for tumor therapy. *Int J Cancer* 136: 476–486. doi:10.1002/ijc.29007

Cheng M, Santich BH, Xu H, Ahmed M, Huse M, Cheung NK. 2016. Successful engineering of a highly potent single-chain variable-fragment (scFv) bispecific antibody to target disialoganglioside (GD2) positive tumors. *Oncoimmunology* 5: e1168557. doi:10.1080/2162402X.2016.1168557

Cheung NK, Landmeier B, Neely J, Nelson AD, Abramowsky C, Ellery S, Adams RB, Miraldi F. 1986. Complete tumor ablation with iodine 131-radiolabeled disialoganglioside GD2-specific monoclonal antibody against human neuroblastoma xenografted in nude mice. *J Natl Cancer Inst* 77: 739–745. doi:10.1093/jnci/77.3.739

Cheung NK, Modak S, Lin Y, Guo H, Zanzonico P, Chung J, Zuo Y, Sanderson J, Wilbert S, Theodore LJ, et al. 2004. Single-chain Fv-streptavidin substantially improved therapeutic index in multistep targeting directed at disialoganglioside GD2. *J Nucl Med* 45: 867–877.

Cheung NK, Cheung IY, Kushner BH, Ostrovnaya I, Chamberlain E, Kramer K, Modak S. 2012a. Murine anti-GD2 monoclonal antibody 3F8 combined with granulocyte-macrophage colony-stimulating factor and 13-*cis*-retinoic acid in high-risk patients with stage 4 neuroblastoma in first remission. *J Clin Oncol* 30: 3264–3270. doi:10.1200/JCO.2011.41.3807

Cheung NK, Guo H, Hu J, Tassev DV, Cheung IY. 2012b. Humanizing murine IgG3 anti-GD2 antibody m3F8 substantially improves antibody-dependent cell-mediated cytotoxicity while retaining targeting in vivo. *Oncoimmunology* 1: 477–486. doi:10.4161/onci.19864

Cheung IY, Kushner BH, Modak S, Basu EM, Roberts SS, Cheung NV. 2017. Phase I trial of anti-GD2 monoclonal antibody hu3F8 plus GM-CSF: impact of body weight, immunogenicity and anti-GD2 response on pharmacokinetics and survival. *Oncoimmunology* 6: e1358331. doi:10.1080/2162402X.2017.1358331

Cheung IY, Cheung NV, Modak S, Mauguen A, Feng Y, Basu E, Roberts SS, Ragupathi G, Kushner BH. 2021. Survival impact of anti-GD2 antibody response in a phase II ganglioside vaccine trial among patients with high-risk neuroblastoma with prior disease progression. *J Clin Oncol* 39: 215–226. doi:10.1200/JCO.20.01892

Cheung IY, Mauguen A, Modak S, Ragupathi G, Basu EM, Roberts SS, Kushner BH, Cheung NK. 2023a. Effect of oral β-glucan on antibody response to ganglioside vaccine in patients with high-risk neuroblastoma: a phase 2 randomized clinical trial. *JAMA Oncol* 9: 242–250. doi:10.1001/jamaoncol.2022.5999

Cheung NKV, Kramer K, Modak S, Kushner BH, Ahmed M, Santich B, Cheal S, Larson S. 2023b. Case study #2: disialoganglioside GD2 as a target for radiopharmaceutical therapy. In *Radiopharmaceutical therapy* (ed. Bodei L, Lewis JS, Zeglis BM), pp. 225–252. Springer, Cham, Switzerland.

Choi SH, Myers J, Tomchuck S, Bonner M, Eid S, Kingsley D, VanHeyst K, Kim SJ, Kim BG, Huang AY. 2023. Oral TGF-βR1 inhibitor Vactosertib promotes osteosarcoma regression by targeting tumor proliferation and enhancing antitumor immunity. *Res Sq* doi:10.21203/rs.3.rs-2709282/v1

Chu Y, Nayyar G, Jiang S, Rosenblum JM, Soon-Shiong P, Safrit JT, Lee DA, Cairo MS. 2021. Combinatorial immunotherapy of N-803 (IL-15 superagonist) and dinutuximab with ex vivo expanded natural killer cells significantly enhances in vitro cytotoxicity against GD2+ pediatric solid tumors and in vivo survival of xenografted immunodeficient NSG mice. *J Immunother Cancer* 9: e002267. doi:10.1136/jitc-2020-002267

Cordoba S, Onuoha S, Thomas S, Pignataro DS, Hough R, Ghorashian S, Vora A, Bonney D, Veys P, Rao K, et al. 2021. CAR T cells with dual targeting of CD19 and CD22 in pediatric and young adult patients with relapsed or refractory B cell acute lymphoblastic leukemia: a phase 1 trial. *Nat Med* 27: 1797–1805. doi:10.1038/s41591-021-01497-1

Del Bufalo F, De Angelis B, Caruana I, Del Baldo G, De Ioris MA, Serra A, Mastronuzzi A, Cefalo MG, Pagliara D, Amicucci M, et al. 2023. GD2-CART01 for relapsed or refractory high-risk neuroblastoma. *N Engl J Med* 388: 1284–1295. doi:10.1056/NEJMoa2210859

De Mattos-Arruda L, Siravegna G. 2021. How to use liquid biopsies to treat patients with cancer. *ESMO Open* 6: 100060. doi:10.1016/j.esmoop.2021.100060

DiJoseph JF, Armellino DC, Boghaert ER, Khandke K, Dougher MM, Sridharan L, Kunz A, Hamann PR, Gorovits B, Udata C, et al. 2004. Antibody-targeted chemotherapy with CMC-544: a CD22-targeted immunoconjugate of calicheamicin for the treatment of B-lymphoid malignancies. *Blood* 103: 1807–1814. doi:10.1182/blood-2003-07-2466

Dobrenkov K, Cheung NK. 2014. GD2-targeted immunotherapy and radioimmunotherapy. *Semin Oncol* 41: 589–612. doi:10.1053/j.seminoncol.2014.07.003

Du H, Hirabayashi K, Ahn S, Kren NP, Montgomery SA, Wang X, Tiruthani K, Mirlekar B, Michaud D, Greene K, et al. 2019. Antitumor responses in the absence of toxicity in solid tumors by targeting B7-H3 via chimeric antigen receptor T cells. *Cancer Cell* **35:** 221–237.e8. doi:10.1016/j.ccell.2019.01.002

Duan LJ, Wang Q, Zhang C, Yang DX, Zhang XY. 2022. Potentialities and challenges of mRNA vaccine in cancer immunotherapy. *Front Immunol* **13:** 923647. doi:10.3389/fimmu.2022.923647

Espinosa-Cotton M, Cheung NV. 2022. Bispecific antibodies for the treatment of neuroblastoma. *Pharmacol Ther* **237:** 108241. doi:10.1016/j.pharmthera.2022.108241

Fathi E, Farahzadi R, Sheervalilou R, Sanaat Z, Vietor I. 2020. A general view of CD33$^+$ leukemic stem cells and CAR-T cells as interesting targets in acute myeloblatsic leukemia therapy. *Blood Res* **55:** 10–16. doi:10.5045/br.2020.55.1.10

Federico SM, Naranjo A, Zhang F, Marachelian A, Desai AV, Shimada H, Braunstein SE, Tinkle CL, Yanik GA, Asgharzadeh S, et al. 2022. A pilot induction regimen incorporating dinutuximab and sargramostim for the treatment of newly diagnosed high-risk neuroblastoma: a report from the Children's Oncology Group. *J Clin Oncol* **40:** 10003. doi:10.1200/JCO.2022.40.16_suppl.10003

Finn RS, Qin S, Ikeda M, Galle PR, Ducreux M, Kim TY, Kudo M, Breder V, Merle P, Kaseb AO, et al. 2020. Atezolizumab plus bevacizumab in unresectable hepatocellular carcinoma. *N Engl J Med* **382:** 1894–1905. doi:10.1056/NEJMoa1915745

Fischer J, Paret C, El Malki K, Alt F, Wingerter A, Neu MA, Kron B, Russo A, Lehmann N, Roth L, et al. 2017. CD19 isoforms enabling resistance to CART-19 immunotherapy are expressed in B-ALL patients at initial diagnosis. *J Immunother* **40:** 187–195. doi:10.1097/CJI.0000000000000169

Frey NV. 2022. Approval of brexucabtagene autoleucel for adults with relapsed and refractory acute lymphocytic leukemia. *Blood* **140:** 11–15. doi:10.1182/blood.2021014892

Fry TJ, Shah NN, Orentas RJ, Stetler-Stevenson M, Yuan CM, Ramakrishna S, Wolters P, Martin S, Delbrook C, Yates B, et al. 2018. CD22-targeted CAR T cells induce remission in B-ALL that is naive or resistant to CD19-targeted CAR immunotherapy. *Nat Med* **24:** 20–28. doi:10.1038/nm.4441

Furman WL, McCarville B, Shulkin BL, Davidoff A, Krasin M, Hsu CW, Pan H, Wu J, Brennan R, Bishop MW, et al. 2022. Improved outcome in children with newly diagnosed high-risk neuroblastoma treated with chemoimmunotherapy: updated results of a phase II study using hu14.18K322A. *J Clin Oncol* **40:** 335–344. doi:10.1200/JCO.21.01375

Gaballa MR, Banerjee P, Milton DR, Jiang X, Ganesh C, Khazal S, Nandivada V, Islam S, Kaplan M, Daher M, et al. 2022. Blinatumomab maintenance after allogeneic hematopoietic cell transplantation for B-lineage acute lymphoblastic leukemia. *Blood* **139:** 1908–1919. doi:10.1182/blood.2021013290

Garaventa A, Bellagamba O, Lo Piccolo MS, Milanaccio C, Lanino E, Bertolazzi L, Villavecchia GP, Cabria M, Scopinaro G, Claudiani F, et al. 1999. 131I-metaiodobenzylguanidine (131I-MIBG) therapy for residual neuroblastoma: a mono-institutional experience with 43 patients. *Br J Cancer* **81:** 1378–1384. doi:10.1038/sj.bjc.6694223

Ge R, Huang GM. 2023. Targeting transforming growth factor β signaling in metastatic osteosarcoma. *J Bone Oncol* **43:** 100513. doi:10.1016/j.jbo.2023.100513

Ghaffari Laleh N, Loeffler CML, Grajek J, Staňková K, Pearson AT, Muti HS, Trautwein C, Enderling H, Poleszczuk J, Kather JN. 2022. Classical mathematical models for prediction of response to chemotherapy and immunotherapy. *PLoS Comput Biol* **18:** e1009822. doi:10.1371/journal.pcbi.1009822

Glowniak JV, Kilty JE, Amara SG, Hoffman BJ, Turner FE. 1993. Evaluation of metaiodobenzylguanidine uptake by the norepinephrine, dopamine and serotonin transporters. *J Nucl Med* **34:** 1140–1146.

Granger MM, Naranjo A, Bagatell R, DuBois SG, McCune JS, Tenney SC, Weiss BD, Mosse YP, Asgharzadeh S, Grupp SA, et al. 2021. Myeloablative busulfan/melphalan consolidation following induction chemotherapy for patients with newly diagnosed high-risk neuroblastoma: Children's Oncology Group trial ANBL12P1. *Transplant Cell Ther* **27:** 490.e1–490.e8. doi:10.1016/j.jctc.2021.03.006

Han X, Wang Y, Wei J, Han W. 2019. Multi-antigen-targeted chimeric antigen receptor T cells for cancer therapy. *J Hematol Oncol* **12:** 128. doi:10.1186/s13045-019-0813-7

Heczey A, Louis CU, Savoldo B, Dakhova O, Durett A, Grilley B, Liu H, Wu MF, Mei Z, Gee A, et al. 2017. CAR T cells administered in combination with lymphodepletion and PD-1 inhibition to patients with neuroblastoma. *Mol Ther* **25:** 2214–2224. doi:10.1016/j.ymthe.2017.05.012

Hernandez R, Erbe A, Gerhardt D, Dennin J, Massey C, Barnhart T, Engle J, Hammer B, Sondel P. 2020. GD2/B7-H3 bispecific antibodies for next-generation neuroblastoma treatment. *J Nucl Med* **61:** 376.

Hogan LE, Brown PA, Ji L, Xu X, Devidas M, Bhatla T, Borowitz MJ, Raetz EA, Carroll A, Heerema NA, et al. 2023. Children's Oncology Group AALL1331: phase III trial of blinatumomab in children, adolescents, and young adults with low-risk B-cell ALL in first relapse. *J Clin Oncol* **41:** 4118–4129. doi:10.1200/JCO.22.02200

Holland EM, Molina JC, Dede K, Moyer D, Zhou T, Yuan CM, Wang HW, Stetler-Stevenson M, Mackall C, Fry TJ, et al. 2022. Efficacy of second CAR-T (CART2) infusion limited by poor CART expansion and antigen modulation. *J Immunother Cancer* **10:** e004483. doi:10.1136/jitc-2021-004483

Hu Y, Zhou Y, Zhang M, Ge W, Li Y, Yang L, Wei G, Han L, Wang H, Zhang X, et al. 2020. The safety and efficacy of a CRISPR/Cas9-engineered universal CAR-T cell product (CTA101) in patients with relapsed/refractory B-cell acute lymphoblastic leukemia. *Blood* **136:** 52. doi:10.1182/blood-2020-142262

Hu Y, Zhou Y, Zhang M, Ge W, Li Y, Yang L, Wei G, Han L, Wang H, Yu S, et al. 2021. CRISPR/Cas9-engineered universal CD19/CD22 dual-targeted CAR-T cell therapy for relapsed/refractory B-cell acute lymphoblastic leukemia. *Clin Cancer Res* **27:** 2764–2772. doi:10.1158/1078-0432.CCR-20-3863

Huibregtse KE, Vo KT, DuBois SG, Fetzko S, Neuhaus J, Batra V, Maris JM, Weiss B, Marachelian A, Yanik GA, et al. 2016. Incidence and risk factors for secondary malignancy in patients with neuroblastoma after treatment with ^{131}I-metaiodobenzylguanidine. *Eur J Cancer* **66:** 144–152. doi:10.1016/j.ejca.2016.07.017

Igarashi Y, Sasada T. 2020. Cancer vaccines: toward the next breakthrough in cancer immunotherapy. *J Immunol Res* **2020:** 5825401. doi:10.1155/2020/5825401

Irurzun-Arana I, McDonald TO, Tróconiz IF, Michor F. 2020. Pharmacokinetic profiles determine optimal combination treatment schedules in computational models of drug resistance. *Cancer Res* **80:** 3372–3382. doi:10.1158/0008-5472.CAN-20-0056

Izadpanah A, Mohammadkhani N, Masoudnia M, Ghasemzad M, Saeedian A, Mehdizadeh H, Poorebrahim M, Ebrahimi M. 2023. Update on immune-based therapy strategies targeting cancer stem cells. *Cancer Med* **12:** 18960. doi:10.1002/cam4.6520

Jain R. 2005. Normalization of tumor vasculature: an emerging concept in antiangiogenic therapy. *Science* **307:** 58–62. doi:10.1126/science.1104819

Jain RK. 2009. A new target for tumor therapy. *N Engl J Med* **360:** 2669–2671. doi:10.1056/NEJMcibr0902054

Jin Y, Lorvik KB, Jin Y, Beck C, Sike A, Persiconi I, Kvaløy E, Saatcioglu F, Dunn C, Kyte JA. 2022. Development of STEAP1 targeting chimeric antigen receptor for adoptive cell therapy against cancer. *Mol Ther Oncolytics* **26:** 189–206. doi:10.1016/j.omto.2022.06.007

Joshi S. 2020. Targeting the tumor microenvironment in neuroblastoma: recent advances and future directions. *Cancers (Basel)* **12:** 2057. doi:10.3390/cancers12082057

Kantarjian H, Thomas D, Jorgensen J, Jabbour E, Kebriaei P, Rytting M, York S, Ravandi F, Kwari M, Faderl S, et al. 2012. Inotuzumab ozogamicin, an anti-CD22-calecheamicin conjugate, for refractory and relapsed acute lymphocytic leukaemia: a phase 2 study. *Lancet Oncol* **13:** 403–411. doi:10.1016/S1470-2045(11)70386-2

Kholodenko IV, Kalinovsky DV, Doronin II, Deyev SM, Kholodenko RV. 2018. Neuroblastoma origin and therapeutic targets for immunotherapy. *J Immunol Res* **2018:** 7394268. doi:10.1155/2018/7394268

Korkaya H, Wicha MS. 2013. HER2 and breast cancer stem cells: more than meets the eye. *Cancer Res* **73:** 3489–3493. doi:10.1158/0008-5472.CAN-13-0260

Kramer K, Humm JL, Souweidane MM, Zanzonico PB, Dunkel IJ, Gerald WL, Khakoo Y, Yeh SD, Yeung HW, Finn RD, et al. 2007. Phase I study of targeted radioimmunotherapy for leptomeningeal cancers using intra-Ommaya 131-I-3F8. *J Clin Oncol* **25:** 5465–5470. doi:10.1200/JCO.2007.11.1807

Krishnadas DK, Shapiro T, Lucas K. 2013. Complete remission following decitabine/dendritic cell vaccine for relapsed neuroblastoma. *Pediatrics* **131:** e336–e341. doi:10.1542/peds.2012-0376

Krishnadas DK, Shusterman S, Bai F, Diller L, Sullivan JE, Cheerva AC, George RE, Lucas KG. 2015. A phase I trial combining decitabine/dendritic cell vaccine targeting MAGE-A1, MAGE-A3 and NY-ESO-1 for children with relapsed or therapy-refractory neuroblastoma and sarcoma. *Cancer Immunol Immunother* **64:** 1251–1260. doi:10.1007/s00262-015-1731-3

Kushner BH, Cheung IY, Modak S, Kramer K, Ragupathi G, Cheung NK. 2014. Phase I trial of a bivalent gangliosides vaccine in combination with β-glucan for high-risk neuroblastoma in second or later remission. *Clin Cancer Res* **20:** 1375–1382. doi:10.1158/1078-0432.CCR-13-1012

Kushner BH, Cheung IY, Modak S, Basu EM, Roberts SS, Cheung NK. 2018a. Humanized 3F8 anti-G$_{D2}$ monoclonal antibody dosing with granulocyte-macrophage colony-stimulating factor in patients with resistant neuroblastoma: a phase 1 clinical trial. *JAMA Oncol* **4:** 1729–1735. doi:10.1001/jamaoncol.2018.4005

Kushner BH, Cheung IY, Modak S, Basu EM, Roberts SS, Cheung NK. 2018b. Phase I trial of humanized 3F8 anti-G$_{D2}$ monoclonal antibody and GM-CSF in patients with resistant neuroblastoma. *JAMA Oncol* **4:** 1729–1735. doi:10.1001/jamaoncol.2018.4005

Ladenstein R, Pötschger U, Valteau-Couanet D, Luksch R, Castel V, Yaniv I, Laureys G, Brock P, Michon JM, Owens C, et al. 2018. Interleukin 2 with anti-GD2 antibody ch14.18/CHO (dinutuximab β) in patients with high-risk neuroblastoma (HR-NBL1/SIOPEN): a multicentre, randomised, phase 3 trial. *Lancet Oncol* **19:** 1617–1629. doi:10.1016/S1470-2045(18)30578-3

Larrosa C, Mora J, Cheung NK. 2023. Global impact of monoclonal antibodies (mAbs) in children: a focus on anti-GD2. *Cancers (Basel)* **15:** 3729. doi:10.3390/cancers15143729

Larson SM, Pentlow KS, Volkow ND, Wolf AP, Finn RD, Lambrecht RM, Graham MC, Di Resta G, Bendriem B, Daghighian F, et al. 1992. PET scanning of iodine-124-3F9 as an approach to tumor dosimetry during treatment planning for radioimmunotherapy in a child with neuroblastoma. *J Nucl Med* **33:** 2020–2023.

Lee DW, Kochenderfer JN, Stetler-Stevenson M, Cui YK, Delbrook C, Feldman SA, Fry TJ, Orentas R, Sabatino M, Shah NN, et al. 2015. T cells expressing CD19 chimeric antigen receptors for acute lymphoblastic leukaemia in children and young adults: a phase 1 dose-escalation trial. *Lancet* **385:** 517–528. doi:10.1016/S0140-6736(14)61403-3

Li AM, Hucks GE, Dinofia AM, Seif AE, Teachey DT, Baniewicz D, Callahan C, Fasano C, McBride B, Gonzalez V, et al. 2018. Checkpoint inhibitors augment CD19-directed chimeric antigen receptor (CAR) T cell therapy in relapsed B-cell acute lymphoblastic leukemia. *Blood* **132:** 556–556. doi:10.1182/blood-2018-99-112572

Li S, Liu M, Do MH, Chou C, Stamatiades EG, Nixon BG, Shi W, Zhang X, Li P, Gao S, et al. 2020. Cancer immunotherapy via targeted TGF-β signalling blockade in T$_H$ cells. *Nature* **587:** 121–125. doi:10.1038/s41586-020-2850-3

Li G, Wang H, Wu H, Chen J. 2022. B7-H3-targeted CAR-T cell therapy for solid tumors. *Int Rev Immunol* **41:** 625–637. doi:10.1080/08830185.2022.2102619

Liu RB, Engels B, Arina A, Schreiber K, Hyjek E, Schietinger A, Binder DC, Butz E, Krausz T, Rowley DA, et al. 2012. Densely granulated murine NK cells eradicate large solid tumors. *Cancer Res* **72:** 1964–1974.

Liu RB, Engels B, Schreiber K, Ciszewski C, Schietinger A, Schreiber H, Jabri B. 2013. IL-15 in tumor microenvironment causes rejection of large established tumors by T cells in a noncognate T cell receptor-dependent manner. *Proc Natl Acad Sci* **110:** 8158–8163. doi:10.1073/pnas.1301022110

Liu S, Deng B, Yin Z, Lin Y, An L, Liu D, Pan J, Yu X, Chen B, Wu T, et al. 2021. Combination of CD19 and CD22 CAR-T cell therapy in relapsed B-cell acute lymphoblastic leukemia after allogeneic transplantation. *Am J Hematol* **96:** 671–679. doi:10.1002/ajh.26160

Locatelli F, Zugmaier G, Mergen N, Bader P, Jeha S, Schlegel PG, Bourquin JP, Handgretinger R, Brethon B, Rossig C, et al. 2020. Blinatumomab in pediatric patients with relapsed/refractory acute lymphoblastic leukemia: results of the RIALTO trial, an expanded access study. *Blood Cancer J* **10:** 77. doi:10.1038/s41408-020-00342-x

Locatelli F, Zugmaier G, Mergen N, Bader P, Jeha S, Schlegel PG, Bourquin JP, Handgretinger R, Brethon B, Rössig C, et al. 2022. Blinatumomab in pediatric relapsed/refractory B-cell acute lymphoblastic leukemia: RIALTO expanded access study final analysis. *Blood Adv* **6:** 1004–1014. doi:10.1182/bloodadvances.2021005579

Löffler A, Kufer P, Lutterbüse R, Zettl F, Daniel PT, Schwenkenbecher JM, Riethmüller G, Dörken B, Bargou RC. 2000. A recombinant bispecific single-chain antibody, CD19 × CD3, induces rapid and high lymphoma-directed cytotoxicity by unstimulated T lymphocytes. *Blood* **95:** 2098–2103. doi:10.1182/blood.V95.6.2098

Long AH, Morgenstern DA, Leruste A, Bourdeaut F, Davis KL. 2022. Checkpoint immunotherapy in pediatrics: here, gone, and back again. *Am Soc Clin Oncol Educ Book* **42:** 781–794. doi:10.1200/EDBK_349799

Lorentzen CL, Haanen JB, Met O, Svane IM. 2022. Clinical advances and ongoing trials of mRNA vaccines for cancer treatment. *Lancet Oncol* **23:** e450–e458. doi:10.1016/S1470-2045(22)00372-2

Loria R, Vici P, Di Lisa FS, Soddu S, Maugeri-Sacca M, Bon G. 2022. Cross-resistance among sequential cancer therapeutics: an emerging issue. *Front Oncol* **12:** 877380. doi:10.3389/fonc.2022.877380

Majzner RG, Theruvath JL, Nellan A, Heitzeneder S, Cui Y, Mount CW, Rietberg SP, Linde MH, Xu P, Rota C, et al. 2019. CAR T cells targeting B7-H3, a pan-cancer antigen, demonstrate potent preclinical activity against pediatric solid tumors and brain tumors. *Clin Cancer Res* **25:** 2560–2574. doi:10.1158/1078-0432.CCR-18-0432

Mantovani A, Marchesi F, Jaillon S, Garlanda C, Allavena P. 2021. Tumor-associated myeloid cells: diversity and therapeutic targeting. *Cell Mol Immunol* **18:** 566–578. doi:10.1038/s41423-020-00613-4

Markham A. 2021. Naxitamab: first approval. *Drugs* **81:** 291–296. doi:10.1007/s40265-021-01467-4

Maude SL, Frey N, Shaw PA, Aplenc R, Barrett DM, Bunin NJ, Chew A, Gonzalez VE, Zheng Z, Lacey SF, et al. 2014. Chimeric antigen receptor T cells for sustained remissions in leukemia. *N Engl J Med* **371:** 1507–1517. doi:10.1056/NEJMoa1407222

Maude SL, Laetsch TW, Buechner J, Rives S, Boyer M, Bittencourt H, Bader P, Verneris MR, Stefanski HE, Myers GD, et al. 2018. Tisagenlecleucel in children and young adults with B-cell lymphoblastic leukemia. *N Engl J Med* **378:** 439–448. doi:10.1056/NEJMoa1709866

McNeer JL, Rau RE, Gupta S, Maude SL, O'Brien MM. 2020. Cutting to the front of the line: immunotherapy for childhood acute lymphoblastic leukemia. *Am Soc Clin Oncol Educ Book* **40:** e132–e143. doi:10.1200/EDBK_278171

Michon J, Perdereau B, Brixy F, Moutel S, Fridman WH, Teillaud JL. 1995. In vivo targeting of human neuroblastoma xenograft by anti-GD2/anti-FcγRI (CD64) bispecific antibody. *Eur J Cancer* **31A:** 631–636. doi:10.1016/0959-8049(95)00013-9

Miederer M, McDevitt MR, Borchardt P, Bergman I, Kramer K, Cheung NK, Scheinberg DA. 2004. Treatment of neuroblastoma meningeal carcinomatosis with intrathecal application of α-emitting atomic nanogenerators targeting disialo-ganglioside GD2. *Clin Cancer Res* **10:** 6985–6992. doi:10.1158/1078-0432.CCR-04-0859

Modak S, Kramer K, Pandit-Taskar N. 2014. Radioimmunotherapy of neuroblastoma. In *Therapeutic nuclear medicine* (ed. Baum RP), pp. 629–638. Springer, Berlin.

Modak S, Kushner BH, Basu E, Roberts SS, Cheung NK. 2017. Combination of bevacizumab, irinotecan, and temozolomide for refractory or relapsed neuroblastoma: results of a phase II study. *Pediatr Blood Cancer* **64:** 10.1002/pbc.26448. doi:10.1002/pbc.26448

Mody R, Yu AL, Naranjo A, Zhang FF, London WB, Shulkin BL, Parisi MT, Servaes SE, Diccianni MB, Hank JA, et al. 2020. Irinotecan, temozolomide, and dinutuximab with GM-CSF in children with refractory or relapsed neuroblastoma: a report from the Children's Oncology Group. *J Clin Oncol* **38:** 2160–2169. doi:10.1200/JCO.20.00203

Moghimi B, Muthugounder S, Jambon S, Tibbetts R, Hung L, Bassiri H, Hogarty MD, Barrett DM, Shimada H, Asgharzadeh S. 2021. Preclinical assessment of the efficacy and specificity of GD2-B7H3 SynNotch CAR-T in metastatic neuroblastoma. *Nat Commun* **12:** 511. doi:10.1038/s41467-020-20785-x

Mora J, Chan GCF, Morgenstern DA, Nysom K, Bear M, Tornøe K, Sørensen PS, Kushner BH. 2022. Naxitamab (NAX) treatment for refractory/relapsed (R/R) high-risk neuroblastoma (HR-NB): response data and efficacy in patient (pt) subgroups. *J Clin Oncol* **40:** e22019. doi:10.1200/JCO.2022.40.16_suppl.e22019

Mora J, Chan GC, Morgenstern DA, Nysom K, Bear MK, Tornøe K, Kushner BH. 2023. Outpatient administration of naxitamab in combination with granulocyte-macrophage colony-stimulating factor in patients with refractory and/or relapsed high-risk neuroblastoma: management of adverse events. *Cancer Rep* **6:** e1627. doi:10.1002/cnr2.1627

Moreno L, Rubie H, Varo A, Le Deley MC, Amoroso L, Chevance A, Garaventa A, Gambart M, Bautista F, Valteau-Couanet D, et al. 2017. Outcome of children with relapsed or refractory neuroblastoma: a meta-analysis of ITCC/SIOPEN European phase II clinical trials. *Pediatr Blood Cancer* **64:** 25–31. doi:10.1002/pbc.26192

Nelde A, Rammensee HG, Walz JS. 2021. The peptide vaccine of the future. *Mol Cell Proteomics* **20:** 100022. doi:10.1074/mcp.R120.002309

O'Brien MM, Ji L, Shah NN, Rheingold SR, Bhojwani D, Yuan CM, Xu X, Yi JS, Harris AC, Brown PA, et al. 2022. Phase II trial of inotuzumab ozogamicin in children and adolescents with relapsed or refractory B-cell acute lymphoblastic leukemia: Children's Oncology Group protocol AALL1621. *J Clin Oncol* **40:** 956–967. doi:10.1200/JCO.21.01693

O'Brien MM, McNeer JL, Rheingold SR, Devidas M, Chen Z, Bhojwani D, Ramsey LB, Agrawal A, Wood BL, Suh E, et al. 2023. A phase 3 trial of inotuzumab ozogamicin for high-risk B-ALL: second safety phase results from Children's Oncology Group AALL1732. *J Clin Oncol* **41:** 10016–10016. doi:10.1200/JCO.2023.41.16_suppl.10016

Ollauri-Ibáñez C, Astigarraga I. 2021. Use of antiangiogenic therapies in pediatric solid tumors. *Cancers (Basel)* **13**: 253. doi:10.3390/cancers13020253

Owonikoko TK, Moser JC, Yoon J, Slotkin EK, Dowlati A, Ma VT, Düring M, Sveistrup J. 2023. Phase 1 trial of GD2-SADA: ^{177}Lu-DOTA drug complex in patients with recurrent or refractory metastatic solid tumors known to express GD2 including small cell lung cancer (SCLC), sarcoma, and malignant melanoma. *J Clin Oncol* **41**: TPS3162. doi:10.1200/JCO.2023.41.16_suppl.TPS3162

Pan J, Niu Q, Deng B, Liu S, Wu T, Gao Z, Liu Z, Zhang Y, Qu X, Zhang Y, et al. 2019. CD22 CAR T-cell therapy in refractory or relapsed B acute lymphoblastic leukemia. *Leukemia* **33**: 2854–2866. doi:10.1038/s41375-019-0488-7

Park JA, Cheung NV. 2017. Limitations and opportunities for immune checkpoint inhibitors in pediatric malignancies. *Cancer Treat Rev* **58**: 22–33. doi:10.1016/j.ctrv.2017.05.006

Park JA, Cheung NV. 2020a. Targets and antibody formats for immunotherapy of neuroblastoma. *J Clin Oncol* **38**: 1836–1848. doi:10.1200/JCO.19.01410

Park JA, Cheung NV. 2020b. GD2 or HER2 targeting T cell engaging bispecific antibodies to treat osteosarcoma. *J Hematol Oncol* **13**: 172. doi:10.1186/s13045-020-01012-y

Park JA, Cheung NV. 2022. Overcoming tumor heterogeneity by ex vivo arming of T cells using multiple bispecific antibodies. *J Immunother Cancer* **10**: e003771. doi:10.1136/jitc-2021-003771

Park JR, Kreissman SG, London WB, Naranjo A, Cohn SL, Hogarty MD, Tenney SC, Haas-Kogan D, Shaw PJ, Kraveka JM, et al. 2019. Effect of tandem autologous stem cell transplant vs single transplant on event-free survival in patients with high-risk neuroblastoma: a randomized clinical trial. *J Am Med Assoc* **322**: 746–755. doi:10.1001/jama.2019.11642

Park JA, Wang L, Cheung NV. 2021. Modulating tumor infiltrating myeloid cells to enhance bispecific antibody-driven T cell infiltration and anti-tumor response. *J Hematol Oncol* **14**: 142. doi:10.1186/s13045-021-01156-5

Park JA, Espinosa-Cotton M, Guo HF, Monette S, Cheung NV. 2023. Targeting tumor vasculature to improve antitumor activity of T cells armed ex vivo with T cell engaging bispecific antibody. *J Immunother Cancer* **11**: e006680. doi:10.1136/jitc-2023-006680

Polishchuk AL, Dubois SG, Haas-Kogan D, Hawkins R, Matthay KK. 2011. Response, survival, and toxicity after iodine-131-metaiodobenzylguanidine therapy for neuroblastoma in preadolescents, adolescents, and adults. *Cancer* **117**: 4286–4293. doi:10.1002/cncr.25987

Qin F, Xia F, Chen H, Cui B, Feng Y, Zhang P, Chen J, Luo M. 2021. A guide to nucleic acid vaccines in the prevention and treatment of infectious diseases and cancers: from basic principles to current applications. *Front Cell Dev Biol* **9**: 633776. doi:10.3389/fcell.2021.633776

Ragupathi G, Livingston PO, Hood C, Gathuru J, Krown SE, Chapman PB, Wolchok JD, Williams LJ, Oldfield RC, Hwu WJ. 2003. Consistent antibody response against ganglioside GD2 induced in patients with melanoma by a GD2 lactone-keyhole limpet hemocyanin conjugate vaccine plus immunological adjuvant QS-21. *Clin Cancer Res* **9**: 5214–5220.

Ramakrishna S, Kaczanowska S, Murty T, Contreras CF, Merchant M, Glod J, Gutierrez N, Alimadadi A, Stroncek D, Highfill S, et al. 2022. Abstract CT142: GD2.Ox40.CD28.z CAR T cell trial in neuroblastoma and osteosarcoma. *Cancer Res* **82**: CT142. doi:10.1158/1538-7445.AM2022-CT142

Richman SA, Nunez-Cruz S, Moghimi B, Li LZ, Gershenson ZT, Mourelatos Z, Barrett DM, Grupp SA, Milone MC. 2018. High-affinity GD2-specific CAR T cells induce fatal encephalitis in a preclinical neuroblastoma model. *Cancer Immunol Res* **6**: 36–46. doi:10.1158/2326-6066.CIR-17-0211

Rousseau RF, Haight AE, Hirschmann-Jax C, Yvon ES, Rill DR, Mei Z, Smith SC, Inman S, Cooper K, Alcoser P, et al. 2003. Local and systemic effects of an allogeneic tumor cell vaccine combining transgenic human lymphotactin with interleukin-2 in patients with advanced or refractory neuroblastoma. *Blood* **101**: 1718–1726. doi:10.1182/blood-2002-08-2493

Ruf P, Jäger M, Ellwart J, Wosch S, Kusterer E, Lindhofer H. 2004. Two new trifunctional antibodies for the therapy of human malignant melanoma. *Int J Cancer* **108**: 725–732. doi:10.1002/ijc.11630

Russell SJ, Barber GN. 2018. Oncolytic viruses as antigen-agnostic cancer vaccines. *Cancer Cell* **33**: 599–605. doi:10.1016/j.ccell.2018.03.011

Russell HV, Strother D, Mei Z, Rill D, Popek E, Biagi E, Yvon E, Brenner M, Rousseau J. 2007. Phase I trial of vaccination with autologous neuroblastoma tumor cells genetically modified to secrete IL-2 and lymphotactin. *J Immunother* **30**: 227–233. doi:10.1097/01.cji.0000211335.14385.57

Russell SJ, Peng KW, Bell JC. 2012. Oncolytic virotherapy. *Nat Biotechnol* **30**: 658–670. doi:10.1038/nbt.2287

Sait S, Modak S. 2017. Anti-GD2 immunotherapy for neuroblastoma. *Expert Rev Anticancer Ther* **17**: 889–904. doi:10.1080/14737140.2017.1364995

Saltz LB, Clarke S, Díaz-Rubio E, Scheithauer W, Figer A, Wong R, Koski S, Lichinitser M, Yang TS, Rivera F, et al. 2008. Bevacizumab in combination with oxaliplatin-based chemotherapy as first-line therapy in metastatic colorectal cancer: a randomized phase III study. *J Clin Oncol* **26**: 2013–2019. doi:10.1200/JCO.2007.14.9930

Sandhu KS, Macias A, Del Real M, Beltran AL, Kim YS, Zhang J, Palmer J, Robbins M, Loomis R, Akhtari M, et al. 2022. Interim results of a phase 1/2 study of pembrolizumab combined with blinatumomab in patients with relapsed/refractory (r/r) ALL. *Blood* **140**: 8985–8986. doi:10.1182/blood-2022-170279

Santich BH, Cheal SM, Ahmed M, McDevitt MR, Ouerfelli O, Yang G, Veach DR, Fung EK, Patel M, Burnes Vargas D, et al. 2021. A self-assembling and disassembling (SADA) bispecific antibody (BsAb) platform for curative two-step pretargeted radioimmunotherapy. *Clin Cancer Res* **27**: 532–541. doi:10.1158/1078-0432.CCR-20-2150

Scarborough JA, McClure E, Anderson P, Dhawan A, Durmaz A, Lessnick SL, Hitomi M, Scott JG. 2020. Identifying states of collateral sensitivity during the evolution of therapeutic resistance in Ewing's sarcoma. *iScience* **23**: 101293. doi:10.1016/j.isci.2020.101293

Shah NN, Highfill SL, Shalabi H, Yates B, Jin J, Wolters PL, Ombrello A, Steinberg SM, Martin S, Delbrook C, et al.

2020. CD4/CD8 T-cell selection affects chimeric antigen receptor (CAR) T-cell potency and toxicity: updated results from a phase I anti-CD22 CAR T-cell trial. *J Clin Oncol* **38**: 1938–1950. doi:10.1200/JCO.19.03279

Shah BD, Ghobadi A, Oluwole OO, Logan AC, Boissel N, Cassaday RD, Leguay T, Bishop MR, Topp MS, Tzachanis D, et al. 2021a. KTE-X19 for relapsed or refractory adult B-cell acute lymphoblastic leukaemia: phase 2 results of the single-arm, open-label, multicentre ZUMA-3 study. *Lancet* **398**: 491–502. doi:10.1016/S0140-6736(21)01222-8

Shah BD, Bishop MR, Oluwole OO, Logan AC, Baer MR, Donnellan WB, O'Dwyer KM, Holmes H, Arellano ML, Ghobadi A, et al. 2021b. KTE-X19 anti-CD19 CAR T-cell therapy in adult relapsed/refractory acute lymphoblastic leukemia: ZUMA-3 phase 1 results. *Blood* **138**: 11–22. doi:10.1182/blood.2020009098

Shah NN, Lee DW, Yates B, Yuan CM, Shalabi H, Martin S, Wolters PL, Steinberg SM, Baker EH, Delbrook CP, et al. 2021c. Long-term follow-up of CD19-CAR T-cell therapy in children and young adults with B-ALL. *J Clin Oncol* **39**: 1650–1659. doi:10.1200/JCO.20.02262

Shalabi H, Qin H, Su A, Yates B, Wolters PL, Steinberg SM, Ligon JA, Silbert S, DéDé K, Benzaoui M, et al. 2022. CD19/22 CAR T cells in children and young adults with B-ALL: phase 1 results and development of a novel bicistronic CAR. *Blood* **140**: 451–463. doi:10.1182/blood.2022015795

Shilyansky J, Jacobs P, Doffek K, Sugg SL. 2007. Induction of cytolytic T lymphocytes against pediatric solid tumors in vitro using autologous dendritic cells pulsed with necrotic primary tumor. *J Pediatr Surg* **42**: 54–61. doi:10.1016/j.jpedsurg.2006.09.008

Shor B, Gerber HP, Sapra P. 2015. Preclinical and clinical development of inotuzumab-ozogamicin in hematological malignancies. *Mol Immunol* **67**: 107–116. doi:10.1016/j.molimm.2014.09.014

Singh N, Frey NV, Engels B, Barrett DM, Shestova O, Ravikumar P, Cummins KD, Lee YG, Pajarillo R, Chun I, et al. 2021. Antigen-independent activation enhances the efficacy of 4-1BB-costimulated CD22 CAR T cells. *Nat Med* **27**: 842–850. doi:10.1038/s41591-021-01326-5

Singh S, Bhardwaj M, Sen A, Nambiyar K, Ahuja A. 2023. Cancer stem cell markers—CD133 and CD44—in paediatric solid tumours: a study of immunophenotypic expression and correlation with clinicopathological parameters. *Indian J Surg Oncol* **14**: 113–121. doi:10.1007/s13193-022-01626-3

Siravegna G, Marsoni S, Siena S, Bardelli A. 2017. Integrating liquid biopsies into the management of cancer. *Nat Rev Clin Oncol* **14**: 531–548. doi:10.1038/nrclinonc.2017.14

Slingluff CL Jr. 2011. The present and future of peptide vaccines for cancer: single or multiple, long or short, alone or in combination? *Cancer J* **17**: 343–350. doi:10.1097/PPO.0b013e318233e5b2

Sotillo E, Barrett DM, Black KL, Bagashev A, Oldridge D, Wu G, Sussman R, Lanauze C, Ruella M, Gazzara MR, et al. 2015. Convergence of acquired mutations and alternative splicing of *CD19* enables resistance to CART-19 immunotherapy. *Cancer Discov* **5**: 1282–1295. doi:10.1158/2159-8290.CD-15-1020

Sterner RC, Sterner RM. 2021. CAR-T cell therapy: current limitations and potential strategies. *Blood Cancer J* **11**: 69. doi:10.1038/s41408-021-00459-7

Sun Y, Su Y, Wang Y, Liu N, Li Y, Chen J, Qiao Z, Niu J, Hu J, Zhang B, et al. 2021. CD19 CAR-T cells with membrane-bound IL-15 for B-cell acute lymphoblastic leukemia after failure of CD19 and CD22 CAR-T cells: case report. *Front Immunol* **12**: 728962. doi:10.3389/fimmu.2021.728962

Suzuki M, Cheung NK. 2015. Disialoganglioside GD2 as a therapeutic target for human diseases. *Expert Opin Ther Targets* **19**: 349–362. doi:10.1517/14728222.2014.986459

Takebe N, Harris PJ, Warren RQ, Ivy SP. 2011. Targeting cancer stem cells by inhibiting wnt, notch, and hedgehog pathways. *Nat Rev Clin Oncol* **8**: 97–106. doi:10.1038/nrclinonc.2010.196

Tapia-Galisteo A, Álvarez-Vallina L, Sanz L. 2023. Bi- and trispecific immune cell engagers for immunotherapy of hematological malignancies. *J Hematol Oncol* **16**: 83. doi:10.1186/s13045-023-01482-w

Tian M, Cheuk AT, Wei JS, Abdelmaksoud A, Chou HC, Milewski D, Kelly MC, Song YK, Dower CM, Li N, et al. 2022. An optimized bicistronic chimeric antigen receptor against GPC2 or CD276 overcomes heterogeneous expression in neuroblastoma. *J Clin Invest* **132**: e155621. doi:10.1172/JCI155621

Verneris MR, Wagner JE. 2007. Recent developments in cell-based immune therapy for neuroblastoma. *J Neuroimmune Pharmacol* **2**: 134–139. doi:10.1007/s11481-007-9065-3

Vishweshwaraiah YL, Dokholyan NV. 2022. mRNA vaccines for cancer immunotherapy. *Front Immunol* **13**: 1029069. doi:10.3389/fimmu.2022.1029069

von Stackelberg A, Locatelli F, Zugmaier G, Handgretinger R, Trippett TM, Rizzari C, Bader P, O'Brien MM, Brethon B, Bhojwani D, et al. 2016. Phase I/phase II study of blinatumomab in pediatric patients with relapsed/refractory acute lymphoblastic leukemia. *J Clin Oncol* **34**: 4381–4389. doi:10.1200/JCO.2016.67.3301

Wang H, Kaur G, Sankin AI, Chen F, Guan F, Zang X. 2019a. Immune checkpoint blockade and CAR-T cell therapy in hematologic malignancies. *J Hematol Oncol* **12**: 59. doi:10.1186/s13045-019-0746-1

Wang L, Hoseini SS, Xu H, Ponomarev V, Cheung NK. 2019b. Silencing Fc domains in T cell-engaging bispecific antibodies improves T-cell trafficking and antitumor potency. *Cancer Immunol Res* **7**: 2013–2024. doi:10.1158/2326-6066.CIR-19-0121

Wang X, Dong Z, Awuah D, Chang WC, Cheng WA, Vyas V, Cha SC, Anderson AJ, Zhang T, Wang Z, et al. 2022. CD19/BAFF-R dual-targeted CAR T cells for the treatment of mixed antigen-negative variants of acute lymphoblastic leukemia. *Leukemia* **36**: 1015–1024. doi:10.1038/s41375-021-01477-x

Wayne AS, Huynh V, Hijiya N, Rouce RH, Brown PA, Krueger J, Kitko CL, Ziga ED, Hermiston ML, Richards MK, et al. 2023. Three-year results from phase I of ZUMA-4: KTE-X19 in pediatric relapsed/refractory acute lymphoblastic leukemia. *Haematologica* **108**: 747–760. doi:10.3324/haematol.2022.280678

Wieduwilt MJ, Yin J, Kour O, Teske R, Stock W, Byrd K, Doucette K, Mangan J, Masters GA, Mims AS, et al. 2023. S117: chemotherapy-free treatment with inotuzumab ozogamicin and blinatumomab for older adults with newly diagnosed, Ph-negative, CD22-positive, B-cell acute lymphoblastic leukemia: Alliance A041703. *Hemasphere* **7**: e08838b7. doi:10.1097/01.HS9.0000967380

Wilson JS, Gains JE, Moroz V, Wheatley K, Gaze MN. 2014. A systematic review of 131I-meta iodobenzylguanidine molecular radiotherapy for neuroblastoma. *Eur J Cancer* **50**: 801–815. doi:10.1016/j.ejca.2013.11.016

Wu ZL, Schwartz E, Seeger R, Ladisch S. 1986. Expression of GD2 ganglioside by untreated primary human neuroblastomas. *Cancer Res* **46**: 440–443.

Xu H, Cheng M, Guo H, Chen Y, Huse M, Cheung NK. 2015. Retargeting T cells to GD2 pentasaccharide on human tumors using bispecific humanized antibody. *Cancer Immunol Res* **3**: 266–277. doi:10.1158/2326-6066.CIR-14-0230-T

Yang L, Shi P, Zhao G, Xu J, Peng W, Zhang J, Zhang G, Wang X, Dong Z, Chen F, et al. 2020. Targeting cancer stem cell pathways for cancer therapy. *Signal Transduct Target Ther* **5**: 8. doi:10.1038/s41392-020-0110-5

Yankelevich M, Kondadasula SV, Thakur A, Buck S, Cheung NK, Lum LG. 2012. Anti-CD3 × anti-GD2 bispecific antibody redirects T-cell cytolytic activity to neuroblastoma targets. *Pediatr Blood Cancer* **59**: 1198–1205. doi:10.1002/pbc.24237

Yankelevich M, Modak S, Chu R, Lee DW, Thakur A, Cheung NKV, Lum LG. 2019. Phase I study of OKT3 x hu3F8 bispecific antibody (GD2Bi) armed T cells (GD2BATs) in GD2-positive tumors. *J Clin Oncol* **37**: 2533. doi:10.1200/JCO.2019.37.15_suppl.2533

Yeh SD, Larson SM, Burch L, Kushner BH, Laquaglia M, Finn R, Cheung NK. 1991. Radioimmunodetection of neuroblastoma with iodine-131-3F8: correlation with biopsy, iodine-131-metaiodobenzylguanidine and standard diagnostic modalities. *J Nucl Med* **32**: 769–776.

Yerrabelli RS, He P, Fung EK, Kramer K, Zanzonico PB, Humm JL, Guo H, Pandit-Taskar N, Larson SM, Cheung NV. 2021. Intraommaya compartmental radioimmunotherapy using ^{131}I-omburtamab—pharmacokinetic modeling to optimize therapeutic index. *Eur J Nucl Med Mol Imaging* **48**: 1166–1177. doi:10.1007/s00259-020-05050-z

Yu AL, Gilman AL, Ozkaynak MF, London WB, Kreissman SG, Chen HX, Smith M, Anderson B, Villablanca JG, Matthay KK, et al. 2010. Anti-GD2 antibody with GM-CSF, interleukin-2, and isotretinoin for neuroblastoma. *N Engl J Med* **363**: 1324–1334. doi:10.1056/NEJMoa0911123

Yu Z, Pestell TG, Lisanti MP, Pestell RG. 2012. Cancer stem cells. *Int J Biochem Cell Biol* **44**: 2144–2151. doi:10.1016/j.biocel.2012.08.022

Yu J, Sun H, Cao W, Song Y, Jiang Z. 2022. Research progress on dendritic cell vaccines in cancer immunotherapy. *Exp Hematol Oncol* **11**: 3. doi:10.1186/s40164-022-00257-2

Zhang M, Xu H. 2023. Peptide-assembled nanoparticles targeting tumor cells and tumor microenvironment for cancer therapy. *Front Chem* **11**: 1115495. doi:10.3389/fchem.2023.1115495

Zhao S, Wang C, Lu P, Lou Y, Liu H, Wang T, Yang S, Bao Z, Han L, Liang X, et al. 2021. Switch receptor T3/28 improves long-term persistence and antitumor efficacy of CAR-T cells. *J Immunother Cancer* **9**: e003176. doi:10.1136/jitc-2021-003176

Zirngibl F, Ivasko SM, Grunewald L, Klaus A, Schwiebert S, Ruf P, Lindhofer H, Astrahantseff K, Andersch L, Schulte JH, et al. 2021. GD2-directed bispecific trifunctional antibody outperforms dinutuximab β in a murine model for aggressive metastasized neuroblastoma. *J Immunother Cancer* **9**: e002923. doi:10.1136/jitc-2021-002923

Acute Promyelocytic Leukemia, Retinoic Acid, and Arsenic: A Tale of Dualities

Domitille Rérolle,[1,2,3] Hsin-Chieh Wu,[1,2,3] and Hugues de Thé[1,2,3,4]

[1]Center for Interdisciplinary Research in Biology (CIRB), Collège de France, CNRS, INSERM, Université PSL, 75005 Paris, France

[2]INSERM U944, CNRS UMR7212, GenCellDis, Institut de Recherche Saint-Louis, Université Paris Cité, 75010 Paris, France

[3]Chaire d'Oncologie Cellulaire et Moléculaire, Collège de France, Université PSL, 75005 Paris, France

[4]Service d'Hématologie Biologique, AP-HP, Hôpital St. Louis, 75010 Paris, France

Correspondence: hugues.dethe@inserm.fr

Acute promyelocytic leukemia (APL) is driven by the promyelocytic leukemia (PML)/retinoic acid receptor α (RARA) fusion oncoprotein. Over the years, it has emerged as a model system to understand how this simple (and sometimes sole) genetic alteration can transform hematopoietic progenitors through the acquisition of dominant-negative properties toward both transcriptional control by nuclear receptors and PML-mediated senescence. The fortuitous identification of two drugs, arsenic trioxide (ATO) and all-*trans*-retinoic acid (ATRA), that respectively bind PML and RARA to initiate PML/RARA degradation, has allowed an unprecedented dissection of the cellular and molecular mechanisms involved in patients' cure by the ATO/ATRA combination. This analysis has unraveled the dual and complementary roles of RARA and PML in both APL initiation and cure by the ATRA/ATO combination. We discuss how some of the features unraveled by APL studies may be more broadly applicable to some other forms of leukemia. In particular, the functional synergy between drugs that promote differentiation and those that initiate apoptosis/senescence to impede self-renewal could pave the way to novel curative combinations.

Based on clinical and cytological explorations, acute promyelocytic leukemia (APL) was first identified as a specific clinical entity of rapid and unfavorable evolution: or, "the most malignant form of acute leukemia" (Hillestad 1957). APL was later associated with the presence of a specific chromosomal translocation, t(15,17), present in virtually all patients (Rowley et al. 1977), later found to drive the promyelocytic leukemia (PML)/ retinoic acid receptor α (RARA) fusion gene, paving the way to the molecular pathogenesis of this condition. Although APL can occur in children, it cannot be considered as a disease of children, neither as a cancer whose physiopathology is directly linked to developmental biology. Actually, APL can occur at every age, with an almost constant incidence (Vickers et al. 2000). Some reports have suggested that its clinical course is more severe in children (Iland et al. 2023). Some very rare familial cases were reported, but the biological reasons underlying these observations remain unknown. APL is an important model disease, deserving an article in this collection as a unique example of cure by targeted therapies.

Recent reviews have addressed APL genetics (Geoffroy and de Thé 2020). Briefly, APL and related diseases are associated with chromosomal translocations always involving a retinoic acid receptor (RAR). In the immense majority of cases, RARA is involved, but, recently, some very rare cases of atypical APLs were shown to harbor retinoic acid receptor γ (RARG) (Wu and Gao 2023) or, more rarely, retinoic acid receptor β (RARB) fusions (Osumi et al. 2018; Zhu et al. 2023). Most RARA fusions involve PML, but a variety of other variant RARA fusions were described, the most common of which is PLZF (Licht et al. 1995). Moreover, viral insertions within the RARA gene, presumably driving its overexpression, may drive an APL-like disease (Astolfi et al. 2021; Chen et al. 2022). The fact that the incidence of APL is essentially constant with age argues that it is driven by a single rate-limiting dominant genetic hit. Nevertheless, many cooperating oncogenic activations were associated with disease progression (FLT3) or relapse (WT1) (Akagi et al. 2009; Wartman et al. 2011; Madan et al. 2016; Lehmann-Che et al. 2018). Critically, the APL genome is remarkably stable when compared to most other tumors.

The cornerstones of APL therapy are arsenic trioxide (ATO) and all-*trans*-retinoic acid (ATRA), two targeted therapies that were discovered largely by chance and now cure over 95% of patients without DNA-damaging drugs (Estey et al. 2006; Lo-Coco et al. 2013, 2016). The physiopathology of the disease and its response to therapy have been deciphered with an unprecedented level of molecular details (Lallemand-Breitenbach et al. 2012; Dos Santos et al. 2013; de Thé et al. 2017). Here, we propose to review recent studies on the biology of APL oncogenesis, and show how the dual targeting of its PML and RARA moieties contributes to yield a leukemia cure. We also highlight how these mechanisms may be more broadly involved in leukemia biology.

PML/RARA, THE MASTER DRIVER OF "CLASSIC" APLs

The constant involvement of a RAR in the various fusion proteins demonstrates that deregulated retinoic acid signaling plays a key role in the initiation of APL. RARs are transcription factors whose activity is regulated by a group of vitamin A–derived ligands, retinoids. They control the differentiation of many tissues and exert potent effects on stem cells, including hematopoietic stem cells (Fig. 1; Cabezas-Wallscheid et al. 2017; Schönberger et al. 2022). Studies combining gene inactivation and treatment with RARA-specific retinoids have clearly demonstrated the role of RARA in myeloid differentiation (Kastner et al. 2001; Walkley et al. 2002). Expression of dominant-negative RARA defective for transcriptional activation has pro-oncogenic properties in many cellular systems, including hematopoietic progenitors or hepatocytes, as well as in specific forms of breast cancers (Tsai et al. 1992; Tsai and Collins 1993; Yanagi-

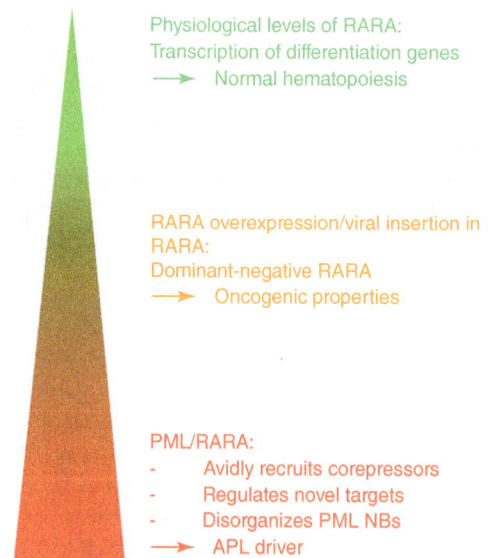

Physiological levels of RARA:
Transcription of differentiation genes
→ Normal hematopoiesis

RARA overexpression/viral insertion in RARA:
Dominant-negative RARA
→ Oncogenic properties

PML/RARA:
- Avidly recruits corepressors
- Regulates novel targets
- Disorganizes PML NBs
→ APL driver

Figure 1. Biology of acute promyelocytic leukemia (APL) oncogenesis. In physiological conditions, retinoic acid receptor α (RARA) is essential for normal hematopoiesis. However, RARA has oncogenic properties when mutated or overexpressed, and the fusion of RARA with promyelocytic leukemia (PML) drives APL through further transcription deregulation and disorganization of PML nuclear bodies (NBs).

Cite this article as *Cold Spring Harb Perspect Med* doi: 10.1101/cshperspect.a041582

tani et al. 2004; Khetchoumian et al. 2007; Tan et al. 2015). In several of these biological systems, mere overexpression of RARA has the same consequences (Fig. 1; Du et al. 1999). Collectively, this suggests that RARA-mediated target gene repression is important for the initiation of APL and more broadly, for transformation.

Modulation of nuclear receptor-mediated transcription is regulated by their ligand-dependent association with protein complexes that regulate epigenetic status (including DNA and histone modifications), transcriptional initiation, and elongation. One of these key complexes is the nuclear receptor corepressor/silencing mediator of retinoic acid (N-Cor/SMRT), which avidly binds RARA and, presumably, even more avidly RARA fusions, because of their ability to self-dimerize (Fig. 2; Lin and Evans 2000; Minucci et al. 2000). PML-enforced PML–RARA/RXRA dimerization also results in a greatly extended repertoire of the oncoprotein DNA-binding sites (Kamashev et al. 2004; Martens et al. 2010). It was thus suggested that the bases for APL initiation relied on basal transcriptional repression of RARA targets, but also other nuclear receptor target genes, controlling myeloid differentiation and/or self-renewal (Fig. 1). However, the overexpression of normal RARA can contribute to myeloid malignancies ex vivo

or in vivo (Du et al. 2000; Astolfi et al. 2021; Chen et al. 2022), suggesting that deregulation of RARs signaling is absolutely central, while the deregulation of de novo targets resulting from PML/RARA dimerization may be an additional contributor to malignant transformation. Recent studies have demonstrated that for some target genes, PML/RARA can be an activator (Tan et al. 2021), although whether this feature is essential for transformation is unknown. Finally, whether the deregulated expression of a small number of master RARA target genes suffices to recapitulate APL initiation remains to be clarified. Identification of these key downstream effectors would bear considerable importance for our understanding of leukemia biology, as some may be shared with other types of acute myelogenous leukemias (AMLs).

IS DIMERIZATION THE ONLY CONTRIBUTION OF PML TO TRANSFORMATION BY PML/RARA?

At least 98% of APL-associated fusion proteins involve PML. Either the emergence of these fusions is greatly facilitated by some topological features of the PML and RARA genes, or PML has a key role in oncogenesis. This may reflect either a central contribution of PML to the gain

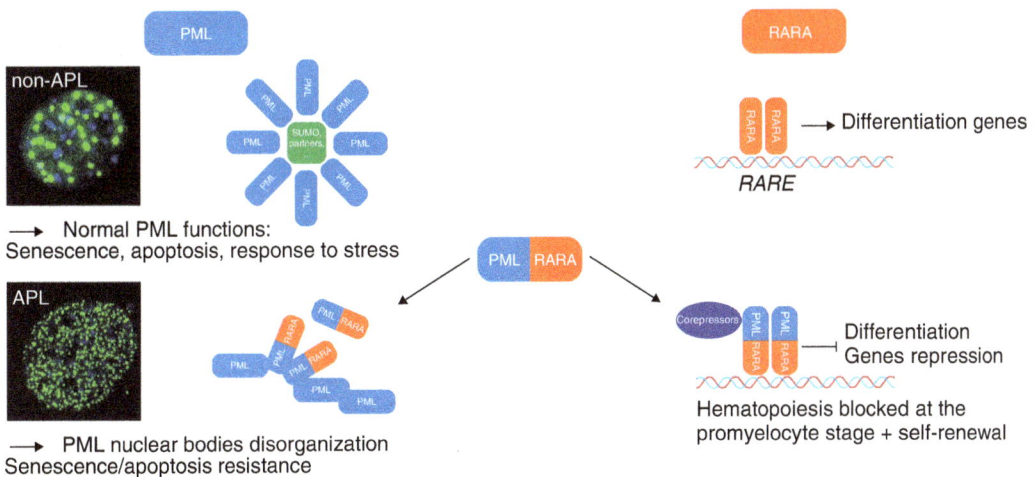

Figure 2. Promyelocytic leukemia (PML)/retinoic acid receptor α (RARA): a double hit oncoprotein. The oncofusion protein disorganizes both PML and RARA structure/function. PML nuclear bodies (NBs) are disorganized into a microspeckled pattern and lose their function, and RARA target genes are repressed due to the recruitment of corepressors, which will block acute promyelocytic leukemia (APL) cells at the promyelocyte stage.

of functions of PML/RARA, or a critical contribution of PML allele loss to the oncogenic process. Experiments in transgenic mice demonstrated that self-dimerizing RARA inefficiently promoted the development of an APL-like disease (Sternsdorf et al. 2006). Thus, it is likely that PML provides some specific features to PML/RARA that favor transformation (Occhionorelli et al. 2011). Actually, fusion of PML to RARB or RARG may also promote transformation (Marinelli et al. 2007). Those might relate to the ability of PML to promote sumoylation (Sahin et al. 2014; Tessier et al. 2022) as posttranslational modification that was repeatedly involved in transcriptional repression (Garcia-Dominguez and Reyes 2009). In that respect, a sumoylation-defective PML/RARA mutant is impaired in its transformation abilities (Zhu et al. 2005).

Importantly, many studies have shown that PML has some features of a tumor-suppressor gene (Salomoni and Pandolfi 2002; Bernardi and Pandolfi 2007). Yet, PML may also promote the survival of some cancer cells and is almost never deleted or mutated in cancer evolution (Carracedo et al. 2012). PML is a gene that has fascinated cell biologists by its ability to nucleate subnuclear domains, PML nuclear bodies (NBs), that play a key role in stress responses and posttranslational modifications of an ever-growing list of partner proteins (Lallemand-Breitenbach and de Thé 2010, 2018). In multiple cellular systems, PML is required for senescence induction (Hsu and Kao 2018; Patra and Müller 2021). Importantly, PML/RARA disrupts the assembly of PML NBs, yielding a "microspeckled" distribution that can be used for APL diagnosis (Fig. 2; Daniel et al. 1993; Dyck et al. 1994, 1995; Koken et al. 1994). Pointing to a role of PML and NB disruption in APL pathogenesis, APL initiation is modestly accelerated in *Pml* absence (Rego et al. 2001; unpubl. data). Thus, PML function is altered during APL pathogenesis and this likely contributes to leukemogenesis. The actual mechanistic links between PML NB formation and tumor-suppressive properties remain debated. The first proposed mechanism was control of P53 function, most likely through the ability of NBs to recruit P53 and most of the enzymes en-

forcing its posttranslational modifications (Pearson et al. 2000; Matt and Hofmann 2018). Yet, PML may also drive P53-independent senescence (Mallette et al. 2004) and profoundly affects mitochondrial functions (Carracedo et al. 2012; Ito et al. 2012). Finally, sumoylation was found to exert prosenescent functions (Bischof and Dejean 2007). This may be promoted by PML NBs, which control global sumoylation (Sahin et al. 2014; Tessier et al. 2022).

Another topic of interest is the actual cell of origin of APL (Guibal et al. 2009; Wojiski et al. 2009). Studies in mouse models or in ex vivo transformation systems have suggested that PML/RARA can promote the proliferation of stem cells, not only myeloid progenitors (Welch et al. 2011). Careful analysis of preleukemic APL mice and their transition to frank APL could provide important insights into the emergence of APL, although some mice/human specificities may complicate this analysis. Altogether, the fusion of PML to RARA yields both novel properties of transcriptional deregulation and disruption of PML NBs (Fig. 2; de Thé et al. 2017). Mechanistically, some studies have suggested a direct role of PML in the acquisition of the transcriptional repression phenotype of PML/RARA (Occhionorelli et al. 2011). Conversely, PML/RARA-associated complexes were implicated in the inhibition of NB assembly (Shima et al. 2013). The PML/RARA fusion exemplifies a remarkable model in which transformation can be driven by a single dominant alteration that disrupts pathways controlled by each of its constitutive moieties.

APL THERAPIES, FROM EMPIRICISM TO CURE BY TARGETED DRUGS

The history of the successive APL therapies was reviewed elsewhere and will be only summarized here to show how they contributed to sequential changes in paradigms, illuminating APL therapy response and ultimately driving its cure. Therapeutically, APL was shown to be sensitive to high-dose anthracyclines in the early 1970s, although the latter often aggravated the bleeding diathesis and resulted in a very high incidence of early deaths (Bernard et al. 1973). Yet, up to 20%

of patients could be cured by chemotherapy alone.

The ability of retinoids to differentiate many cell lines, including AML ones (Breitman et al. 1980, 1981), led to the first human trials of ATRA, which induced rapid in vivo differentiation of the leukemic cells and showed clearance of the disease, culminating in complete remissions (Fig. 3; Huang et al. 1988; Castaigne et al. 1990; Chomienne et al. 1990). This was the first success of "differentiation therapy" (Warrell et al. 1993; Degos et al. 1995). The success of retinoic acid treatment initiated the cloning of PML/RARA and the first physiopathological models wherein ATRA reversed the transcriptional repression of PML/RARA-silenced genes, thereby initiating differentiation of leukemic cells (de Thé et al. 1990, 1991). Differentiated APL cells (granulocytes) being very short-lived, terminal APL cell differentiation drives disease

clearance. Yet, in most cases, patients rapidly relapsed, sometimes bearing on-target mutations on PML/RARA that impeded transcriptional reactivation by ATRA, formally demonstrating that ATRA is a PML/RARA-targeted therapy (Gallagher 2002). Bypassing these relapses, a combination of retinoic acid and anthracycline allowed APL eradication and could cure two-thirds of patients, so the introduction of ATRA was an actual therapeutic revolution (Fenaux et al. 1994).

Yet, the devil is in the details and it progressively appeared that several pieces did not fit in the puzzle. Not all retinoids have clinical efficacy, despite their ability to reactivate transcription and some rare variant RARA fusions were clinically insensitive to ATRA (Licht et al. 1995). Murine models clearly demonstrated that differentiation and APL regression could be uncoupled, genetically and pharmacologically (Koken

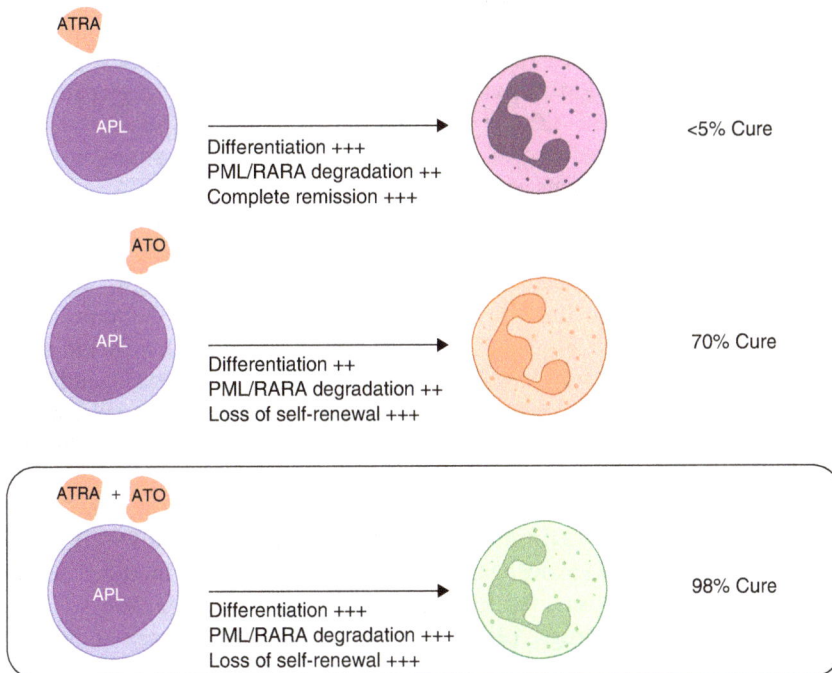

Figure 3. All-*trans*-retinoic acid (ATRA) and arsenic trioxide (ATO): a synergy to drive 98% cure in acute promyelocytic leukemia (APL) patients. Fewer than 5% of patients treated with single-agent ATRA are cured. Indeed, ATRA mainly induces transient complete remissions. ATO can cure up to 70% APL patients, due to its ability to target promyelocytic leukemia (PML) and trigger loss of self-renewal. Finally, the ATRA + ATO combination can lead up to 95% cures; the two molecules synergize as they target the retinoic acid receptor α (RARA) and PML moieties independently (see also Fig. 4).

et al. 1999; Nasr et al. 2008; Ablain et al. 2013). Subsequent studies in other models clearly established that differentiation of AML cells may not be the irreversible process that had once been foreseen (McKenzie et al. 2019), so that differentiation through transcriptional control was likely not the only driver of ATRA therapy.

Further understanding of APL therapy came from another clinical revolution: the exquisite sensitivity of APLs to ATO (Chen et al. 1996, 1997). Clinically, ATO may definitively cure up to 70% of APL patients as a single agent, including those that had become resistant to ATRA (Mathews et al. 2006, 2010). Thus, ATO is clinically far more efficient than ATRA. Interestingly, ATO primarily drives differentiation in vivo, although some apoptosis was also noted ex vivo (Chen et al. 1996, 1997). Critically, ATO directly targets PML/RARA, enforcing its degradation (Fig. 3; Chen et al. 1997; Zhu et al. 1997; Muller et al. 1998; Lallemand-Breitenbach et al. 2001; for review, see Zhu et al. 2002). PML/RARA degradation restores PML NB formation. Targeting of PML/RARA is enforced through its PML moiety, since ATO dramatically alters PML NBs and also degrades normal PML proteins (Zhu et al. 1997). Subsequent studies have dissected, in great molecular detail, the mechanisms of ATO binding onto PML and its consequences with respect to cell biology and biochemistry (Lallemand-Breitenbach et al. 2001, 2008; Jeanne et al. 2010; Zhang et al. 2010; Lallemand-Breitenbach and de Thé 2018; Bercier et al. 2023). Actually, ATRA also degrades PML/RARA and RARA through a molecular mechanism involving the UBR5 ubiquitin E3 ligase (Zhu et al. 1999; Kopf et al. 2000; Bruck et al. 2009; Tsai et al. 2023). Thus, both active APL drugs target each of the constitutive moieties of the PML/RARA fusion to initiate its degradation (Fig. 4). Following the PML/RARA example, many strategies (such as proteolysis-targeting chimeric drugs [PROTACs]) were implemented to initiate degradation of other driver oncogenes (Dale et al. 2021).

In principle, the degradation of a driver oncogene is the best way to inactivate it. Indeed, the dominant-negative effects of PML/RARA on retinoic acid signaling and NB assembly should disappear. Actually, in several mice models of Myc-

induced leukemogenesis, extinction of oncogene expression precipitates apoptosis or senescence to drive short-lived remissions (Felsher and Bishop 1999; Jain et al. 2002). In other AML models, functional inhibition of oncogenes can drive terminal differentiation (Wang et al. 2013; Amatangelo et al. 2017; de Thé 2018). This raised an important question with respect to ATRA activity: is PML/RARA degradation sufficient for differentiation and/or APL clearance or is transcriptional reactivation of PML/RARA and RARA of any importance for clinical responses? While some of these questions remain unsettled, the use of mouse models has provided some important answers. In murine APL models, the absence of RARA did not alter the initial onset of differentiation (Kogan et al. 2000). Analyses of "uncoupled" retinoids that efficiently initiate RARA and PML/RARA-dependent transcription, but not degradation, demonstrated that these drugs are much less potent than ATRA to initiate APL clearance in vivo, despite yielding similar differentiation (Ablain et al. 2013). Genetic and pharmacological studies demonstrated that PML/RARA degradation is essential for ATRA response (Nasr et al. 2008, 2009; Ablain et al. 2013). Transcriptional activation through RARA or PML/RARA may still be of some importance. Yet, that ATO induces APL differentiation in vivo with similar kinetics and efficiency as ATRA suggests that PML/RARA loss suffices to initiate this process (Vitaliano-Prunier et al. 2014). One should stress that in APL, ATRA is almost never curative on its own. This may in fact reflect some pharmacodynamic issues, as liposome-mediated delivery of ATRA resulted in some long-term remissions (Douer et al. 2001; Tsimberidou et al. 2006), possibly reflecting a more efficient PML/RARA degradation. Yet, note that the master genes silenced by altered RARs to initiate APL self-renewal and differentiation arrest remain to be identified.

Importantly, mouse models clearly showed that normal PML is absolutely required for clinical response to either ATRA or ATO in vivo (Ablain et al. 2014; unpubl. data). Thus, therapy-induced NB reformation is an essential determinant of APL response (Rérolle and de Thé 2023). Nuclear body restoration may be the key difference be-

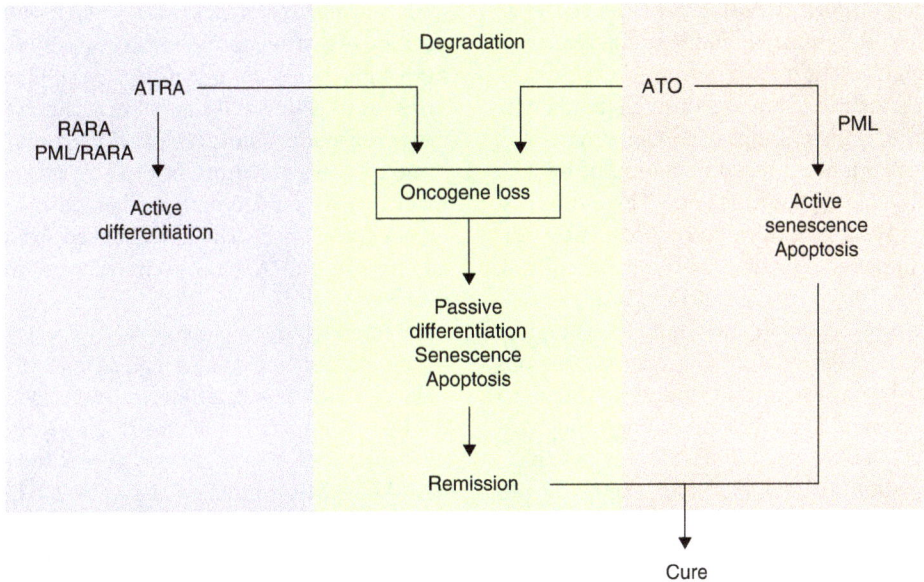

Figure 4. Mechanisms of acute promyelocytic leukemia (APL) cure by all-*trans*-retinoic acid (ATRA) + arsenic trioxide (ATO). ATRA or ATO induces promyelocytic leukemia (PML)/retinoic acid receptor α (RARA) degradation. Oncoprotein loss initiates passive differentiation, apoptosis, and senescence. ATRA may also actively induce APL cell differentiation, while ATO activates PML-induced senescence and apoptosis, allowing patient cure.

tween mere oncogene extinction, driving short-term remission, and long-term response or cure (Fig. 4). In the case of APL, PML/RARA loss allows PML NB reformation, which has additional growth-suppressive function independently from restored transcriptional control of retinoid targets. Indeed, PML is a master gene of senescence, at least in part through NB formation. Moreover, since ATO enforces PML NB formation independently from PML/RARA degradation, ATO-enhanced PML NB targeting is likely to be an important contributor to ATO clinical efficacy (Fig. 4). Similar to PML overexpression-driven senescence, the molecular mechanisms through which PML NB reformation upon PML/RARA destruction is required for full APL response remain imperfectly understood. PML is required for activation of P53 upon ATRA and/or ATO treatments. Moreover, in this context, PML is more important than P53 in driving responses. This could be mediated by PML global effects on sumoylation, a posttranslational modification shown to favor senescence and to be tightly linked to chemotherapy sensitivity in AMLs (Bossis et al. 2014; Tessier

et al. 2022). Critically, PML mutations were discovered in therapy-resistant patients for whom PML/RARA remained normal (Lehmann-Che et al. 2014; Iaccarino et al. 2016), a situation that we recently recapitulated in murine models (unpubl. data), genetically demonstrating the critical requirement of normal PML for ATO response.

Analysis of these different responses to therapies highlights an important caveat regarding therapeutic responses in AML. In APL, the latter involves both differentiation and loss of self-renewal. In contrast to initial beliefs, the two may be somehow disentangled, as some self-renewal may persist even in terminally differentiated cells (Nasr et al. 2008; de Thé 2018; McKenzie et al. 2019). Clearly, loss of PML/RARA-mediated repression releases a differentiation program that is critical for the initial tumor debulking. A different, complementary, PML-mediated loss of self-renewal/senescence program is also required for the full therapeutic effect (Ito et al. 2008). Thus, PML NB reformation can be "passive," following PML/RARA degradation by ATRA, or "active" following ATO-enforced NB biogenesis through targeting of normal PML (Fig. 4; Bercier et al. 2023). ATRA primarily induces differentiation through transcriptional control; ATO primarily induces loss of self-renewal through its ability to target PML (Figs. 3 and 4; Bercier et al. 2023). This explains why the combination of the two drugs is highly synergistic (Figs. 3 and 4), as first shown in animal models and subsequently validated in patients (Lallemand-Breitenbach et al. 1999; Lo-Coco et al. 2013, 2016). Mechanistically, this frontline combination induces synergistic PML/RARA degradation (because the degrons activated by ATRA or ATO binding are different) and maximal PML NB reformation through direct ATO-targeting of PML. The ATRA/ATO combination is now the gold standard, with more than 95% cures, at least in patients enrolled in clinical trials (Lehmann et al. 2017).

Overall, the APL model has highlighted a remarkable series of dualities: fusion of PML to RARA as the master gene of transformation, PML/RARA degradation through ATRA and ATO binding to its PML and RARA moieties, activation of differentiation, and loss of self-renewal. In times of immuno-oncology, whether APL eradication is purely a cell-autonomous process or whether the latter requires the immune system, as demonstrated in single-agent ATRA therapy (Westervelt et al. 2002; Padua et al. 2003; Robin et al. 2006), remains unsettled.

BROADENING THE LESSONS FROM THE APL MODEL TO OTHER AMLs?

What do 30 years of APL research provide as insights into other malignancies? At the level of pathogenesis, deregulated retinoic acid signaling has been observed in several other malignancies, notably human breast cancers and other AMLs or murine models of hepatocarcinoma (Khetchoumian et al. 2007; Tan et al. 2015). From a therapeutic point of view, AMLs that express high levels of RARA were shown to be intrinsically ATRA-sensitive (McKeown et al. 2017). This observation is in line with ex vivo experiments demonstrating that mere RARA overexpression initiates ATRA-reversible immortalization of murine hematopoietic progenitors (Du et al. 1999; Zhu et al. 2007). Mechanisms involved are currently uncharacter-

ized, but may resemble studies wherein overexpression of thyroid hormone receptors mimicked expression of its dominant-negative mutant *v-ErbA* for its ability to immortalize avian erythrocytic progenitors (Bauer et al. 1998). Importantly, some clinical translations of these findings were made in AML patients unfit for chemotherapy for whom a clinical benefit was obtained from the addition of ATRA to conventional treatments (Lübbert et al. 2020; de Botton et al. 2023). Of note, several studies demonstrated that enforcing cell cycle arrest of AML cells renders them susceptible to ATRA-induced differentiation (Boutzen et al. 2016; Mugoni et al. 2019). Finally, dominant-negative RARA mutations were observed in non-APL AMLs treated by high-dose ATRA for suspicion of APL (Zhao et al. 2019), genetically demonstrating a strong RARA-dependent selective pressure of ATRA treatment. While the RARA downstream effector genes are unknown, they most likely overlap with PML/RARA-repressed genes that initiate APL. Thus, these novel ATRA/AML connections could open some important translational opportunities in a disease that remains incurable in patients unfit for chemotherapy.

PML has also rich connections with human pathogenesis. It may be involved in other chromosomal translocations (Kurahashi et al. 2011) and is tightly linked to many viral infections, through its ability to act as a first-line innate response (Scherer and Stamminger 2016). PML was also tightly linked to the natural history of many epithelial cancers, massively induced at the initial stages of transformation (possibly as a reflection of interferon [IFN] induction), but lost at an advanced stage, perhaps reflecting bypass of a senescence checkpoint (Koken et al. 1995; Gurrieri et al. 2004). This is actually very reminiscent of the activation of DNA damage response upon oncogenic activation (Bartkova et al. 2005; Gorgoulis et al. 2005). However, in P53-mutant tumors, PML was proposed to have prosurvival functions, possibly linked to the effects of PML on mitochondrial fitness (Carracedo et al. 2012; Ito et al. 2012). In some settings, PML expression in tumors is tightly correlated to survival. The absolute requirement of PML for APL response to ATRA or ATO begged the question of its role in the pathogenesis or therapeutic response of other hematological malignancies. In AMLs associated with mutation of the NPM1 chaperone, mutant NPM1 impedes PML NB formation (El Hajj et al. 2015; Martelli et al. 2015). Oxidative stress induced by Actinomycin D restored PML NB formation and the latter was required for AML cell lines to respond through senescence induction (Gionfriddo et al. 2021; Wu et al. 2021). PML expression levels are primarily tuned by IFN signaling (Stadler et al. 1995). Myeloproliferative neoplasms (MPNs) driven by an activating mutation in Jak2 are often treated with IFN, which induces a slow clearance of transformed cells (Kiladjian et al. 2008). PML NB biogenesis is dependent on PML protein expression, PML assembly upon ATO treatment or ROS exposure, and PML degradation by a number of well-identified proteolytic pathways (Gamell et al. 2014; Niwa-Kawakita et al. 2017; Lallemand-Breitenbach and de Thé 2018). Thus, maximal NB formation can be enforced by a combination of IFN and ATO (Quignon et al. 1998). Should PML NBs be one of the downstream effectors of IFN activity in MPN, its activity should be enhanced by ATO. A dramatic synergy was indeed observed in murine models of the disease for MPN clearance and leukemia-initiating cell eradication. Importantly, the latter was dependent on *Pml* presence (Dagher et al. 2021). How general is the PML dependency for response to cancer therapy awaits further studies. Given the role of PML to control P53 signaling and the importance of the latter in chemotherapy response, it would be particularly interesting to explore Pml dependency on chemotherapy response in murine cancer models. Similarly, a structure–function analysis of PML role in therapy response could orient as to which of the many proposed functions of PML (sumoylation control, mitochondrial fitness, antioxidant, P53 control, etc.) is actually responsible for these effects.

CONCLUDING REMARKS: TOWARD A BIOLOGY OF CURE?

APL is currently one of the only examples of cure by targeted therapies that has now become the clinical gold standard. As such, it has allowed an unprecedented mechanistic exploration of the molecular requirements for disease eradication. It was long thought that dissection of APL pathogenesis and therapy response would only be relevant to APL. Yet, recent studies have shown that deregulated RARA signaling (and ATRA sensitivity) is more broadly implicated in AML than the sole APL setting (McKeown et al. 2017). Similarly, PML contributes to therapy response for several other leukemia/therapy pairs (Rérolle and de Thé 2023). In fact, studies of ovarian cancer response to platinum-based chemotherapy have also suggested a positive impact of PML expression, at least in part through mitochondrial control (Gentric et al. 2019). More broadly, PML is a key regulator of sumoylation (Tessier et al. 2022), a posttranslational modification essential to multiple forms of stress responses, including senescence induction (Bischof and Dejean 2007). Mechanistically, how PML arbitrates post-therapy cell fate remains to be determined.

Beyond the targeting of retinoid signaling by ATRA and PML/sumoylation by ATO, APL studies could enlighten and broaden the dual contributions of differentiation and senescence in cancer therapy (Fig. 4). In APL, differentiation is insufficient for cure per se, but most likely contributes to cure through both transient debulking and some reduction in self-renewal. Retinoids are not the only drugs that may initiate leukemia differentiation (Wang et al. 2013; de Thé 2018). Similarly, senescence and cell cycle arrest may be triggered by many drugs. The synergism between differentiation and senescence may constitute a model for future curative associations in AMLs, and perhaps more broadly in cancers, showing how their dual effects discovered in APL could cooperate in other conditions.

ACKNOWLEDGMENTS

Work in the authors' laboratory is supported by grants from the ERC (PML-Therapy, ADG-785917), as well as Institut National du Cancer (PLBio INCA). H.d.T. received consulting fees from SYROS, but the authors have no other conflict of interests.

REFERENCES

Ablain J, Leiva M, Peres L, Fonsart J, Anthony E, de Thé H. 2013. Uncoupling RARA transcriptional activation and degradation clarifies the bases for APL response to therapies. *J Exp Med* **210**: 647–653. doi:10.1084/jem.20122337

Ablain J, Rice K, Soilihi H, de Reynies A, Minucci S, de Thé H. 2014. Activation of a promyelocytic leukemia-tumor protein 53 axis underlies acute promyelocytic leukemia cure. *Nat Med* **20**: 167–174. doi:10.1038/nm.3441

Akagi T, Shih LY, Kato M, Kawamata N, Yamamoto G, Sanada M, Okamoto R, Miller CW, Liang DC, Ogawa S, et al. 2009. Hidden abnormalities and novel classification of t (15;17) acute promyelocytic leukemia (APL) based on genomic alterations. *Blood* **113**: 1741–1748. doi:10.1182/blood-2007-12-130260

Amatangelo MD, Quek L, Shih A, Stein EM, Roshal M, David MD, Marteyn B, Farnoud NR, de Botton S, Bernard OA, et al. 2017. Enasidenib induces acute myeloid leukemia cell differentiation to promote clinical response. *Blood* **130**: 732–741. doi:10.1182/blood-2017-04-779447

Astolfi A, Masetti R, Indio V, Bertuccio SN, Messelodi D, Rampelli S, Leardini D, Carella M, Serravalle S, Libri V, et al. 2021. Torque teno mini virus as a cause of childhood acute promyelocytic leukemia lacking PML/RARA fusion. *Blood* **138**: 1773–1777. doi:10.1182/blood.2021011677

Bartkova J, Hořejší Z, Koed K, Krämer A, Tort F, Zieger K, Guldberg P, Sehested M, Nesland JM, Lukas C, et al. 2005. DNA damage response as a candidate anti-cancer barrier in early human tumorigenesis. *Nature* **434**: 864–870. doi:10.1038/nature03482

Bauer A, Mikulits W, Lagger G, Stengl G, Brosch G, Beug H. 1998. The thyroid hormone receptor functions as a ligand-operated developmental switch between proliferation and differentiation of erythroid progenitors. *EMBO J* **17**: 4291–4303. doi:10.1093/emboj/17.15.4291

Bercier P, Wang QQ, Zang N, Zhang J, Yang C, Maimaitiyiming Y, Abou-Ghali M, Berthier C, Wu C, Niwa-Kawakita M, et al. 2023. Structural basis of PML/RARA oncoprotein targeting by arsenic unravels a cysteine rheostat controlling PML body assembly and function. *Cancer Discov* **13**: 2548–2565. doi:10.1158/2159-8290.CD-23-0453

Bernard J, Weil M, Boiron M, Jacquillat C, Flandrin G, Gemon MF. 1973. Acute promyelocytic leukemia: results of treatment by daunorubicin. *Blood* **41**: 489–496. doi:10.1182/blood.V41.4.489.489

Bernardi R, Pandolfi PP. 2007. Structure, dynamics and functions of promyelocytic leukaemia nuclear bodies. *Nat Rev Mol Cell Biol* **8**: 1006–1016. doi:10.1038/nrm2277

Bischof O, Dejean A. 2007. SUMO is growing senescent. *Cell Cycle* **6**: 677–681. doi:10.4161/cc.6.6.4021

Bossis G, Sarry JE, Kifagi C, Ristic M, Saland E, Vergez F, Salem T, Boutzen H, Baik H, Brockly F, et al. 2014. The ROS/SUMO axis contributes to the response of acute myeloid leukemia cells to chemotherapeutic drugs. *Cell Rep* **7**: 1815–1823. doi:10.1016/j.celrep.2014.05.016

Boutzen H, Saland E, Larrue C, de Toni F, Gales L, Castelli FA, Cathebas M, Zaghdoudi S, Stuani L, Kaoma T, et al. 2016. Isocitrate dehydrogenase 1 mutations prime the all-*trans* retinoic acid myeloid differentiation pathway in acute myeloid leukemia. *J Exp Med* **213**: 483–497. doi:10.1084/jem.20150736

Breitman TR, Selonick SE, Collins SJ. 1980. Induction of differentiation of the human promyelocytic leukemia cell line (HL-60) by retinoic acid. *Proc Natl Acad Sci* **77**: 2936–2940. doi:10.1073/pnas.77.5.2936

Breitman TR, Collins SJ, Keene BR. 1981. Terminal differentiation of human promyelocytic leukemic cells in primary culture in response to retinoic acid. *Blood* **57**: 1000–1004. doi:10.1182/blood.V57.6.1000.1000

Bruck N, Vitoux D, Ferry C, Duong V, Bauer A, de Thé H, Rochette-Egly C. 2009. A coordinated phosphorylation cascade initiated by p38MAPK/MSK1 directs RARα to target promoters. *EMBO J* **28**: 34–47. doi:10.1038/emboj.2008.256

Cabezas-Wallscheid N, Buettner F, Sommerkamp P, Klimmeck D, Ladel L, Thalheimer FB, Pastor-Flores D, Roma LP, Renders S, Zeisberger P, et al. 2017. Vitamin A-retinoic acid signaling regulates hematopoietic stem cell dormancy. *Cell* **169**: 807–823.e19. doi:10.1016/j.cell.2017.04.018

Carracedo A, Weiss D, Leliaert AK, Bhasin M, de Boer VC, Laurent G, Adams AC, Sundvall M, Song SJ, Ito K, et al. 2012. A metabolic prosurvival role for PML in breast cancer. *J Clin Invest* **122**: 3088–3100. doi:10.1172/JCI62129

Castaigne S, Chomienne C, Daniel MT, Ballerini P, Berger R, Fenaux P, Degos L. 1990. All-*trans* retinoic acid as a differentiation therapy for acute promyelocytic leukemia. I: Clinical results. *Blood* **76**: 1704–1709. doi:10.1182/blood.V76.9.1704.1704

Chen GQ, Zhu J, Shi XG, Ni JH, Zhong HJ, Si GY, Jin XL, Tang W, Li XS, Xong SM, et al. 1996. In vitro studies on cellular and molecular mechanisms of arsenic trioxide (As_2O_3) in the treatment of acute promyelocytic leukemia: As_2O_3 induces NB4 cell apoptosis with downregulation of Bcl-2 expression and modulation of PML-RAR α/PML proteins. *Blood* **88**: 1052–1061. doi:10.1182/blood.V88.3.1052.1052

Chen GQ, Shi XG, Tang W, Xiong SM, Zhu J, Cai X, Han ZG, Ni JH, Shi GY, Jia PM, et al. 1997. Use of arsenic trioxide (As_2O_3) in the treatment of acute promyelocytic leukemia (APL). I: As_2O_3 exerts dose-dependent dual effects on APL cells. *Blood* **89**: 3345–3353.

Chen X, Wang F, Zhou X, Zhang Y, Cao P, Ma X, Yuan L, Fang J, Liu M, Liu M, et al. 2022. Torque teno mini virus driven childhood acute promyelocytic leukemia: the third case report and sequence analysis. *Front Oncol* **12**: 1074913. doi:10.3389/fonc.2022.1074913

Chomienne C, Ballerini P, Balitrand N, Daniel MT, Fenaux P, Castaigne S, Degos L. 1990. All-*trans* retinoic acid in acute promyelocytic leukemias. II: In vitro studies: structure-function relationship. *Blood* **76**: 1710–1717. doi:10.1182/blood.V76.9.1710.1710

Dagher T, Maslah N, Edmond V, Cassinat B, Vainchenker W, Giraudier S, Pasquier F, Verger E, Niwa-Kawakita M, Lallemand-Breitenbach V, et al. 2021. JAK2V617F myeloproliferative neoplasm eradication by a novel interferon/arsenic therapy involves PML. *J Exp Med* **218**: e20201268. doi:10.1084/jem.20201268

Dale B, Cheng M, Park KS, Kaniskan HU, Xiong Y, Jin J. 2021. Advancing targeted protein degradation for cancer therapy. *Nat Rev Cancer* **21**: 638–654. doi:10.1038/s41568-021-00365-x

Daniel MT, Koken M, Romagné O, Barbey S, Bazarbachi A, Stadler M, Guillemin M-C, Degos L, Chomienne C, de Thé H. 1993. PML protein expression in hematopoietic and acute promyelocytic leukemia cells. *Blood* 82: 1858–1867. doi:10.1182/blood.V82.6.1858.1858

de Botton S, Cluzeau T, Vigil CE, Cook R, Rousselot P, Rizzieri DA, Liesveld JL, Fenaux P, Braun T, Banos A, et al. 2023. Targeting *RARA* overexpression with tamibarotene, a potent and selective RARα agonist, is a novel approach in AML. *Blood Adv* 7: 1858–1870. doi:10.1182/bloodadvances.2022008806

Degos L, Dombret H, Chomienne C, Daniel MT, Miclea JM, Chastang C, Castaigne S, Fenaux P. 1995. All-*trans* retinoic acid as a differentiating agent in the treatment of acute promyelocytic leukemia. *Blood* 85: 2643–2653. doi:10.1182/blood.V85.10.2643.bloodjournal85102643

de Thé H. 2018. Differentiation therapy revisited. *Nat Rev Cancer* 18: 117–127. doi:10.1038/nrc.2017.103

de Thé H, Chomienne C, Lanotte M, Degos L, Dejean A. 1990. The t(15;17) translocation of acute promyelocytic leukemia fuses the retinoic acid receptor α gene to a novel transcribed locus. *Nature* 347: 558–561. doi:10.1038/347558a0

de Thé H, Lavau C, Marchio A, Chomienne C, Degos L, Dejean A. 1991. The PML-RARα fusion mRNA generated by the t(15;17) translocation in acute promyelocytic leukemia encodes a functionally altered RAR. *Cell* 66: 675–684. doi:10.1016/0092-8674(91)90113-D

de Thé H, Pandolfi PP, Chen Z. 2017. Acute promyelocytic leukemia: a paradigm for oncoprotein-targeted cure. *Cancer Cell* 32: 552–560. doi:10.1016/j.ccell.2017.10.002

Dos Santos GA, Kats L, Pandolfi PP. 2013. Synergy against PML-RARa: targeting transcription, proteolysis, differentiation, and self-renewal in acute promyelocytic leukemia. *J Exp Med* 210: 2793–2802. doi:10.1084/jem.20131121

Douer D, Estey E, Santillana S, Bennett JM, Lopez-Bernstein G, Boehm K, Williams T. 2001. Treatment of newly diagnosed and relapsed acute promyelocytic leukemia with intravenous liposomal all-*trans* retinoic acid. *Blood* 97: 73–80. doi:10.1182/blood.V97.1.73

Du C, Redner RL, Cooke MP, Lavau C. 1999. Overexpression of wild-type retinoic acid receptor α (RARα) recapitulates retinoic acid-sensitive transformation of primary myeloid progenitors by acute promyelocytic leukemia RARα-fusion genes. *Blood* 94: 793–802. doi:10.1182/blood.V94.2.793

Du C, Fang M, Li Y, Li L, Wang X. 2000. Smac, a mitochondrial protein that promotes cytochrome c-dependent caspase activation by eliminating IAP inhibition. *Cell* 102: 33–42. doi:10.1016/S0092-8674(00)00008-8

Dyck JA, Maul GG, Miller WH, Chen JD, Kakizuka A, Evans RM. 1994. A novel macromolecular structure is a target of the promyelocyte-retinoic acid receptor oncoprotein. *Cell* 76: 333–343. doi:10.1016/0092-8674(94)90340-9

Dyck JA, Warrell RP, Evans RM, Miller WH. 1995. Rapid diagnosis of acute promyelocytic leukemia by immunohistochemical localization of PML/RARα protein. *Blood* 86: 862–867. doi:10.1182/blood.V86.3.862.862

El Hajj H, Dassouki Z, Berthier C, Raffoux E, Ades L, Legrand O, Hleihel R, Sahin U, Tawil N, Salameh A, et al. 2015. Retinoic acid and arsenic trioxide trigger degradation of mutated NPM1, resulting in apoptosis of AML cells. *Blood* 125: 3447–3454. doi:10.1182/blood-2014-11-612416

Estey E, Garcia-Manero G, Ferrajoli A, Faderl S, Verstovsek S, Jones D, Kantarjian H. 2006. Use of all-*trans* retinoic acid plus arsenic trioxide as an alternative to chemotherapy in untreated acute promyelocytic leukemia. *Blood* 107: 3469–3473. doi:10.1182/blood-2005-10-4006

Felsher DW, Bishop JM. 1999. Reversible tumorigenesis by MYC in hematopoietic lineages. *Mol Cell* 4: 199–207. doi:10.1016/S1097-2765(00)80367-6

Fenaux P, Chastang C, Chomienne C, Degos L, Degos L; European APL Group. 1994. Tretinoin with chemotherapy in newly diagnosed acute promyelocytic leukaemia. *Lancet* 343: 1033. doi:10.1016/S0140-6736(94)90151-1

Gallagher RE. 2002. Retinoic acid resistance in acute promyelocytic leukemia. *Leukemia* 16: 1940–1958. doi:10.1038/sj.leu.2402719

Gamell C, Jan Paul P, Haupt Y, Haupt S. 2014. PML tumour suppression and beyond: therapeutic implications. *FEBS Lett* 588: 2653–2662. doi:10.1016/j.febslet.2014.02.007

Garcia-Dominguez M, Reyes JC. 2009. SUMO association with repressor complexes, emerging routes for transcriptional control. *Biochim Biophys Acta* 1789: 451–459. doi:10.1016/j.bbagrm.2009.07.001

Gentric G, Kieffer Y, Mieulet V, Goundiam O, Bonneau C, Nemati F, Hurbain I, Raposo G, Popova T, Stern MH, et al. 2019. PML-regulated mitochondrial metabolism enhances chemosensitivity in human ovarian cancers. *Cell Metab* 29: 156–173.e10. doi:10.1016/j.cmet.2018.09.002

Geoffroy MC, de Thé H. 2020. Classic and variants APLs, as viewed from a therapy response. *Cancers (Basel)* 12: 967. doi:10.3390/cancers12040967

Gionfriddo I, Brunetti L, Mezzasoma F, Milano F, Cardinali V, Ranieri R, Venanzi A, Pierangeli S, Vetro C, Spinozzi G, et al. 2021. Dactinomycin induces complete remission associated with nucleolar stress response in relapsed/refractory NPM1-mutated AML. *Leukemia* 35: 2552–2562. doi:10.1038/s41375-021-01192-7

Gorgoulis VG, Vassiliou LV, Karakaidos P, Zacharatos P, Kotsinas A, Liloglou T, Venere M, Ditullio RA Jr, Kastrinakis NG, Levy B, et al. 2005. Activation of the DNA damage checkpoint and genomic instability in human precancerous lesions. *Nature* 434: 907–913. doi:10.1038/nature03485

Guibal FC, Alberich-Jorda M, Hirai H, Ebralidze A, Levantini E, Di Ruscio A, Zhang P, Santana-Lemos BA, Neuberg D, Wagers AJ, et al. 2009. Identification of a myeloid committed progenitor as the cancer-initiating cell in acute promyelocytic leukemia. *Blood* 114: 5415–5425. doi:10.1182/blood-2008-10-182071

Gurrieri C, Capodieci P, Bernardi R, Scaglioni PP, Nafa K, Rush LJ, Verbel DA, Cordon-Cardo C, Pandolfi PP. 2004. Loss of the tumor suppressor PML in human cancers of multiple histologic origins. *J Natl Cancer Inst* 96: 269–279. doi:10.1093/jnci/djh043

Hillestad LK. 1957. Acute promyelocytic leukemia. *Acta Med Scand* 159: 189–194. doi:10.1111/j.0954-6820.1957.tb00124.x

Hsu KS, Kao HY. 2018. PML: regulation and multifaceted function beyond tumor suppression. *Cell Biosci* 8: 5. doi:10.1186/s13578-018-0204-8

Huang M, Ye Y, Chen R, Chai J, Lu J, Zhoa L, Gu L, Wang Z. 1988. Use of all-*trans* retinoic acid in the treatment of acute promyelocytic leukaemia. *Blood* **72:** 567–572. doi:10.1182/blood.V72.2.567.567

Iaccarino L, Ottone T, Divona M, Cicconi L, Cairoli R, Voso MT, Lo-Coco F. 2016. Mutations affecting both the rearranged and the unrearranged *PML* alleles in refractory acute promyelocytic leukaemia. *Br J Haematol* **172:** 909–913. doi:10.1111/bjh.13910

Iland HJ, Russell N, Dillon R, Schuh AC, Tedjaseputra A, Wei A, Khwaja A, Knapper S, Lane SW, Reynolds J, et al. 2023. Characteristics and outcomes of patients with acute promyelocytic leukemia and extreme hyperleukocytosis at presentation. *Blood Adv* **7:** 2580–2585. doi:10.1182/bloodadvances.2022007126

Ito K, Bernardi R, Morotti A, Matsuoka S, Saglio G, Ikeda Y, Rosenblatt J, Avigan DE, Teruya-Feldstein J, Pandolfi PP. 2008. PML targeting eradicates quiescent leukaemia-initiating cells. *Nature* **453:** 1072–1078. doi:10.1038/nature07016

Ito K, Carracedo A, Weiss D, Arai F, Ala U, Avigan DE, Schafer ZT, Evans RM, Suda T, Lee CH, et al. 2012. A PML–PPAR-δ pathway for fatty acid oxidation regulates hematopoietic stem cell maintenance. *Nat Med* **18:** 1350–1358. doi:10.1038/nm.2882

Jain M, Arvanitis C, Chu K, Dewey W, Leonhardt E, Trinh M, Sundberg CD, Bishop JM, Felsher DW. 2002. Sustained loss of a neoplastic phenotype by brief inactivation of *MYC*. *Science* **297:** 102–104. doi:10.1126/science.1071489

Jeanne M, Lallemand-Breitenbach V, Ferhi O, Koken M, Le Bras M, Duffort S, Peres L, Berthier C, Soilihi H, Raught B, et al. 2010. PML/RARA oxidation and arsenic binding initiate the antileukemia response of As$_2$O$_3$. *Cancer Cell* **18:** 88–98. doi:10.1016/j.ccr.2010.06.003

Kamashev DE, Vitoux D, De Thé H. 2004. PML–RARA-RXR oligomers mediate retinoid and rexinoid/cAMP cross-talk in acute promyelocytic leukemia cell differentiation. *J Exp Med* **199:** 1163–1174. doi:10.1084/jem.20032226

Kastner P, Lawrence HJ, Waltzinger C, Ghyselinck NB, Chambon P, Chan S. 2001. Positive and negative regulation of granulopoiesis by endogenous RARα. *Blood* **97:** 1314–1320. doi:10.1182/blood.V97.5.1314

Khetchoumian K, Teletin M, Tisserand J, Mark M, Herquel B, Ignat M, Zucman-Rossi J, Cammas F, Lerouge T, Thibault C, et al. 2007. Loss of *Trim24* (Tif1α) gene function confers oncogenic activity to retinoic acid receptor α. *Nat Genet* **39:** 1500–1506. doi:10.1038/ng.2007.15

Kiladjian JJ, Cassinat B, Chevret S, Turlure P, Cambier N, Roussel M, Bellucci S, Grandchamp B, Chomienne C, Fenaux P. 2008. Pegylated interferon-alfa-2a induces complete hematologic and molecular responses with low toxicity in polycythemia vera. *Blood* **112:** 3065–3072. doi:10.1182/blood-2008-03-143537

Kogan SC, Hong SH, Shultz DB, Privalsky ML, Bishop JM. 2000. Leukemia initiated by PMLRARα: the PML domain plays a critical role while retinoic acid–mediated transactivation is dispensable. *Blood* **95:** 1541–1550. doi:10.1182/blood.V95.5.1541.005k28_1541_1550

Koken MHM, Puvion-Dutilleul F, Guillemin MC, Viron A, Linares-Cruz G, Stuurman N, de Jong L, Szostecki C, Calvo F, Chomienne C, et al. 1994. The t(15;17) translocation alters a nuclear body in a retinoic acid-reversible fashion.

EMBO J **13:** 1073–1083. doi:10.1002/j.1460-2075.1994.tb06356.x

Koken MHM, Linares-Cruz G, Quignon F, Viron A, Chelbi-Alix MK, Sobczak-Thépot J, Juhlin L, Degos L, Calvo F, de Thé H. 1995. The PML growth-suppressor has an altered expression in human oncogenesis. *Oncogene* **10:** 1315–1324.

Koken MHM, Daniel M-T, Gianni M, Zelent A, Licht J, Buzyn A, Minard P, Degos L, Varet B, de Thé H. 1999. Retinoic acid, but not arsenic trioxide, degrades the PLZF/RARα fusion protein, without inducing terminal differentiation or apoptosis, in a RA-therapy resistant t(11;17)(q23;q21) APL patient. *Oncogene* **18:** 1113–1118. doi:10.1038/sj.onc.1202414

Kopf E, Plassat JL, Vivat V, de Thé H, Chambon P, Rochette-Egly C. 2000. Dimerization with retinoid X receptors and phosphorylation modulate the retinoic acid-induced degradation of retinoic acid receptors α and γ through the ubiquitin-proteasome pathway. *J Biol Chem* **275:** 33280–33288. doi:10.1074/jbc.M002840200

Kurahashi S, Hayakawa F, Miyata Y, Yasuda T, Minami Y, Tsuzuki S, Abe A, Naoe T. 2011. PAX5-PML acts as a dual dominant-negative form of both PAX5 and PML. *Oncogene* **30:** 1822–1830. doi:10.1038/onc.2010.554

Lallemand-Breitenbach V, de Thé H. 2010. PML nuclear bodies. *Cold Spring Harb Perspect Biol* **2:** a000661. doi:10.1101/cshperspect.a000661

Lallemand-Breitenbach V, de Thé H. 2018. PML nuclear bodies: from architecture to function. *Curr Opin Cell Biol* **52:** 154–161. doi:10.1016/j.ceb.2018.03.011

Lallemand-Breitenbach V, Guillemin M-C, Janin A, Daniel M-T, Degos L, Kogan SC, Bishop JM, de Thé H. 1999. Retinoic acid and arsenic synergize to eradicate leukemic cells in a mouse model of acute promyelocytic leukemia. *J Exp Med* **189:** 1043–1052. doi:10.1084/jem.189.7.1043

Lallemand-Breitenbach V, Zhu J, Puvion F, Koken M, Honoré N, Doubeikovsky A, Duprez E, Pandolfi PP, Puvion E, Freemont P, et al. 2001. Role of promyelocytic leukemia (PML) sumolation in nuclear body formation, 11S proteasome recruitment, and As$_2$O$_3$-induced PML or PML/retinoic acid receptor α degradation. *J Exp Med* **193:** 1361–1372. doi:10.1084/jem.193.12.1361

Lallemand-Breitenbach V, Jeanne M, Benhenda S, Nasr R, Lei M, Peres L, Zhou J, Zhu J, Raught B, de Thé H. 2008. Arsenic degrades PML or PML–RARα through a SUMO-triggered RNF4/ubiquitin-mediated pathway. *Nat Cell Biol* **10:** 547–555. doi:10.1038/ncb1717

Lallemand-Breitenbach V, Zhu J, Chen Z, de Thé H. 2012. Curing APL through PML/RARA degradation by As$_2$O$_3$. *Trends Mol Med* **18:** 36–42. doi:10.1016/j.molmed.2011.10.001

Lehmann S, Deneberg S, Antunovic P, Rangert-Derolf A, Garelius H, Lazarevic V, Myhr-Eriksson K, Möllgård L, Uggla B, Wahlin A, et al. 2017. Early death rates remain high in high-risk APL: update from the Swedish Acute Leukemia Registry 1997–2013. *Leukemia* **31:** 1457–1459. doi:10.1038/leu.2017.71

Lehmann-Che J, Bally C, de Thé H. 2014. Therapy resistance in APL. *New Engl J Med* **371:** 1170–1172. doi:10.1056/NEJMc1409040

Lehmann-Che J, Bally C, Letouzé E, Berthier C, Yuan H, Jollivet F, Ades L, Cassinat B, Hirsch P, Pigneux A, et al.

2018. Dual origin of relapses in retinoic-acid resistant acute promyelocytic leukemia. *Nat Commun* **9:** 2047. doi:10.1038/s41467-018-04384-5

Licht JD, Chomienne C, Goy A, Chen A, Scott AA, Head DR, Michaux JL, Wu Y, DeBlasio A, Miller WH Jr, et al. 1995. Clinical and molecular characterization of a rare syndrome of acute promyelocytic leukemia associated with translocation (11;17). *Blood* **85:** 1083–1094. doi:10.1182/blood .V85.4.1083.bloodjournal8541083

Lin R, Evans R. 2000. Acquisition of oncogenic potential by RAR chimeras in acute promyelocytic leukemia through formation of homodimers. *Mol Cell* **5:** 821–830. doi:10 .1016/S1097-2765(00)80322-6

Lo-Coco F, Avvisati G, Vignetti M, Thiede C, Orlando SM, Iacobelli S, Ferrara F, Fazi P, Cicconi L, Di Bona E, et al. 2013. Retinoic acid and arsenic trioxide for acute promyelocytic leukemia. *N Engl J Med* **369:** 111–121. doi:10.1056/ NEJMoa1300874

Lo-Coco F, Di Donato L, Gimema, Schlenk RF; German–Austrian Acute Myeloid Leukemia Study Group and Study Alliance Leukemia. 2016. Targeted therapy alone for acute promyelocytic leukemia. *N Engl J Med* **374:** 1197–1198. doi:10.1056/NEJMc1513710

Lübbert M, Grishina O, Schmoor C, Schlenk RF, Jost E, Crysandt M, Heuser M, Thol F, Salih HR, Schittenhelm MM, et al. 2020. Valproate and retinoic acid in combination with decitabine in elderly nonfit patients with acute myeloid leukemia: results of a multicenter, randomized, 2 × 2, phase II trial. *J Clin Oncol* **38:** 257–270. doi:10.1200/JCO .19.01053

Madan V, Shyamsunder P, Han L, Mayakonda A, Nagata Y, Sundaresan J, Kanojia D, Yoshida K, Ganesan S, Hattori N, et al. 2016. Comprehensive mutational analysis of primary and relapse acute promyelocytic leukemia. *Leukemia* **30:** 1672–1681. doi:10.1038/leu.2016.69

Mallette FA, Goumard S, Gaumont-Leclerc MF, Moiseeva O, Ferbeyre G. 2004. Human fibroblasts require the Rb family of tumor suppressors, but not p53, for PML-induced senescence. *Oncogene* **23:** 91–99. doi:10.1038/sj.onc.1206886

Marinelli A, Bossi D, Pelicci PG, Minucci S. 2007. A redundant oncogenic potential of the retinoic receptor (RAR) α, β and γ isoforms in acute promyelocytic leukemia. *Leukemia* **21:** 647–650.

Martelli MP, Gionfriddo I, Mezzasoma F, Milano F, Pierangeli S, Mulas F, Pacini R, Tabarrini A, Pettirossi V, Rossi R, et al. 2015. Arsenic trioxide and all-*trans* retinoic acid target NPM1 mutant oncoprotein levels and induce apoptosis in NPM1-mutated AML cells. *Blood* **125:** 3455–3465. doi:10.1182/blood-2014-11-611459

Martens JH, Brinkman AB, Simmer F, Francoijs KJ, Nebbioso A, Ferrara F, Altucci L, Stunnenberg HG. 2010. PML-RARα/RXR alters the epigenetic landscape in acute promyelocytic leukemia. *Cancer Cell* **17:** 173–185. doi:10 .1016/j.ccr.2009.12.042

Mathews V, George B, Lakshmi KM, Viswabandya A, Bajel A, Balasubramanian P, Shaji RV, Srivastava VM, Srivastava A, Chandy M. 2006. Single-agent arsenic trioxide in the treatment of newly diagnosed acute promyelocytic leukemia: durable remissions with minimal toxicity. *Blood* **107:** 2627–2632. doi:10.1182/blood-2005-08-3532

Mathews V, George B, Chendamarai E, Lakshmi KM, Desire S, Balasubramanian P, Viswabandya A, Thirugnanam R,

Abraham A, Shaji RV, et al. 2010. Single-agent arsenic trioxide in the treatment of newly diagnosed acute promyelocytic leukemia: long-term follow-up data. *J Clin Oncol* **28:** 3866–3871. doi:10.1200/JCO.2010.28.5031

Matt S, Hofmann TG. 2018. Crosstalk between p53 modifiers at PML bodies. *Mol Cell Oncol* **5:** e1074335. doi:10.1080/ 23723556.2015.1074335

McKenzie MD, Ghisi M, Oxley EP, Ngo S, Cimmino L, Esnault C, Liu R, Salmon JM, Bell CC, Ahmed N, et al. 2019. Interconversion between tumorigenic and differentiated states in acute myeloid leukemia. *Cell Stem Cell* **25:** 258–272.e9. doi:10.1016/j.stem.2019.07.001

McKeown MR, Corces MR, Eaton ML, Fiore C, Lee E, Lopez JT, Chen MW, Smith D, Chan SM, Koenig JL, et al. 2017. Superenhancer analysis defines novel epigenomic subtypes of non-APL AML, including an RARα dependency targetable by SY-1425, a potent and selective RARα agonist. *Cancer Discov* **7:** 1136–1153. doi:10.1158/2159-8290 .CD-17-0399

Minucci S, Maccarana M, Cioce M, De Luca P, Gelmetti V, Segalla S, Di Croce L, Giavara S, Matteucci C, Gobbi A, et al. 2000. Oligomerization of RAR and AML1 transcription factors as a novel mechanism of oncogenic activation. *Mol Cell* **5:** 811–820. doi:10.1016/S1097-2765(00)80321-4

Mugoni V, Panella R, Cheloni G, Chen M, Pozdnyakova O, Stroopinsky D, Guarnerio J, Monteleone E, Lee JD, Mendez L, et al. 2019. Vulnerabilities in mIDH2 AML confer sensitivity to APL-like targeted combination therapy. *Cell Res* **29:** 446–459. doi:10.1038/s41422-019-0162-7

Muller S, Matunis MJ, Dejean A. 1998. Conjugation with the ubiquitin-related modifier SUMO-1 regulates the partitioning of PML within the nucleus. *EMBO J* **17:** 61–70. doi:10.1093/emboj/17.1.61

Nasr R, Guillemin MC, Ferhi O, Soilihi H, Peres L, Berthier C, Rousselot P, Robledo-Sarmiento M, Lallemand-Breitenbach V, Gourmel B, et al. 2008. Eradication of acute promyelocytic leukemia-initiating cells through PML-RARA degradation. *Nat Med* **14:** 1333–1342. doi:10.1038/nm .1891

Nasr R, Lallemand-Breitenbach V, Zhu J, Guillemin MC, de Thé H. 2009. Therapy-induced *PML/RARA* proteolysis and acute promyelocytic leukemia cure. *Clin Cancer Res* **15:** 6321–6326. doi:10.1158/1078-0432.CCR-09-0209

Niwa-Kawakita M, Ferhi O, Soilihi H, Le Bras M, Lallemand-Breitenbach V, de Thé H. 2017. PML is a ROS sensor activating p53 upon oxidative stress. *J Exp Med* **214:** 3197–3206. doi:10.1084/jem.20160301

Occhionorelli M, Santoro F, Pallavicini I, Gruszka A, Moretti S, Bossi D, Viale A, Shing D, Ronzoni S, Muradore I, et al. 2011. The self-association coiled-coil domain of PML is sufficient for the oncogenic conversion of the retinoic acid receptor (RAR) α. *Leukemia* **25:** 814–820. doi:10.1038/leu .2011.18

Osumi T, Tsujimoto SI, Tamura M, Uchiyama M, Nakabayashi K, Okamura K, Yoshida M, Tomizawa D, Watanabe A, Takahashi H, et al. 2018. Recurrent *RARB* translocations in acute promyelocytic leukemia lacking *RARA* translocation. *Cancer Res* **78:** 4452–4458. doi:10.1158/0008-5472 .CAN-18-0840

Padua RA, Larghero J, Robin M, le Pogam C, Schlageter MH, Muszlak S, Fric J, West R, Rousselot P, Phan TH, et al. 2003. PML-RARA-targeted DNA vaccine induces protective immunity in a mouse model of leukemia. *Nat Med* **9:** 1413–1417. doi:10.1038/nm949

Patra U, Müller S. 2021. A tale of usurpation and subversion: SUMO-dependent integrity of promyelocytic leukemia nuclear bodies at the crossroad of infection and immunity. *Front Cell Dev Biol* **9:** 696234. doi:10.3389/fcell.2021.696234

Pearson M, Carbone R, Sebastiani C, Cioce M, Fagioli M, Saito S, Higashimoto Y, Appella E, Minucci S, Pandolfi PP, et al. 2000. PML regulates p53 acetylation and premature senescence induced by oncogenic Ras. *Nature* **406:** 207–210. doi:10.1038/35018127

Quignon F, de Bels F, Koken M, Feunteun J, Ameisen JC, de Thé H. 1998. PML induces a novel caspase-independent cell death process. *Nat Genet* **20:** 259–265. doi:10.1038/3068

Rego EM, Wang ZG, Peruzzi D, He LZ, Cordon-Cardo C, Pandolfi PP. 2001. Role of promyelocytic leukemia (PML) protein in tumor suppression. *J Exp Med* **193:** 521–530. doi:10.1084/jem.193.4.521

Rérolle D, de Thé H. 2023. The PML hub: an emerging actor of leukemia therapies. *J Exp Med* **220:** e20221213. doi:10.1084/jem.20221213

Robin M, Andreu-Gallien J, Schlageter MH, Bengoufa D, Guillemot I, Pokorna K, Robert C, Larghero J, Rousselot P, Raffoux E, et al. 2006. Frequent antibody production against RARα in both APL mice and patients. *Blood* **108:** 1972–1974. doi:10.1182/blood-2006-03-013177

Rowley JD, Golomb HM, Dougherty C. 1977. 15/17 translocation, a consistent chromosomal change in acute promyelocytic leukaemia. *Lancet* **309:** 549–550. doi:10.1016/S0140-6736(77)91415-5

Sahin U, Ferhi O, Jeanne M, Benhenda S, Berthier C, Jollivet F, Niwa-Kawakita M, Faklaris O, Setterblad N, de Thé H, et al. 2014. Oxidative stress-induced assembly of PML nuclear bodies controls sumoylation of partner proteins. *J Cell Biol* **204:** 931–945. doi:10.1083/jcb.201305148

Salomoni P, Pandolfi PP. 2002. The role of PML in tumor suppression. *Cell* **108:** 165–170. doi:10.1016/S0092-8674(02)00626-8

Scherer M, Stamminger T. 2016. Emerging role of PML nuclear bodies in innate immune signaling. *J Virol* **90:** 5850–5854. doi:10.1128/JVI.01979-15

Schönberger K, Obier N, Romero-Mulero MC, Cauchy P, Mess J, Pavlovich PV, Zhang YW, Mitterer M, Rettkowski J, Lalioti ME, et al. 2022. Multilayer omics analysis reveals a non-classical retinoic acid signaling axis that regulates hematopoietic stem cell identity. *Cell Stem Cell* **29:** 131–148. e10. doi:10.1016/j.stem.2021.10.002

Shima Y, Honma Y, Kitabayashi I. 2013. PML-RARα and its phosphorylation regulate pml oligomerization and HIPK2 stability. *Cancer Res* **73:** 4278–4288. doi:10.1158/0008-5472.CAN-12-3814

Stadler M, Chelbi-Alix MK, Koken MHM, Venturini L, Lee C, Saïb A, Quignon F, Pelicano L, Guillemin M-C, Schindler C, et al. 1995. Transcriptional induction of the PML growth suppressor gene by interferons is mediated through an ISRE and a GAS element. *Oncogene* **11:** 2565–2573.

Sternsdorf T, Phan VT, Maunakea ML, Ocampo CB, Sohal J, Silletto A, Galimi F, Le Beau MM, Evans RM, Kogan SC. 2006. Forced retinoic acid receptor α homodimers prime

mice for APL-like leukemia. *Cancer Cell* **9:** 81–94. doi:10.1016/j.ccr.2005.12.030

Tan J, Ong CK, Lim WK, Ng CC, Thike AA, Ng LM, Rajasegaran V, Myint SS, Nagarajan S, Thangaraju S, et al. 2015. Genomic landscapes of breast fibroepithelial tumors. *Nat Genet* **47:** 1341–1345. doi:10.1038/ng.3409

Tan Y, Wang X, Song H, Zhang Y, Zhang R, Li S, Jin W, Chen SJ, Fang H, Chen Z, et al. 2021. A PML/RARα direct target atlas redefines transcriptional deregulation in acute promyelocytic leukemia. *Blood* **137:** 1503–1516. doi:10.1182/blood.2020005698

Tessier S, Ferhi O, Geoffroy MC, González-Prieto R, Canat A, Quentin S, Pla M, Niwa-Kawakita M, Bercier P, Rérolle D, et al. 2022. Exploration of nuclear body-enhanced sumoylation reveals that PML represses 2-cell features of embryonic stem cells. *Nat Commun* **13:** 5726. doi:10.1038/s41467-022-33147-6

Tsai S, Collins SJ. 1993. A dominant negative retinoic acid receptor blocks neutrophil differentiation at the promyelocyte stage. *Proc Natl Acad Sci* **90:** 7153–7157. doi:10.1073/pnas.90.15.7153

Tsai S, Bartelmez S, Heyman R, Damm K, Evans R, Collins SJ. 1992. A mutated retinoic acid receptor-α exhibiting dominant-negative activity alters the lineage development of a multipotent hematopoietic cell line. *Genes Dev* **6:** 2258–2269. doi:10.1101/gad.6.12a.2258

Tsai JM, Aguirre JD, Li YD, Brown J, Focht V, Kater L, Kempf G, Sandoval B, Schmitt S, Rutter JC, et al. 2023. UBR5 forms ligand-dependent complexes on chromatin to regulate nuclear hormone receptor stability. *Mol Cell* **83:** 2753–2767.e10. doi:10.1016/j.molcel.2023.06.028

Tsimberidou AM, Tirado-Gomez M, Andreeff M, O'Brien S, Kantarjian H, Keating M, Lopez-Berestein G, Estey E. 2006. Single-agent liposomal all-*trans* retinoic acid can cure some patients with untreated acute promyelocytic leukemia: an update of the University of Texas M. D. Anderson Cancer Center Series. *Leuk Lymphoma* **47:** 1062–1068. doi:10.1080/10428190500463932

Vickers M, Jackson G, Taylor P. 2000. The incidence of acute promyelocytic leukemia appears constant over most of a human lifespan, implying only one rate limiting mutation. *Leukemia* **14:** 722–726. doi:10.1038/sj.leu.2401722

Vitaliano-Prunier A, Halftermeyer J, Ablain J, de Reynies A, Peres L, Le Bras M, Metzger D, de Thé H. 2014. Clearance of PML/RARA-bound promoters suffice to initiate APL differentiation. *Blood* **124:** 3772–3780. doi:10.1182/blood-2014-03-561852

Walkley CR, Yuan YD, Chandraratna RA, McArthur GA. 2002. Retinoic acid receptor antagonism in vivo expands the numbers of precursor cells during granulopoiesis. *Leukemia* **16:** 1763–1772. doi:10.1038/sj.leu.2402625

Wang F, Travins J, DeLaBarre B, Penard-Lacronique V, Schalm S, Hansen E, Straley K, Kernytsky A, Liu W, Gliser C, et al. 2013. Targeted inhibition of mutant IDH2 in leukemia cells induces cellular differentiation. *Science* **340:** 622–626. doi:10.1126/science.1234769

Warrell R, de Thé H, Wang Z, Degos L. 1993. Acute promyelocytic leukemia. *New Engl J Med* **329:** 177–189. doi:10.1056/NEJM199307153290307

Wartman LD, Larson DE, Xiang Z, Ding L, Chen K, Lin L, Cahan P, Klco JM, Welch JS, Li C, et al. 2011. Sequencing a mouse acute promyelocytic leukemia genome reveals ge-

Cite this article as *Cold Spring Harb Perspect Med* doi: 10.1101/cshperspect.a041582

netic events relevant for disease progression. *J Clin Invest* **121:** 1445–1455. doi:10.1172/JCI45284

Welch JS, Yuan W, Ley TJ. 2011. PML-RARA can increase hematopoietic self-renewal without causing a myeloproliferative disease in mice. *J Clin Invest* **121:** 1636–1645. doi:10.1172/JCI42953

Westervelt P, Pollock JL, Oldfather KM, Walter MJ, Ma MK, Williams A, DiPersio JF, Ley TJ. 2002. Adaptive immunity cooperates with liposomal all-*trans*-retinoic acid (ATRA) to facilitate long-term molecular remissions in mice with acute promyelocytic leukemia. *Proc Natl Acad Sci* **99:** 9468–9473. doi:10.1073/pnas.132657799

Wojiski S, Guibal FC, Kindler T, Lee BH, Jesneck JL, Fabian A, Tenen DG, Gilliland DG. 2009. PML–RARα initiates leukemia by conferring properties of self-renewal to committed promyelocytic progenitors. *Leukemia* **23:** 1462–1471. doi:10.1038/leu.2009.63

Wu D, Gao R. 2023. Acute myeloid leukemia with NUP98:: RARG resembling acute promyelocytic leukemia accompanying ARID1B gene mutation. *Hematology* **28:** 2227495. doi:10.1080/16078454.2023.2227495

Wu HC, Rérolle D, Berthier C, Hleihel R, Sakamoto T, Quentin S, Benhenda S, Morganti C, Wu CC, Conte L, et al. 2021. Actinomycin D targets NPM1c-primed mitochondria to restore PML-driven senescence in AML therapy. *Cancer Discov* **11:** 3198–3213. doi:10.1158/2159-8290 .CD-21-0177

Yanagitani A, Yamada S, Yasui S, Shimomura T, Murai R, Murawaki Y, Hashiguchi K, Kanbe T, Saeki T, Ichiba M, et al. 2004. Retinoic acid receptor α dominant negative form causes steatohepatitis and liver tumors in transgenic mice. *Hepatology* **40:** 366–375. doi:10.1002/hep.20335

Zhang XW, Yan XJ, Zhou ZR, Yang FF, Wu ZY, Sun HB, Liang WX, Song AX, Lallemand-Breitenbach V, Jeanne M, et al. 2010. Arsenic trioxide controls the fate of the PML-RARα oncoprotein by directly binding PML. *Science* **328:** 240–243. doi:10.1126/science.1183424

Zhao J, Liang JW, Xue HL, Shen SH, Chen J, Tang YJ, Yu LS, Liang HH, Gu LJ, Tang JY, et al. 2019. The genetics and clinical characteristics of children morphologically diagnosed as acute promyelocytic leukemia. *Leukemia* **33:** 1387–1399. doi:10.1038/s41375-018-0338-z

Zhu J, Koken MHM, Quignon F, Chelbi-Alix MK, Degos L, Wang ZY, Chen Z, de Thé H. 1997. Arsenic-induced PML targeting onto nuclear bodies: implications for the treatment of acute promyelocytic leukemia. *Proc Natl Acad Sci* **94:** 3978–3983. doi:10.1073/pnas.94.8.3978

Zhu J, Gianni M, Kopf E, Honoré N, Chelbi-Alix M, Koken M, Quignon F, Rochette-Egly C, de Thé H. 1999. Retinoic acid induces proteasome-dependent degradation of retinoic acid receptor α (RARα) and oncogenic RARα fusion proteins. *Proc Natl Acad Sci* **96:** 14807–14812. doi:10 .1073/pnas.96.26.14807

Zhu J, Chen Z, Lallemand-Breitenbach V, de Thé H. 2002. How acute promyelocytic leukaemia revived arsenic. *Nat Rev Cancer* **2:** 705–714. doi:10.1038/nrc887

Zhu J, Zhou J, Peres L, Riaucoux F, Honoré N, Kogan S, de Thé H. 2005. A sumoylation site in PML/RARA is essential for leukemic transformation. *Cancer Cell* **7:** 143–153. doi:10 .1016/j.ccr.2005.01.005

Zhu J, Nasr R, Pérès L, Riaucoux-Lormière F, Honoré N, Berthier C, Kamashev D, Zhou J, Vitoux D, Lavau C, et al. 2007. RXR is an essential component of the oncogenic PML/RARA complex in vivo. *Cancer Cell* **12:** 23–35. doi:10.1016/j.ccr.2007.06.004

Zhu HH, Qin YZ, Zhang ZL, Liu YJ, Wen LJ, You MJ, Zhang C, Such E, Luo H, Yuan HJ, et al. 2023. A global study for acute myeloid leukemia with *RARG* rearrangement. *Blood Adv* **7:** 2972–2982. doi:10.1182/bloodadvances .2022008364

Principles in the Development of Contemporary Treatment of Childhood Malignancies: The First 75 Years

Katie A. Greenzang and Stephen E. Sallan

Department of Pediatric Oncology, Dana-Farber Cancer Institute; Division of Pediatric Hematology/Oncology, Boston Children's Hospital; Harvard Medical School, Boston, Massachusetts 02115, USA

Correspondence: katie_greenzang@dfci.harvard.edu

Over the last 75 years, pediatric cancer has gone from nearly universally fatal, to having a >80% chance of long-term survival. Below we share highlights in this 75-year history, beginning with the "birth" of chemotherapy in treating childhood leukemia, through the development of multiagent chemotherapy, risk-stratified therapy, the use of molecular strategies in diagnosis and treatment, and adapting treatment to the needs of particularly vulnerable patient groups such as adolescents and young adults (AYAs). While pediatric leukemia treatment demonstrates the ever-improving cures achieved through iterative incorporation of novel discoveries, this experience is contrasted with that of osteosarcoma, where scientific advances made over recent decades have yet to be translated into meaningful improvements in long-term survival. We conclude with a brief overview of current areas of focus, including precision medicine, immunotherapy, and other treatment advancements, yet describe the need to couple these scientific breakthroughs with consideration of equitable access and evaluation of the long-term impacts of these "newer" therapies in survivorship. Substantial further work is needed to achieve our goal of curing all children with cancer as harmlessly as possible.

There have been tremendous gains in survival over the last 75 years of pediatric cancer treatment. In the 1940s, most childhood cancers were considered fatal at the time of diagnosis, and now >80% of children with cancer will go on to be long-term survivors (Siegel et al. 2020). This marked improvement in outcomes has been driven by the development and refinement of multiagent chemotherapy regimens, novel diagnostic and treatment approaches, and through iterative cooperative group trials and collaborations within and across specialties and areas of expertise.

Before the mid-twentieth century, the only curative treatment for childhood cancers was surgery. Yet, this approach was unsuccessful in treating the most common malignancies, namely, leukemias and brain tumors. Even apparently localized extracranial solid tumors, such as soft tissue and bone sarcomas, were rarely cured with surgery alone. Routine radiation therapy, both as an adjunct to surgery of localized and regionally spread solid tumors, or used as primary therapy for brain tumors and some lymphomas, began to flourish in the late 1960s. For example, during the 1970s, a high percentage of children with non-hematogenously disseminated childhood Hodgkin lymphoma, stages 1–3, were cured by radiation therapy. During that era, technological advancements, especially the introduction of megavoltage machines that replaced outdated orthovoltage equipment, enhanced therapeutic capacity and simultaneously diminished late-occurring toxicities. Nonetheless, the long-term sequelae of radiation eventually led to practice-changing approaches, namely, the routine use of chemotherapy for nearly all patients and the simultaneous reduction of radiation doses.

Since its "birth" in treating childhood leukemia in 1948 (Farber et al. 1948), chemotherapy, by definition a systemic modality, as compared with surgery or radiation, became and remained the mainstay for the treatment of childhood cancers. Principles pertaining to the use of cytotoxic chemotherapeutic drugs evolved rapidly in the 1960s, when a small number of drugs became the therapeutic workhorses across a multitude of diseases. As of 2023, the majority of children with a malignant disease receive some chemotherapy. Best exemplified by therapeutic advances in the treatment of acute lymphoblastic leukemia (ALL) (see below), the early and prevailing principles of chemotherapy include the routine use of combinations of drugs, especially those with minimally overlapping side effects and different mechanisms of action. While obvious in the treatment of leukemias and metastatic solid tumors, as well as most lymphomas, the use of chemotherapy in most "localized" solid tumors is predicated on the untreated "natural history" of the disease, namely, that local control with surgery and/or radiation therapy is not curative and thus unrecognized disseminated disease is present at or near the time that the tumor is discovered. Hence the development of the use of systemic chemotherapy in addition to local control measures such as surgery or radiation, through neoadjuvant chemotherapy, the use of chemotherapy before initiation of local controlling modalities, and postlocal control adjuvant chemotherapy. We use the evolution of treatment of two solid tumors, osteosarcoma and Wilms tumor, as prime examples (see below).

While a marked improvement in the cure of all types of childhood cancers (Fig. 1) best exemplifies the therapeutic value of combination chemotherapy, the acute and chronic toxicities of chemotherapy, of magnitudes varying from transient and self-healing to long-term and life-threatening, always accompany its use. Therefore, risk-stratified care to allow reduction of therapy for those diseases with excellent outcomes, or introduction of novel, hopefully less toxic, targeted agents, are recent approaches and areas of current research to continue to optimize cure while trying to avoid or minimize drug-associated toxicities.

ALL: FROM FATAL TO CURATIVE: SINGLE AGENT TO COMBINATION DRUG CHEMOTHERAPY

The seminal publication pertaining to the early treatment of childhood acute leukemia "In the beginning …" appeared June 3, 1948 in the *New England Journal of Medicine* (Farber et al. 1948). Using aminopterin, supplied by a biochemist at Lederle Laboratories, Yellapragada Subbarow, Dr. Sidney Farber and colleagues at Boston Children's Hospital reported transient antileukemia responses to this antifolate, thus ushering in the era of single-agent chemotherapy. Around that time, a methyl derivative of aminopterin, methotrexate, also discovered by Subbarow, was found to be less toxic than aminopterin. Since then, methotrexate has been used universally in the treatment of childhood ALL for the past 75 years.

Why an antifolate? In the decade preceding Farber's observation, folic acid and vitamin B12 each had been found to cure previously fatal blood disorders (macrocytic anemia of pregnancy and pernicious anemia) that, at least under the microscope, appeared similar to ALL. Farber had tried both of those drugs for children with leukemia without success. In fact, he reported that treatment with folic acid resulted in an "accelerated phase" of ALL (Farber et al. 1948). That observation, and his knowledge that Subbarow, a colleague, was developing antifolates, led to the eventual use of aminopterin.

Cite this article as *Cold Spring Harb Perspect Med* doi: 10.1101/cshperspect.a041634

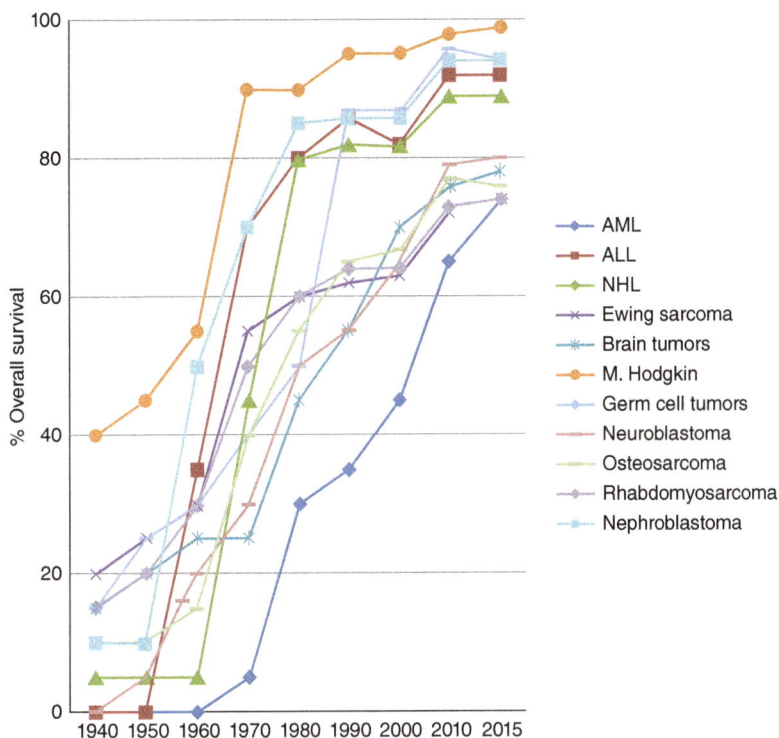

Figure 1. Seventy-five years of improvement in survival in pediatric malignancies (Burdach et al. 2018). Increase in survival rates in Germany. Two-year survival until 1980, 5-year survival from 1980 (Siegel et al. 2020). (This figure is reprinted, with permission, from Burdach et al. 2018 under the terms of the Creative Commons Attribution 4.0 International License, creativecommons.org/licenses/by/4.0.)

Unfortunately, the use of a single drug, even one as active as methotrexate, led to transient disease responses, but never cure. That reality was paramount in the thinking of two young internists, Drs. Emil Frei III and Emil J. Freireich, who from the mid-1950s into the mid-1960s worked together at the Clinical Center at the National Cancer Institute (NCI) Leukemia Service. This center, led by Dr. Gordon Zubrod, was charged with curing childhood ALL, a universally fatal disease. Because 95% of their patients suffered early deaths from hemorrhage and infection, their immediate first goals were to develop blood product support and antibiotic regimens to minimize infectious complications. Not until they had substantially overcome those early complications could they even begin to evaluate antileukemia therapies. Drs. Frei and Freireich, who were not only colleagues, but also became close personal friends, readily related that many voices both inside and outside of the National Institutes of Health (NIH) viewed the Clinical Center with skepticism, disparagement, disdain, and even accusations of unethical practice. They attributed the strength of support from both their leaders, like Dr. Zubrod, and the parents of their patients, as significant sources of encouragement to pursue what they recognized was achievable: curing this previously incurable disease.

During that era, available antileukemia agents included aminopterin, methotrexate, 6-mercaptopurine (6-MP), prednisone, and, somewhat later, vincristine. Frei and Freireich extrapolated data from mouse leukemia models developed by Howard Skipper and Frank Schabel Jr. at the Southern Research Institute in Birmingham, Alabama (Schabel et al. 1961) and sorted out the most effective drug combinations. The use of more than a single drug originated from experience with antibiotics and drug resistance for infectious diseases,

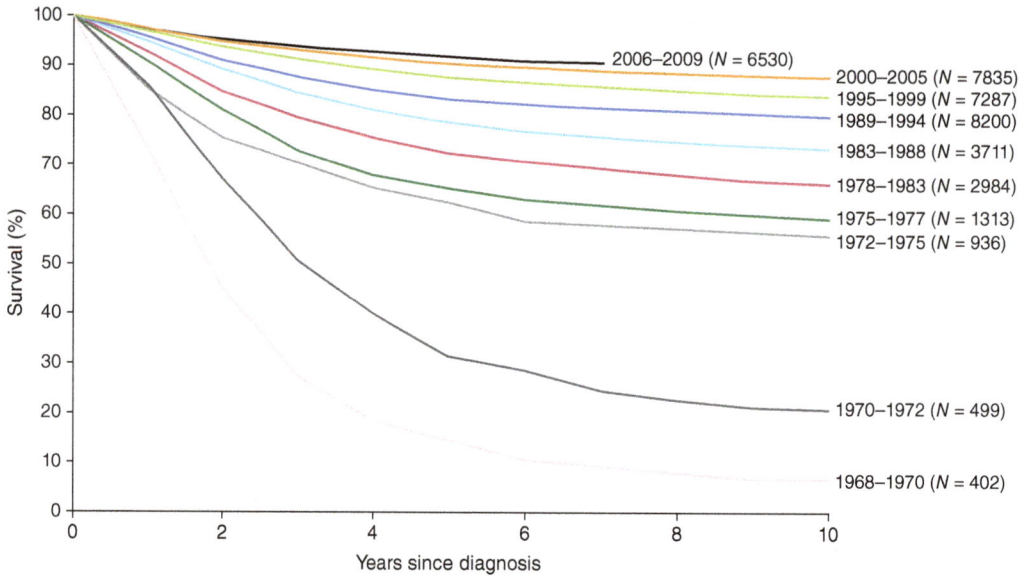

Figure 2. Fifty years of improvement in survival of childhood acute lymphoblastic leukemia (ALL). Overall survival among children with ALL who were enrolled in Children's Cancer Group and Children's Oncology Group Clinical Trials, 1968–2009. (This figure is reprinted, with permission, from Hunger and Mullighan 2015.)

specifically tuberculosis. Conceptionally, Frei and Freireich optimized individual drug doses and schedules that delivered maximal antileukemia doses and nonoverlapping toxicities, for example, adding nonmyelosuppressive combinations (vincristine and prednisone) with full doses of myelosuppressive drugs (methotrexate and 6-MP). Initially, they studied two-drug combinations, then three drugs, and with the arrival of vincristine, a four-drug combination: first VAMP (with aminopterin), then POMP (same agents, except with methotrexate replacing aminopterin). By 1964, they had witnessed their first cures and initiated multicenter clinical protocols, thus ushering in the era of combination chemotherapy (Frei et al. 1965). From the early 1960s until today, with rare exceptions, combination chemotherapy has remained essential for the successful treatment of multiple malignant diseases.

Highlights of 60 Years of Combination Chemotherapy in ALL

By the late 1960s, most ALL relapses were in the central nervous system (CNS), necessitating a fresh, more targeted approach toward the treatment of the CNS (Fig. 2). Although commonly, and incorrectly, called "prophylactic" CNS treatments, autopsy evidence from early deaths demonstrated extant CNS leukemia involvement was nearly universal (Price and Johnson 1973). Again, the infectious disease models, especially the use of intrathecal antibiotics used in the treatment of tuberculosis, led to early experience with intrathecal antileukemia drugs such as methotrexate, an intervention with known efficacy in the treatment of symptomatic CNS leukemia. Attempts to treat the CNS as a "local" phenomenon led to the use of cranial and craniospinal radiation, starting at low doses and doubling until control was achieved. CNS treatment early in the course of therapy became routine. These critical interventions markedly increased overall cure rates (Aur et al. 1971). However, these CNS-directed treatments, particularly radiation, had numerous severe short- and long-term consequences including vertebral marrow suppression, necessitating lowering of systemic chemotherapy doses, dose-related neurocognitive toxicity, and risk of second malignant neoplasms (Neglia et al. 1991). Ultimately, the use of intrathecal chemotherapy permitted the replacement of spinal radiation. Eventually, intrathecal therapy, with or without high doses of systemic methotrexate, proved highly successful, resulting in progressively less use of any radiation.

In the late 1960s, significant additions to the ALL armamentarium included the anthracyclines, doxorubicin, and daunorubicin, as well as the therapeutic enzyme L-asparaginase. Other active agents with more limited roles included cyclophosphamide, etoposide, and cytarabine.

Until the late 1970s, all children received identical therapy. As biomarkers that could predict the likelihood of treatment response were identified, investigators introduced the concept of "risk-stratified" therapy. Initially, treatments were based on patients' age and white blood cell (WBC) count at the time of diagnosis, criteria that became standard over the past 40 years (Miller et al. 1974). Around that same time, differences in treatment outcomes based on immunologic cell surface markers became apparent, again resulting in different "risk-based" treatments (Sallan et al. 1980). The earliest different treatment approaches were for those with T-cell ALL and "non-T-cell" ALL, the latter soon recognized as B-lineage ALL. Subsequent technological advances in the 1980s permitted treatment differences based upon chromosomal mutations, in the 1990s based on gene expression patterns, and eventually based upon measurement of minimal residual disease. By the early twenty-first century, the discovery of precisely targeted therapies, such as tyrosine kinase inhibitors (TKI) in Philadephia-positive ALL, and immunotherapies in relapsed ALL, such as CAR-T cells, bispecific T-cell engagers (e.g., blinatumomab), and antibody–drug conjugates (e.g., inotuzumab ozogamicin), opened entirely new therapeutic approaches.

Adolescent and Young Adult Acute Lymphoblastic Leukemia

The Child is the father of the Man.
                    —William Wordsworth

Around the turn of the twenty-first century, serendipity played a role in what led to the near-doubling of cures of ALL in the adolescent and young adult (AYA) population, particularly those 18–30 years of age. By coincidence, two University of Chicago leukemia investigators, Drs. Wendy Stock, an adult medical oncologist, and James Nachman, a pediatric oncologist, each became the Principal Investigator of national, cooperative group ALL trials with overlapping age eligibility criteria. As both trials enrolled patients 16–20 years of age with newly diagnosed ALL, this afforded an unanticipated opportunity to ascertain whether differences between the adult and pediatric treatment approaches resulted in outcome differences for this age group. They did! Event-free survival for the same AYA population treated on the pediatric or adult trial was 63% versus 34%, respectively (Stock et al. 2008). Interestingly, very similar pediatric versus adult protocol outcomes for AYA populations with overlapping ages were reported from trials around the world (Boissel et al. 2003; Muffly and Curran 2019). Subsequent prospective "pediatric inspired" protocols eventually became the standard of care for Philadelphia chromosome-negative AYA ALL and within a decade had doubled the likelihood of cure for that population (Stock et al. 2019). Similar to childhood ALL outcomes, differences in AYA patients included prognostic factors such as postinduction minimal residual disease and Philadelphia chromosome mutations. Explanations for the earlier differences between pediatric and adult approaches to ALL included differences in the biology of the leukemia (e.g., more Philadelphia chromosome-positive and Philadelphia-like and less ETV6/RUNX1 in AYA), more comorbidities in AYA, and perhaps some cultural differences between pediatric and adult oncologist approaches to the rigidity of chemotherapy scheduling and oversight of adherence (Schiffer 2003; Boissel and Baruchel 2018). One fundamental difference in the current approach to childhood, as opposed to AYA, ALL pertains to the role of stem cell transplantation in first remission in Philadelphia-positive ALL: A standard of care only for AYA patients.

These strides in outcomes for AYA patients with leukemia highlight a current area of focus across oncology—working to close the outcomes gap for AYAs with cancer by ensuring optimal access to clinical trials and tailoring treatment and supportive care for this particularly vulnerable population (Hudson and Bhatia 2024). While many other pediatric and AYA malignancies have shown improvements in outcomes that have paralleled the story of ALL, there are several malignancies that have shown a lack of improvement in outcomes in recent decades, and others where a gap remains between pediatric outcomes and those of AYAs (Keegan et al. 2024). As the incidence of cancer in AYAs is increasing for several malignancies (Bleyer 2023), continued focus on treatment approaches that incorporate the distinct needs and issues facing AYAs is paramount.

TREATMENT PROGRESS IN THE CARE OF CHILDREN WITH SOLID TUMORS

Principles of pediatric cancer treatment that began in leukemia as described above, specifically the use of multiagent chemotherapy in combination with local treatment approaches of radiation or surgery, risk-stratified therapy, and identification of novel agents, all contributed to the development of current treatment approaches in pediatric solid tumors, although with some key distinctions. Here, we note two representative tumors, Wilms tumor and osteosarcoma, which highlight both the marked improvement in outcomes for some solid tumors, and the ongoing challenges and as yet unanswered questions.

Wilms Tumor

Although originally described in the late nineteenth century by a surgeon, Max Wilms, during his pathology training (Wilms 1899; Raffensperger 2015), the history of the treatment of Wilms tumor had its origins in the 1930s and 1940s in large part by Boston Children's Hospital surgeons Drs. William Edwards Ladd and Robert Gross (Ladd 1938; Randolph 1998). Contrary to the classic surgical approach to the kidney by a flank incision, Ladd initiated an anterior abdominal approach that mitigated the risk of tumor rupture, and, in an era of limited imaging, allowed palpation for involved ipsilateral lymph nodes and permitted the opening of Gerota's fascia and exploration of the contralateral kidney (Ladd 1938). Surgery alone was curative in only ~30% of children (Randolph 1998). Gross, in conjunction with a radiologist, Dr. Edwin Neu-

hauser, combined two local modalities (surgery and radiation therapy) and by 1950 had improved cure rates to 47% (Randolph 1998). In the mid-1950s, Sidney Farber introduced the routine use of a third therapeutic modality, chemotherapy (initially with actinomycin D), increasing cure rates to ~65% and opening the modern era of trimodal therapy for Wilms tumor (Farber 1966).

By the late 1960s, both vincristine and actinomycin were recognized as equally effective drugs for this disease. At that time the National Wilms Tumor Study group was formed to conduct large-scale clinical trials, the first of which compared single-agent chemotherapy, with either actinomycin D or vincristine, to combination chemotherapy with both drugs (Wolff 1975). Analogous to prior experience with ALL, combination chemotherapy demonstrated superior outcomes, with cure rates eventually >90% in subsequent clinical trials, enabled by increasingly risk-stratified treatment, sometimes adding other agents, and through adaptation of therapy based on initial treatment response (Dome et al. 2015). Importantly, biological differences and the extent of involvement of the tumors, as well as strategies to minimize therapy-related toxicities, all became critical factors in optimizing therapy for individual children.

Osteosarcoma

The story of osteosarcoma therapy in some ways begins like that of ALL, but diverges in several important aspects. Osteosarcoma, the most common bone tumor in children and adolescents, was first described in the early 1800s. Before the 1970s, surgical resection through amputation was the mainstay of therapy. Yet, despite this aggressive local tumor control, more than 50% of patients experienced distant metastatic disease within 6 months of surgery, and >80% of patients experienced relapse, suggesting the presence of micrometastatic disease at diagnosis in the vast majority of patients (Link et al. 1986; Isakoff et al. 2015). The definitive randomized control trial comparing surgery alone to surgery followed by multiagent chemotherapy was published in the *New England Journal of Medicine* in 1986 and

Cite this article as *Cold Spring Harb Perspect Med* doi: 10.1101/cshperspect.a041634

found 17% 2-year relapse-free survival in the surgery-only control group, as compared with 66% in the group that received adjuvant chemotherapy (Link et al. 1986). Concurrently, the idea of neoadjuvant chemotherapy (or induction therapy) delivered before local control surgery, in addition to adjuvant chemotherapy (or consolidation therapy) after surgery was being developed in part to support surgical advances, specifically limb salvage, by allowing more time preoperatively for the development of more advanced prostheses. The theoretic advantages to neoadjuvant chemotherapy included immediate treatment of micrometastatic disease, increased opportunity for limb-sparing surgery after some tumor shrinkage, and the ability to evaluate response to therapy to guide future chemotherapeutics and determine those at greatest risk of suboptimal outcomes (Rosen et al. 1982). As the Pediatric Oncology Group study comparing neoadjuvant versus adjuvant chemotherapy found similar results between groups (Goorin et al. 2003), preoperative chemotherapy followed by surgery and then further adjuvant chemotherapy became the standard of care given the advantages described above.

While these initial advances in the use of multiagent chemotherapy and surgery markedly improved survival, further attempts to improve outcomes have been unsuccessful, and treatment has largely remained the same over recent decades (Isakoff et al. 2015). The most utilized three-drug regimen for both induction and consolidation therapy consists of MAP, or methotrexate, doxorubicin (adriamycin), and cisplatin. The traditional approach of "more is better" has not proven true in osteosarcoma, where the expansion to five chemotherapeutic agents in place of the traditional three has only added toxicities without increasing survival (Gill and Gorlick 2021). Despite tremendous progress in understanding the biology of osteosarcoma, many of these discoveries have yet to be translated into therapeutic advances that improve outcomes.

Local control of osteosarcoma has evolved tremendously over the decades, to simultaneously optimize complete surgical resection with clean margins, in addition to improving functional and quality-of-life outcomes. Whereas amputation was the primary approach historically, many current patients are candidates for limb preservation or for rotationplasty or tibial turnabout, where an uninvolved section of the limb is used in the reconstruction to preserve joint function and growth, particularly in growing children (Harrison et al. 2018). Surgery remains a mainstay of treatment for metastatic disease as well, particularly for pulmonary metastatic disease.

CURRENT AND FUTURE DIRECTIONS: FROM EMPIRICISM TO PRECISION MEDICINE AND IMMUNOTHERAPIES

The first 50 years of the history of the field as exemplified by the three clinical examples above culminated in the end of the era of enlightened empiricism. Over the most recent 25 years, few new cytotoxic chemotherapeutic drugs have impacted outcomes for children with cancer. Instead, advancements have been driven by technology and biology, from improved understanding of the specific molecular drivers of disease and development of associated targeted treatments, to molecular measures of the extent of disease (e.g., circulating tumor DNA) (Shulman et al. 2018; Stankunaite et al. 2022; Kjær et al. 2023), as well as molecular measures of treatment efficacy (e.g., minimal residual disease) to help further align the amount of treatment required for individual patients.

Specific, molecularly targeted drugs, so-called precision medicine, heralded the twenty-first century of childhood cancer therapy. The poster child for the field was the TKI, imatinib. Its addition to chemotherapy for Philadelphia-positive ALL significantly improved survival for affected children (Schultz et al. 2014), obviated the need for stem cell transplantation for those with CML (Ata et al. 2023), and played an essential role in some solid tumors as well (e.g., gastrointestinal stromal tumor) (Andrzejewska et al. 2022). Within the first decade of TKI use, second, third, and even fourth generations of the drugs became available, often manifesting better outcomes than their predecessors.

Immunotherapies, another game changer of the twenty-first century, began making their mark early in the 2010s when the first patient with relapsed ALL was successfully treated with chimeric antigen receptor T cells (CAR-T cells) (Grupp et al. 2013). Subsequent trials proved the efficacy of CAR-T-cell therapy, mostly in adult hematologic malignancies such as lymphoma, and multiple myeloma, and future directions of research include the adoption of CAR-T-cell therapies for solid tumors (Zappa et al. 2023). Around the same time, bispecific T-cell engager drugs, specifically blinatumomab, demonstrated its clinical efficacy in relapsed B-cell ALL (Gore et al. 2018), and within a decade the drug had begun to find a place for some children with newly diagnosed ALL (van der Sluis et al. 2023). A third type of immunotherapy, so-called antibody–drug conjugates (e.g., inotuzumab) also proved efficacious in the treatment of relapsed ALL (Stokke and Bhojwani 2021). Collectively, these three immunotherapies have changed the face of treatment in ALL and are likely to soon have much wider applicability. Whereas therapeutic monoclonal antibodies have found widespread use in many adult malignancies, their use in children has been more limited. That said, rituximab is routinely used in non-Hodgkin lymphomas, dinutuximab in neuroblastoma, and emapalumab in hemophagocytic lymphohistiocytosis (Larrosa et al. 2023).

In addition to molecular diagnosis and therapeutics, technological advances have also advanced our ability to image and treat many diseases, and allowed more precisely targeted radiation treatments. For example, intensity-modulated radiotherapy (IMRT), clearly facilitated by continuously improving imaging, including the routine use of CT scans and MRI, delivers more precise dosing to target areas while sparing normal tissue damage. In the early twenty-first century, PET–CT further enhanced target delineation and for many diseases, like Hodgkin lymphoma, also became the principal imaging modality to evaluate early therapeutic responses. For other diseases, especially brain tumors, both proton beam radiation and stereotactic radiosurgery further enhanced target location and resulted in less radiation toxicity for many children. Similarly, metaiodobenzylguanidine (MIBG), long used as an imaging modality in neuroblastoma when labeled with iodine-131 (131I-MIBG), is

efficacious as a more targeted treatment modality in relapsed and high-risk disease (Weiss et al. 2021). These advances have further contributed to optimal risk-stratified treatment, and attempts to decrease the morbidity and late effects of toxic therapies.

Systematic evaluation of long-term survivors of childhood malignancies demonstrates that virtually every child experiences some toxicity of treatment. For some children, the toxicities, be they mild or severe, are transient. For many others, toxicities are life-long, often compromising health, well-being, and quality of life, and in too many individuals, toxicities may be fatal (Oeffinger et al. 2006; Hudson et al. 2013; Bhakta et al. 2017). Increased knowledge of the long-term impacts of treatment, such as the risk of anthracycline-related cardiotoxicity, or alkylator-associated infertility, has led to important strides in the prevention and screening for these toxicities through the use of cardioprotectants such as dexrazoxane (Lipshultz et al. 2004), early oncofertility counseling leading to increased rates of sperm and oocyte banking (Ligon et al. 2023), and more recently, studies to identify those at greatest genetic risk of late effects (Clemens et al. 2018). Additionally, the development of specialized survivorship care centers that provide multidisciplinary and risk-based screening and care specific to the needs of childhood cancer survivors has contributed greatly to care, outcomes, and research. Thus, the goal of all contemporary treatment regimens includes balancing the maximal likelihood of cure and minimal likelihood of harm. While there is hope that more targeted novel therapies will have fewer "off-target" short- and long-term toxicities, the toxicities and late effects of many novel therapies remain to be seen and warrant further study.

Recent research has also elucidated the critical role of social determinants of health—nonmedical factors that influence health outcomes such as income, education, access to care, discrimination, and food insecurity—and pediatric cancer outcomes. Despite highly protocolized care typically delivered in subspecialty pediatric cancer centers, multiple studies have found that children experiencing poverty or resource limitations, and those from minoritized backgrounds, experience worse cancer outcomes (Bona et al. 2021; Tran et al. 2022; Gupta et al. 2023). Recognition of these outcome disparities is only a first step. Current and future research seeks to understand the biological and social underpinnings of these disparities, to ensure equitable access to care, clinical trials, and therapeutic discoveries, and to develop health equity interventions all with a goal of eliminating outcome disparities (Winestone et al. 2023).

Clearly, there is substantial work to be done to advance knowledge and cures while decreasing toxicities and improving short- and long-term quality of life for all children with cancer. Areas of future research include optimizing the duration of therapy for each child to minimize the length of treatment for those who can be cured in less time. Similarly, determining when therapy can be safely deintensified for those cancers with excellent rates of cure must also be pursued. Presumably, some children can be cured with shorter than conventional courses of treatment, or less intensive therapy, but who? And how does one individualize treatment duration or intensity? For example, we cure most children with acute promyelocytic leukemia after only 4–6 months of chemotherapy, some children with acute myelogenous leukemia with similar durations of chemotherapy, yet nearly all children with ALL receive at least 2 years of treatment. The optimal duration or intensity of therapy questions can be ethically and emotionally challenging for providers and parents alike when trying to balance the risk of relapse against the risk of further harm. Continued work developing novel therapeutic approaches and targeted therapies is clearly crucial and can lead to opportunities to deintensify therapy. Additional areas of future discovery include ensuring that therapeutic advances are equitably available and applied to all patients including historically marginalized and minoritized patients, AYAs, and those in rural or resource-limited locations. This will require ongoing national and international cooperative group collaboration and research, accelerating the process from discovery to approval of novel agents, particularly for pediatric usage, as well as patient and family partnerships. Throughout all of this, we must continue to study and improve patient experiences of treatment and survivorship to discover the long-term sequelae of novel therapies and to ensure our scientific advances align with patient priorities and needs and improve quality of life in addition to survival.

While the key discoveries and lessons learned from the first 75 years of pediatric cancer treatment outlined above have dramatically improved rates of cure and quality of cure for pediatric cancers, substantial further work is needed to achieve our goal of curing all children with cancer. This history sets the table for the remainder of this collection, the contents of which will provide the ingredients, recipes, and creativity for the next 25 years. Ultimately, we hope these discoveries will cure all children with cancer and do so as harmlessly as possible.

REFERENCES

Andrzejewska M, Czarny J, Derwich K. 2022. Latest advances in the management of pediatric gastrointestinal stromal tumors. *Cancers (Basel)* **14:** 4989. doi:10.3390/cancers 14204989

Ata F, Benkhadra M, Ghasoub R, Fernyhough LJ, Omar NE, Nashwan AJ, Aldapt MB, Mushtaq K, Kassem NA, Yassin MA. 2023. Tyrosine kinase inhibitors in pediatric chronic myeloid leukemia: a focused review of clinical trials. *Front Oncol* **13:** 1285346. doi:10.3389/fonc.2023.1285346

Aur RJ, Simone J, Hustu HO, Walters T, Borella L, Pratt C, Pinkel D. 1971. Central nervous system therapy and combination chemotherapy of childhood lymphocytic leukemia. *Blood* **37:** 272–281. doi:10.1182/blood.v37.3.272.272

Bhakta N, Liu Q, Ness KK, Baassiri M, Eissa H, Yeo F, Chemaitilly W, Ehrhardt MJ, Bass J, Bishop MW, et al. 2017. The cumulative burden of surviving childhood cancer: an initial report from the St Jude Lifetime Cohort Study (SJLIFE). *Lancet* **390:** 2569–2582. doi:10.1016/s0140-6736(17)31610-0

Bleyer A. 2023. Increasing cancer in adolescents and young adults: cancer types and causation implications. *J Adolesc Young Adult Oncol* **12:** 285–296. doi:10.1089/jayao.2022.0134

Boissel N, Baruchel A. 2018. Acute lymphoblastic leukemia in adolescents and young adults: treat as adults or as children? *Blood* **132:** 351–361. doi:10.1182/blood-2018-02-778530

Boissel N, Auclerc MF, Lhéritier V, Perel Y, Thomas X, Leblanc T, Rousselot P, Cayuela JM, Gabert J, Fegueux N, et al. 2003. Should adolescents with acute lymphoblastic leukemia be treated as old children or young adults? Comparison of the French FRALLE-93 and LALA-94 trials. *J Clin Oncol* **21:** 774–780. doi:10.1200/jco.2003.02.053

Bona K, Li Y, Winestone LE, Getz KD, Huang YS, Fisher BT, Desai AV, Richardson T, Hall M, Naranjo A, et al. 2021. Poverty and targeted immunotherapy: survival in Children's Oncology Group Clinical Trials for high-risk neuroblastoma. *J Natl Cancer Inst* **113**: 282–291. doi:10.1093/jnci/djaa107

Burdach SEG, Westhoff MA, Steinhauser MF, Debatin KM. 2018. Precision medicine in pediatric oncology. *Mol Cell Pediatr* **5**: 6. doi:10.1186/s40348-018-0084-3

Clemens E, van der Kooi ALF, Broer L, van Dulmen-den Broeder E, Visscher H, Kremer L, Tissing W, Loonen J, Ronckers CM, Pluijm SMF, et al. 2018. The influence of genetic variation on late toxicities in childhood cancer survivors: a review. *Crit Rev Oncol Hematol* **126**: 154–167. doi:10.1016/j.critrevonc.2018.04.001

Dome JS, Graf N, Geller JI, Fernandez CV, Mullen EA, Spreafico F, Van den Heuvel-Eibrink M, Pritchard-Jones K. 2015. Advances in Wilms tumor treatment and biology: progress through international collaboration. *J Clin Oncol* **33**: 2999–3007. doi:10.1200/jco.2015.62.1888

Farber S. 1966. Chemotherapy in the treatment of leukemia and Wilms' tumor. *J Am Med Assoc* **198**: 826–836.

Farber S, Diamond LK, Mercer RD, Sylvester RF, Wolff JA. 1948. Temporary remissions in acute leukemia in children produced by folic acid antagonist, 4-aminopteroyl-glutamic acid (aminopterin). *N Engl J Med* **238**: 787–793. doi:10.1056/nejm194806032382301

Frei E III, Karon M, Levin RH, Freireich EJ, Taylor RJ, Hananian J, Selawry O, Holland JF, Hoogstraten B, Wolman IJ, et al. 1965. The effectiveness of combinations of antileukemic agents in inducing and maintaining remission in children with acute leukemia. *Blood* **26**: 642–656.

Gill J, Gorlick R. 2021. Advancing therapy for osteosarcoma. *Nat Rev Clin Oncol* **18**: 609–624. doi:10.1038/s41571-021-00519-8

Goorin AM, Schwartzentruber DJ, Devidas M, Gebhardt MC, Ayala AG, Harris MB, Helman LJ, Grier HE, Link MP, Pediatric Oncology Group. 2003. Presurgical chemotherapy compared with immediate surgery and adjuvant chemotherapy for nonmetastatic osteosarcoma: Pediatric Oncology Group Study POG-8651. *J Clin Oncol* **21**: 1574–1580. doi:10.1200/jco.2003.08.165

Gore L, Locatelli F, Zugmaier G, Handgretinger R, O'Brien MM, Bader P, Bhojwani D, Schlegel PG, Tuglus CA, von Stackelberg A. 2018. Survival after blinatumomab treatment in pediatric patients with relapsed/refractory B-cell precursor acute lymphoblastic leukemia. *Blood Cancer J* **8**: 80. doi:10.1038/s41408-018-0117-0

Grupp SA, Kalos M, Barrett D, Aplenc R, Porter D, Rheingold SR, Teachey DT, Chew A, Hauck B, Wright F, et al. 2013. Chimeric antigen receptor-modified T cells for acute lymphoid leukemia. *N Engl J Med* **368**: 1509–1518. doi:10.1056/nejmoa1215134

Gupta S, Dai Y, Chen Z, Winestone LE, Teachey DT, Bona K, Aplenc R, Rabin KR, Zweidler-McKay P, Carroll AJ, et al. 2023. Racial and ethnic disparities in childhood and young adult acute lymphocytic leukaemia: secondary analyses of eight Children's Oncology Group cohort trials. *Lancet Haematol* **10**: e129–e141. doi:10.1016/s2352-3026(22)00371-4

Harrison DJ, Geller DS, Gill JD, Lewis VO, Gorlick R. 2018. Current and future therapeutic approaches for osteosarcoma. *Expert Rev Anticancer Ther* **18**: 39–50. doi:10.1080/14737140.2018.1413939

Hudson MM, Bhatia S. 2024. Mind the gap: a multiprong approach to minimizing the gap in outcomes among ado-lescents and young adults with cancer. *J Clin Oncol* **42**: 617–620. doi:10.1200/jco.23.02240

Hudson MM, Ness KK, Gurney JG, Mulrooney DA, Chemaitilly W, Krull KR, Green DM, Armstrong GT, Nottage KA, Jones KE, et al. 2013. Clinical ascertainment of health outcomes among adults treated for childhood cancer. *J Am Med Assoc* **309**: 2371–2381. doi:10.1001/jama.2013.6296

Hunger SP, Mullighan CG. 2015. Acute lymphoblastic leukemia in children. *N Engl J Med* **373**: 1541–1552. doi:10.1056/NEJMra1400972

Isakoff MS, Bielack SS, Meltzer P, Gorlick R. 2015. Osteosarcoma: current treatment and a collaborative pathway to success. *J Clin Oncol* **33**: 3029–3035. doi:10.1200/jco.2014.59.4895

Keegan THM, Abrahão R, Alvarez EM. 2024. Survival trends among adolescents and young adults diagnosed with cancer in the United States: comparisons with children and older adults. *J Clin Oncol* **42**: 630–641. doi:10.1200/jco.23.01367

Kjær EKR, Vase CB, Rossing M, Ahlborn LB, Hjalgrim LL. 2023. Detection of circulating tumor-derived material in peripheral blood of pediatric sarcoma patients: a systematic review. *Transl Oncol* **34**: 101690. doi:10.1016/j.tranon.2023.101690

Ladd WE. 1938. Embryoma of the kidney (Wilms' tumor). *Ann Surg* **108**: 885–902. doi:10.1097/00000658-193811000-00010

Larrosa C, Mora J, Cheung NK. 2023. Global impact of monoclonal antibodies (mAbs) in children: a focus on anti-GD2. *Cancers (Basel)* **15**: 3729. doi:10.3390/cancers15143729

Ligon JA, Hayashi M, Ciampa D, Kramer C, Guastella A, Fuchs RJ, Herati AS, Christianson MS, Chen AR. 2023. A multidisciplinary pediatric oncofertility team improves fertility preservation and counseling across 7 years. *Cancer Rep* **6**: e1753. doi:10.1002/cnr2.1753

Link MP, Goorin AM, Miser AW, Green AA, Pratt CB, Belasco JB, Pritchard J, Malpas JS, Baker AR, Kirkpatrick JA, et al. 1986. The effect of adjuvant chemotherapy on relapse-free survival in patients with osteosarcoma of the extremity. *N Engl J Med* **314**: 1600–1606. doi:10.1056/nejm198606193142502

Lipshultz SE, Rifai N, Dalton VM, Levy DE, Silverman LB, Lipsitz SR, Colan SD, Asselin BL, Barr RD, Clavell LA, et al. 2004. The effect of dexrazoxane on myocardial injury in doxorubicin-treated children with acute lymphoblastic leukemia. *N Engl J Med* **351**: 145–153. doi:10.1056/nejmoa035153

Miller DR, Sonley M, Karon M, Breslow N, Hammond D. 1974. Additive therapy in the maintenance of remission in acute lymphoblastic leukemia of childhood: the effect of the initial leukocyte count. *Cancer* **34**: 508–517. doi:10.1002/1097-0142(197409)34:3<508::aid-cncr2820340306>3.0.co;2-t

Muffly L, Curran E. 2019. Pediatric-inspired protocols in adult acute lymphoblastic leukemia: are the results bearing fruit? *Hematology* **2019**: 17–23. doi:10.1182/hematology.2019000009

Neglia JP, Meadows AT, Robison LL, Kim TH, Newton WA, Ruymann FB, Sather HN, Hammond GD. 1991. Second neoplasms after acute lymphoblastic leukemia in childhood. *N Engl J Med* **325**: 1330–1336. doi:10.1056/nejm199111073251902

Oeffinger KC, Mertens AC, Sklar CA, Kawashima T, Hudson MM, Meadows AT, Friedman DL, Marina N, Hobbie W, Kadan-Lottick NS, et al. 2006. Chronic health conditions in adult survivors of childhood cancer. *N Engl J Med* **355:** 1572–1582. doi:10.1056/nejmsa060185

Price RA, Johnson WW. 1973. The central nervous system in childhood leukemia. I: The arachnoid. *Cancer* **31:** 520–533. doi:10.1002/1097-0142(197303)31:3<520::aid-cncr2820310306>3.0.co;2-2

Raffensperger J. 2015. Max Wilms and his tumor. *J Pediatr Surg* **50:** 356–359. doi:10.1016/j.jpedsurg.2014.10.054

Randolph CJ. 1998. Treatment of mixed tumors of the kidney in childhood, by Robert E. Gross, MD, and Edward B.D. Neuhauser, MD, Pediatrics, 1950;6:843–852. *Pediatrics* **102:** 209–210.

Rosen G, Caparros B, Huvos AG, Kosloff C, Nirenberg A, Cacavio A, Marcove RC, Lane JM, Mehta B, Urban C. 1982. Preoperative chemotherapy for osteogenic sarcoma: selection of postoperative adjuvant chemotherapy based on the response of the primary tumor to preoperative chemotherapy. *Cancer* **49:** 1221–1230. doi:10.1002/1097-0142(19820315)49:6<1221::aid-cncr2820490625>3.0.co;2-e

Sallan SE, Ritz J, Pesando J, Gelber R, O'Brien C, Hitchcock S, Coral F, Schlossman SF. 1980. Cell surface antigens: prognostic implications in childhood acute lymphoblastic leukemia. *Blood* **55:** 395–402.

Schabel FM, Montgomery JA, Skipper HE, Laster WR, Thomson JR. 1961. Experimental evaluation of potential anticancer agents. I: Quantitative therapeutic evaluation of certain purine analogs. *Cancer Res* **21:** 690–699.

Schiffer CA. 2003. Differences in outcome in adolescents with acute lymphoblastic leukemia: a consequence of better regimens? Better doctors? Both? *J Clin Oncol* **21:** 760–761. doi:10.1200/jco.2003.11.116

Schultz KR, Carroll A, Heerema NA, Bowman WP, Aledo A, Slayton WB, Sather H, Devidas M, Zheng HW, Davies SM, et al. 2014. Long-term follow-up of imatinib in pediatric Philadelphia chromosome-positive acute lymphoblastic leukemia: Children's Oncology Group Study AALL0031. *Leukemia* **28:** 1467–1471. doi:10.1038/leu.2014.30

Shulman DS, Klega K, Imamovic-Tuco A, Clapp A, Nag A, Thorner AR, Van Allen E, Ha G, Lessnick SL, Gorlick R, et al. 2018. Detection of circulating tumour DNA is associated with inferior outcomes in Ewing sarcoma and osteosarcoma: a report from the Children's Oncology Group. *Br J Cancer* **119:** 615–621. doi:10.1038/s41416-018-0212-9

Siegel DA, Richardson LC, Henley SJ, Wilson RJ, Dowling NF, Weir HK, Tai EW, Buchanan Lunsford N. 2020. Pediatric cancer mortality and survival in the United States, 2001-2016. *Cancer* **126:** 4379–4389. doi:10.1002/cncr.33080

Stankunaite R, George SL, Gallagher L, Jamal S, Shaikh R, Yuan L, Hughes D, Proszek PZ, Carter P, Pietka G, et al. 2022. Circulating tumour DNA sequencing to determine therapeutic response and identify tumour heterogeneity in patients with paediatric solid tumours. *Eur J Cancer* **162:** 209–220. doi:10.1016/j.ejca.2021.09.042

Stock W, La M, Sanford B, Bloomfield CD, Vardiman JW, Gaynon P, Larson RA, Nachman J, Children's Cancer Group, Cancer and Leukemia Group B studies. 2008. What determines the outcomes for adolescents and young adults with acute lymphoblastic leukemia treated on cooperative group protocols? A comparison of Children's Cancer Group and Cancer and Leukemia Group B studies. *Blood* **112:** 1646–1654. doi:10.1182/blood-2008-01-130237

Stock W, Luger SM, Advani AS, Yin J, Harvey RC, Mullighan CG, Willman CL, Fulton N, Laumann KM, Malnassy G, et al. 2019. A pediatric regimen for older adolescents and young adults with acute lymphoblastic leukemia: results of CALGB 10403. *Blood* **133:** 1548–1559. doi:10.1182/blood-2018-10-881961

Stokke JL, Bhojwani D. 2021. Antibody–drug conjugates for the treatment of acute pediatric leukemia. *J Clin Med* **10:** 3556. doi:10.3390/jcm10163556

Tran YH, Coven SL, Park S, Mendonca EA. 2022. Social determinants of health and pediatric cancer survival: a systematic review. *Pediatr Blood Cancer* **69:** e29546. doi:10.1002/pbc.29546

van der Sluis IM, de Lorenzo P, Kotecha RS, Attarbaschi A, Escherich G, Nysom K, Stary J, Ferster A, Brethon B, Locatelli F, et al. 2023. Blinatumomab added to chemotherapy in infant lymphoblastic leukemia. *N Engl J Med* **388:** 1572–1581. doi:10.1056/nejmoa2214171

Weiss BD, Yanik G, Naranjo A, Zhang FF, Fitzgerald W, Shulkin BL, Parisi MT, Russell H, Grupp S, Pater L, et al. 2021. A safety and feasibility trial of ^{131}I-MIBG in newly diagnosed high-risk neuroblastoma: a Children's Oncology Group study. *Pediatr Blood Cancer* **68:** e29117. doi:10.1002/pbc.29117

Wilms M. 1899. *Die Mischgeschwülste der Niere*. A. Georgi, Leipzig, Germany.

Winestone LE, Beauchemin MP, Bona K, Kahn J, Prasad P, Robles JM, Velez MC, Diversity and Health Disparities Committee. 2023. Children's Oncology Group's 2023 blueprint for research: diversity and health disparities. *Pediatr Blood Cancer* **70:** e30592. doi:10.1002/pbc.30592

Wolff JA. 1975. Advances in the treatment of Wilms' tumor. *Cancer* **35:** 901–904. doi:10.1002/1097-0142(197503)35:3<901::aid-cncr2820350707>3.0.co;2-l

Zappa E, Vitali A, Anders K, Molenaar JJ, Wienke J, Künkele A. 2023. Adoptive cell therapy in paediatric extracranial solid tumours: current approaches and future challenges. *Eur J Cancer* **194:** 113347. doi:10.1016/j.ejca.2023.113347

Cancer Therapies Targeting Cellular Metabolism

Benjamin Morris[1] and Alejandro Gutierrez[2]

[1]Biomedical and Biological Sciences Program, Harvard Medical School, Boston, Massachusetts 02115, USA
[2]Division of Molecular Oncology, St. Jude Children's Research Hospital, Memphis, Tennessee 38105, USA

Correspondence: alejandro.gutierrez@stjude.org

Cancer is caused by mutations that drive aberrant growth, proliferation, and invasion, thus overriding regulatory mechanisms that normally link these processes to organismal needs and cellular physiology. This imposes demands for the production of energy and biomass and for survival in microenvironments that are often nonphysiologic and nutrient-poor, which are met by rewiring of cellular metabolism. The resultant dependence of tumor cells on altered metabolism can induce sensitivity to specific metabolic perturbations that can be exploited for cancer therapy. Some cancers are caused by mutations that impart a novel function to metabolic enzymes, leading to the production of a tumor-promoting metabolite that is dispensable in normal cells, representing an ideal therapeutic target. Tumors can also exploit metabolic regulation of cellular immunity to evade antitumor immune responses, and deciphering this biology has revealed potential targets for therapeutic intervention. Here, we discuss a number of illustrative examples highlighting the therapeutic potential and the challenges of targeting metabolism for cancer therapy.

More than 100 years ago, Otto Warburg made the observation that tumor cells have a high rate of glycolysis relative to respiration, even in the presence of oxygen (aerobic glycolysis). In so doing, Warburg described cancer as a disease of altered cellular metabolism for the first time (Warburg 1925). The early interpretation that this reflected impaired respiration has been supplanted by the contemporary view that most cancers have high rates of glycolysis with normal respiration, a strategy that provides glycolytic intermediates to fuel biomass production (Pavlova et al. 2022). Nevertheless, the core tenets of the initial observation remain true—altered cellular metabolism is a defining feature of cancer.

Metabolic alterations in tumor cells represent attractive targets for cancer therapy for several reasons. Metabolic flux in cancer cells is typically quite different from that in normal cells, and this altered flux is often essential for the tumor cell's survival. However, metabolic pathways also have important roles in normal cells, and some therapies designed to target cancer metabolism have shown unacceptable toxicity. Any successful drug must be specific enough for cancer cells to minimize off-target effects on normal tissue. Additionally, the therapeutic ef-

A 1C metabolism

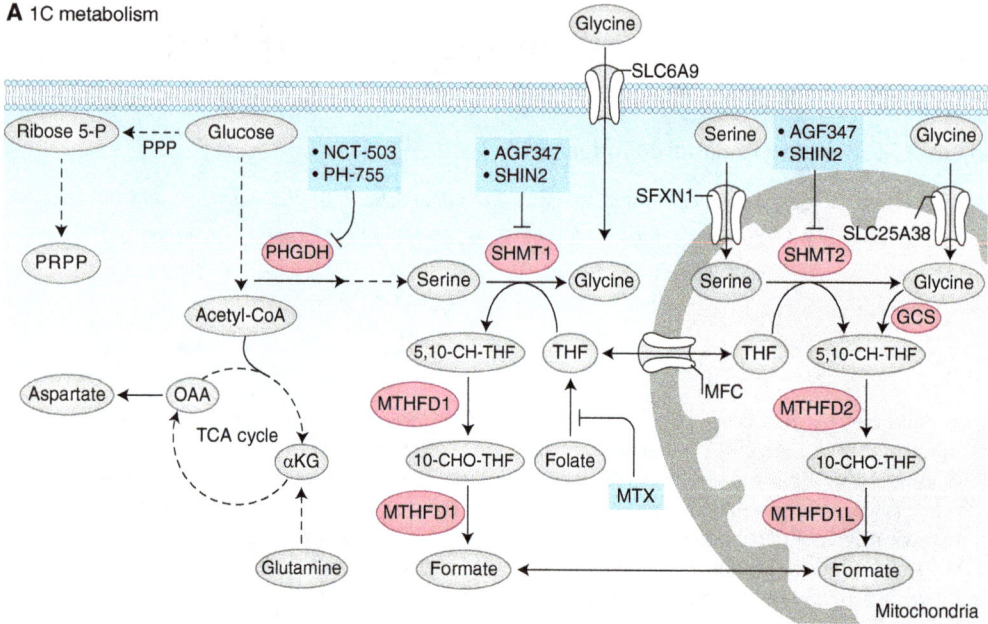

B Purine synthesis

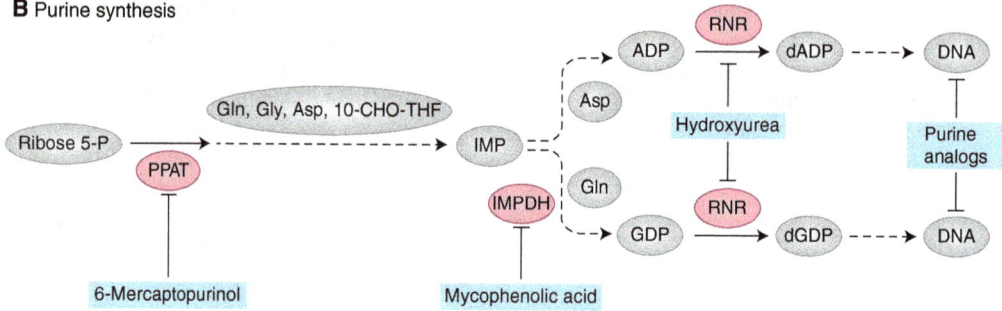

C Pyrimidine synthesis

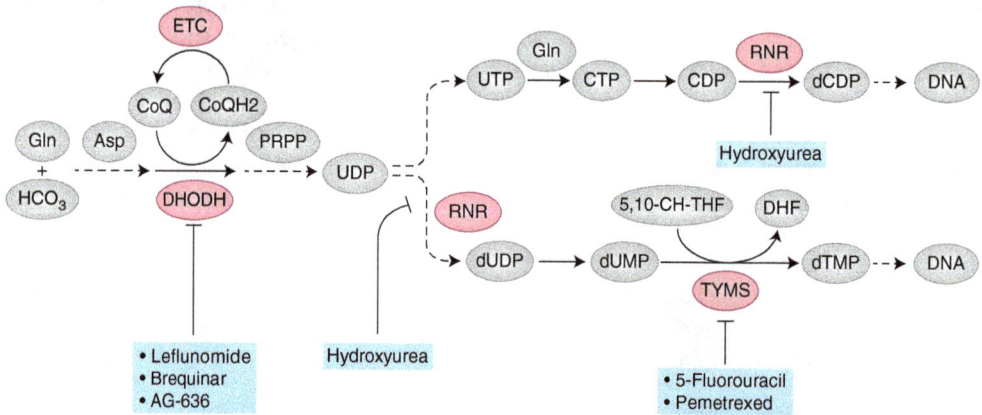

Figure 1. (*See following page for legend.*)

 Cite this article as *Cold Spring Harb Perspect Med* doi: 10.1101/cshperspect.a041657

fect must be sufficiently robust to overcome the remarkable plasticity of many cancers, which can alter their reliance on specific pathways to survive when these are blocked. A successful therapy targeting cellular metabolism must, therefore, be sufficiently specific for metabolic changes in the cancer cells to minimize deleterious side effects and must target a pathway that is so essential for those cells that they are unable to survive when it is disrupted. Although this is a difficult balance to strike, metabolic factors are targeted by many successful cancer therapies and remain the focus of ongoing efforts for therapeutic development.

The field of cancer metabolism is too broad to be covered in a single paper, but we will focus here on several illustrative examples of both the successes and challenges of targeting cellular metabolism as a cancer treatment. (For further reading, we point the reader to several thorough reviews addressing specific aspects of this field: Vander Heiden and DeBerardinis 2017; Elia and Haigis 2021; Martínez-Reyes and Chandel 2021; Meacham et al. 2022; Pavlova et al. 2022; Stine et al. 2022; Brunner and Finley 2023; Xiao et al. 2023.)

ANTIMETABOLITES: TARGETING DE NOVO NUCLEOTIDE SYNTHESIS

Although Warburg's initial discovery focused on constitutive aerobic glycolysis in cancer, clinical trials of glycolysis inhibitors have shown significant toxicity and limited clinical efficacy (Landau et al. 1958; Singh et al. 2005; Voss et al. 2018). Thus, if a favorable therapeutic index exists for glycolysis inhibition as a cancer therapeutic, it has not been obvious from the clinical approaches tested to date. Rather, the first breakthrough leveraging metabolic cancer therapies was made by Farber and colleagues in the 1940s. These investigators found that a synthetic folate analog, aminopterin, was able to induce remission in children with acute lymphoblastic leukemia (ALL), which was previously untreatable (Farber and Diamond 1948). This work led to the development of antimetabolites, a then-novel class of drugs that interfere with nucleoside synthesis.

Antimetabolites are broadly divided into two classes: antifolates, which interfere with the folate cycle, and nucleoside analogs, which mimic the structure of endogenous nucleosides. Nucleoside analogs primarily exert their cytotoxicity by inhibiting the function of nucleoside-metabolizing enzymes or by directly integrating themselves into newly synthesized nucleotide strands, resulting in abnormal DNA or RNA molecules. Today, antimetabolites have been developed to target multiple stages of nucleoside synthesis (Fig. 1). Some notable examples include 6-mercaptopurine and 6-thioguanine, which after intracellular phosphorylation, block the activity of 5-phosphoribosyl-1-pyrophosphatase aminotransferase, preventing the

Figure 1. Inhibitors of nucleotide synthesis. (*A*) One-carbon (1C) metabolism. Nucleotide synthesis uses ribose 5-P produced by the pentose phosphate pathway (PPP) and aspartate produced from oxaloacetate, as well as glycine and tetrahydrofolate methyl donors—methylene-THF (5,10-CH-THF) and formyl-THF (10-CHO-THF)—arising from 1C metabolism linked to the conversion of serine to glycine. 1C metabolism occurs in both the cytoplasm and mitochondria. (*B*) Purine synthesis is a multistep, multienzyme pathway that uses ribose 5-P, glutamine (Gln), glycine (Gly), aspartate (Asp), and 10-CHO-THF to make inosine monophosphate (IMP). IMP is converted to ADP and GDP, which can then be converted to dADP and dGDP. (*C*) Pyrimidine synthesis is a multistep process that uses phosphoribosyl pyrophosphate (PRPP) as a scaffold to produce UDP from glutamine, carbonate, and aspartate. Ribonucleotide reductase (RNR) converts UDP to dUDP and CDP to dCDP. Thymidylate synthase (TYMS), which can be inhibited by the clinical agents pemetrexed or 5-fluorouracil (5-FU), converts dUMP to dTMP. (CoQH2) Reduced coenzyme Q, (DHF) dihydrofolate, (DHODH) dihydroorotate dehydrogenase, (ETC) electron transport chain, (GCS) glycine cleavage system, (HCO3) bicarbonate, (IMPDH) inosine monophosphate dehydrogenase, (MTHFD1) methylenetetrahydrofolate dehydrogenase 1, (MTHFD1L) methylenetetrahydrofolate dehydrogenase 1-like, (MTHFD2) methylenetetrahydrofolate dehydrogenase 2, (MTX) methotrexate, (PHGDH) phosphoglycerate dehydrogenase, (PPAT) phosphoribosyl pyrophosphate amidotransferase, (RNR) ribonucleotide reductase, (SHMT1) serine hydroxymethyl transferase 1. (Reprinted with permission from Stine et al. 2022.)

first step of purine synthesis. Pyrimidine synthesis is also targetable—brequinar and leflunomide inhibit the upstream enzyme dihydroorotate dehydrogenase (DHODH), while capecitabine and 5-fluorouracil inhibit thymidylate synthase to limit the availability of thymidine. Antifolates, including aminopterin and the closely related methotrexate, structurally resemble folate and inhibit one-carbon-dependent reactions, including de novo nucleotide synthesis, as well as the synthesis of several other molecules including amino acids, phospholipids, and polyamines.

The traditional explanation for the sensitivity of cancer cells to antimetabolites is based on proliferation rate. Since cancer cells divide at a faster rate than many normal cells, they have higher rates of DNA replication and, therefore, increased reliance on de novo nucleotide synthesis. Interfering with these processes with antimetabolites has thus proven an effective way to treat cancer. These inhibitors have achieved clinical success for a wide range of cancers: 6-mercaptopurine and 6-thioguanine for leukemias, 5-fluorouracil and capecitabine for gastrointestinal and other carcinomas, and pemetrexed for mesothelioma and non-small-cell lung cancer (Stine et al. 2022). Inhibition of pyrimidine synthesis with brequinar, leflunomide, or related drugs has not had the same clinical impact (Maroun et al. 1993; Moore et al. 1993) but is Food and Drug Administration (FDA)-approved for the treatment of rheumatoid arthritis due to its immunosuppressive effects (Strand et al. 1999) and remains of interest for cancer therapy (Sykes et al. 2016; Cuthbertson et al. 2020). Antifolates have also been highly successful clinically. Current antifolates include methotrexate and, more recently, pemetrexed and pralatrexate, all of which are clinically approved for at least one cancer type and have improved efficacy and/or decreased toxicity compared to their precursor, aminopterin (Thiersch 1949; Marchi and O'Connor 2012; Visentin et al. 2012).

However, challenges arise when interfering with a process as integral to general cellular function as nucleotide synthesis. Cancer cells are not alone in their increased requirements for nucleotide synthesis, and antimetabolites thus decrease the viability of other fast-growing cells as well; these include hair, bone marrow, and skin (Zwart et al. 2023). Antimetabolite resistance is also a notable concern, as cells are able to alter their reliance on inhibited pathways or up-regulate inhibited proteins to stoichiometrically overcome drug treatment (Schimke et al. 1978; Goker et al. 1995; Ishizaka et al. 1995; Kinsella and Smith 1998). The challenge of overcoming resistance mechanisms while avoiding severe off- or on-target toxicities, a constant in almost all areas of cancer therapeutics, necessitates the continued optimization of dosing strategies and rational combinations for existing therapies in addition to the development of novel antimetabolite drug candidates.

ASPARAGINASE: AMINO ACID HOMEOSTASIS AS A THERAPEUTIC VULNERABILITY

Of the 20 amino acids, nine are unable to be synthesized by humans and are therefore classified as essential, while 11 are nonessential or semiessential, as the body can, under homeostatic conditions, synthesize these from other molecules. Nonessential amino acids can become available for cells through several mechanisms: anabolically via de novo synthesis, catabolically through the degradation of proteins or other molecules, or by uptake from environmental sources (Palm and Thompson 2017). Critically, cancer cells have increased amino acid requirements to support the doubling of biomass required with each cell division. These requirements are met, in part, by up-regulation of amino acid biosynthesis pathways (Lieu et al. 2020). Understanding the alterations in amino acid metabolism associated with cancer thus enables the identification of specific vulnerabilities to amino acid starvation, which can be therapeutically targeted.

One such vulnerability is asparagine starvation. Asparagine is considered a nonessential amino acid because it is not required in the diet: it can be synthesized, along with glutamate, by asparagine synthetase (ASNS) from aspartate and glutamine (Lomelino et al. 2017). ASNS is widely expressed throughout the body, suggesting that most tissues can meet their asparagine requirements via de novo synthesis (Milman and Cooney 1974; Chen et al. 2004). However, many cancers are unable to meet their asparagine requirements via de novo synthesis, enabling the depletion of asparagine with the enzyme asparaginase as a viable therapeutic strategy. The anticancer activity of asparaginase was first discovered following the observation that injection of guinea pig serum-induced regression of mouse lymphomas in vivo, an effect traced to an asparaginase activity in guinea pigs that is lacking in mice and humans (Kidd 1953; Broome 1963). Soon after, a highly efficient asparaginase derived from *Escherichia coli* entered clinical trials for various tumors, showing the highest efficacy in leukemia (Clarkson et al. 1970; Tallal et al. 1970; Clavell et al. 1986). Asparaginase remains widely used and highly effective for the treatment of ALL, and is also clinically used for other cancer types, including lymphoma and acute myeloid leukemias (AMLs) (Clavell et al. 1986; Pession et al. 2005; DeAngelo et al. 2015).

Mechanistically, asparaginase deaminates—and thus depletes—free asparagine. In the simplest explanation, ASNS-low cancer cells are sensitive to asparaginase because they are unable to tolerate asparagine depletion, whereas normal cells are ASNS-high and tolerate asparaginase treatment by synthesizing asparagine de novo. Indeed, ASNS is up-regulated upon amino acid depletion by the so-called integrated stress response (Pakos-Zebrucka et al. 2016), and up-regulation of ASNS has been linked to asparaginase resistance (Horowitz et al. 1968; Hutson et al. 1997; Stams et al. 2005; Williams et al. 2020). However, ASNS expression alone does not fully explain variability in asparaginase response (Stams et al. 2003; Holleman et al. 2004; Appel et al. 2006). Clinical asparaginases also have some glutaminase activity (Narta et al. 2007), and glutamine depletion increases cellular dependence on asparagine (Pavlova et al. 2018). However, the importance of glutamine depletion for asparaginase response is context-dependent and incompletely understood (Nguyen et al. 2018; Chan et al. 2019).

Several recent findings have begun to shed light on mechanisms underlying the cellular response to asparaginase. In human cells, the only known enzymatic utilization of asparagine is for protein synthesis, as studies to date have revealed no detectable use of asparagine for the production of other metabolites (Pavlova et al. 2018; Sullivan et al. 2018). However, intracellular asparagine is also required as an amino acid exchange factor, enabling the uptake of other amino acids that are important for cell growth, including serine, histidine, and glutamine (Krall et al. 2016). Asparaginase treatment thus diminishes the availability of asparagine as a precursor for protein synthesis, and as an exchange factor for the import of other amino acids. Asparaginase triggers apoptotic cell death in drug-sensitive tumor cells (Holleman et al. 2003; Lee et al. 2019), which presumably reflects, at least in part, the loss of protein synthesis capacity, although the precise molecular events that mediate asparaginase-induced cell death remain to be fully defined.

Relapse following asparaginase-based cancer therapies is a major clinical problem, and several mechanisms of asparaginase resistance have been identified (Fig. 2; Davidsen and Sullivan 2020). One asparaginase-resistance mechanism is through an increased reliance on catabolic sources of amino acid production. Some asparaginase-resistant leukemias and colorectal cancers have been shown to up-regulate protein degradation through the ubiquitin-proteasome pathway in response to asparagine starvation (Hinze et al. 2022). This adaptive response can be blocked to reverse asparaginase resistance by activation of a noncanonical (β-catenin-independent) branch of Wnt signaling termed Wnt-dependent stabilization of proteins, or by pharmacologic inhibition of the kinase GSK3α (Hinze et al. 2019, 2020). Autophagy, an alternative pathway of protein degradation, has also been implicated in asparaginase resistance in leukemia and in normal cells (Takahashi et al. 2017; Hinze et al. 2020). Another independent screen showed that asparaginase resistance in some tumors requires the aspartate and glutamate transporter SLC1A3 to maintain intracellular aspartate concentrations; aspartate is a sub-

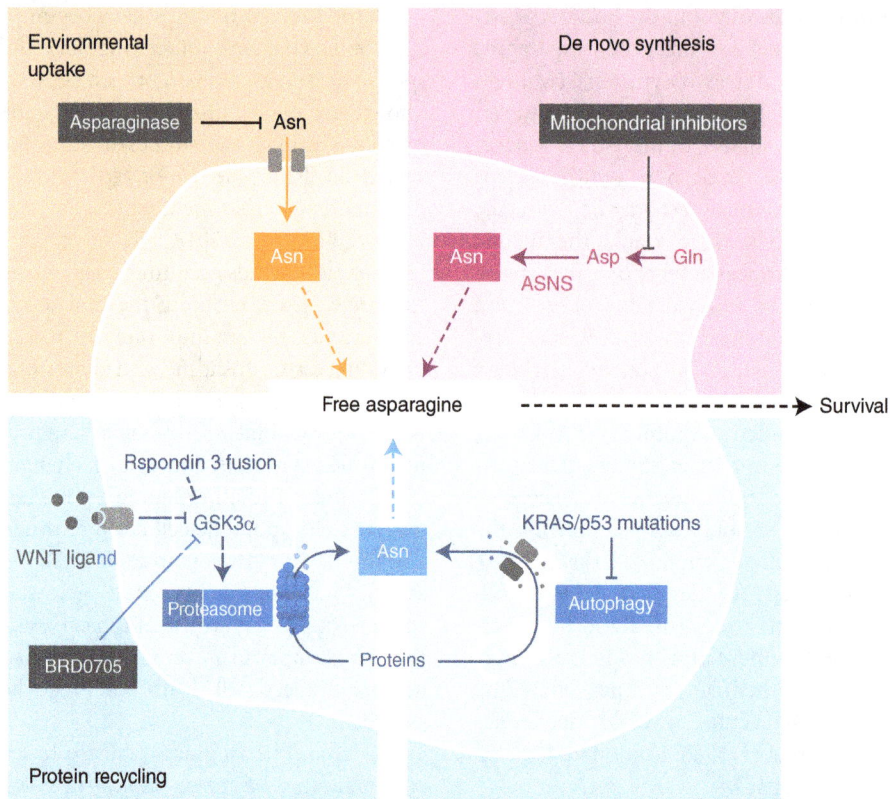

Figure 2. Mechanisms of asparaginase resistance. Asparaginase eliminates asparagine available in the environment. Mitochondrial inhibition limits the conversion of glutamine to aspartate, thereby depleting the substrate for asparagine synthesis. KRAS/TP53 mutations inhibit autophagy, forcing cancer cells to rely on proteasomal degradation for recycling protein into free asparagine. BRD0705, WNT ligands, and Rspondin 3 fusion activate WNT/STOP, thereby inhibiting proteasomal degradation required for cancer cells to survive asparagine depletion. (Asn) Asparagine, (ASNS) asparagine synthetase, (Asp) aspartate, and (Gln) glutamine. (Reprinted with permission from Davidsen and Sullivan 2020.)

strate of ASNS and is poorly cell-permeable when this transporter is lacking (Sun et al. 2019). Preclinically, simultaneously targeting the asparaginase-resistance factors GSK3α or SLC1A3 alongside asparaginase treatment induces asparaginase sensitivity (Sun et al. 2019; Hinze et al. 2020, 2022). These synthetic lethal interactions might enable novel treatment strategies for asparaginase-resistant cancers.

CELL SIGNALING PATHWAYS AND ONCOGENIC MUTATIONS

Perhaps predictably, many oncogenic mutations in signaling proteins and transcription factors alter cellular metabolism. Aberrant activation of the PI3K/AKT/mTOR signaling axis is among the most common aberrations in human cancer (Hoxhaj and Manning 2020). This pathway regulates many aspects of cell growth, including nutrient import, amino acid synthesis, glycolysis, and transcriptional signals promoting cell growth and proliferation. The PI3K/AKT/mTOR pathway is therefore an attractive target for therapeutic intervention to target signaling and metabolic pathways that are specifically dysregulated in cancer. Despite extensive efforts, PI3K pathway inhibitors have struggled to achieve the clinical success that preclinical experiments suggested might be possible (Janku et al. 2018). However, several insights from disappointing clinical trials are helping increase the probability of success for current-generation PI3K inhibitors. Hyperglycemia is a common on-target effect of inhibition of glycolysis; early detection and management strategies, including the administration of antidiabetes drugs, can help enable continued treatment (Tankova et al. 2022). Targeting PI3K is also challenging due to its numerous different isoforms, each of which has slightly different structures and functions. Improving the specificity and selectivity of PI3K inhibitors will help decrease off-target effects, while using isoform-specific inhibitors for cancers driven by mutations in just one PI3K isoform may provide a safer and more effective therapeutic alternative, as long as compensatory activation of other isoforms can be avoided (Venkatesan et al. 2010; Juric et al. 2015, 2018; Shapiro et al. 2015). Finally, promising results have been reported from combining PI3K/AKT/mTOR inhibition with other cancer therapeutics, suggesting the possibility of rational drug combinations to overcome resistance mechanisms, especially as advances in precision oncology help to identify better predictive biomarkers for drug responses (Baselga et al. 2012, 2017; Curigliano et al. 2023).

NEOMORPHIC IDH MUTATIONS: TARGETING PRODUCTION OF AN ONCOGENIC METABOLITE

In recent years, mutant isocitrate dehydrogenase (IDH1 and IDH2) proteins have emerged as targets to disrupt cancer-specific metabolism. IDH proteins normally catalyze the conversion of isocitrate to α-ketoglutarate (also known as 2-oxoglutarate) as part of the tricarboxylic (TCA) cycle (Przybyla-Zawislak et al. 1999). Genomewide analyses identified recurrent analogous missense mutations of IDH1 (affecting amino acid R132) or IDH2 (R140 or R172) in glioma and AML (Parsons et al. 2008; Mardis et al. 2009; Yan et al. 2009; Patel et al. 2011). The fact that these were heterozygous missense mutations recurrently involving specific amino acids immediately raised the possibility that these were gain-of-function mutations. Mechanistic

analysis revealed that these cancer-associated IDH1/2 mutations impart a neomorphic enzymatic activity that catalyzes the reduction of α-ketoglutarate to the R-enantiomer of 2-hydroxyglutarate (R-2HG) (Fig. 3; Dang et al. 2009; Gross et al. 2010). R-2HG is present at low levels in normal cells, but at millimolar concentrations in IDH mutant tumors (Choi et al. 2012). Strikingly, R-2HG is an oncometabolite that is sufficient for oncogenic transformation (Koivunen et al. 2012; Losman et al. 2013). R-2HG has several oncogenic effects, including activation of the α-ketoglutarate-dependent prolyl hydroxylase EGLN, as well as inhibition of other α-ketoglutarate-dependent tumor suppressors such as TET family proteins (Losman et al. 2020).

The novel function imparted by IDH mutations, which are selectively present in the cancer cells of these patients, made these compelling targets for the development of cancer-specific therapeutics. Further strengthening the therapeutic rationale was that transformation by R-2HG is reversible following the removal of this oncometabolite (Losman et al. 2013). This prompted the development of inhibitors of mutant IDH proteins, the first of which was enasidenib, a selective inhibitor of mutant IDH2 that efficiently reduces levels of R-2HG and tumor growth in relevant preclinical models (Yen et al. 2017). This was soon followed by ivosidenib, an inhibitor of mutated IDH1 that similarly inhibits R-2HG production and impairs tumor viability in preclinical models (Popovici-Muller et al. 2018). In early clinical trials, enasidenib showed both safety and efficacy in relapsed/refractory IDH2-mutant AML (Stein et al. 2017), and a recent phase 3 clinical trial supports its clinical activity in AML (de Botton et al. 2023). Ivosidenib has likewise achieved clinical success, showing meaningful clinical activity in IDH1-mutant cholangiocarcinoma (Lowery et al. 2019; Abou-Alfa et al. 2020). Ivosidenib is also efficacious against other cancer types, including IDH1-mutated AML (Montesinos et al. 2022). Brain tumors can be challenging to treat due to the low penetration of drugs across the blood–brain barrier (BBB), but a combination IDH1/2 inhibitor with favorable BBB permeability (vor-

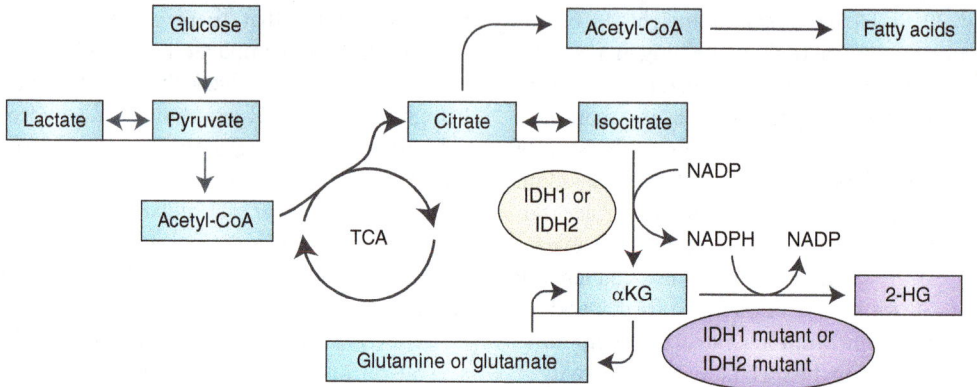

Figure 3. Functions of normal and mutant isocitrate dehydrogenase (IDH) proteins in cellular metabolism. Certain somatic mutations at crucial arginine residues in IDH1 (which is cytoplasmic) and IDH2 (which is mitochondrial) are common early driver mutations in glioma and acute myeloid leukemia (AML). These mutations are unusual because they cause the gain of a novel enzymatic activity. Instead of isocitrate being converted to α-ketoglutarate (αKG) with the production of reduced nicotinamide adenine dinucleotide phosphate (NADPH), αKG is converted to 2-hydroxyglutarate (2-HG) with the consumption of NADPH. 2-HG builds up to high levels in tumor cells and tissues of affected patients and supports tumor progression by a mechanism that is yet to be determined. (TCA) Tricarboxylic acid cycle. (Reprinted with permission from Cairns et al. 2011.)

asidenib) was developed (Konteatis et al. 2020), and it has since shown efficacy against IDH1/2 mutant low-grade gliomas (Mellinghoff et al. 2023).

As is the case with virtually all known cancer therapies, the emergence of resistance to inhibitors of mutant IDH proteins has been a clinical problem. IDH inhibitor resistance often arises via secondary point mutations that restore the production of R-2HG, highlighting that these tumors are under strong selective pressure to maintain R-2HG production. In the case of the IDH2 inhibitor enasidenib, these mutations can occur in *trans* and affect the dimerization domain of the wild-type (and not the mutant) IDH2 allele. Expression of the IDH2 dimerization domain mutant alone does not produce R-2HG, but when coexpressed with the oncogenic IDH2 mutation, allows continued production of R-2HG even in the presence of enasidenib (Intlekofer et al. 2018). In the case of the IDH1 inhibitor ivosidenib, resistance often arises via second-site mutations of the same allele that result in both impaired binding by the inhibitor and increased efficiency of R-2HG production (Reinbold et al. 2022). Alternatively, comuta-

tions in other genes with well-established roles in AML pathogenesis, including *DNMT3A*, *NPM1*, *SRSF2*, *ASXL1*, *RUNX1*, *NRAS*, and *TP53* can provide alternative resistance mechanisms that obviate the need for R-2HG for AML survival (Choe et al. 2020). Thus, while IDH GOF inhibitors represent an impressive recent development in the clinical targeting of cancer metabolism, further work is needed to mitigate resistance and maximize the curative potential of IDH1/2 inhibitor-based cancer therapeutics.

BEYOND CANCER: TARGETING IMMUNOMETABOLISM TO IMPROVE ANTITUMOR IMMUNITY

The immune system serves as an important line of defense against cancer development (Hanahan 2022). Consequently, stimulating cellular antitumor immunity through inhibition of immune checkpoint molecules such as CTLA4, PD1 (also known as PDCD1), or PDL1 (CD274), or using chimeric antigen receptors (CARs) or bispecific antibodies to reprogram cellular immunity to tumor antigens, have become highly effective treatments for several tumor types (Park and Cheung 2024). However, these approaches fail to induce durable responses in many patients. One attractive strategy to improve antitumor immunity is focused on so-called immunometabolism, or the metabolic regulation of immune responses. Upon activation, T cells and natural killer (NK) cells require environmental nutrients to generate the biomass and energy required for the proliferative burst that characterizes an effective immune response (Pearce 2010; Im et al. 2016). Restriction of nutrients in the tumor microenvironment (TME) has emerged as a key mechanism underlying resistance to antitumor immunity (Chang et al. 2015). As with cancer cell–intrinsic metabolism, the number of therapeutically promising immunometabolic pathways is too large to comprehensively cover here. We will, therefore, focus on one illustrative example: arginine. (For further reading, we refer the reader to several excellent recent reviews: Elia and Haigis 2021; Corrado and Pearce 2022; Møller et al. 2022; Han et al. 2023.)

Arginine is a semiessential amino acid that is synthesized by arginosuccinate synthase 1 (ASS1). However, many cancers lose expression of ASS1, becoming reliant on external arginine for survival and growth (Allen et al. 2014; Bean et al. 2016; Geiger et al. 2016; Locke et al. 2016; Hajaj et al. 2024). In contrast, arginine is often depleted from the TME, an effect that can be mediated in party by the production of arginase by normal myeloid cells in the TME (Rodriguez et al. 2004). The case of arginine highlights a challenge in targeting cancer metabolism. At first glance, the high prevalence of ASS1 deficiency in cancer might suggest that, like asparagine, arginine depletion might be an effective strategy to target cancer-specific amino acid vulnerability while sparing healthy tissue. However, there is an additional consequence of the arginine-deficient conditions of the TME: suppression of T-cell activation. Thus, although arginine promotes cancer growth (Al-Koussa et al. 2020), increasing arginine levels can actually decrease cancer progression due to increased antitumor immunity (Geiger et al. 2016).

It has long been known that arginine deprivation impairs T-cell responses (Brittenden et al. 1994; Ochoa et al. 2001; Choi et al. 2009), and more recent preclinical data have shown that providing excess arginine improves anticancer T-cell responses (Geiger et al. 2016). Taken together, these and other findings suggest an important role for arginine in establishing an effective and durable antitumor immune response. Indeed, dietary supplementation of arginine improves survival and prevents tumor growth in mouse models of cancer (Geiger et al. 2016; Canale et al. 2021). In both mice and humans, circulating arginine levels are predictive of immunotherapy efficacy, further highlighting the critical role of arginine in antitumor immunity (Peyraud et al. 2022).

From a therapeutic perspective, increasing arginine levels in the TME is achievable, but the dosing regimens used in experimental models are challenging for human patients. An inventive preclinical approach was used to genetically engineer a bacterium (*E. coli* Nissle 1917) to home to tumors and convert ammonia into arginine (Canale et al. 2021). An alternative way to increase arginine availability for T cells in the TME is to inhibit its degradation using arginase I inhibitors, and early-stage clinical trials have shown promising results in patients with advanced cancer (Naing et al. 2019; Javle et al. 2021).

Several alternative approaches to target immunometabolic pathways are also emerging as part of therapeutics leveraging CAR-T or CAR-NK cells, in which immune cells are isolated, transduced ex vivo with a CAR that redirects these to tumor antigens, and subsequently reinjected into a patient. It is possible to select or engineer CAR cells with altered metabolism during ex vivo cellular reprogramming (Kawalekar et al. 2016; Sukumar et al. 2016; Klebanoff et al. 2017). Genome engineering has also been applied to introduce permanent changes in genes that regulate CAR metabolism to improve therapeutic efficacy (Atkins et al. 2020; Narayan et al. 2022). Strategies are also emerging to modulate the activity of tumor-infiltrating lymphocytes, including modulation of lactate (Sonveaux et al. 2008; Brand et al. 2016; Cascone et al. 2018), glutamine (Tannir et al. 2018), and acetate (Balmer et al. 2016).

OXIDATIVE AND REDUCTIVE STRESS: EMERGING TARGETS IN CANCER METABOLISM

The therapeutic potential of altering redox homeostasis (the balance between reductive and oxidative reactions in the cell) has also been studied in the context of physiologically healthy and malignant cells. Oxidative agents, including reactive oxygen species (ROS) and reactive nitrogen species (RNS), are generally increased in cancer cells, and to a certain threshold, this can promote cancer by acting as second messengers in cell signaling pathways and creating an immunosuppressive TME (Nakamura and Takada 2021). However, too much ROS/RNS can be toxic, so many cancers also express high levels of antioxidants, including glutathione and thioredoxin, to ensure their own survival (Chao et al. 1992; Baker et al. 1997). Ongoing studies have led to the identification of cancer-specific vulnerabilities to the perturbation of redox stress pathways that can be exploited therapeutically.

Oxidative stress is produced, directly or indirectly, by most conventional chemotherapeutics and radiation, and is thought to be one mechanism by which these agents induce cancer cell death (Hayes et al. 2020). However, early efforts to directly target ROS pathways had limited clinical benefit due to toxicity and limited responses in patients (Hayes et al. 2020). Ongoing preclinical and early-stage clinical studies continue to identify new ways to induce oxidative stress specifically in cancer cells, including inhibition of glutaminase and/or FLT3 (Gregory et al. 2016, 2018; Rashmi et al. 2020). A systematic approach characterizing responses to a panel of anticancer drugs known to regulate ROS levels sought to identify mechanisms by which cancer cells sense and respond to altered ROS levels (Zhang et al. 2023). This study revealed that these compounds modify reactive cysteine residues on a large number of cellular proteins. Additionally, CHK1 was identified as a nuclear ROS sensor that inhibits mitochondrial translation during oxidative stress, thus alleviating ROS accumulation and enabling cell survival.

A CHK1 inhibitor already in clinical trials for leukemia (Daud et al. 2015; Webster et al. 2017) was furthermore shown to reduce nuclear ROS levels and synergize with the ROS-inducing small molecule auranofin (Abdalbari and Telleria 2021; Zhang et al. 2023). Taken together, this study is an illustrative example of the value of integrated, systemic approaches to identify novel drug targets to combat resistance.

Although less well-studied than oxidative stress, reductive stress in cancer has recently received more attention (Ge et al. 2024). Just as too many oxidative agents can disrupt cellular homeostasis, an accumulation of reducing molecules (including NADH or NADPH) and/or antioxidants can result in an excessively reductive environment and ultimately cell death (Fig. 4). One key regulator of cellular redox state is the transcription factor NRF2, which activates a broad antioxidant gene expression program upon inhibition of its negative regulator KEAP1 in conditions of oxidative stress (Weiss-Sadan et al. 2023). While activating mutations in NRF2 or inactivating mutations of KEAP1 are recurrent in some cancers, aberrant activation of this pathway can also hinder cell growth in certain conditions, as some tumors are reliant on KEAP1-mediated repression of NRF2 for growth in vitro and in vivo (DeBlasi et al. 2023; Weiss-Sadan et al. 2023). Potential therapeutic strategies are emerging for cancers with mutations of the KEAP1-NRF2 pathway. For instance, growth of *KEAP1*-mutant cancers is inhibited in preclinical models by inhibition of mitochondrial complex 1 by IACS-010759 (Molina et al. 2018; Weiss-Sadan et al. 2023), although complex 1 inhibition does have significant systemic toxicity (Yap et al. 2023). Loss of KEAP1 and aberrant activation of NRF2 also trigger hypersensitivity to inhibition of glutaminase, providing an alternative potential therapeutic approach to target these tumors (Romero et al. 2017). Directly targeting KEAP1 might also be a potential therapeutic strategy to regulate cellular redox state, with a KEAP1 activator recently entering a phase I clinical trial for advanced solid tumors (NCT05954312).

Cite this article as *Cold Spring Harb Perspect Med* doi: 10.1101/cshperspect.a041657

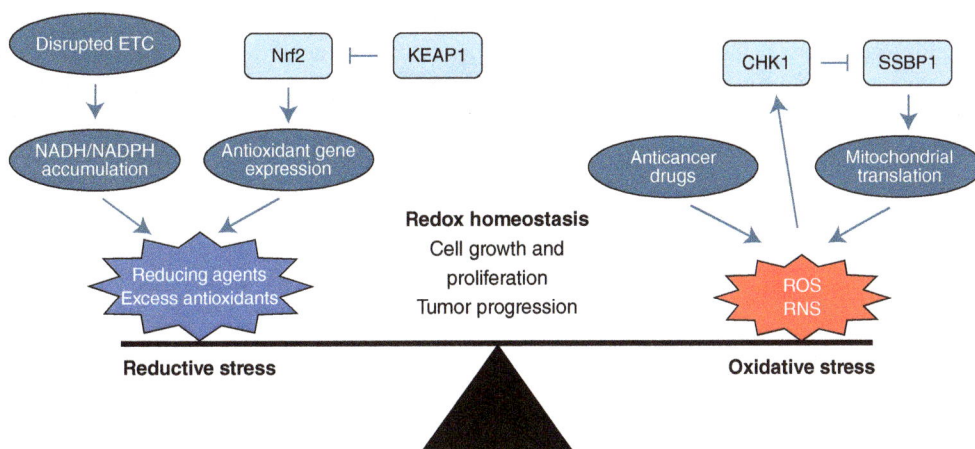

Figure 4. Redox homeostasis in cancer. Under conditions of redox homeostasis, reductive and oxidative stress are balanced, which promotes cancer growth and proliferation. Accumulation of reactive oxygen species, including H_2O_2 and $O_2\bullet$ as well as reactive nitrogen species, can be driven by both mitochondrial translation and anticancer small molecules such as auranofin, cisplatin, or doxorubicin. These are sensed by checkpoint kinase 1 (CHK1), which inhibits the mitochondrial single-stranded DNA-binding protein 1 (SSBP1) to decrease mitochondrial translation and enable resistance to ROS-inducing small molecules. Reductive stress can also induce cell death or dysfunction and is primarily driven by the accumulation of reducing agents such as NADH or NADPH or overexpression of antioxidants. A primary cause for NADH/NADPH accumulation is disruption of the electron transport chain (ETC) through either pharmacological inhibition, genetic mutations, or hypoxic conditions. The transcription factor Nrf2 drives the expression of antioxidant genes, including ALDH3A1, and is inhibited by KEAP1 to maintain homeostasis.

WHY ARE SOME METABOLIC CANCER THERAPIES SUCCESSFUL WHILE OTHERS FAIL?

As described in this work, several approaches targeting metabolic vulnerabilities of cancer cells have had remarkable clinical impact. Yet other seemingly rational approaches, such as inhibition of glycolysis based on Warburg's seminal studies, showed intolerable toxicity at doses that failed to induce meaningful clinical responses. Why this apparent discrepancy? The clinical impact of any cancer therapeutic is dependent on the degree to which it triggers differential toxicity to cancer versus normal cells (the so-called therapeutic index). Thus, the ability of normal tissues and organs to tolerate a therapeutic agent is as important for the clinical success of a therapeutic as its toxicity to cancer cells. However, our ability to identify therapeutic vulnerabilities in tumor cells has advanced much more rapidly than our ability to assess

the toxicity of these approaches to normal tissues and organs. Indeed, many vital functions remain very difficult to model in experimental systems suitable for medium- or even low-throughput screening. An additional challenge is the relative difficulty of assessing therapeutic efficacy in experimental models that fully recapitulate the organismal physiology and heterogeneity of human patient populations. Nevertheless, the development of increasingly sophisticated experimental models of human cancer promises to improve our ability to identify transformative new therapies.

Conceptually, the most tumor-selective cancer therapeutic is one that targets a factor that is uniquely relevant in tumor cells, as exemplified by inhibitors of oncometabolite production by mutant IDH proteins. While we speculate that oncometabolite-driven oncogenesis will prove to be more common than currently recognized, most cancers lack such a neomorphic oncogenic factor that is dispensable in normal cells. Instead, most therapies that can cure patients in the clinic target factors that are important for all cells, such as nucleic acids and amino acids for example. Why are these approaches clinically successful at all? The aberrant proliferation of cancer cells is undoubtedly part of the answer. Indeed, biomass must double with every cell division, so a high proliferation rate imposes considerable metabolic demands for biomass production. The oncogenic mutations that drive uncontrolled growth and proliferation in cancer cells uncouple these processes from regulation by nutrient availability and cellular physiology, thus disabling adaptive responses that allow normal cells to survive during metabolic perturbations. Finally, the ability of most, if not all, therapies to cure cancer requires the induction of an effective antitumor immune response. This may be facilitated by the induction of cell death in a subset of cancer cells expressing mutation-induced neoantigens, leading to antitumor immunity against cancer cells expressing the same neoantigens that survived the initial therapeutic insult. Powerful tolerance mechanisms that evolved to protect normal cells from autoimmunity may prevent the development of immunity to normal cells, even if these are killed to the same degree as tumor cells.

CONCLUDING REMARKS

Dating back to Warburg, metabolism has been one of the most well-studied aspects of cancer biology. Meanwhile, our appreciation for the diverse range of processes controlled by metabolism—ranging from initial transformation and growth to immunosuppression and metastasis—continues to grow. Thus, metabolic pathways remain among the most clinically viable targets for cancer therapy. Yet cancer is, in many ways, natural selection "gone wrong"—a mutational process that promotes organismal evolution can also increase short-term fitness in an individual cell, allowing it to outcompete its neighbors and grow uncontrollably, to the ultimate detriment of the organism. It is, therefore, not surprising that cancers can often evolve in the face of treatment-induced selective pressure to develop resistance. Indeed, patients receiving many of the therapies discussed above often suffer from relapse following treatment. There is, therefore, an urgent need to define novel therapeutic approaches to overcome resistance and maximize the curative potential of cancer therapies. Modern, unbiased screening approaches promise to further expedite the process of new target discovery through the systematic characterization of changes in signaling pathways, metabolite production, and other processes during tumor progression and treatment (Li et al. 2019). Furthermore, developing a greater understanding of metabolic adaptations in normal cells—including immune cells, stem cells, and specific tissues of relevance, such as the pancreas or adipose tissue—will aid in the development of therapies that more potently and selectively eliminate cancer cells. It is thus an exciting time for researching cellular or cancer metabolism, with a wealth of previous research to build from and a recent acceleration of findings.

REFERENCES

*Reference is also in this subject collection.

Abdalbari FH, Telleria CM. 2021. The gold complex auranofin: new perspectives for cancer therapy. *Discov Onc* 12: 42. doi:10.1007/s12672-021-00439-0

Abou-Alfa GK, Macarulla T, Javle MM, Kelley RK, Lubner SJ, Adeva J, Cleary JM, Catenacci DV, Borad MJ, Bridgewater J, et al. 2020. Ivosidenib in IDH1-mutant, chemotherapy-refractory cholangiocarcinoma (ClarIDHy): a multicentre, randomised, double-blind, placebo-controlled, phase 3 study. *Lancet Oncol* 21: 796–807. doi:10.1016/S1470-2045(20)30157-1

Al-Koussa H, El Mais N, Maalouf H, Abi-Habib R, El-Sibai M. 2020. Arginine deprivation: a potential therapeutic for cancer cell metastasis? A review. *Cancer Cell Int* 20: 150. doi:10.1186/s12935-020-01232-9

Allen MD, Luong P, Hudson C, Leyton J, Delage B, Ghazaly E, Cutts R, Yuan M, Syed N, Lo Nigro C, et al. 2014. Prognostic and therapeutic impact of argininosuccinate synthetase 1 control in bladder cancer as monitored longitudinally by PET imaging. *Cancer Res* 74: 896–907. doi:10.1158/0008-5472.CAN-13-1702

Appel IM, den Boer ML, Meijerink JPP, Veerman AJP, Reniers NCM, Pieters R. 2006. Up-regulation of asparagine synthetase expression is not linked to the clinical response L-asparaginase in pediatric acute lymphoblastic leukemia. *Blood* 107: 4244–4249. doi:10.1182/blood-2005-06-2597

Atkins RM, Menges MA, Bauer A, Turner JG, Locke FL. 2020. Metabolically flexible CAR T cells (mfCAR-T), with constitutive expression of PGC-1α resistant to post translational modifications, exhibit superior survival and function in vitro. *Blood* **136**: 30. doi:10.1182/blood-2020-143217

Baker A, Payne CM, Briehl MM, Powis G. 1997. Thioredoxin, a gene found overexpressed in human cancer, inhibits apoptosis in vitro and in vivo. *Cancer Res* **57**: 5162–5167.

Balmer ML, Ma EH, Bantug GR, Grählert J, Pfister S, Glatter T, Jauch A, Dimeloe S, Slack E, Dehio P, et al. 2016. Memory CD8$^+$ T cells require increased concentrations of acetate induced by stress for optimal function. *Immunity* **44**: 1312–1324. doi:10.1016/j.immuni.2016.03.016

Baselga J, Campone M, Piccart M, Burris HA, Rugo HS, Sahmoud T, Noguchi S, Gnant M, Pritchard KI, Lebrun F, et al. 2012. Everolimus in postmenopausal hormone-receptor-positive advanced breast cancer. *N Engl J Med* **366**: 520–529. doi:10.1056/NEJMoa1109653

Baselga J, Im SA, Iwata H, Cortés J, De Laurentiis M, Jiang Z, Arteaga CL, Jonat W, Clemons M, Ito Y, et al. 2017. Buparlisib plus fulvestrant versus placebo plus fulvestrant in postmenopausal, hormone receptor-positive, HER2-negative, advanced breast cancer (BELLE-2): a randomised, double-blind, placebo-controlled, phase 3 trial. *Lancet Oncol* **18**: 904–916. doi:10.1016/S1470-2045(17)30376-5

Bean GR, Kremer JC, Prudner BC, Schenone AD, Yao JC, Schultze MB, Chen DY, Tanas MR, Adkins DR, Bomalaski J, et al. 2016. A metabolic synthetic lethal strategy with arginine deprivation and chloroquine leads to cell death in ASS1-deficient sarcomas. *Cell Death Dis* **7**: e2406–e2406. doi:10.1038/cddis.2016.232

Brand A, Singer K, Koehl GE, Kolitzus M, Schoenhammer G, Thiel A, Matos C, Bruss C, Klobuch S, Peter K, et al. 2016. LDHA-associated lactic acid production blunts tumor immunosurveillance by T and NK cells. *Cell Metab* **24**: 657–671. doi:10.1016/j.cmet.2016.08.011

Brittenden J, Park KG, Heys SD, Ross C, Ashby J, Ak A-S, Eremin O. 1994. L-arginine stimulates host defenses in patients with breast cancer. *Surgery* **115**: 205–212.

Broome JD. 1963. Evidence that the L-asparaginase of guinea pig serum is responsible for its antilymphoma effects. I: Properties of the L-asparaginase of guinea pig serum in relation to those of the antilymphoma substance. *J Exp Med* **118**: 99–120. doi:10.1084/jem.118.1.99

Brunner JS, Finley LWS. 2023. Metabolic determinants of tumour initiation. *Nat Rev Endocrinol* **19**: 134–150. doi:10.1038/s41574-022-00773-5

Cairns RA, Harris IS, Mak TW. 2011. Regulation of cancer cell metabolism. *Nat Rev Cancer* **11**: 85–95. doi:10.1038/nrc2981

Canale FP, Basso C, Antonini G, Perotti M, Li N, Sokolovska A, Neumann J, James MJ, Geiger S, Jin W, et al. 2021. Metabolic modulation of tumours with engineered bacteria for immunotherapy. *Nature* **598**: 662–666. doi:10.1038/s41586-021-04003-2

Cascone T, McKenzie JA, Mbofung RM, Punt S, Wang Z, Xu C, Williams LJ, Wang Z, Bristow CA, Carugo A, et al. 2018. Increased tumor glycolysis characterizes immune resistance to adoptive T cell therapy. *Cell Metab* **27**: 977–987.e4. doi:10.1016/j.cmet.2018.02.024

Chan M, Gravel M, Bramoullé A, Bridon G, Avizonis D, Shore GC, Roulston A. 2014. Synergy between the NAMPT inhibitor GMX1777(8) and pemetrexed in non-small cell lung cancer cells is mediated by PARP

activation and enhanced NAD consumption. *Cancer Res* **74**: 5948–5954. doi:10.1158/0008-5472.CAN-14-0809

Chan WK, Horvath TD, Tan L, Link T, Harutyunyan KG, Pontikos MA, Anishkin A, Du D, Martin LA, Yin E, et al. 2019. Glutaminase activity of L-asparaginase contributes to durable preclinical activity against acute lymphoblastic leukemia. *Mol Cancer Ther* **18**: 1587–1592. doi:10.1158/1535-7163.MCT-18-1329

Chang CH, Qiu J, O'Sullivan D, Buck MD, Noguchi T, Curtis JD, Chen Q, Gindin M, Gubin MM, van der Windt GJW, et al. 2015. Metabolic competition in the tumor microenvironment is a driver of cancer progression. *Cell* **162**: 1229–1241. doi:10.1016/j.cell.2015.08.016

Chao CC, Huang YT, Ma CM, Chou WY, Lin-Chao S. 1992. Overexpression of glutathione S-transferase and elevation of thiol pools in a multidrug-resistant human colon cancer cell line. *Mol Pharmacol* **41**: 69–75.

Chen H, Pan YX, Dudenhausen EE, Kilberg MS. 2004. Amino acid deprivation induces the transcription rate of the human asparagine synthetase gene through a timed program of expression and promoter binding of nutrient-responsive basic region/leucine zipper transcription factors as well as localized histone acetylation. *J Biol Chem* **279**: 50829–50839. doi:10.1074/jbc.M409173200

Choe S, Wang H, DiNardo CD, Stein EM, de Botton S, Roboz GJ, Altman JK, Mims AS, Watts JM, Pollyea DA, et al. 2020. Molecular mechanisms mediating relapse following ivosidenib monotherapy in IDH1-mutant relapsed or refractory AML. *Blood Adv* **4**: 1894–1905. doi:10.1182/bloodadvances.2020001503

Choi BS, Martinez-Falero IC, Corset C, Munder M, Modolell M, Müller I, Kropf P. 2009. Differential impact of l-arginine deprivation on the activation and effector functions of T cells and macrophages. *J Leukoc Biol* **85**: 268–277. doi:10.1189/jlb.0508310

Choi C, Ganji SK, DeBerardinis RJ, Hatanpaa KJ, Rakheja D, Kovacs Z, Yang XL, Mashimo T, Raisanen JM, Marin-Valencia I, et al. 2012. 2-Hydroxyglutarate detection by magnetic resonance spectroscopy in IDH-mutated patients with gliomas. *Nat Med* **18**: 624–629. doi:10.1038/nm.2682

Clarkson B, Krakoff I, Burchenal J, Karnofsky D, Golbey R, Dowling M, Oettgen H, Lipton A. 1970. Clinical results of treatment with *E. coli* L-asparaginase in adults with leukemia, lymphoma, and solid tumors. *Cancer* **25**: 279–305. doi:10.1002/1097-0142(197002)25:2<279::AID-CNCR2820250205>3.0.CO;2-7

Clavell LA, Gelber RD, Cohen HJ, Hitchcock-Bryan S, Cassady JR, Tarbell NJ, Blattner SR, Tantravahi R, Leavitt P, Sallan SE. 1986. Four-agent induction and intensive asparaginase therapy for treatment of childhood acute lymphoblastic leukemia. *N Engl J Med* **315**: 657–663. doi:10.1056/NEJM198609113151101

Corrado M, Pearce EL. 2022. Targeting memory T-cell metabolism to improve immunity. *J Clin Invest* **132**: e148546. doi:10.1172/JCI148546

Curigliano G, Shapiro GI, Kristeleit RS, Abdul Razak AR, Leong S, Alsina M, Giordano A, Gelmon KA, Stringer-Reasor E, Vaishampayan UN, et al. 2023. A phase 1B open-label study of gedatolisib (PF-05212384) in combination with other anti-tumour agents for patients with advanced solid tumours and triple-negative breast cancer. *Br J Cancer* **128:** 30–41. doi:10.1038/s41416-022-02025-9

Cuthbertson CR, Guo H, Kyani A, Madak JT, Arabzada Z, Neamati N. 2020. The dihydroorotate dehydrogenase inhibitor brequinar is synergistic with ENT1/2 Inhibitors. *ACS Pharmacol Transl Sci* **3:** 1242–1252. doi:10.1021/acsptsci.0c00124

Dang L, White DW, Gross S, Bennett BD, Bittinger MA, Driggers EM, Fantin VR, Jang HG, Jin S, Keenan MC, et al. 2009. Cancer-associated IDH1 mutations produce 2-hydroxyglutarate. *Nature* **462:** 739–744. doi:10.1038/nature08617

Daud AI, Ashworth MT, Strosberg J, Goldman JW, Mendelson D, Springett G, Venook AP, Loechner S, Rosen LS, Shanahan F, et al. 2015. Phase I dose-escalation trial of checkpoint kinase 1 inhibitor MK-8776 as monotherapy and in combination with gemcitabine in patients with advanced solid tumors. *JCO* **33:** 1060–1066. doi:10.1200/JCO.2014.57.5027

Davidsen K, Sullivan LB. 2020. Free asparagine or die: cancer cells require proteasomal protein breakdown to survive asparagine depletion. *Cancer Discov* **10:** 1632–1634. doi:10.1158/2159-8290.CD-20-1251

DeAngelo DJ, Stevenson KE, Dahlberg SE, Silverman LB, Couban S, Supko JG, Amrein PC, Ballen KK, Seftel MD, Turner AR, et al. 2015. Long-term outcome of a pediatric-inspired regimen used for adults aged 18–50 years with newly diagnosed acute lymphoblastic leukemia. *Leukemia* **29:** 526–534. doi:10.1038/leu.2014.229

DeBlasi JM, Falzone A, Caldwell S, Prieto-Farigua N, Prigge JR, Schmidt EE, Chio IIC, Karreth FA, DeNicola GM. 2023. Distinct Nrf2 signaling thresholds mediate lung tumor initiation and progression. *Cancer Res* **83:** 1953–1967. doi:10.1158/0008-5472.CAN-22-3848

de Botton S, Montesinos P, Schuh AC, Papayannidis C, Vyas P, Wei AH, Ommen H, Semochkin S, Kim HJ, Larson RA, et al. 2023. Enasidenib vs conventional care in older patients with late-stage mutant-*IDH2* relapsed/refractory AML: a randomized phase 3 trial. *Blood* **141:** 156–167. doi:10.1182/blood.2021014901

Elia I, Haigis MC. 2021. Metabolites and the tumour microenvironment: from cellular mechanisms to systemic metabolism. *Nat Metab* **3:** 21–32. doi:10.1038/s42255-020-00317-z

Farber S, Diamond LK. 1948. Temporary remissions in acute leukemia in children produced by folic acid antagonist, 4-aminopteroyl-glutamic acid. *N Engl J Med* **238:** 787–793. doi:10.1056/NEJM194806032382301

Ge M, Papagiannakopoulos T, Bar-Peled L. 2024. Reductive stress in cancer: coming out of the shadows. *Trends Cancer* **10:** 103–112. doi:10.1016/j.trecan.2023.10.002

Geiger R, Rieckmann JC, Wolf T, Basso C, Feng Y, Fuhrer T, Kogadeeva M, Picotti P, Meissner F, Mann M, et al. 2016. L-arginine modulates T cell metabolism and enhances survival and anti-tumor activity. *Cell* **167:** 829–842.e13. doi:10.1016/j.cell.2016.09.031

Goker E, Waltham M, Kheradpour A, Trippett T, Mazumdar M, Elisseyeff Y, Schnieders B, Steinherz P, Tan C, Berman E. 1995. Amplification of the dihydrofolate reductase gene is a mechanism of acquired resistance to methotrexate in patients with acute lymphoblastic leukemia and is correlated with p53 gene mutations. *Blood* **86:** 677–684. doi:10.1182/blood.V86.2.677.bloodjournal862677

Gregory MA, D'Alessandro A, Alvarez-Calderon F, Kim J, Nemkov T, Adane B, Rozhok AI, Kumar A, Kumar V, Pollyea DA, et al. 2016. ATM-G6PD-driven redox metabolism promotes FLT3 inhibitor resistance in acute myeloid leukemia. *Proc Natl Acad Sci* **113:** E6669–E6678. doi:10.1073/pnas.1603876113

Gregory MA, Nemkov T, Reisz JA, Zaberezhnyy V, Hansen KC, D'Alessandro A, DeGregori J. 2018. Glutaminase inhibition improves FLT3 inhibitor therapy for acute myeloid leukemia. *Exp Hematol* **58:** 52–58. doi:10.1016/j.exphem.2017.09.007

Gross S, Cairns RA, Minden MD, Driggers EM, Bittinger MA, Jang HG, Sasaki M, Jin S, Schenkein DP, Su SM, et al. 2010. Cancer-associated metabolite 2-hydroxyglutarate accumulates in acute myelogenous leukemia with isocitrate dehydrogenase 1 and 2 mutations. *J Exp Med* **207:** 339–344. doi:10.1084/jem.20092506

Hajaj E, Pozzi S, Erez A. 2024. From the inside out: exposing the roles of urea cycle enzymes in tumors and their micro and macro environments. *Cold Spring Harb Perspect Med* **14:** a041538. doi:10.1101/cshperspect.a041538

Han S, Georgiev P, Ringel AE, Sharpe AH, Haigis MC. 2023. Age-associated remodeling of T cell immunity and metabolism. *Cell Metab* **35:** 36–55. doi:10.1016/j.cmet.2022.11.005

Hanahan D. 2022. Hallmarks of cancer: new dimensions. *Cancer Discov* **12:** 31–46. doi:10.1158/2159-8290.CD-21-1059

Hayes JD, Dinkova-Kostova AT, Tew KD. 2020. Oxidative stress in cancer. *Cancer Cell* **38:** 167–197. doi:10.1016/j.ccell.2020.06.001

Hinze L, Pfirrmann M, Karim S, Degar J, McGuckin C, Vinjamur D, Sacher J, Stevenson KE, Neuberg DS, Orellana E, et al. 2019. Synthetic lethality of wnt pathway activation and asparaginase in drug-resistant acute leukemias. *Cancer Cell* **35:** 664–676.e7. doi:10.1016/j.ccell.2019.03.004

Hinze L, Labrosse R, Degar J, Han T, Schatoff EM, Schreek S, Karim S, McGuckin C, Sacher JR, Wagner F, et al. 2020. Exploiting the therapeutic interaction of WNT pathway activation and asparaginase for colorectal cancer therapy. *Cancer Discov* **10:** 1690–1705. doi:10.1158/2159-8290.CD-19-1472

Hinze L, Schreek S, Zeug A, Ibrahim NK, Fehlhaber B, Loxha L, Cinar B, Ponimaskin E, Degar J, McGuckin C, et al. 2022. Supramolecular assembly of GSK3α as a cellular response to amino acid starvation. *Mol Cell* **82:** 2858–2870.e8. doi:10.1016/j.molcel.2022.05.025

Holleman A, den Boer ML, Kazemier KM, Janka-Schaub GE, Pieters R. 2003. Resistance to different classes of drugs is associated with impaired apoptosis in childhood acute lymphoblastic leukemia. *Blood* **102:** 4541–4546. doi:10.1182/blood-2002-11-3612

Holleman A, Cheok MH, den Boer ML, Yang W, Veerman AJP, Kazemier KM, Pei D, Cheng C, Pui CH, Relling MV, et al. 2004. Gene-expression patterns in drug-resistant acute lymphoblastic leukemia cells and response to treatment. *N Engl J Med* **351:** 533–542. doi:10.1056/NEJMoa033513

Horowitz B, Madras BK, Meister A, Old LJ, Boyse EA, Stockert E. 1968. Asparagine synthetase activity of mouse leukemias. *Science* 160: 533–535. doi:10.1126/science.160.3827.533

Hoxhaj G, Manning BD. 2020. The PI3K-AKT network at the interface of oncogenic signalling and cancer metabolism. *Nat Rev Cancer* 20: 74–88. doi:10.1038/s41568-019-0216-7

Hutson RG, Kitoh T, Moraga Amador DA, Cosic S, Schuster SM, Kilberg MS. 1997. Amino acid control of asparagine synthetase: relation to asparaginase resistance in human leukemia cells. *Am J Physiol* 272: C1691–C1699. doi:10.1152/ajpcell.1997.272.5.C1691

Im SJ, Hashimoto M, Gerner MY, Lee J, Kissick HT, Burger MC, Shan Q, Hale JS, Lee J, Nasti TH, et al. 2016. Defining CD8$^+$ T cells that provide the proliferative burst after PD-1 therapy. *Nature* 537: 417–421. doi:10.1038/nature19330

Intlekofer AM, Shih AH, Wang B, Nazir A, Rustenburg AS, Albanese SK, Patel M, Famulare C, Correa FM, Takemoto N, et al. 2018. Acquired resistance to IDH inhibition through *trans* or *cis* dimer-interface mutations. *Nature* 559: 125–129. doi:10.1038/s41586-018-0251-7

Ishizaka Y, Chernov MV, Burns CM, Stark GR. 1995. p53-dependent growth arrest of REF52 cells containing newly amplified DNA. *Proc Natl Acad Sci* 92: 3224–3228. doi:10.1073/pnas.92.8.3224

Janku F, Yap TA, Meric-Bernstam F. 2018. Targeting the PI3K pathway in cancer: are we making headway? *Nat Rev Clin Oncol* 15: 273–291. doi:10.1038/nrclinonc.2018.28

Javle MM, Bridgewater JA, Gbolahan OB, Jungels C, Cho MT, Papadopoulos KP, Thistlethwaite FC, Canon JLR, Cheng L, Ioannidis S, et al. 2021. A phase I/II study of safety and efficacy of the arginase inhibitor INCB001158 plus chemotherapy in patients (Pts) with advanced biliary tract cancers. *JCO* 39: 311–311. doi:10.1200/JCO.2021.39.3_suppl.311

Juric D, Castel P, Griffith M, Griffith OL, Won HH, Ellis H, Ebbesen SH, Ainscough BJ, Ramu A, Iyer G, et al. 2015. Convergent loss of PTEN leads to clinical resistance to a PI(3)Kα inhibitor. *Nature* 518: 240–244. doi:10.1038/nature13948

Juric D, Rodon J, Tabernero J, Janku F, Burris HA, Schellens JHM, Middleton MR, Berlin J, Schuler M, Gil-Martin M, et al. 2018. Phosphatidylinositol 3-kinase α–selective inhibition with alpelisib (BYL719) in *PIK3CA*-altered solid tumors: results from the first-in-human study. *JCO* 36: 1291–1299. doi:10.1200/JCO.2017.72.7107

Kawalekar OU, O'Connor RS, Fraietta JA, Guo L, McGettigan SE, Posey AD, Patel PR, Guedan S, Scholler J, Keith B, et al. 2016. Distinct signaling of coreceptors regulates specific metabolism pathways and impacts memory development in CAR T cells. *Immunity* 44: 380–390. doi:10.1016/j.immuni.2016.01.021

Kidd JG. 1953. Regression of transplanted lymphomas induced in vivo by means of normal guinea pig serum. I: Course of transplanted cancers of various kinds in mice and rats given guinea pig serum, horse serum, or rabbit serum. *J Exp Med* 98: 565–582. doi:10.1084/jem.98.6.565

Kinsella AR, Smith D. 1998. Tumor resistance to antimetabolites. *Gen Pharmacol* 30: 623–626. doi:10.1016/S0306-3623(97)00383-2

Klebanoff CA, Crompton JG, Leonardi AJ, Yamamoto TN, Chandran SS, Eil RL, Sukumar M, Vodnala SK, Hu J, Ji Y, et al. 2017. Inhibition of AKT signaling uncouples T cell differentiation from expansion for receptor-engineered adoptive immunotherapy. *JCI Insight* 2: e95103. doi:10.1172/jci.insight.95103

Koivunen P, Lee S, Duncan CG, Lopez G, Lu G, Ramkissoon S, Losman JA, Joensuu P, Bergmann U, Gross S, et al. 2012. Transformation by the (R)-enantiomer of 2-hydroxyglutarate linked to EGLN activation. *Nature* 483: 484–488. doi:10.1038/nature10898

Konteatis Z, Artin E, Nicolay B, Straley K, Padyana AK, Jin L, Chen Y, Narayaraswamy R, Tong S, Wang F, et al. 2020. Vorasidenib (AG-881): a first-in-class, brain-penetrant dual inhibitor of mutant IDH1 and 2 for treatment of glioma. *ACS Med Chem Lett* 11: 101–107. doi:10.1021/acsmedchemlett.9b00509

Krall AS, Xu S, Graeber TG, Braas D, Christofk HR. 2016. Asparagine promotes cancer cell proliferation through use as an amino acid exchange factor. *Nat Commun* 7: 11457. doi:10.1038/ncomms11457

Landau BR, Laszlo J, Stengle J, Burk D. 1958. Certain metabolic and pharmacologic effects in cancer patients given infusions of 2-deoxy-D-glucose. *J Natl Cancer Inst* 21: 485–494.

Lee JK, Kang S, Wang X, Rosales JL, Gao X, Byun HG, Jin Y, Fu S, Wang J, Lee KY. 2019. HAP1 loss confers l-asparaginase resistance in ALL by downregulating the calpain-1-Bid-caspase-3/12 pathway. *Blood* 133: 2222–2232. doi:10.1182/blood-2018-12-890236

Li H, Ning S, Ghandi M, Kryukov GV, Gopal S, Deik A, Souza A, Pierce K, Keskula P, Hernandez D, et al. 2019. The landscape of cancer cell line metabolism. *Nat Med* 25: 850–860. doi:10.1038/s41591-019-0404-8

Lieu EL, Nguyen T, Rhyne S, Kim J. 2020. Amino acids in cancer. *Exp Mol Med* 52: 15–30. doi:10.1038/s12276-020-0375-3

Locke M, Ghazaly E, Freitas MO, Mitsinga M, Lattanzio L, Lo Nigro C, Nagano A, Wang J, Chelala C, Szlosarek P, et al. 2016. Inhibition of the polyamine synthesis pathway is synthetically lethal with loss of argininosuccinate synthase 1. *Cell Rep* 16: 1604–1613. doi:10.1016/j.celrep.2016.06.097

Lomelino CL, Andring JT, McKenna R, Kilberg MS. 2017. Asparagine synthetase: function, structure, and role in disease. *J Biol Chem* 292: 19952–19958. doi:10.1074/jbc.R117.819060

Losman JA, Looper RE, Koivunen P, Lee S, Schneider RK, McMahon C, Cowley GS, Root DE, Ebert BL, Kaelin WG. 2013. (R)-2-Hydroxyglutarate is sufficient to promote leukemogenesis and its effects are reversible. *Science* 339: 1621–1625. doi:10.1126/science.1231677

Losman JA, Koivunen P, Kaelin WG. 2020. 2-Oxoglutarate-dependent dioxygenases in cancer. *Nat Rev Cancer* 20: 710–726. doi:10.1038/s41568-020-00303-3

Lowery MA, Burris HA, Janku F, Shroff RT, Cleary JM, Azad NS, Goyal L, Maher EA, Gore L, Hollebecque A, et al. 2019. Safety and activity of ivosidenib in patients with IDH1-mutant advanced cholangiocarcinoma: a phase 1 study. *Lancet Gastroenterol Hepatol* **4**: 711–720. doi:10 .1016/S2468-1253(19)30189-X

Marchi E, O'Connor OA. 2012. Safety and efficacy of pralatrexate in the treatment of patients with relapsed or refractory peripheral T-cell lymphoma. *Ther Adv Hematol* **3**: 227–235. doi:10.1177/2040620712445330

Mardis ER, Ding L, Dooling DJ, Larson DE, McLellan MD, Chen K, Koboldt DC, Fulton RS, Delehaunty KD, McGrath SD, et al. 2009. Recurring mutations found by sequencing an acute myeloid leukemia genome. *N Engl J Med* **361**: 1058–1066. doi:10.1056/NEJMoa0903840

Maroun J, Ruckdeschel J, Natale R, Morgan R, Dallaire B, Sisk R, Gyves J. 1993. Multicenter phase II study of brequinar sodium in patients with advanced lung cancer. *Cancer Chemother Pharmacol* **32**: 64–66. doi:10.1007/ BF00685878

Martínez-Reyes I, Chandel NS. 2021. Cancer metabolism: looking forward. *Nat Rev Cancer* **21**: 669–680. doi:10 .1038/s41568-021-00378-6

Meacham CE, DeVilbiss AW, Morrison SJ. 2022. Metabolic regulation of somatic stem cells in vivo. *Nat Rev Mol Cell Biol* **23**: 428–443. doi:10.1038/s41580-022-00462-1

Mellinghoff IK, van den Bent MJ, Blumenthal DT, Touat M, Peters KB, Clarke J, Mendez J, Yust-Katz S, Welsh L, Mason WP, et al. 2023. Vorasidenib in IDH1- or IDH2-mutant low-grade glioma. *N Engl J Med* **389**: 589–601. doi:10.1056/NEJMoa2304194

Milman HA, Cooney DA. 1974. The distribution of l-asparagine synthetase in the principal organs of several mammalian and avian species. *Biochem J* **142**: 27–35. doi:10 .1042/bj1420027

Molina JR, Sun Y, Protopopova M, Gera S, Bandi M, Bristow C, McAfoos T, Morlacchi P, Ackroyd J, Agip ANA, et al. 2018. An inhibitor of oxidative phosphorylation exploits cancer vulnerability. *Nat Med* **24**: 1036–1046. doi:10 .1038/s41591-018-0052-4

Møller SH, Hsueh PC, Yu YR, Zhang L, Ho PC. 2022. Metabolic programs tailor T cell immunity in viral infection, cancer, and aging. *Cell Metab* **34**: 378–395. doi:10.1016/j .cmet.2022.02.003

Montesinos P, Recher C, Vives S, Zarzycka E, Wang J, Bertani G, Heuser M, Calado RT, Schuh AC, Yeh SP, et al. 2022. Ivosidenib and azacitidine in IDH1-mutated acute myeloid leukemia. *N Engl J Med* **386**: 1519–1531. doi:10 .1056/NEJMoa2117344

Moore M, Maroun J, Robert F, Natale R, Neidhart J, Dallaire B, Sisk R, Gyves J. 1993. Multicenter phase II study of brequinar sodium in patients with advanced gastrointestinal cancer. *Invest New Drugs* **11**: 61–65. doi:10.1007/ BF00873913

Naing A, Bauer T, Papadopoulos KP, Rahma O, Tsai F, Garralda E, Naidoo J, Pai S, Gibson MK, Rybkin I, et al. 2019. Phase I study of the arginase inhibitor INCB001158 (1158) alone and in combination with pembrolizumab (PEM) in patients (Pts) with advanced/metastatic (adv/ met) solid tumours. *Ann Oncol* **30**: v160. doi:10.1093/ annonc/mdz244.002

Nakamura H, Takada K. 2021. Reactive oxygen species in cancer: current findings and future directions. *Cancer Sci* **112**: 3945–3952. doi:10.1111/cas.15068

Narayan V, Barber-Rotenberg JS, Jung IY, Lacey SF, Rech AJ, Davis MM, Hwang WT, Lal P, Carpenter EL, Maude SL,

et al. 2022. PSMA-targeting TGFβ-insensitive armored CAR T cells in metastatic castration-resistant prostate cancer: a phase 1 trial. *Nat Med* **28**: 724–734. doi:10 .1038/s41591-022-01726-1

Narta UK, Kanwar SS, Azmi W. 2007. Pharmacological and clinical evaluation of L-asparaginase in the treatment of leukemia. *Crit Rev Oncol Hematol* **61**: 208–221. doi:10 .1016/j.critrevonc.2006.07.009

Nguyen HA, Su Y, Zhang JY, Antanasijevic A, Caffrey M, Schalk AM, Liu L, Rondelli D, Oh A, Mahmud DL, et al. 2018. A novel l-asparaginase with low l-glutaminase co-activity is highly efficacious against both T- and B-cell acute lymphoblastic leukemias in vivo. *Cancer Res* **78**: 1549–1560. doi:10.1158/0008-5472.CAN-17-2106

Ochoa JB, Strange J, Kearney P, Gellin G, Endean E, Fitzpatrick E. 2001. Effects of L-arginine on the proliferation of T lymphocyte subpopulations. *JPEN J Parenter Enteral Nutr* **25**: 23–29. doi:10.1177/014860710102500123

Pakos-Zebrucka K, Koryga I, Mnich K, Ljujic M, Samali A, Gorman AM. 2016. The integrated stress response. *EMBO Rep* **17**: 1374–1395. doi:10.15252/embr.20164 2195

Palm W, Thompson CB. 2017. Nutrient acquisition strategies of mammalian cells. *Nature* **546**: 234–242. doi:10 .1038/nature22379

* Park JA, Cheung NKV. 2024. Immunotherapies for childhood cancer. *Cold Spring Harb Perspect Med*: a041574. doi:10.1101/cshperspect.a041574

Parsons DW, Jones S, Zhang X, Lin JCH, Leary RJ, Angenendt P, Mankoo P, Carter H, Siu IM, Gallia GL, et al. 2008. An integrated genomic analysis of human glioblastoma multiforme. *Science* **321**: 1807–1812. doi:10.1126/ science.1164382

Patel KP, Ravandi F, Ma D, Paladugu A, Barkoh BA, Medeiros LJ, Luthra R. 2011. Acute myeloid leukemia with *IDH1* or *IDH2* mutation: frequency and clinicopathologic features. *Am J Clin Pathol* **135**: 35–45. doi:10.1309/ AJCPD7NR2RMNQDVF

Pavlova NN, Hui S, Ghergurovich JM, Fan J, Intlekofer AM, White RM, Rabinowitz JD, Thompson CB, Zhang J. 2018. As extracellular glutamine levels decline, asparagine becomes an essential amino acid. *Cell Metab* **27**: 428–438. e5. doi:10.1016/j.cmet.2017.12.006

Pavlova NN, Zhu J, Thompson CB. 2022. The hallmarks of cancer metabolism: still emerging. *Cell Metab* **34**: 355–377. doi:10.1016/j.cmet.2022.01.007

Pearce EL. 2010. Metabolism in T cell activation and differentiation. *Curr Opin Immunol* **22**: 314–320. doi:10.1016/j .coi.2010.01.018

Pession A, Valsecchi MG, Masera G, Kamps WA, Magyarosy E, Rizzari C, van Wering ER, Lo Nigro L, van der Does A, Locatelli F, et al. 2005. Long-term results of a randomized trial on extended use of high dose L-asparaginase for standard risk childhood acute lymphoblastic leukemia. *J Clin Oncol* **23**: 7161–7167. doi:10.1200/JCO.2005.11.411

Peyraud F, Guégan JP, Bodet D, Nafia I, Fontan L, Auzanneau C, Cousin S, Roubaud G, Cabart M, Chomy F, et al. 2022. Circulating L-arginine predicts the survival of cancer patients treated with immune checkpoint inhibitors. *Ann Oncol* **33**: 1041–1051. doi:10.1016/j.annonc.2022.07 .001

Popovici-Muller J, Lemieux RM, Artin E, Saunders JO, Salituro FG, Travins J, Cianchetta G, Cai Z, Zhou D, Cui D, et al. 2018. Discovery of AG-120 (Ivosidenib): a first-in-class mutant IDH1 inhibitor for the treatment of IDH1 mutant cancers. *ACS Med Chem Lett* **9**: 300–305. doi:10.1021/acsmedchemlett.7b00421

Przybyla-Zawislak B, Gadde DM, Ducharme K, McCammon MT. 1999. Genetic and biochemical interactions involving tricarboxylic acid cycle (TCA) function using a collection of mutants defective in all TCA cycle genes. *Genetics* **152**: 153–166. doi:10.1093/genetics/152.1.153

Rashmi R, Jayachandran K, Zhang J, Menon V, Muhammad N, Zahner M, Ruiz F, Zhang S, Cho K, Wang Y, et al. 2020. Glutaminase inhibitors induce thiol-mediated oxidative stress and radiosensitization in treatment-resistant cervical cancers. *Mol Cancer Ther* **19**: 2465–2475. doi:10.1158/1535-7163.MCT-20-0271

Reinbold R, Hvinden IC, Rabe P, Herold RA, Finch A, Wood J, Morgan M, Staudt M, Clifton IJ, Armstrong FA, et al. 2022. Resistance to the isocitrate dehydrogenase 1 mutant inhibitor ivosidenib can be overcome by alternative dimer-interface binding inhibitors. *Nat Commun* **13**: 4785. doi:10.1038/s41467-022-32436-4

Rodriguez PC, Quiceno DG, Zabaleta J, Ortiz B, Zea AH, Piazuelo MB, Delgado A, Correa P, Brayer J, Sotomayor EM, et al. 2004. Arginase I production in the tumor microenvironment by mature myeloid cells inhibits T-cell receptor expression and antigen-specific T-cell responses. *Cancer Res* **64**: 5839–5849. doi:10.1158/0008-5472.CAN-04-0465

Romero R, Sayin VI, Davidson SM, Bauer MR, Singh SX, LeBoeuf SE, Karakousi TR, Ellis DC, Bhutkar A, Sánchez-Rivera FJ, et al. 2017. Keap1 loss promotes Kras-driven lung cancer and results in dependence on glutaminolysis. *Nat Med* **23**: 1362–1368. doi:10.1038/nm.4407

Schimke RT, Kaufman RJ, Alt FW, Kellems RF. 1978. Gene amplification and drug resistance in cultured murine cells. *Science* **202**: 1051–1055. doi:10.1126/science.715457

Shapiro GI, Bell-McGuinn KM, Molina JR, Bendell JC, Spicer J, Kwak EL, Pandya S, Millham R, Borzillo G, Pierce K, et al. 2015. First-in-human study of PF-05212384 (PKI-587), a small-molecule, intravenous, dual inhibitor of PI3K and mTOR in patients with advanced cancer. *Clin Cancer Res* **21**: 1888–1895. doi:10.1158/1078-0432.CCR-14-1306

Singh D, Banerji AK, Dwarakanath BS, Tripathi RP, Gupta JP, Mathew TL, Ravindranath T, Jain V. 2005. Optimizing cancer radiotherapy with 2-deoxy-d-glucose dose escalation studies in patients with glioblastoma multiforme. *Strahlenther Onkol* **181**: 507–514. doi:10.1007/s00066-005-1320-z

Sonveaux P, Végran F, Schroeder T, Wergin MC, Verrax J, Rabbani ZN, Saedeleer CJD, Kennedy KM, Diepart C, Jordan BF, et al. 2008. Targeting lactate-fueled respiration selectively kills hypoxic tumor cells in mice. *J Clin Invest* **118**: 3930–3942. doi:10.1172/JCI36843

Stams WAG, den Boer ML, Beverloo HB, Meijerink JPP, Stigter RL, van Wering ER, Janka-Schaub GE, Slater R, Pieters R. 2003. Sensitivity to L-asparaginase is not associated with expression levels of asparagine synthetase in t

(12;21)⁺ pediatric ALL. *Blood* **101**: 2743–2747. doi:10.1182/blood-2002-08-2446

Stams WAG, den Boer ML, Holleman A, Appel IM, Beverloo HB, van Wering ER, Janka-Schaub GE, Evans WE, Pieters R. 2005. Asparagine synthetase expression is linked with l-asparaginase resistance in TEL-AML1–negative but not TEL-AML1–positive pediatric acute lymphoblastic leukemia. *Blood* **105**: 4223–4225. doi:10.1182/blood-2004-10-3892

Stein EM, DiNardo CD, Pollyea DA, Fathi AT, Roboz GJ, Altman JK, Stone RM, DeAngelo DJ, Levine RL, Flinn IW, et al. 2017. Enasidenib in mutant IDH2 relapsed or refractory acute myeloid leukemia. *Blood* **130**: 722–731. doi:10.1182/blood-2017-04-779405

Stine ZE, Schug ZT, Salvino JM, Dang CV. 2022. Targeting cancer metabolism in the era of precision oncology. *Nat Rev Drug Discov* **21**: 141–162. doi:10.1038/s41573-021-00339-6

Strand V, Cohen S, Schiff M, Weaver A, Fleischmann R, Cannon G, Fox R, Moreland L, Olsen N, Furst D, et al. 1999. Treatment of active rheumatoid arthritis with leflunomide compared with placebo and methotrexate. *Arch Intern Med* **159**: 2542–2550. doi:10.1001/archinte.159.21.2542

Sukumar M, Liu J, Mehta GU, Patel SJ, Roychoudhuri R, Crompton JG, Klebanoff CA, Ji Y, Li P, Yu Z, et al. 2016. Mitochondrial membrane potential identifies cells with enhanced stemness for cellular therapy. *Cell Metab* **23**: 63–76. doi:10.1016/j.cmet.2015.11.002

Sullivan LB, Luengo A, Danai LV, Bush LN, Diehl FF, Hosios AM, Lau AN, Elmiligy S, Malstrom S, Lewis CA, et al. 2018. Aspartate is an endogenous metabolic limitation for tumour growth. *Nat Cell Biol* **20**: 782–788. doi:10.1038/s41556-018-0125-0

Sun J, Nagel R, Zaal EA, Ugalde AP, Han R, Proost N, Song JY, Pataskar A, Burylo A, Fu H, et al. 2019. SLC1A3 contributes to L-asparaginase resistance in solid tumors. *EMBO J* **38**: e102147. doi:10.15252/embj.2019102147

Sykes DB, Kfoury YS, Mercier FE, Wawer MJ, Law JM, Haynes MK, Lewis TA, Schajnovitz A, Jain E, Lee D, et al. 2016. Inhibition of dihydroorotate dehydrogenase overcomes differentiation blockade in acute myeloid leukemia. *Cell* **167**: 171–186.e15. doi:10.1016/j.cell.2016.08.057

Takahashi H, Inoue J, Sakaguchi K, Takagi M, Mizutani S, Inazawa J. 2017. Autophagy is required for cell survival under L-asparaginase-induced metabolic stress in acute lymphoblastic leukemia cells. *Oncogene* **36**: 4267–4276. doi:10.1038/onc.2017.59

Tallal L, Tan C, Oettgen H, Wollner N, McCarthy M, Helson L, Burchenal J, Karnofsky D, Murphy ML. 1970. E. coli L-asparaginase in the treatment of leukemia and solid tumors in 131 children. *Cancer* **25**: 306–320. doi:10.1002/1097-0142(197002)25:2<306::AID-CNCR2820250206>3.0.CO;2-H

Tankova T, Senkus E, Beloyartseva M, Borštnar S, Catrinoiu D, Frolova M, Hegmane A, Janež A, Krnić M, Lengyel Z, et al. 2022. Management strategies for hyperglycemia associated with the α-selective PI3K inhibitor alpelisib for the treatment of breast cancer. *Cancers (Basel)* **14**: 1598. doi:10.3390/cancers14071598

Tannir NM, Motzer RJ, Agarwal N, Liu PY, Whiting SH, O'Keeffe B, Tran X, Fiji GP, Escudier B. 2018. CANTATA: a randomized phase 2 study of CB-839 in combination with cabozantinib vs. placebo with cabozantinib in patients with advanced/metastatic renal cell carcinoma. *JCO* **36**: TPS4601–TPS4601. doi:10.1200/JCO.2018.36.15_suppl.TPS4601

Thiersch JB. 1949. Bone-marrow changes in man after treatment with aminopterin, amethopterin, and aminoanfol; with special reference to megaloblastosis and tumor remission. *Cancer* **2**: 877–883. doi:10.1002/1097-0142 (194909)2:5<877::AID-CNCR2820020520>3.0.CO;2-0

Vander Heiden MG, DeBerardinis RJ. 2017. Understanding the intersections between metabolism and cancer biology. *Cell* **168**: 657–669. doi:10.1016/j.cell.2016.12.039

Venkatesan AM, Dehnhardt CM, Delos Santos E, Chen Z, Dos Santos O, Ayral-Kaloustian S, Khafizova G, Brooijmans N, Mallon R, Hollander I, et al. 2010. Bis(morpholino-1,3,5-triazine) derivatives: potent adenosine 5′-triphosphate competitive phosphatidylinositol-3-kinase/mammalian target of rapamycin inhibitors: discovery of compound 26 (PKI-587), a highly efficacious dual inhibitor. *J Med Chem* **53**: 2636–2645. doi:10.1021/jm901830p

Visentin M, Zhao R, Goldman ID. 2012. The antifolates. *Hematol Oncol Clin North Am* **26**: 629–648. doi:10.1016/j.hoc.2012.02.002

Voss M, Lorenz NI, Luger AL, Steinbach JP, Rieger J, Ronellenfitsch MW. 2018. Rescue of 2-deoxyglucose side effects by ketogenic diet. *Int J Mol Sci* **19**: 2462. doi:10.3390/ijms19082462

Warburg O. 1925. The metabolism of carcinoma cells. *J Cancer Res* **9**: 148–163. doi:10.1158/jcr.1925.148

Webster JA, Tibes R, Morris L, Blackford AL, Litzow M, Patnaik M, Rosner GL, Gojo I, Kinders R, Wang L, et al. 2017. Randomized phase II trial of cytosine arabinoside with and without the CHK1 inhibitor MK-8776 in relapsed and refractory acute myeloid leukemia. *Leuk Res* **61**: 108–116. doi:10.1016/j.leukres.2017.09.005

Weiss-Sadan T, Ge M, Hayashi M, Gohar M, Yao CH, de Groot A, Harry S, Carlin A, Fischer H, Shi L, et al. 2023. NRF2 activation induces NADH-reductive stress, providing a metabolic vulnerability in lung cancer. *Cell Metab* **35**: 487–503.e7. doi:10.1016/j.cmet.2023.01.012

Williams RT, Guarecuco R, Gates LA, Barrows D, Passarelli MC, Carey B, Baudrier L, Jeewajee S, La K, Prizer B, et al. 2020. ZBTB1 regulates asparagine synthesis and leukemia cell response to L-asparaginase. *Cell Metab* **31**: 852–861. e6. doi:10.1016/j.cmet.2020.03.008

Xiao Y, Yu TJ, Xu Y, Ding R, Wang YP, Jiang YZ, Shao ZM. 2023. Emerging therapies in cancer metabolism. *Cell Metab* **35**: 1283–1303. doi:10.1016/j.cmet.2023.07.006

Yan H, Parsons DW, Jin G, McLendon R, Rasheed BA, Yuan W, Kos I, Batinic-Haberle I, Jones S, Riggins GJ, et al. 2009. *IDH1* and *IDH2* mutations in gliomas. *N Engl J Med* **360**: 765–773. doi:10.1056/NEJMoa0808710

Yap TA, Daver N, Mahendra M, Zhang J, Kamiya-Matsuoka C, Meric-Bernstam F, Kantarjian HM, Ravandi F, Collins ME, Francesco MED, et al. 2023. Complex I inhibitor of oxidative phosphorylation in advanced solid tumors and acute myeloid leukemia: phase I trials. *Nat Med* **29**: 115–126. doi:10.1038/s41591-022-02103-8

Yen K, Travins J, Wang F, David MD, Artin E, Straley K, Padyana A, Gross S, DeLaBarre B, Tobin E, et al. 2017. AG-221, a first-in-class therapy targeting acute myeloid leukemia harboring oncogenic *IDH2* mutations. *Cancer Discov* **7**: 478–493. doi:10.1158/2159-8290.CD-16-1034

Zhang J, Simpson CM, Berner J, Chong HB, Fang J, Ordulu Z, Weiss-Sadan T, Possemato AP, Harry S, Takahashi M, et al. 2023. Systematic identification of anticancer drug targets reveals a nucleus-to-mitochondria ROS-sensing pathway. *Cell* **186**: 2361–2379.e25. doi:10.1016/j.cell .2023.04.026

Zwart NRK, Franken MD, Tissing WJE, Lubberman FJE, McKay JA, Kampman E, Kok DE. 2023. Folate, folic acid, and chemotherapy-induced toxicities: a systematic literature review. *Crit Rev Oncol Hematol* **188**: 104061. doi:10.1016/j.critrevonc.2023.104061

Cite this article as *Cold Spring Harb Perspect Med* doi: 10.1101/cshperspect.a041657

Adoptive Cell Therapy for Pediatric Solid Tumors

Amy B. Hont and Catherine M. Bollard

Center for Cancer and Immunology Research, Children's National Hospital, Washington, D.C. 20010, USA

Correspondence: ahoughte@childrensnational.org

Patients with relapsed or refractory pediatric solid tumors have limited therapeutic options with little to no appreciable improvements in outcomes in over two decades. Adoptive cell therapy (ACT) is a promising, targeted option for patients with the potential to minimize acute and long-term toxicities. In this review, we (1) characterize the development and manufacture different ACT approaches used for pediatric solid tumors, and (2) discuss the obstacles when targeting and treating solid tumors. The outcomes of the clinical applications of the various cell therapy products are also reviewed along with the future potential, including novel product development and combination therapies. In sum, this review serves as a comprehensive review of the clinical trial results evaluating the safety, feasibility, and efficacy of novel cell therapy products in the clinic for the treatment of pediatric solid tumors and seeks to provide new insights regarding ACT successes, failures, and challenges to benefit a rapidly expanding immunotherapy field.

Patients with relapsed or refractory solid tumors have poor outcomes. Cancer is a leading cause of death in children age 0–14 years, second only to accidents (Siegel et al. 2018), and solid tumors account for more than half of these deaths (Siegel et al. 2020). Solid tumors make up one-third of all cancers in children age 0–1, and more than half of cancers for adolescents age 15–19 years (Allen-Rhoades et al. 2018; American Cancer Society 2018). While select patients with localized or low-risk disease have good outcomes with current treatment strategies (American Cancer Society 2018), a vast majority of patients with relapsed/refractory solid tumors have a dismal prognosis, often with <25% of patients surviving at 5 years despite aggressive, toxic therapies (Ceschel et al. 2006; Kim et al.

2008). Relapse remains the most common cause of death (Ward et al. 2014; Malogolowkin et al. 2017), while survivors have serious long-term sequelae related to therapy. The development of effective and targeted therapies is therefore critical for this patient population.

Adoptive cell therapy (ACT) offers the opportunity to harness the power of the immune system in a targeted manner to eradicate malignancy and has proven to be safe and feasible, although challenges to clinical efficacy persist. Immune evasion mechanisms, particularly antigen loss, and an unfavorable tumor microenvironment (TME) (Majzner and Mackall 2018) cultivated by solid tumors have resulted in only modest responses thus far (Park et al. 2007; Pule et al. 2008; Robbins et al. 2011; Heczey and Louis

2013; Lamers et al. 2013; Zhang et al. 2016; Heczey et al. 2017; Hont et al. 2019; Kulczycka et al. 2023). While ACT in the form of the recently Food and Drug Administration (FDA)-approved tumor-infiltrating lymphocyte (TIL) therapy has demonstrated remarkable outcomes for metastatic or refractory melanoma (Rosenberg et al. 1986, 2008, 2011; Dudley et al. 2002, 2008), translation to pediatric solid tumors has been limited by the fact that pediatric tumors are less immunogenic (Downing et al. 2012; Casey and Cheung 2020). To date, chimeric antigen receptor (CAR)-T-cell products, native or engineered T-cell receptor (TCR) products, natural killer (NK) cell products, and TILs have produced varying levels of success in pediatric solid tumors. Several advances in product generation and insights to antitumor activity in vivo have however led to promising approaches in investigation in active clinical trials.

This review will address the development and manufacture of ACTs for pediatric solid tumors, describe the obstacles to clinical efficacy in vivo and propose future directions of ACT for the treatment of pediatric solid tumors

CONSIDERATIONS FOR TARGETING PEDIATRIC SOLID TUMORS WITH ACT

Target Identification

Identifying immunogenic targets is crucial when designing effective ACT strategies targeting pediatric solid tumors. These antigens must be widely expressed on the tumor with minimal expression in healthy tissues to optimize "on-target, on-tumor" activity and minimize "off-target" and/or "off-tumor" toxicities. Such targets can be uniform for a specific cell therapy product (e.g., CD19 CAR-T for B-cell malignancies), as a one-size-fits-most approach. This has the advantage of standardized production, facilitating product characterization, and enhanced clarity of the risks associated with the antigen-specific T-cell population. Conversely, target antigens may be individualized per patient, similar to the TIL approach, in which tumor-reactive T cells are cultured from a patient's tumor to be reinfused at clinically

relevant numbers, typically with cytokine support. As a result, there is less known about the antigen-specificity of TIL T-cell population, but the approach carries the advantage of generating and expanding a T-cell population with an antitumor response specific to the patient's own tumor. Additionally, the T cells may target both known tumor-associated antigens (TAAs) or neoantigens, providing a comprehensive, personalized product.

Efforts to identify immunogenic antigens expressed on pediatric solid tumors continue to grow the list of targets, although no one antigen has provided the key to cell therapy success to date. The cancer testis antigen (CTA) group, so-called for their identification on malignant cells as well as reproductive organs such as testis and placenta, is a group of antigens long-identified on solid tumors (as well as hematologic malignancies). This group is wide ranging and includes TAAs such as the MAGE family, GAGE family, SSX family, preferentially expressed antigen in melanoma (PRAME), and NY-ESO-1 (Ishida et al. 1996; Dalerba et al. 2001; Chi et al. 2002; Naka et al. 2002; Oberthuer et al. 2004; Epping and Bernards 2006; Hunder et al. 2008; Robbins et al. 2011; Toledo et al. 2011; Brenne et al. 2012; Pollack et al. 2012; Tan et al. 2012; Weide et al. 2012; Krishnadas et al. 2013; Yao et al. 2014; Roszik et al. 2017). Oncofetal antigens are a similar group with shared overexpression on fetal and cancer tissues, including glypican-3 (GPC3) (Kinoshita et al. 2015; Moek et al. 2018; Ortiz et al. 2019; Zheng et al. 2022). Another frequently investigated group of antigens is comprised of the products of overexpressed genes that convey a survival or metastatic benefit. These include survivin and other members of the inhibitors of apoptosis proteins (IAPs), Wilms tumor 1 (WT1), SOX2, HER2, GD2, and B7-H3. Broadly, the vast majority of these antigens have limited expression in healthy tissues (Shinozawa et al. 2000; Scharnhorst et al. 2001; Yin 2011; Chau and Hastie 2012), are up-regulated on solid tumors (Tamm et al. 1998; Shinozawa et al. 2000; Lee and Haber 2001; Takamizawa et al. 2001; Carpentieri et al. 2002; Altieri 2003; Oberthuer et al. 2004; Caldas et al. 2006; Coughlin et al. 2006; Fukuda and Pelus 2006; Hylander et al. 2006; Barbolina et al. 2008; Brenne et al. 2012; Yang et al. 2012; Brett et al. 2013; Kim et al. 2014; Kopp and Katsanis 2016; Li et al. 2016; Roszik et al. 2017; Zayed and Petersen 2018; Majzner et al. 2019; Wang et al. 2020; Hingorani et al. 2022), and typically associate with more aggressive disease, chemotherapy resistance, and decreased survival (Takamizawa et al. 2001; Tran et al. 2002; Tamm et al. 2004; Segal et al. 2005; Coughlin et al. 2006; Epping and Bernards 2006; Pennati et al. 2007; Basta-Jovanovic et al. 2012; Tan et al. 2012; Qi et al. 2015; Xie et al. 2016). Therefore, multiple potential targets have been identified in pediatric cancers but targeting of such antigens can be challenging due to the panoply of immune evasion strategies these tumors employ, including antigen loss, major histocompatibility complex (MHC) down-regulation, and the immune suppressive TME.

Antigen Escape and MHC Loss/Down-Regulation

A crucial consideration when targeting pediatric cancer is the potent immune escape mechanism of antigen loss. This has proven to be particularly challenging in pediatric solid tumors, which display heterogenous antigen expression and varying degrees of antigen density, thus are more easily able to down-regulate targeted antigens to escape tumor-directed cytolytic activity. The simultaneous targeting of multiple antigens, as in native TCR products such as TAA-specific T cells (TAA-T) (Chapuis et al. 2013, 2016; Hont et al. 2019; Lulla et al. 2020, 2021; Vasileiou et al. 2021; Kinoshita et al. 2022) and TILs (Topalian et al. 1987; Rosenberg et al. 1998; Dudley et al. 2001, 2002, 2008; Rosenberg et al. 2008, 2011; Stevanović et al. 2015; Goff et al. 2016), has the potential to alleviate the problem of antigen loss. These products have also demonstrated the ability to induce antigen spreading, through exposure to new TAAs and recruitment of an endogenous antitumor response (Bollard et al. 2014; Gulley et al. 2017; Hont et al. 2019; Brossart 2020).

Down-regulation of the MHC presents an additional potent immune evasion strategy to prevent attack by T cells recognizing the tumor via the native TCR in a human leukocyte antigen (HLA)-restricted manner. It has been posited that IFN-γ secreted by ACTs can potentially up-regulate MHC in vivo. However, the use of CAR-T can overcome this limitation and, while technically challenging, CAR-T directed to target multiple antigens simultaneously are in progress with mixed preclinical results (Hegde et al. 2016; Yang et al. 2021).

The continued observation of antigen loss and MHC down-regulation is indicative that there will likely need to be a multimodal approach to reduce tumor burden, up-regulate antigen and MHC expression, and facilitate T-cell activity. These findings highlight the importance of obtaining follow-up tumor tissue to monitor for this specific phenomenon (MHC and antigen loss/down-regulation) post-ACT when feasible and safe for patients.

Overcoming the Immune Suppressive Tumor Microenvironment

The immunosuppressive TME can lead not only to immune escape (MHC down-regulation and antigen escape, as above) but can directly contribute to the limited persistence of cell therapy products in vivo. While physical barriers (e.g., the blood–brain barrier for brain tumors) can obstruct the migration of infused cell therapy products to their target, tumors utilize several strategies to impede antitumor cytotoxicity. The TME is composed of a complex and dynamic milieu cultivated to promote tumor growth and progression (Table 1).

Pediatric solid tumors are described as a group as immune "deserts" or "cold," meaning that tumor mutational burden and immunogenic potential is low (Downing et al. 2012; Casey and Cheung 2020). As such, an important aspect of creating successful immune cell therapy in pediatric solid tumors is transformation of a cold or immune suppressive TME to a hot or immunogenic TME (Terry et al. 2020), which can be accomplished through polarization of the cell infiltrate to immune activated or exposure of

Table 1. The immune evasion tactics of the tumor microenvironment

TME component		Mechanism of immune evasion	Mitigating approaches
Antigen escape and major histocompatibility complex (MHC) down-regulation		Down-regulation of targeted antigen in MHC-independent therapies (e.g., chimeric antigen receptor [CAR]-T cells) or of MHC complex in MHC-restricted cytolysis (e.g., tumor-associated antigen T cell [TAA-T] products)	Multiantigen targeting, epigenetic modifiers or inflammatory cytokines (e.g., IFN-γ) to induce antigen/MHC expression
Cellular milieu	Regulatory T cells (Tregs)	Secretion of/excess inhibitory cytokines (IL-10, TGF-β, adenosine) to suppress natural killer (NK) and T-cell activation; decreased IFN-γ, IL-2; dendritic cell (DC) tolerance (Li et al. 2020; Scott et al. 2021)	Repolarization of cells to inflammatory phenotype (e.g., M1 tumor-associated macrophages [TAMs]) (Boutilier and Elsawa 2021) Cytokine-directed strategies to mitigate downstream effects of immune suppressive cell population (see cytokine milieu below) Immune checkpoint blockade (Laumont et al. 2022) CD40 agonist antibodies (Morrison et al. 2020; Vonderheide 2020) Targeted therapies to modulate immune cell phenotype (e.g., JAK inhibitors, ATRA, vitamin D) (Sherman et al. 2014; Biffi et al. 2019; Chen et al. 2021)
	Myeloid-derived suppressor cells (MDSCs)	TAM precursor, promote immunosuppressive cytokines (IL-10, TGF-β, adenosine), depletion of T-cell nutrients (Kumar et al. 2016)	
	TAMs (M2)	Secretion of IL-10 (Zhang et al. 2012), perpetuated by IL-4 and TGF-β; promotion of angiogenesis, tumor progression/metastasis (Zhang et al. 2012; Anderson and Simon 2020; Pittet et al. 2022)	
	Tumor-associated neutrophils (TANs)	N2 TANs associated with worse prognosis (Templeton et al. 2014; Coffelt et al. 2016; Masucci et al. 2019)	
	B cells	B regulatory function induces tolerance, secretion of immunosuppressive cytokines (IL-10, TGF-β, adenosine) (Anderson and Simon 2020; Laumont et al. 2022; Laumont and Nelson 2023)	
	Cancer-associated fibroblasts (CAFs)	Secretion of immunosuppressive cytokines (IL-10, TGF-β, IL-6, IL-9), vascular endothelial growth factor (VEGF) remodeling of extracellular matrix, angiogenesis, metabolic competition, exosomes (Xing et al. 2010; Chen et al. 2021)	
Cytokine/ enzyme milieu	TGF-β	Known to hinder the immune response, promotes Treg and other immune cell differentiation, antagonizes DC differentiation (Hanks et al. 2013; Pickup et al. 2013; Caja and Vannucci 2015)	Modification of cell therapy products to render resistance and/or sinking of cytokines (Yvon et al. 2017; Bollard et al. 2018; Fukumura et al. 2018; Burga et al. 2019) Antibodies (neutralizing or blocking), decoy receptors, enzymatic inhibitors, or tyrosine kinase inhibitors to decrease production or interfere with
	IL-10	Can directly impair antigen presentation and antigen-presenting cells (APCs), decreases Th1 cytokines (Itakura et al. 2011; Sato et al. 2011)	

Continued

 Cite this article as *Cold Spring Harb Perspect Med* doi: 10.1101/cshperspect.a041636

Table 1. *Continued*

TME component	Mechanism of immune evasion	Mitigating approaches
IL-8	Induces epithelial-to-mesenchymal transition (EMT), promotes angiogenesis, recruits immune suppressive cells, induces chemoresistance (Fousek et al. 2021)	signaling (Finley and Popel 2013; Munn and Mellor 2016; Fox et al. 2018; Liu et al. 2018; Anderson and Simon 2020; Fousek et al. 2021; Sosnowska et al. 2021; Ramji et al. 2022)
VEGF	Induces angiogenesis, promotes migration and metastasis (Finley and Popel 2013), enhances quantity and function of Treg, MDSC, and TAM population, inhibits effector T-cell function and DC priming (Fukumura et al. 2018)	
Indoleamine-2,3-dioxygenase (IDO)	Anti-inflammatory, associated with tolerogenic response in antigen-specific T cells and APCs, Treg and MDSC activation, induced by and supporting expression of immune checkpoints (Munn and Mellor 2016; Liu et al. 2018)	
Arginase	Depletes arginine to attenuate effector T-cell response, impair activation by APCs, and induce Tregs (Rodriguez et al. 2004; Grzywa et al. 2020)	
Exosomes	Microvesicles within TME that promote tumor progression, angiogenesis, and metastasis; promote CAFs, attenuate effector T-cell response; transfer immune checkpoint expression (Roma-Rodrigues et al. 2014; Li and Nabet 2019; Wu et al. 2019; Anderson and Simon 2020; Jin et al. 2022)	Inhibition of exosome formation and release (Huang et al. 2019; Poggio et al. 2019; Kim et al. 2022); may be used as synthetic drug delivery systems (Roma-Rodrigues et al. 2014; Wu et al. 2019; Jin et al. 2022)
Immune checkpoint expression	Expression of immune checkpoints (CTLA-4, PD-1, PD-L1, LAG-3, TIM-3) on the tumor and/or effector cell surface induce anergy and exhaustion	Immune checkpoint blockade (Postow et al. 2015; Kyi and Postow 2016; Long et al. 2022)

antigens to support antigen spread and ongoing immune response (Gulley et al. 2017). The infiltrating immune cell population in particular can secrete cytokines, chemokines, and enzymes to promote tumor survival, progression, and metastasis, including transforming growth factor β (TGF-β), vascular endothelial growth factor (VEGF), IL-10, IL-8, indoleamine-2,3-dioxygenase (IDO), adenosine, and arginase (Tormoen et al. 2018; Labani-Motlagh et al. 2020). This immunosuppressive cell population includes regulatory T cells (Tregs), myeloid-de-rived suppressor cells (MDSCs), tumor-associated macrophages (TAMs), tumor-associated neutrophils (TANs), cancer-associated fibroblasts (CAFs), and regulatory B cells (Bregs) to support tumor growth and metastasis (Yang et al. 2020; Scott et al. 2021), which in turn maintains the survival and proliferation of these cells, perpetuating an immunosuppressive, self-sustaining, pro-TME (Kumar et al. 2016; Yang et al. 2020). A higher proportion of these cells within the TME are associated with a worse prognosis, while polarization to an inflammatory phenotype (e.g., CD8[+] cytotoxic T [Tc] cells, CD4[+] helper T [Th1] cells, B cells, TANs, and TAMs), can be positive prognostic indicators (Anderson and Simon 2020; Boutilier and El-sawa 2021; Downs-Canner et al. 2022; Pittet et al. 2022), likely further supported by cytokine secretion (e.g., IFN-γ, IL-2) and implicating a potential mechanism of overcoming the TME. For example, macrophages and neutrophils are plastic and able to be polarized to an M1/N1 phenotype through Th1 signals (e.g., tumor ne-crosis factor [TNF], IFN-γ, IL-12), while Th2 signals (TGF-β, IL-4, IL-13) and hypoxia sup-port M2/N2 polarization (Fridlender et al. 2009; Boutilier and Elsawa 2021; Linde et al. 2023). Similarly, the TME can subvert the activation of dendritic cells (DCs) through improper prim-ing to induce a tolerogenic response to TAAs and enhance Treg polarization (Labani-Mot-lagh et al. 2020). Approaches to various compo-nents of the TME are summarized in Table 1.

The cumulative effect of the TME to pro-mote tumor survival is manifested in other ways as well. On the cell surface, tumors can regulate the expression of proteins to decrease activation (i.e., MHC down-regulation or Fas) or promote immune cell inactivation. Immune checkpoints such as programmed death-ligand 1 (PD-L1) on the tumor surface induce T-cell anergy and exhaustion, and cytokines in the TME can further promote exhaustion via up-regulation of PD-1, CTLA-4 LAG-3, TIM-3 on the T-cell surface (Table 1). The high energy rate and metabolic plasticity of tumors lead to a rel-atively hypoxic, low-nutrient environment in which immune cells are inhibited, and tumors are able to thrive (Tang et al. 2021). These effects all require consideration and, when possible, mitigation when creating cell therapies. Modifi-cation of ACT products with cytokine receptors that are either resistant to immunosuppressive tactics or promote T-cell activation upon bind-ing will likely play an important role in over-coming the TME (Bollard et al. 2018). These modified immune products can (1) lead to sink-ing of the cytokine from the TME, depriving the tumor its effects, (2) provide a homing mecha-nism for infused immune cells to traffic to tu-mors more efficiently, and (3) translate the nat-

urally immunosuppressive signal of the secreted cytokine into a neutral (and thus resistant) or activation signal to further promote adoptively transferred cell persistence and antitumor cyto-toxicity.

Location and Homing

Outside of creating a barrier through the immu-nosuppressive TME, tumors can be difficult to target through means of poor or inadequate trafficking of infused cell therapies to the site(s) of solid tumors. Modification of cell therapy through enhanced or artificial expression of chemokine receptors (Moon et al. 2011; Wang et al. 2021, 2023) or local delivery of cell therapy (Adusumilli et al. 2014; Sagnella et al. 2022) can optimize tumor penetration and enhance clini-cal responses (Craddock et al. 2010; Mirzaei et al. 2017). Tumors may also be transformed through low-dose radiation to expose neoantigens and enhance the secretion of chemokines attracting T cells to the tumor site (Morotti et al. 2021).

GENERATION OF ACT PRODUCTS

ACT products require different manufacturing strategies depending on mechanism of cytolytic activity (see Fig. 1). The difference in approach has subsequent implications on the antitumor function in vivo. The various manufacturing strategies are summarized below.

Tumor-Infiltrating Lymphocytes (TILs)

The therapeutic potential of ACT for the treat-ment of cancers was established in the 1980s by Rosenberg et al. (1986), who identified the pro-cess of using TILs for relapsed or refractory mel-anoma. This manufacturing process, which has recently garnered FDA approval, requires the dissection of a tumor specimen that is then cul-tured in medium with high-dose IL-2 (6000 IU/mL) to expand antigen-specific T cells (Dudley et al. 2003). Products are selected by tumor-spe-cific activity after coculture with a tumor speci-men and expanded to clinically significant num-bers. TIL products are infused into patients after a preparatory regimen, which can include total body irradiation and/or lymphodepleting che-

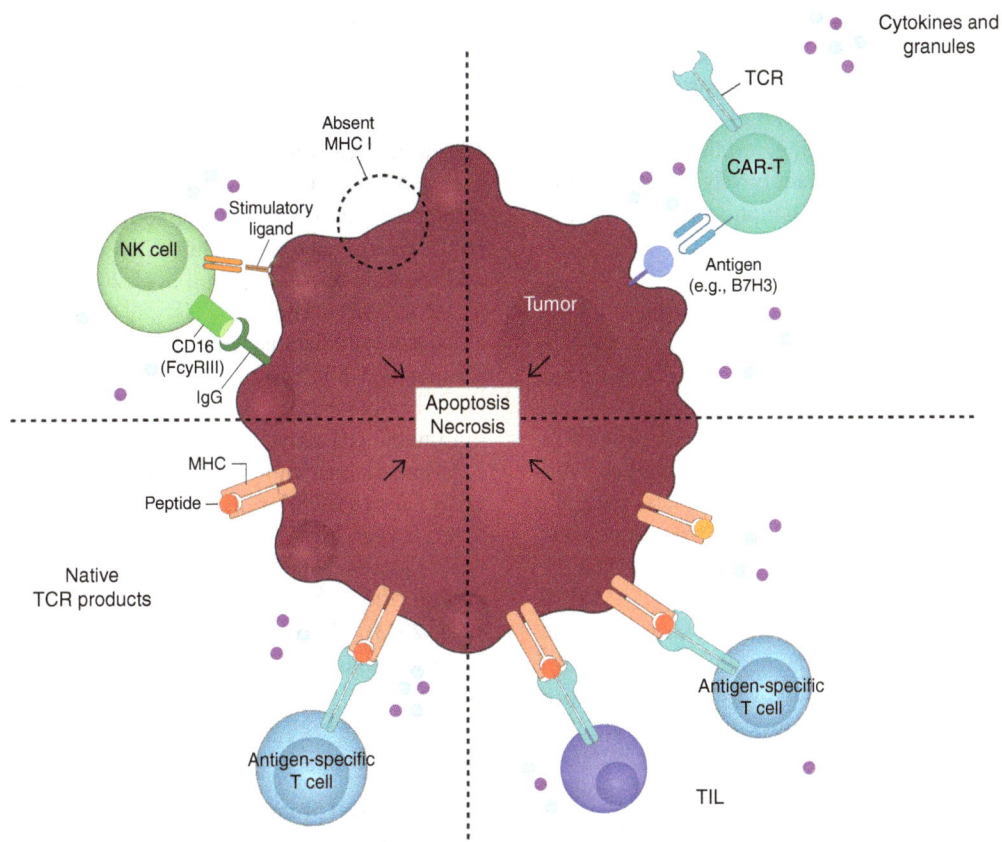

Figure 1. Summary of differing approaches to adoptive cell therapy for pediatric solid tumors including (1) Chimeric antigen receptor (CAR)-T products with the CAR targeting a surface antigen; (2) tumor-infiltrating lymphocyte (TIL) products, with increased heterogeneity and specificity to wide variety of antigens; (3) native T-cell receptor (TCR) products, with ability to target intracellular and extracellular antigens; and (4) natural killer (NK) cell products, independent of major histocompatibility complex (MHC). Cytokines (IFN-γ and TNF-α) and granules (perforin, granzyme B) released upon activation of adoptively transferred cell therapy, leading to tumor apoptosis and necrosis.

motherapy along with serial administration of IL-2 (Rosenberg et al. 2011). Given the difference in immunobiology of the TME in pediatric solid tumors, including the immune cell infiltrate, advances have made slower progress in the pediatric population. However, recent preclinical studies are encouraging (Lussier et al. 2015; Ollé Hurtado et al. 2019; Hensel et al. 2022).

CAR-T

The generation of CAR-T products has evolved with significant gains since the first generation in 1989 and the early 1990s (Gross et al. 1989;

Eshhar et al. 1993), after identifying the need for costimulatory domain. The first clinical application in the late 2000s demonstrated the potential for this cell product to eradicate relapsed/refractory cancer, most successfully used in diffuse large B-cell lymphoma and pediatric B-cell acute lymphoblastic leukemia with FDA approvals for these disease indications (Brentjens et al. 2011, 2013). In general, to manufacture CAR-T products, T cells from a patient or healthy donor are activated and expanded nonspecifically in vitro and transduced to express an artificial receptor for the desired target. Subsequent iterations, or generations, of CAR-T products have incorporated enhanced stimulatory domains to support T-cell function and persistence. More recently, new approaches that secrete supporting cytokines (TRUCKs) upon stimulation are being evaluated. Typically, CAR-T products are adoptively transferred after a fludarabine-based lymphodepleting chemotherapy (Amini et al. 2022).

TCR-Specific Products

Native αβTCR products, or multi-TAA-T products are expanded ex vivo similar to the approach used for virus-specific T cells (VSTs) (Hont et al. 2022). Because circulating TAA-T cells are present at low frequencies naturally, this typically requires repeated stimulations with antigen-presenting cells (APCs) to either (1) stimulate and expand antigen-specific T cells to clinically relevant numbers, or (2) select products via expression of activation signals (IFN-γ, 41BB) or streptamers followed by ex vivo expansion in the presence of cytokines (Becker et al. 2001; Peggs et al. 2011; Freimüller et al. 2015; Lee et al. 2020; Jiang et al. 2023). T-cell products can also be genetically modified to express the TCRα and β chains that confer a predesignated specificity (Shafer et al. 2022). This requires precise identification of immunogenic, preserved epitopes and the HLA restriction.

Off-the-Shelf Products

ACT products may be optimized using healthy (allogeneic) donors. Autologous cell therapy products have demonstrated an good safety profile outside of cytokine release syndrome (CRS), but patients are often subject to other salvage therapies while awaiting generation and release testing. Autologous immune cells may also be predisposed to an anergic or exhausted response to targeted antigens that would preclude in vivo efficacy. Hence, the ability to treat patients with a partially HLA-matched allogeneic product or an HLA-agnostic product would allow for "on demand" treatment at the ideal timing with an optimized product. Third-party multi-VSTs have demonstrated safety and efficacy including in the cancer setting (e.g., EBV⁺ lymphomas)

(Leen et al. 2013; Tzannou et al. 2017), suggesting that this approach could be feasible and safe in patients with non-viral-associated malignancies using a similar multiantigen-specific T-cell approach targeting nonviral TAAs. Off-the-shelf CAR-T cells have been investigated with strategies (genetic engineering, cell selection) to reduce risks of graft-versus-host-disease (GVHD) and immune rejection (Depil et al. 2020; Aparicio et al. 2023). Allogeneic NK cell therapy has also been well tolerated, given the ability of KIR-mismatched NK cell to eliminate host APCs; however, they may have a multifactorial role in immune/T-cell activation (Chu et al. 2022).

The cellular composition, specifically the ratio of CD4 to CD8 T cells, in cell therapy products has been explored for CD19 CAR-T products (e.g., lisocabtagene maraleucel or liso-cel) (Turtle et al. 2016; Ogasawara et al. 2022; Galli et al. 2023). However, the importance of this approach is not as well developed for ACT products targeting pediatric solid tumors.

Other Cell Types

While αβT-cell products comprise the majority of ACT in clinical use and are the focus of this review, other cell types are being investigated for targeted ACT against cancer. While a comprehensive review of these non-T-cell products lies outside the scope of this review, these approaches are briefly summarized as follows: NK cell and invariant NK T (iNKT)-cell products can be expanded ex vivo from patient or healthy donor blood with the use of supportive cytokines and in some cases feeder cells to be infused in patients (Chu et al. 2022; Delfanti et al. 2022; Xiao et al. 2022). Similarly, γδT-cell products are also a potentially safe, translatable product (Yazdanifar et al. 2020; Ma et al. 2023). Finally, investigators have generated CAR-NK cell products with promising results, in a process similar to CAR-T-cell product generation (Chu et al. 2022; Xiao et al. 2022).

APPROACH TO ACT IN CLINICAL USE FOR PEDIATRIC SOLID TUMORS

Lymphodepleting chemotherapy creates a favorable environment for homeostatic proliferation and enhanced persistence in vivo (Williams et al. 2007) and has been well tolerated (Dudley et al. 2002; Rüttinger et al. 2007; Rosenberg et al. 2008). In combination with adoptive cell transfer, lymphodepleting chemotherapy has been associated with improved responses and durable remissions (Rosenberg et al. 1986; Muranski et al. 2006; Dudley et al. 2008; Hay and Turtle 2017). Optimization with lymphodepleting chemotherapy may promote better clinical responses and further prolong progression-free survival (PFS) by supporting T-cell persistence and altering the TME (Rosenberg et al. 1986, 2008; Dudley et al. 2002, 2008; Muranski et al. 2006; Rüttinger et al. 2007; Williams et al. 2007; Brentjens et al. 2011; Hay and Turtle 2017). Historically, these lymphodepleting regimens have been implemented and optimized in the setting of CAR-T for hematologic malignancies, primarily consisting of fludarabine and cyclophosphamide administered over 3–5 days prior to

adoptive cell transfer. Lymphodepleting chemotherapy has also been highly beneficial and well tolerated in the context of TIL therapy for melanoma (Dudley et al. 2002; Muranski et al. 2006; Rosenberg et al. 2011). However, such approaches have not been as robust outside of these settings. Therefore, there may be a unique lymphodepleting cocktail that would benefit patients receiving ACT for pediatric solid tumors and this remains an area to be explored.

Solid tumors present several challenges in vivo (as described herein) for the successful usage of ACT. So far, there have been limited improvements in outcomes for high-risk patients with solid tumors. In pediatric patients with solid tumors, the most frequently investigated ACT approaches have been CAR-T and TCR products (Guzman et al. 2023), although effective targets have been challenging to identify. Table 2 summarizes the differences in these approaches, while Table 3 details the clinical results. Because

Table 2. Summary of adoptive cell therapy products

Cell product	Pro	Con
Tumor-infiltrating lymphocytes (TILs)	Polyfunctional, multiantigen specific, personalized product	Unpredictable reproducibility in the isolation and expansion of TILs, specifically in pediatric solid tumors; relatively unknown product specificity; major histocompatibility complex (MHC)-restricted
Chimeric antigen receptor (CAR)-T products	Established manufacturing process with success in r/r hematologic malignancies Non-MHC-restricted	Classically single antigen targeting with high rate of antigen loss Dependent on the expression of surface protein and antigen density Toxicity (cytokine release syndrome [CRS], neurotoxicity, pseudoprogression) Time to generate product
αβTCR products	Multiantigen specificity Target intracellular and extracellular antigens Favorable safety profile Enhancement of antitumor immunity	Less robust antitumor response MHC-restricted Time to generate product Potential T-cell anergy or exhaustion to tumor-associated antigens (TAAs)
Natural killer (NK) cell products	Non-MHC-restricted	Potential for off-target toxicity Inhibited by MHC I expression
Off-the-shelf products	Rapid administration Less anergic/exhausted immune cell population	Potential for graft-versus-host-disease (GVHD) Smaller memory cell population May need to consider human leukocyte antigen (HLA) restriction

Table 3. Clinical experience of cell therapy for pediatric solid tumors

Cell product type	Product/diagnosis	Patient population	Clinical response	Safety/toxicity and other clinical activity
Chimeric antigen receptor (CAR)-T products	GD2 CAR-T to treat diffuse midline glioma administered IV and ICV (Majzner et al. 2022)	Four patients (interim report); two children, two adults	Prolonged overall survival: three of four patients experienced clinical and radiographic improvement (Majzner et al. 2022)	Grade 1–3 CRS, tumor inflammation-associated neurotoxicity; managed with intensive supportive care
	HER2 CAR-T to treat pediatric brain tumors administered locoregionally (Vitanza et al. 2021)	Three patients (interim report); one child, two adults	1 SD, 2 PD (Vitanza et al. 2021)	No dose-limiting toxicities (DLTs); overall safe with clinical evidence of inflammation (fever, headache, pain)
	HER2 CAR-T to treat pediatric solid tumors (Ahmed et al. 2015)	19 patients; 12 children, seven adults	17 evaluable patients: SD ($n = 4$), PD ($n = 13$)	Well tolerated, no DLTs
	GD2 CAR-T for neuroblastoma (Heczey et al. 2017; Straathof et al. 2020; Del Bufalo et al. 2023)	12 patients; 12 children, zero adults	Three patients with tumor regression but no objective clinical responses (Straathof et al. 2020)	Well tolerated with two cases of cytokine release syndrome (CRS) (grade 2–3); no on-target, off-tumor toxicity
		11 patients; nine children, two adults	Four patients infused on cohort 1 (dose escalation CAR-T), SD ($n = 2$), PD ($n = 2$); four patients infused on cohort 2 (lymphodepletion [LD] prior to CAR-T infusion), SD ($n = 1$), PD ($n = 3$); three patients infused on cohort 3 (LD prior to CAR-T infusion with PD-1 inhibitor coadministration), SD ($n = 2$), PD ($n = 1$) (Heczey et al. 2017)	Well tolerated, no DLTs; isolated fever ($n = 5$), fever and neutropenia ($n = 2$); grade 2 CRS ($n = 1$)
		27 patients; 25 children, two adults	CR ($n = 9$), PR ($n = 8$), SD ($n = 5$), PD ($n = 5$) (Del Bufalo et al. 2023)	No DLTs; CRS documented in 20 patients, only one of which required suicide gene activation to clear product
	CE7R CAR-T for neuroblastoma: stable disease (Park et al. 2007)	Six patients; six children, zero adults	Five evaluable patients, PR ($n = 1$), PD ($n = 4$) (Park et al. 2007)	Well tolerated, no on-target, off-tumor toxicity Grade 3 pneumonitis ($n = 1$) and grade 3 cytopenia ($n = 4$) occurred in 12 infusions

Continued

Table 3. *Continued*

Cell product type	Product/diagnosis	Patient population	Clinical response	Safety/toxicity and other clinical activity
CAR-modified antigen-specific T cells	VSTs modified with HER2-CAR to treat glioblastoma (Ahmed et al. 2017)	17 patients; seven children, 10 adults	16 evaluable patients, PR ($n = 1$), SD ($n = 7$), PD ($n = 8$); three patients with SD are alive 24–29 mo postinfusion without disease progression (Ahmed et al. 2017)	Well tolerated, no DLTs; grade 2 headache and/or seizure ($n = 2$)
	EBV-specific T cells modified with GD2-CAR to treat r/r neuroblastoma (Pule et al. 2008)	11 patients; 11 children, zero adults	Eight evaluable patients, CR ($n = 1$); three additional patients with tumor necrosis or regression (Pule et al. 2008)	Well tolerated without safety concerns
αβTCR products	TAA-T (target WT1, PRAME, survivin) for r/r pediatric solid tumors (Hont et al. 2019)	15 patients; 13 children, two adults	SD ($n = 11$), PD ($n = 4$); well tolerated without DLTs (Hont et al. 2019)	Well tolerated, no DLTs or infusion-related adverse events; no incidence of CRS; grade 1 fatigue ($n = 1$) and grade 1 myalgia ($n = 1$)
	NY-ESO-1-specific T cells administered with systemic IL-2 for A*02-positive patients including adolescents and young adults (AYAs) with r/r synovial sarcoma (Robbins et al. 2011)	11 patients with metastatic melanoma; six patients with synovial sarcoma; two young adults (age 19 and 20 yr); 15 adults	Metastatic melanoma, CR ($n = 2$), PR ($n = 3$), PD ($n = 6$); synovial cell sarcoma, PR ($n = 4$), PD ($n = 2$) (Robbins et al. 2011)	Well tolerated, no toxicities associated with transferred cells; transient neutropenia and thrombocytopenia associated with LD; transient toxicities associated with IL-2
Natural killer (NK) cell products	Anti-GD2 antibody in combination with chemotherapy and haploidentical NK cell therapy to treat relapsed/refractory neuroblastoma (Federico et al. 2017)	13 patients (11 received NK cell infusion); 13 children, zero adults	CR ($n = 4$), PR ($n = 4$), SD ($n = 5$); 12 mo OS: 77% (Federico et al. 2017)	Grade 3/4 myelosuppression ($n = 13$); grade 1/2 pain ($n = 13$), likely related to chemotherapy or anti-GD2 antibody
	NK-cell therapy with and without recombinant human IL-15 (rh!L-15) to treat r/r pediatric solid tumors (Segal et al. 2023)	16 patients (median age 16.1 yr)	PR ($n = 3$), SD ($n = 13$) (Segal et al. 2023)	CRS in two patients (one DLT); pericardial tamponade/capillary leak (DLT) in one of four patients receiving rhIL-15

(IV) Intravenous; (ICV) intracerebroventricular.

these are in the context of phase I or II clinical trials, patients often have multiply relapsed/refractory disease, widely variable disease burden, and unique pretreatment histories, making comparison to identify optimal approaches challenging. While these approaches have been broadly found to be safe and well tolerated, translating an effective treatment approach is challenging. Most likely, a successful approach will incorporate modulation of the TME, a multitarget approach, and thoughtful timing of these complex biologics after the patient's last tumor-directed therapy.

FUTURE DIRECTIONS AND CONCLUSIONS

Patients with pediatric solid tumors that have failed current treatment approaches with chemotherapy, surgery, and radiation have dismal outcomes without major advances in recent decades. Outcomes for some blood cancers and melanoma have been transformed by cell therapeutics such as CAR-T targeting antigens expressed widely by the tumor (Maude et al. 2018), and TIL treatments (Rosenberg et al. 2011) with or without checkpoint inhibition (Garon et al. 2015). However, the translation of such successful strategies to pediatric solid tumors has been complicated by numerous factors including heterogenous antigen expression, immune evasion tactics, and an immune-suppressive TME. Notably, the development of targeted antibody therapies has provided better survival outcomes in high-risk neuroblastoma, a complex tumor, in which the therapy has been administered typically in the setting of minimal residual disease at the end of multimodal therapy (Yu et al. 2010). With current research, enhanced cell therapy products, rational combination strategies, and optimal timing of adoptive cell transfer may hold the key to providing future patients with effective treatments and lasting cures.

COMPETING INTEREST STATEMENT

C.M.B. and A.B.H. have intellectual property related to developing gene-engineered cells and antigen-specific T cells for solid tumors.

ACKNOWLEDGMENTS

C.M.B. and A.B.H. are supported by a Cancer Grand Challenges grant awarded by the Cancer Research United Kingdom and the National Cancer Institute of the National Institutes of Health (1OT2CA278700-01). A.B.H. has received support from the Department of Defense Peer Reviewed Cancer Research Program (W81XWH22110259).

REFERENCES

Adusumilli PS, Cherkassky L, Villena-Vargas J, Colovos C, Servais E, Plotkin J, Jones DR, Sadelain M. 2014. Regional delivery of mesothelin-targeted CAR T cell therapy generates potent and long-lasting CD4-dependent tumor immunity. *Sci Transl Med* **6**: 261ra151. doi:10.1126/scitranslmed.3010162

Ahmed N, Brawley VS, Hegde M, Robertson C, Ghazi A, Gerken C, Liu E, Dakhova O, Ashoori A, Corder A, et al. 2015. Human epidermal growth factor receptor 2 (HER2)-specific chimeric antigen receptor-modified T cells for the immunotherapy of HER2-positive sarcoma. *J Clin Oncol* **33**: 1688–1696. doi:10.1200/JCO.2014.58.0225

Ahmed N, Brawley V, Hegde M, Bielamowicz K, Kalra M, Landi D, Robertson C, Gray TL, Diouf O, Wakefield A, et al. 2017. HER2-specific chimeric antigen receptor-modified virus-specific T cells for progressive glioblastoma: a phase 1 dose-escalation trial. *JAMA Oncol* **3**: 1094–1101. doi:10.1001/jamaoncol.2017.0184

Allen-Rhoades W, Whittle SB, Rainusso N. 2018. Pediatric solid tumors in children and adolescents: an overview. *Pediatr Rev* **39**: 444–453. doi:10.1542/pir.2017-0268

Altieri DC. 2003. Survivin, versatile modulation of cell division and apoptosis in cancer. *Oncogene* **22**: 8581–8589. doi:10.1038/sj.onc.1207113

American Cancer Society. 2018. *Cancer facts & figures 2018*. American Cancer Society, Atlanta.

Amini L, Silbert SK, Maude SL, Nastoupil LJ, Ramos CA, Brentjens RJ, Sauter CS, Shah NN, Abou-El-Enein M. 2022. Preparing for CAR T cell therapy: patient selection, bridging therapies and lymphodepletion. *Nat Rev Clin Oncol* **19**: 342–355. doi:10.1038/s41571-022-00607-3

Anderson NM, Simon MC. 2020. The tumor microenvironment. *Curr Biol* **30**: R921–R925. doi:10.1016/j.cub.2020.06.081

Aparicio C, Acebal C, González-Vallinas M. 2023. Current approaches to develop "off-the-shelf" chimeric antigen receptor (CAR)-T cells for cancer treatment: a systematic review. *Exp Hematol Oncol* **12**: 73. doi:10.1186/s40164-023-00435-w

Barbolina MV, Adley BP, Shea LD, Stack MS. 2008. Wilms tumor gene protein 1 is associated with ovarian cancer metastasis and modulates cell invasion. *Cancer* **112**: 1632–1641. doi:10.1002/cncr.23341

Basta-Jovanovic G, Radojevic-Skodric S, Brasanac D, Djuricic S, Milasin J, Bogdanovic L, Opric D, Savin M, Baralic I,

Jovanovic M. 2012. Prognostic value of survivin expression in Wilms tumor. *J BUON* 17: 168–173.

Becker C, Pohla H, Frankenberger B, Schüler T, Assenmacher M, Schendel DJ, Blankenstein T. 2001. Adoptive tumor therapy with T lymphocytes enriched through an IFN-γ capture assay. *Nat Med* 7: 1159–1162. doi:10.1038/nm1001-1159

Biffi G, Oni TE, Spielman B, Hao Y, Elyada E, Park Y, Preall J, Tuveson DA. 2019. IL1-induced JAK/STAT signaling is antagonized by TGFβ to shape CAF heterogeneity in pancreatic ductal adenocarcinoma. *Cancer Discov* 9: 282–301. doi:10.1158/2159-8290.CD-18-0710

Bollard CM, Gottschalk S, Torrano V, Diouf O, Ku S, Hazrat Y, Carrum G, Ramos C, Fayad L, Shpall EJ, et al. 2014. Sustained complete responses in patients with lymphoma receiving autologous cytotoxic T lymphocytes targeting Epstein–Barr virus latent membrane proteins. *J Clin Oncol* 32: 798–808. doi:10.1200/JCO.2013.51.5304

Bollard CM, Tripic T, Cruz CR, Dotti G, Gottschalk S, Torrano V, Dakhova O, Carrum G, Ramos CA, Liu H, et al. 2018. Tumor-specific T-cells engineered to overcome tumor immune evasion induce clinical responses in patients with relapsed Hodgkin lymphoma. *J Clin Oncol* 36: 1128–1139. doi:10.1200/JCO.2017.74.3179

Boutilier AJ, Elsawa SF. 2021. Macrophage polarization states in the tumor microenvironment. *Int J Mol Sci* 22: 6995. doi:10.3390/ijms22136995

Brenne K, Nymoen DA, Reich R, Davidson B. 2012. PRAME (preferentially expressed antigen of melanoma) is a novel marker for differentiating serous carcinoma from malignant mesothelioma. *Am J Clin Pathol* 137: 240–247. doi:10.1309/AJCPGA95KVSAUDMF

Brentjens RJ, Rivière I, Park JH, Davila ML, Wang X, Stefanski J, Taylor C, Yeh R, Bartido S, Borquez-Ojeda O, et al. 2011. Safety and persistence of adoptively transferred autologous CD19-targeted T cells in patients with relapsed or chemotherapy refractory B-cell leukemias. *Blood* 118: 4817–4828. doi:10.1182/blood-2011-04-348540

Brentjens RJ, Davila ML, Riviere I, Park J, Wang X, Cowell LG, Bartido S, Stefanski J, Taylor C, Olszewska M, et al. 2013. CD19-targeted t cells rapidly induce molecular remissions in adults with chemotherapy-refractory acute lymphoblastic leukemia. *Sci Transl Med* 5: 177ra138. doi:10.1126/scitranslmed.3005930

Brett A, Pandey S, Fraizer G. 2013. The Wilms' tumor gene (WT1) regulates E-cadherin expression and migration of prostate cancer cells. *Mol Cancer* 12: 3. doi:10.1186/1476-4598-12-3

Brossart P. 2020. The role of antigen spreading in the efficacy of immunotherapies. *Clin Cancer Res* 26: 4442–4447. doi:10.1158/1078-0432.CCR-20-0305

Burga RA, Yvon E, Chorvinsky E, Fernandes R, Cruz CRY, Bollard CM. 2019. Engineering the TGFβ receptor to enhance the therapeutic potential of natural killer cells as an immunotherapy for neuroblastoma. *Clin Cancer Res* 25: 4400–4412. doi:10.1158/1078-0432.CCR-18-3183

Caja F, Vannucci L. 2015. TGFβ: a player on multiple fronts in the tumor microenvironment. *J Immunotoxicol* 12: 300–307. doi:10.3109/1547691X.2014.945667

Caldas H, Holloway MP, Hall BM, Qualman SJ, Altura RA. 2006. Survivin-directed RNA interference cocktail is a potent suppressor of tumour growth in vivo. *J Med Genet* 43: 119–128. doi:10.1136/jmg.2005.034686

Carpentieri DF, Nichols K, Chou PM, Matthews M, Pawel B, Huff D. 2002. The expression of WT1 in the differentiation of rhabdomyosarcoma from other pediatric small round blue cell tumors. *Mod Pathol* 15: 1080–1086. doi:10.1097/01.MP.0000028646.03760.6B

Casey DL, Cheung NV. 2020. Immunotherapy of pediatric solid tumors: treatments at a crossroads, with an emphasis on antibodies. *Cancer Immunol Res* 8: 161–166. doi:10.1158/2326-6066.CIR-19-0692

Ceschel S, Casotto V, Valsecchi MG, Tamaro P, Jankovic M, Hanau G, Fossati F, Pillon M, Rondelli R, Sandri A, et al. 2006. Survival after relapse in children with solid tumors: a follow-up study from the Italian off-therapy registry. *Pediatr Blood Cancer* 47: 560–566. doi:10.1002/pbc.20726

Chapuis AG, Ragnarsson GB, Nguyen HN, Chaney CN, Pufnock JS, Schmitt TM, Duerkopp N, Roberts IM, Pogosov GL, Ho WY, et al. 2013. Transferred WT1-reactive CD8[+] T cells can mediate antileukemic activity and persist in post-transplant patients. *Sci Transl Med* 5: 174ra127. doi:10.1126/scitranslmed.3004916

Chapuis AG, Roberts IM, Thompson JA, Margolin KA, Bhatia S, Lee SM, Sloan HL, Lai IP, Farrar EA, Wagener F, et al. 2016. T-cell therapy using interleukin-21-primed cytotoxic T-cell lymphocytes combined with cytotoxic T-cell lymphocyte antigen-4 blockade results in long-term cell persistence and durable tumor regression. *J Clin Oncol* 34: 3787–3795. doi:10.1200/JCO.2015.65.5142

Chau YY, Hastie ND. 2012. The role of Wt1 in regulating mesenchyme in cancer, development, and tissue homeostasis. *Trends Genet* 28: 515–524. doi:10.1016/j.tig.2012.04.004

Chen Y, McAndrews KM, Kalluri R. 2021. Clinical and therapeutic relevance of cancer-associated fibroblasts. *Nat Rev Clin Oncol* 18: 792–804. doi:10.1038/s41571-021-00546-5

Chi SN, Cheung NK, Cheung IY. 2002. Expression of SSX-2 and SSX-4 genes in neuroblastoma. *Int J Biol Markers* 17: 219–223. doi:10.1177/172460080201700401

Chu J, Gao F, Yan M, Zhao S, Yan Z, Shi B, Liu Y. 2022. Natural killer cells: a promising immunotherapy for cancer. *J Transl Med* 20: 240. doi:10.1186/s12967-022-03437-0

Coffelt SB, Wellenstein MD, de Visser KE. 2016. Neutrophils in cancer: neutral no more. *Nat Rev Cancer* 16: 431–446. doi:10.1038/nrc.2016.52

Coughlin CM, Fleming MD, Carroll RG, Pawel BR, Hogarty MD, Shan X, Vance BA, Cohen JN, Jairaj S, Lord EM, et al. 2006. Immunosurveillance and survivin-specific T-cell immunity in children with high-risk neuroblastoma. *J Clin Oncol* 24: 5725–5734. doi:10.1200/JCO.2005.05.3314

Craddock JA, Lu A, Bear A, Pule M, Brenner MK, Rooney CM, Foster AE. 2010. Enhanced tumor trafficking of GD2 chimeric antigen receptor T cells by expression of the chemokine receptor CCR2b. *J Immunother* 33: 780–788. doi:10.1097/CJI.0b013e3181ee6675

Dalerba P, Frascella E, Macino B, Mandruzzato S, Zambon A, Rosolen A, Carli M, Ninfo V, Zanovello P. 2001. MAGE, BAGE and GAGE gene expression in human rhabdomyosarcomas. *Int J Cancer* **93**: 85–90. doi:10.1002/ijc.1307

Del Bufalo F, De Angelis B, Caruana I, Del Baldo G, De Ioris MA, Serra A, Mastronuzzi A, Cefalo MG, Pagliara D, Amicucci M, et al. 2023. GD2-CART01 for relapsed or refractory high-risk neuroblastoma. *N Engl J Med* **388**: 1284–1295. doi:10.1056/NEJMoa2210859

Delfanti G, Dellabona P, Casorati G, Fedeli M. 2022. Adoptive immunotherapy with engineered iNKT cells to target cancer cells and the suppressive microenvironment. *Front Med (Lausanne)* **9**: 897750. doi:10.3389/fmed.2022.897750

Depil S, Duchateau P, Grupp SA, Mufti G, Poirot L. 2020. "Off-the-shelf" allogeneic CAR T cells: development and challenges. *Nat Rev Drug Discov* **19**: 185–199. doi:10.1038/s41573-019-0051-2

Downing JR, Wilson RK, Zhang J, Mardis ER, Pui CH, Ding L, Ley TJ, Evans WE. 2012. The Pediatric Cancer Genome Project. *Nat Genet* **44**: 619–622. doi:10.1038/ng.2287

Downs-Canner SM, Meier J, Vincent BG, Serody JS. 2022. B cell function in the tumor microenvironment. *Annu Rev Immunol* **40**: 169–193. doi:10.1146/annurev-immunol-101220-015603

Dudley ME, Wunderlich J, Nishimura MI, Yu D, Yang JC, Topalian SL, Schwartzentruber DJ, Hwu P, Marincola FM, Sherry R, et al. 2001. Adoptive transfer of cloned melanoma-reactive T lymphocytes for the treatment of patients with metastatic melanoma. *J Immunother* **24**: 363–373. doi:10.1097/00002371-200107000-00012

Dudley ME, Wunderlich JR, Robbins PF, Yang JC, Hwu P, Schwartzentruber DJ, Topalian SL, Sherry R, Restifo NP, Hubicki AM, et al. 2002. Cancer regression and autoimmunity in patients after clonal repopulation with antitumor lymphocytes. *Science* **298**: 850–854. doi:10.1126/science.1076514

Dudley ME, Wunderlich JR, Shelton TE, Even J, Rosenberg SA. 2003. Generation of tumor-infiltrating lymphocyte cultures for use in adoptive transfer therapy for melanoma patients. *J Immunother* **26**: 332–342. doi:10.1097/00002371-200307000-00005

Dudley ME, Yang JC, Sherry R, Hughes MS, Royal R, Kammula U, Robbins PF, Huang J, Citrin DE, Leitman SF, et al. 2008. Adoptive cell therapy for patients with metastatic melanoma: evaluation of intensive myeloablative chemoradiation preparative regimens. *J Clin Oncol* **26**: 5233–5239. doi:10.1200/JCO.2008.16.5449

Epping MT, Bernards R. 2006. A causal role for the human tumor antigen preferentially expressed antigen of melanoma in cancer. *Cancer Res* **66**: 10639–10642. doi:10.1158/0008-5472.CAN-06-2522

Eshhar Z, Waks T, Gross G, Schindler DG. 1993. Specific activation and targeting of cytotoxic lymphocytes through chimeric single chains consisting of antibody-binding domains and the γ or ζ subunits of the immunoglobulin and T-cell receptors. *Proc Natl Acad Sci* **90**: 720–724. doi:10.1073/pnas.90.2.720

Federico SM, McCarville MB, Shulkin BL, Sondel PM, Hank JA, Hutson P, Meagher M, Shafer A, Ng CY, Leung W, et al. 2017. A pilot trial of humanized anti-GD2 monoclonal antibody (hu14.18K322A) with chemotherapy and natural killer cells in children with recurrent/refractory neu-

roblastoma. *Clin Cancer Res* **23**: 6441–6449. doi:10.1158/1078-0432.CCR-17-0379

Finley SD, Popel AS. 2013. Effect of tumor microenvironment on tumor VEGF during anti-VEGF treatment: systems biology predictions. *J Natl Cancer Inst* **105**: 802–811. doi:10.1093/jnci/djt093

Fousek K, Horn LA, Palena C. 2021. Interleukin-8: a chemokine at the intersection of cancer plasticity, angiogenesis, and immune suppression. *Pharmacol Ther* **219**: 107692. doi:10.1016/j.pharmthera.2020.107692

Fox E, Oliver T, Rowe M, Thomas S, Zakharia Y, Gilman PB, Muller AJ, Prendergast GC. 2018. Indoximod: an immunometabolic adjuvant that empowers T cell activity in cancer. *Front Oncol* **8**: 370. doi:10.3389/fonc.2018.00370

Freimüller C, Stemberger J, Artwohl M, Germeroth L, Witt V, Fischer G, Tischer S, Eiz-Vesper B, Knippertz I, Dörrie J, et al. 2015. Selection of adenovirus-specific and Epstein–Barr virus-specific T cells with major histocompatibility class I streptamers under good manufacturing practice (GMP)-compliant conditions. *Cytotherapy* **17**: 989–1007. doi:10.1016/j.jcyt.2015.03.613

Fridlender ZG, Sun J, Kim S, Kapoor V, Cheng G, Ling L, Worthen GS, Albelda SM. 2009. Polarization of tumor-associated neutrophil phenotype by TGF-β: "N1" versus "N2" TAN. *Cancer Cell* **16**: 183–194. doi:10.1016/j.ccr.2009.06.017

Fukuda S, Pelus LM. 2006. Survivin, a cancer target with an emerging role in normal adult tissues. *Mol Cancer Ther* **5**: 1087–1098. doi:10.1158/1535-7163.MCT-05-0375

Fukumura D, Kloepper J, Amoozgar Z, Duda DG, Jain RK. 2018. Enhancing cancer immunotherapy using antiangiogenics: opportunities and challenges. *Nat Rev Clin Oncol* **15**: 325–340. doi:10.1038/nrclinonc.2018.29

Galli E, Bellesi S, Pansini I, Di Cesare G, Iacovelli C, Malafronte R, Maiolo E, Chiusolo P, Sica S, Sorà F, et al. 2023. The CD4/CD8 ratio of infused CD19-CAR-T is a prognostic factor for efficacy and toxicity. *Br J Haematol* **203**: 564–570. doi:10.1111/bjh.19117

Garon EB, Rizvi NA, Hui R, Leighl N, Balmanoukian AS, Eder JP, Patnaik A, Aggarwal C, Gubens M, Horn L, et al. 2015. Pembrolizumab for the treatment of non-small-cell lung cancer. *N Engl J Med* **372**: 2018–2028. doi:10.1056/NEJMoa1501824

Goff SL, Dudley ME, Citrin DE, Somerville RP, Wunderlich JR, Danforth DN, Zlott DA, Yang JC, Sherry RM, Kammula US, et al. 2016. Randomized, prospective evaluation comparing intensity of lymphodepletion before adoptive transfer of tumor-infiltrating lymphocytes for patients with metastatic melanoma. *J Clin Oncol* **34**: 2389–2397. doi:10.1200/JCO.2016.66.7220

Gross G, Waks T, Eshhar Z. 1989. Expression of immunoglobulin-T-cell receptor chimeric molecules as functional receptors with antibody-type specificity. *Proc Natl Acad Sci* **86**: 10024–10028. doi:10.1073/pnas.86.24.10024

Grzywa TM, Sosnowska A, Matryba P, Rydzynska Z, Jasinski M, Nowis D, Golab J. 2020. Myeloid cell-derived arginase in cancer immune response. *Front Immunol* **11**: 938. doi:10.3389/fimmu.2020.00938

Gulley JL, Madan RA, Pachynski R, Mulders P, Sheikh NA, Trager J, Drake CG. 2017. Role of antigen spread and distinctive characteristics of immunotherapy in cancer treatment. *J Natl Cancer Inst* **109**: djw261. doi:10.1093/jnci/djw261

Guzman G, Reed MR, Bielamowicz K, Koss B, Rodriguez A. 2023. CAR-T therapies in solid tumors: opportunities and challenges. *Curr Oncol Rep* **25**: 479–489. doi:10.1007/s11912-023-01380-x

Hanks BA, Holtzhausen A, Evans KS, Jamieson R, Gimpel P, Campbell OM, Hector-Greene M, Sun L, Tewari A, George A, et al. 2013. Type III TGF-β receptor downregulation generates an immunotolerant tumor microenvironment. *J Clin Invest* **123**: 3925–3940. doi:10.1172/JCI65745

Hay KA, Turtle CJ. 2017. Chimeric antigen receptor (CAR) T cells: lessons learned from targeting of CD19 in B-cell malignancies. *Drugs* **77**: 237–245. doi:10.1007/s40265-017-0690-8

Heczey A, Louis CU. 2013. Advances in chimeric antigen receptor immunotherapy for neuroblastoma. *Discov Med* **16**: 287–294.

Heczey A, Louis CU, Savoldo B, Dakhova O, Durett A, Grilley B, Liu H, Wu MF, Mei Z, Gee A, et al. 2017. CAR T cells administered in combination with lymphodepletion and PD-1 inhibition to patients with neuroblastoma. *Mol Ther* **25**: 2214–2224. doi:10.1016/j.ymthe.2017.05.012

Hegde M, Mukherjee M, Grada Z, Pignata A, Landi D, Navai SA, Wakefield A, Fousek K, Bielamowicz K, Chow KK, et al. 2016. Tandem CAR T cells targeting HER2 and IL13Rα2 mitigate tumor antigen escape. *J Clin Invest* **126**: 3036–3052. doi:10.1172/JCI83416

Hensel J, Metts J, Gupta A, Ladle BH, Pilon-Thomas S, Mullinax J. 2022. Adoptive cellular therapy for pediatric solid tumors: beyond chimeric antigen receptor-T cell therapy. *Cancer J* **28**: 322–327. doi:10.1097/PPO.0000000000000603

Hingorani P, Zhang W, Zhang Z, Xu Z, Wang WL, Roth ME, Wang Y, Gill JB, Harrison DJ, Teicher BA, et al. 2022. Trastuzumab deruxtecan, antibody-drug conjugate targeting HER2, is effective in pediatric malignancies: a report by the pediatric preclinical testing consortium. *Mol Cancer Ther* **21**: 1318–1325. doi:10.1158/1535-7163.MCT-21-0758

Hont AB, Cruz CR, Ulrey R, O'Brien B, Stanojevic M, Datar A, Albihani S, Saunders D, Hanajiri R, Panchapakesan K, et al. 2019. Immunotherapy of relapsed and refractory solid tumors with ex vivo expanded multiantigen-associated specific cytotoxic T lymphocytes: a phase I study. *J Clin Oncol* **37**: 2349–2359.

Hont AB, Powell AB, Sohai DK, Valdez IK, Stanojevic M, Geiger AE, Chaudhary K, Dowlati E, Bollard CM, Cruz CRY. 2022. The generation and application of antigen-specific T cell therapies for cancer and viral-associated disease. *Mol Ther* **30**: 2130–2152. doi:10.1016/j.ymthe.2022.02.002

Huang L, Hu C, Chao H, Zhang Y, Li Y, Hou J, Xu Z, Lu H, Li H, Chen H. 2019. Drug-resistant endothelial cells facilitate progression, EMT and chemoresistance in nasopharyngeal carcinoma via exosomes. *Cell Signal* **63**: 109385. doi:10.1016/j.cellsig.2019.109385

Hunder NN, Wallen H, Cao J, Hendricks DW, Reilly JZ, Rodmyre R, Jungbluth A, Gnjatic S, Thompson JA, Yee C. 2008. Treatment of metastatic melanoma with autologous CD4⁺ T cells against NY-ESO-1. *N Engl J Med* **358**: 2698–2703. doi:10.1056/NEJMoa0800251

Hylander B, Repasky E, Shrikant P, Intengan M, Beck A, Driscoll D, Singhal P, Lele S, Odunsi K. 2006. Expression of Wilms tumor gene (WT1) in epithelial ovarian cancer. *Gynecol Oncol* **101**: 12–17. doi:10.1016/j.ygyno.2005.09.052

Ishida H, Matsumura T, Salgaller ML, Ohmizono Y, Kadono Y, Sawada T. 1996. MAGE-1 and MAGE-3 or -6 expression in neuroblastoma-related pediatric solid tumors. *Int J Cancer* **69**: 375–380. doi:10.1002/(SICI)1097-0215(19961021)69:5<375::AID-IJC4>3.0.CO;2-2

Itakura E, Huang RR, Wen DR, Paul E, Wünsch PH, Cochran AJ. 2011. IL-10 expression by primary tumor cells correlates with melanoma progression from radial to vertical growth phase and development of metastatic competence. *Mod Pathol* **24**: 801–809. doi:10.1038/modpathol.2011.5

Jiang W, Avdic S, Lee KH, Street J, Castellano-González G, Simms R, Clancy LE, Blennerhassett R, Patrick E, Chan AS, et al. 2023. Persistence of ex vivo expanded tumour and pathogen specific T-cells after allogeneic stem cell transplant for myeloid malignancies (the INTACT study). *Leukemia* **37**: 2330–2333. doi:10.1038/s41375-023-02033-5

Jin M, Zhang S, Wang M, Li Q, Ren J, Luo Y, Sun X. 2022. Exosomes in pathogenesis, diagnosis, and therapy of ischemic stroke. *Front Bioeng Biotechnol* **10**: 980548. doi:10.3389/fbioe.2022.980548

Kim A, Fox E, Warren K, Blaney SM, Berg SL, Adamson PC, Libucha M, Byrley E, Balis FM, Widemann BC. 2008. Characteristics and outcome of pediatric patients enrolled in phase I oncology trials. *Oncologist* **13**: 679–689. doi:10.1634/theoncologist.2008-0046

Kim A, Park EY, Kim K, Lee JH, Shin DH, Kim JY, Park do Y, Lee CH, Sol MY, Choi KU, et al. 2014. Prognostic significance of WT1 expression in soft tissue sarcoma. *World J Surg Oncol* **12**: 214. doi:10.1186/1477-7819-12-214

Kim JH, Lee CH, Baek MC. 2022. Dissecting exosome inhibitors: therapeutic insights into small-molecule chemicals against cancer. *Exp Mol Med* **54**: 1833–1843. doi:10.1038/s12276-022-00898-7

Kinoshita Y, Tanaka S, Souzaki R, Miyoshi K, Kohashi K, Oda Y, Nakatsura T, Taguchi T. 2015. Glypican 3 expression in pediatric malignant solid tumors. *Eur J Pediatr Surg* **25**: 138–144.

Kinoshita H, Cooke KR, Grant M, Stanojevic M, Cruz CR, Keller M, Fortiz MF, Hoq F, Lang H, Barrett AJ, et al. 2022. Outcome of donor-derived TAA-T cell therapy in patients with high-risk or relapsed acute leukemia post allogeneic BMT. *Blood Adv* **6**: 2520–2534. doi:10.1182/bloodadvances.2021006831

Kopp LM, Katsanis E. 2016. Targeted immunotherapy for pediatric solid tumors. *Oncoimmunology* **5**: e1087637. doi:10.1080/2162402X.2015.1087637

Krishnadas DK, Bai F, Lucas KG. 2013. Cancer testis antigen and immunotherapy. *Immunotargets Ther* **2**: 11–19. doi:10.2147/ITT.S35570

Kulczycka M, Derlatka K, Tasior J, Lejman M, Zawitkowska J. 2023. CAR T-cell therapy in children with solid tumors. *J Clin Med* **12**: 2326. doi:10.3390/jcm12062326

Kumar V, Patel S, Tcyganov E, Gabrilovich DI. 2016. The nature of myeloid-derived suppressor cells in the tumor microenvironment. *Trends Immunol* **37:** 208–220. doi:10.1016/j.it.2016.01.004

Kyi C, Postow MA. 2016. Immune checkpoint inhibitor combinations in solid tumors: opportunities and challenges. *Immunotherapy* **8:** 821–837. doi:10.2217/imt-2016-0002

Labani-Motlagh A, Ashja-Mahdavi M, Loskog A. 2020. The tumor microenvironment: a milieu hindering and obstructing antitumor immune responses. *Front Immunol* **11:** 940. doi:10.3389/fimmu.2020.00940

Lamers CH, Sleijfer S, van Steenbergen S, van Elzakker P, van Krimpen B, Groot C, Vulto A, den Bakker M, Oosterwijk E, Debets R, et al. 2013. Treatment of metastatic renal cell carcinoma with CAIX CAR-engineered T cells: clinical evaluation and management of on-target toxicity. *Mol Ther* **21:** 904–912. doi:10.1038/mt.2013.17

Laumont CM, Nelson BH. 2023. B cells in the tumor microenvironment: multi-faceted organizers, regulators, and effectors of anti-tumor immunity. *Cancer Cell* **41:** 466–489. doi:10.1016/j.ccell.2023.02.017

Laumont CM, Banville AC, Gilardi M, Hollern DP, Nelson BH. 2022. Tumour-infiltrating B cells: immunological mechanisms, clinical impact and therapeutic opportunities. *Nat Rev Cancer* **22:** 414–430. doi:10.1038/s41568-022-00466-1

Lee SB, Haber DA. 2001. Wilms tumor and the WT1 gene. *Exp Cell Res* **264:** 74–99. doi:10.1006/excr.2000.5131

Lee KH, Gowrishankar K, Street J, McGuire HM, Luciani F, Hughes B, Singh M, Clancy LE, Gottlieb DJ, Micklethwaite KP, et al. 2020. Ex vivo enrichment of PRAME antigen-specific T cells for adoptive immunotherapy using CD137 activation marker selection. *Clin Transl Immunol* **9:** e1200. doi:10.1002/cti2.1200

Leen AM, Bollard CM, Mendizabal AM, Shpall EJ, Szabolcs P, Antin JH, Kapoor N, Pai SY, Rowley SD, Kebriaei P, et al. 2013. Multicenter study of banked third-party virus-specific T cells to treat severe viral infections after hematopoietic stem cell transplantation. *Blood* **121:** 5113–5123. doi:10.1182/blood-2013-02-486324

Li I, Nabet BY. 2019. Exosomes in the tumor microenvironment as mediators of cancer therapy resistance. *Mol Cancer* **18:** 32. doi:10.1186/s12943-019-0975-5

Li J, Shen J, Wang K, Hornicek F, Duan Z. 2016. The roles of Sox family genes in sarcoma. *Curr Drug Targets* **17:** 1761–1772. doi:10.2174/1389450117666160502145311

Li C, Jiang P, Wei S, Xu X, Wang J. 2020. Regulatory T cells in tumor microenvironment: new mechanisms, potential therapeutic strategies and future prospects. *Mol Cancer* **19:** 116. doi:10.1186/s12943-020-01234-1

Linde IL, Prestwood TR, Qiu J, Pilarowski G, Linde MH, Zhang X, Shen L, Reticker-Flynn NE, Chiu DK, Sheu LY, et al. 2023. Neutrophil-activating therapy for the treatment of cancer. *Cancer Cell* **41:** 356–372.e310. doi:10.1016/j.ccell.2023.01.002

Liu M, Wang X, Wang L, Ma X, Gong Z, Zhang S, Li Y. 2018. Targeting the IDO1 pathway in cancer: from bench to bedside. *J Hematol Oncol* **11:** 100. doi:10.1186/s13045-018-0644-y

Long AH, Morgenstern DA, Leruste A, Bourdeaut F, Davis KL. 2022. Checkpoint immunotherapy in pediatrics: here, gone, and back again. *Am Soc Clin Oncol Educ Book* **42:** 1–14. doi:10.1200/EDBK_349799

Lulla PD, Tzannou I, Vasileiou S, Carrum G, Ramos CA, Kamble R, Wang T, Wu M, Bilgi M, Gee AP, et al. 2020. The safety and clinical effects of administering a multi-antigen-targeted T cell therapy to patients with multiple myeloma. *Sci Transl Med* **12:** eaaz3339. doi:10.1126/scitranslmed.aaz3339

Lulla PD, Naik S, Vasileiou S, Tzannou I, Watanabe A, Kuvalekar M, Lulla S, Carrum G, Ramos CA, Kamble R, et al. 2021. Clinical effects of administering leukemia-specific donor T cells to patients with AML/MDS after allogeneic transplant. *Blood* **137:** 2585–2597. doi:10.1182/blood.2020009471

Lussier DM, O'Neill L, Nieves LM, McAfee MS, Holechek SA, Collins AW, Dickman P, Jacobsen J, Hingorani P, Blattman JN. 2015. Enhanced T-cell immunity to osteosarcoma through antibody blockade of PD-1/PD-L1 interactions. *J Immunother* **38:** 96–106. doi:10.1097/CJI.0000000000000065

Ma L, Feng Y, Zhou Z. 2023. A close look at current γδ T-cell immunotherapy. *Front Immunol* **14:** 1140623. doi:10.3389/fimmu.2023.1140623

Majzner RG, Mackall CL. 2018. Tumor antigen escape from CAR T-cell therapy. *Cancer Discov* **8:** 1219–1226. doi:10.1158/2159-8290.CD-18-0442

Majzner RG, Theruvath JL, Nellan A, Heitzeneder S, Cui Y, Mount CW, Rietberg SP, Linde MH, Xu P, Rota C, et al. 2019. CAR T cells targeting B7-H3, a pan-cancer antigen, demonstrate potent preclinical activity against pediatric solid tumors and brain tumors. *Clin Cancer Res* **25:** 2560–2574. doi:10.1158/1078-0432.CCR-18-0432

Majzner RG, Ramakrishna S, Yeom KW, Patel S, Chinnasamy H, Schultz LM, Richards RM, Jiang L, Barsan V, Mancusi R, et al. 2022. GD2-CAR T cell therapy for H3K27M-mutated diffuse midline gliomas. *Nature* **603:** 934–941. doi:10.1038/s41586-022-04489-4

Malogolowkin MH, Hemmer MT, Le-Rademacher J, Hale GA, Mehta PA, Smith AR, Kitko C, Abraham A, Abdel-Azim H, Dandoy C, et al. 2017. Outcomes following autologous hematopoietic stem cell transplant for patients with relapsed Wilms' tumor: a CIBMTR retrospective analysis. *Bone Marrow Transplant* **52:** 1549–1555. doi:10.1038/bmt.2017.178

Masucci MT, Minopoli M, Carriero MV. 2019. Tumor associated neutrophils. Their role in tumorigenesis, metastasis, prognosis and therapy. *Front Oncol* **9:** 1146. doi:10.3389/fonc.2019.01146

Maude SL, Laetsch TW, Buechner J, Rives S, Boyer M, Bittencourt H, Bader P, Verneris MR, Stefanski HE, Myers GD, et al. 2018. Tisagenlecleucel in children and young adults with B-cell lymphoblastic leukemia. *N Engl J Med* **378:** 439–448. doi:10.1056/NEJMoa1709866

Mirzaei HR, Rodriguez A, Shepphird J, Brown CE, Badie B. 2017. Chimeric antigen receptors T cell therapy in solid tumor: challenges and clinical applications. *Front Immunol* **8:** 1850. doi:10.3389/fimmu.2017.01850

Moek KL, Fehrmann RSN, van der Vegt B, de Vries EGE, de Groot DJA. 2018. Glypican 3 overexpression across a broad spectrum of tumor types discovered with functional genomic mRNA profiling of a large cancer database.

Cite this article as *Cold Spring Harb Perspect Med* doi: 10.1101/cshperspect.a041636

Am J Pathol **188:** 1973–1981. doi:10.1016/j.ajpath.2018.05.014

Moon EK, Carpenito C, Sun J, Wang LC, Kapoor V, Predina J, Powell DJ, Riley JL, June CH, Albelda SM. 2011. Expression of a functional CCR2 receptor enhances tumor localization and tumor eradication by retargeted human T cells expressing a mesothelin-specific chimeric antibody receptor. *Clin Cancer Res* **17:** 4719–4730. doi:10.1158/1078-0432.CCR-11-0351

Morotti M, Albukhari A, Alsaadi A, Artibani M, Brenton JD, Curbishley SM, Dong T, Dustin ML, Hu Z, McGranahan N, et al. 2021. Promises and challenges of adoptive T-cell therapies for solid tumours. *Br J Cancer* **124:** 1759–1776. doi:10.1038/s41416-021-01353-6

Morrison AH, Diamond MS, Hay CA, Byrne KT, Vonderheide RH. 2020. Sufficiency of CD40 activation and immune checkpoint blockade for T cell priming and tumor immunity. *Proc Natl Acad Sci* **117:** 8022–8031. doi:10.1073/pnas.1918971117

Munn DH, Mellor AL. 2016. IDO in the tumor microenvironment: inflammation, counter-regulation, and tolerance. *Trends Immunol* **37:** 193–207. doi:10.1016/j.it.2016.01.002

Muranski P, Boni A, Wrzesinski C, Citrin DE, Rosenberg SA, Childs R, Restifo NP. 2006. Increased intensity lymphodepletion and adoptive immunotherapy—how far can we go? *Nat Clin Pract Oncol* **3:** 668–681. doi:10.1038/ncponc0666

Naka N, Araki N, Nakanishi H, Itoh K, Mano M, Ishiguro S, de Bruijn DR, Myoui A, Ueda T, Yoshikawa H. 2002. Expression of SSX genes in human osteosarcomas. *Int J Cancer* **98:** 640–642. doi:10.1002/ijc.10277

Oberthuer A, Hero B, Spitz R, Berthold F, Fischer M. 2004. The tumor-associated antigen PRAME is universally expressed in high-stage neuroblastoma and associated with poor outcome. *Clin Cancer Res* **10:** 4307–4313. doi:10.1158/1078-0432.CCR-03-0813

Ogasawara K, Lymp J, Mack T, Dell'Aringa J, Huang CP, Smith J, Peiser L, Kostic A. 2022. In vivo cellular expansion of lisocabtagene maraleucel and association with efficacy and safety in relapsed/refractory large B-cell lymphoma. *Clin Pharmacol Ther* **112:** 81–89. doi:10.1002/cpt.2561

Ollé Hurtado M, Wolbert J, Fisher J, Flutter B, Stafford S, Barton J, Jain N, Barone G, Majani Y, Anderson J. 2019. Tumor infiltrating lymphocytes expanded from pediatric neuroblastoma display heterogeneity of phenotype and function. *PLoS ONE* **14:** e0216373. doi:10.1371/journal.pone.0216373

Ortiz MV, Roberts SS, Glade Bender J, Shukla N, Wexler LH. 2019. Immunotherapeutic targeting of GPC3 in pediatric solid embryonal tumors. *Front Oncol* **9:** 108. doi:10.3389/fonc.2019.00108

Park JR, Digiusto DL, Slovak M, Wright C, Naranjo A, Wagner J, Meechoovet HB, Bautista C, Chang WC, Ostberg JR, et al. 2007. Adoptive transfer of chimeric antigen receptor re-directed cytolytic T lymphocyte clones in patients with neuroblastoma. *Mol Ther* **15:** 825–833. doi:10.1038/sj.mt.6300104

Peggs KS, Thomson K, Samuel E, Dyer G, Armoogum J, Chakraverty R, Pang K, Mackinnon S, Lowdell MW. 2011. Directly selected cytomegalovirus-reactive donor

T cells confer rapid and safe systemic reconstitution of virus-specific immunity following stem cell transplantation. *Clin Infect Dis* **52:** 49–57. doi:10.1093/cid/ciq042

Pennati M, Folini M, Zaffaroni N. 2007. Targeting survivin in cancer therapy: fulfilled promises and open questions. *Carcinogenesis* **28:** 1133–1139. doi:10.1093/carcin/bgm047

Pickup M, Novitskiy S, Moses HL. 2013. The roles of TGFβ in the tumour microenvironment. *Nat Rev Cancer* **13:** 788–799. doi:10.1038/nrc3603

Pittet MJ, Michielin O, Migliorini D. 2022. Clinical relevance of tumour-associated macrophages. *Nat Rev Clin Oncol* **19:** 402–421. doi:10.1038/s41571-022-00620-6

Poggio M, Hu T, Pai CC, Chu B, Belair CD, Chang A, Montabana E, Lang UE, Fu Q, Fong L, et al. 2019. Suppression of exosomal PD-L1 induces systemic anti-tumor immunity and memory. *Cell* **177:** 414–427.e413. doi:10.1016/j.cell.2019.02.016

Pollack SM, Li Y, Blaisdell MJ, Farrar EA, Chou J, Hoch BL, Loggers ET, Rodler E, Eary JF, Conrad EU, et al. 2012. NYESO-1/LAGE-1s and PRAME are targets for antigen specific T cells in chondrosarcoma following treatment with 5-Aza-2-deoxycitabine. *PLoS ONE* **7:** e32165. doi:10.1371/journal.pone.0032165

Postow MA, Callahan MK, Wolchok JD. 2015. Immune checkpoint blockade in cancer therapy. *J Clin Oncol* **33:** 1974–1982. doi:10.1200/JCO.2014.59.4358

Pule MA, Savoldo B, Myers GD, Rossig C, Russell HV, Dotti G, Huls MH, Liu E, Gee AP, Mei Z, et al. 2008. Virus-specific T cells engineered to coexpress tumor-specific receptors: persistence and antitumor activity in individuals with neuroblastoma. *Nat Med* **14:** 1264–1270. doi:10.1038/nm.1882

Qi XW, Zhang F, Wu H, Liu JL, Zong BG, Xu C, Jiang J. 2015. Wilms' tumor 1 (WT1) expression and prognosis in solid cancer patients: a systematic review and meta-analysis. *Sci Rep* **5:** 8924. doi:10.1038/srep08924

Ramji K, Grzywa TM, Sosnowska A, Paterek A, Okninska M, Pilch Z, Barankiewicz J, Garbicz F, Borg K, Bany-Laszewicz U, et al. 2022. Targeting arginase-1 exerts antitumor effects in multiple myeloma and mitigates bortezomib-induced cardiotoxicity. *Sci Rep* **12:** 19660. doi:10.1038/s41598-022-24137-1

Robbins PF, Morgan RA, Feldman SA, Yang JC, Sherry RM, Dudley ME, Wunderlich JR, Nahvi AV, Helman LJ, Mackall CL, et al. 2011. Tumor regression in patients with metastatic synovial cell sarcoma and melanoma using genetically engineered lymphocytes reactive with NY-ESO-1. *J Clin Oncol* **29:** 917–924. doi:10.1200/JCO.2010.32.2537

Rodriguez PC, Quiceno DG, Zabaleta J, Ortiz B, Zea AH, Piazuelo MB, Delgado A, Correa P, Brayer J, Sotomayor EM, et al. 2004. Arginase I production in the tumor microenvironment by mature myeloid cells inhibits T-cell receptor expression and antigen-specific T-cell responses. *Cancer Res* **64:** 5839–5849. doi:10.1158/0008-5472.CAN-04-0465

Roma-Rodrigues C, Fernandes AR, Baptista PV. 2014. Exosome in tumour microenvironment: overview of the crosstalk between normal and cancer cells. *Biomed Res Int* **2014:** 179486. doi:10.1155/2014/179486

Rosenberg SA, Spiess P, Lafreniere R. 1986. A new approach to the adoptive immunotherapy of cancer with tumor-infiltrating lymphocytes. *Science* 233: 1318–1321. doi:10.1126/science.3489291

Rosenberg SA, Yang JC, White DE, Steinberg SM. 1998. Durability of complete responses in patients with metastatic cancer treated with high-dose interleukin-2: identification of the antigens mediating response. *Ann Surg* 228: 307–319. doi:10.1097/00000658-199809000-00004

Rosenberg SA, Restifo NP, Yang JC, Morgan RA, Dudley ME. 2008. Adoptive cell transfer: a clinical path to effective cancer immunotherapy. *Nat Rev Cancer* 8: 299–308. doi:10.1038/nrc2355

Rosenberg SA, Yang JC, Sherry RM, Kammula US, Hughes MS, Phan GQ, Citrin DE, Restifo NP, Robbins PF, Wunderlich JR, et al. 2011. Durable complete responses in heavily pretreated patients with metastatic melanoma using T-cell transfer immunotherapy. *Clin Cancer Res* 17: 4550–4557. doi:10.1158/1078-0432.CCR-11-0116

Roszik J, Wang WL, Livingston JA, Roland CL, Ravi V, Yee C, Hwu P, Futreal A, Lazar AJ, Patel SR, et al. 2017. Overexpressed PRAME is a potential immunotherapy target in sarcoma subtypes. *Clin Sarcoma Res* 7: 11. doi:10.1186/s13569-017-0077-3

Rüttinger D, van den Engel NK, Winter H, Schlemmer M, Pohla H, Grützner S, Wagner B, Schendel DJ, Fox BA, Jauch KW, et al. 2007. Adjuvant therapeutic vaccination in patients with non-small cell lung cancer made lymphopenic and reconstituted with autologous PBMC: first clinical experience and evidence of an immune response. *J Transl Med* 5: 43. doi:10.1186/1479-5876-5-43

Sagnella SM, White AL, Yeo D, Saxena P, van Zandwijk N, Rasko JEJ. 2022. Locoregional delivery of CAR-T cells in the clinic. *Pharmacol Res* 182: 106329. doi:10.1016/j.phrs.2022.106329

Sato T, Terai M, Tamura Y, Alexeev V, Mastrangelo MJ, Selvan SR. 2011. Interleukin 10 in the tumor microenvironment: a target for anticancer immunotherapy. *Immunol Res* 51: 170–182. doi:10.1007/s12026-011-8262-6

Scharnhorst V, van der Eb AJ, Jochemsen AG. 2001. WT1 proteins: functions in growth and differentiation. *Gene* 273: 141–161. doi:10.1016/S0378-1119(01)00593-5

Scott EN, Gocher AM, Workman CJ, Vignali DAA. 2021. Regulatory T cells: barriers of immune infiltration into the tumor microenvironment. *Front Immunol* 12: 702726. doi:10.3389/fimmu.2021.702726

Segal NH, Blachere NE, Shiu HY, Leejee S, Antonescu CR, Lewis JJ, Wolchok JD, Houghton AN. 2005. Antigens recognized by autologous antibodies of patients with soft tissue sarcoma. *Cancer Immun* 5: 4.

Segal J, Kaczanowska S, Bernstein D, Zhang N, Dinh A, Somerville R, Highfill S, Stroncek D, Conlon K, Waldmann T, et al. 2023. 621A phase I study of autologous activated NK cells ± rhIL15 in children and young adults with refractory solid tumors. *J ImmunoTher Cancer* 11: A708. doi:10.1136/jitc-2023-SITC2023.0621

Shafer P, Kelly LM, Hoyos V. 2022. Cancer therapy with TCR-engineered T cells: current strategies, challenges, and prospects. *Front Immunol* 13: 835762. doi:10.3389/fimmu.2022.835762

Sherman MH, Yu RT, Engle DD, Ding N, Atkins AR, Tiriac H, Collisson EA, Connor F, Van Dyke T, Kozlov S, et al.

2014. Vitamin D receptor-mediated stromal reprogramming suppresses pancreatitis and enhances pancreatic cancer therapy. *Cell* 159: 80–93. doi:10.1016/j.cell.2014.08.007

Shinozawa I, Inokuchi K, Wakabayashi I, Dan K. 2000. Disturbed expression of the anti-apoptosis gene, survivin, and EPR-1 in hematological malignancies. *Leukemia Res* 24: 965–970. doi:10.1016/S0145-2126(00)00065-5

Siegel RL, Miller KD, Jemal A. 2018. Cancer statistics, 2018. *CA Cancer J Clin* 68: 7–30. doi:10.3322/caac.21442

Siegel DA, Richardson LC, Henley SJ, Wilson RJ, Dowling NF, Weir HK, Tai EW, Buchanan Lunsford N. 2020. Pediatric cancer mortality and survival in the United States, 2001–2016. *Cancer* 126: 4379–4389. doi:10.1002/cncr.33080

Sosnowska A, Chlebowska-Tuz J, Matryba P, Pilch Z, Greig A, Wolny A, Grzywa TM, Rydzynska Z, Sokolowska O, Rygiel TP, et al. 2021. Inhibition of arginase modulates T-cell response in the tumor microenvironment of lung carcinoma. *Oncoimmunology* 10: 1956143. doi:10.1080/2162402X.2021.1956143

Stevanović S, Draper LM, Langhan MM, Campbell TE, Kwong ML, Wunderlich JR, Dudley ME, Yang JC, Sherry RM, Kammula US, et al. 2015. Complete regression of metastatic cervical cancer after treatment with human papillomavirus-targeted tumor-infiltrating T cells. *J Clin Oncol* 33: 1543–1550. doi:10.1200/JCO.2014.58.9093

Straathof K, Flutter B, Wallace R, Jain N, Loka T, Depani S, Wright G, Thomas S, Cheung GW, Gileadi T, et al. 2020. Antitumor activity without on-target off-tumor toxicity of GD2-chimeric antigen receptor T cells in patients with neuroblastoma. *Sci Transl Med* 12: eabd6169. doi:10.1126/scitranslmed.abd6169

Takamizawa S, Scott D, Wen J, Grundy P, Bishop W, Kimura K, Sandler A. 2001. The survivin:fas ratio in pediatric renal tumors. *J Pediatr Surg* 36: 37–42. doi:10.1053/jpsu.2001.20000

Tamm I, Wang Y, Sausville E, Scudiero DA, Vigna N, Oltersdorf T, Reed JC. 1998. IAP-family protein survivin inhibits caspase activity and apoptosis induced by Fas (CD95), Bax, caspases, and anticancer drugs. *Cancer Res* 58: 5315–5320.

Tamm I, Richter S, Oltersdorf D, Creutzig U, Harbott J, Scholz F, Karawajew L, Ludwig WD, Wuchter C. 2004. High expression levels of x-linked inhibitor of apoptosis protein and survivin correlate with poor overall survival in childhood de novo acute myeloid leukemia. *Clin Cancer Res* 10: 3737–3744. doi:10.1158/1078-0432.CCR-03-0642

Tan P, Zou C, Yong B, Han J, Zhang L, Su Q, Yin J, Wang J, Huang G, Peng T, et al. 2012. Expression and prognostic relevance of PRAME in primary osteosarcoma. *Biochem Biophys Res Commun* 419: 801–808. doi:10.1016/j.bbrc.2012.02.110

Tang T, Huang X, Zhang G, Hong Z, Bai X, Liang T. 2021. Advantages of targeting the tumor immune microenvironment over blocking immune checkpoint in cancer immunotherapy. *Signal Transduct Target Ther* 6: 72. doi:10.1038/s41392-020-00449-4

Templeton AJ, McNamara MG, Šeruga B, Vera-Badillo FE, Aneja P, Ocaña A, Leibowitz-Amit R, Sonpavde G, Knox

JJ, Tran B, et al. 2014. Prognostic role of neutrophil-to-lymphocyte ratio in solid tumors: a systematic review and meta-analysis. *J Natl Cancer Inst* **106**: dju124. doi:10.1093/jnci/dju124

Terry RL, Meyran D, Ziegler DS, Haber M, Ekert PG, Trapani JA, Neeson PJ. 2020. Immune profiling of pediatric solid tumors. *J Clin Invest* **130**: 3391–3402. doi:10.1172/JCI137181

Toledo SR, Zago MA, Oliveira ID, Proto-Siqueira R, Okamoto OK, Severino P, Vêncio RZ, Gamba FT, Silva WA, Moreira-Filho CA, et al. 2011. Insights on PRAME and osteosarcoma by means of gene expression profiling. *J Orthop Sci* **16**: 458–466. doi:10.1007/s00776-011-0106-7

Topalian SL, Muul LM, Solomon D, Rosenberg SA. 1987. Expansion of human tumor infiltrating lymphocytes for use in immunotherapy trials. *J Immunol Methods* **102**: 127–141. doi:10.1016/S0022-1759(87)80018-2

Tormoen GW, Crittenden MR, Gough MJ. 2018. Role of the immunosuppressive microenvironment in immunotherapy. *Adv Radiat Oncol* **3**: 520–526. doi:10.1016/j.adro.2018.08.018

Tran J, Master Z, Yu JL, Rak J, Dumont DJ, Kerbel RS. 2002. A role for survivin in chemoresistance of endothelial cells mediated by VEGF. *Proc Natl Acad Sci* **99**: 4349–4354. doi:10.1073/pnas.072586399

Turtle CJ, Hanafi LA, Berger C, Hudecek M, Pender B, Robinson E, Hawkins R, Chaney C, Cherian S, Chen X, et al. 2016. Immunotherapy of non-Hodgkin's lymphoma with a defined ratio of CD8$^+$ and CD4$^+$ CD19-specific chimeric antigen receptor-modified T cells. *Sci Transl Med* **8**: a116. doi:10.1126/scitranslmed.aaf8621

Tzannou I, Papadopoulou A, Naik S, Leung K, Martinez CA, Ramos CA, Carrum G, Sasa G, Lulla P, Watanabe A, et al. 2017. Off-the-shelf virus-specific T cells to treat BK virus, human herpesvirus 6, cytomegalovirus, Epstein–Barr virus, and adenovirus infections after allogeneic hematopoietic stem-cell transplantation. *J Clin Oncol* **35**: 3547–3557. doi:10.1200/JCO.2017.73.0655

Vasileiou S, Lulla PD, Tzannou I, Watanabe A, Kuvalekar M, Callejas WL, Bilgi M, Wang T, Wu MJ, Kamble R, et al. 2021. T-cell therapy for lymphoma using nonengineered multiantigen-targeted T cells is safe and produces durable clinical effects. *J Clin Oncol: JCO* **39**: 1415–1425. doi:10.1200/JCO.20.02224

Vitanza NA, Johnson AJ, Wilson AL, Brown C, Yokoyama JK, Künkele A, Chang CA, Rawlings-Rhea S, Huang W, Seidel K, et al. 2021. Locoregional infusion of HER2-specific CAR T cells in children and young adults with recurrent or refractory CNS tumors: an interim analysis. *Nat Med* **27**: 1544–1552. doi:10.1038/s41591-021-01404-8

Vonderheide RH. 2020. CD40 agonist antibodies in cancer immunotherapy. *Annu Rev Med* **71**: 47–58. doi:10.1146/annurev-med-062518-045435

Wang S, Liu X, Chen Y, Zhan X, Wu T, Chen B, Sun G, Yan S, Xu L. 2020. The role of SOX2 overexpression in prognosis of patients with solid tumors: a meta-analysis and system review. *Medicine (Baltimore)* **99**: e19604. doi:10.1097/MD.0000000000019604

Wang Y, Wang J, Yang X, Yang J, Lu P, Zhao L, Li B, Pan H, Jiang Z, Shen X, et al. 2021. Chemokine receptor CCR2b enhanced anti-tumor function of chimeric antigen recep-

tor T cells targeting mesothelin in a non-small-cell lung carcinoma model. *Front Immunol* **12**: 628906. doi:10.3389/fimmu.2021.628906

Wang J, Wang Y, Pan H, Zhao L, Yang X, Liang Z, Shen X, Zhang J, Yang J, Zhu Y, et al. 2023. Chemokine receptors CCR6 and PD1 blocking scFv E27 enhances anti-EGFR CAR-T therapeutic efficacy in a preclinical model of human non-small cell lung carcinoma. *Int J Mol Sci* **24**: 5424. doi:10.3390/ijms24065424

Ward E, DeSantis C, Robbins A, Kohler B, Jemal A. 2014. Childhood and adolescent cancer statistics, 2014. *CA Cancer J Clin* **64**: 83–103. doi:10.3322/caac.21219

Weide B, Zelba H, Derhovanessian E, Pflugfelder A, Eigentler TK, Di Giacomo AM, Maio M, Aarntzen EH, de Vries IJ, Sucker A, et al. 2012. Functional T cells targeting NY-ESO-1 or Melan-A are predictive for survival of patients with distant melanoma metastasis. *J Clin Oncol* **30**: 1835–1841. doi:10.1200/JCO.2011.40.2271

Williams KM, Hakim FT, Gress RE. 2007. T cell immune reconstitution following lymphodepletion. *Semin Immunol* **19**: 318–330. doi:10.1016/j.smim.2007.10.004

Wu Q, Zhou L, Lv D, Zhu X, Tang H. 2019. Exosome-mediated communication in the tumor microenvironment contributes to hepatocellular carcinoma development and progression. *J Hematol Oncol* **12**: 53. doi:10.1186/s13045-019-0739-0

Xiao J, Zhang T, Gao F, Zhou Z, Shu G, Zou Y, Yin G. 2022. Natural killer cells: a promising kit in the adoptive cell therapy toolbox. *Cancers (Basel)* **14**: 5657. doi:10.3390/cancers14225657

Xie Y, Ma X, Gu L, Li H, Chen L, Li X, Gao Y, Fan Y, Zhang Y, Yao Y, et al. 2016. Prognostic and clinicopathological significance of survivin expression in renal cell carcinoma: a systematic review and meta-analysis. *Sci Rep* **6**: 29794. doi:10.1038/srep29794

Xing F, Saidou J, Watabe K. 2010. Cancer associated fibroblasts (CAFs) in tumor microenvironment. *Front Biosci (Landmark Ed)* **15**: 166–179. doi:10.2741/3613

Yang S, Zheng J, Ma Y, Zhu H, Xu T, Dong K, Xiao X. 2012. Oct4 and Sox2 are overexpressed in human neuroblastoma and inhibited by chemotherapy. *Oncol Rep* **28**: 186–192.

Yang Y, Li C, Liu T, Dai X, Bazhin AV. 2020. Myeloid-derived suppressor cells in tumors: from mechanisms to antigen specificity and microenvironmental regulation. *Front Immunol* **11**: 1371. doi:10.3389/fimmu.2020.01371

Yang Y, McCloskey JE, Yang H, Puc J, Alcaina Y, Vedvyas Y, Gomez Gallegos AA, Ortiz-Sánchez E, de Stanchina E, Min IM, et al. 2021. Bispecific CAR T cells against EpCAM and inducible ICAM-1 overcome antigen heterogeneity and generate superior antitumor responses. *Cancer Immunol Res* **9**: 1158–1174. doi:10.1158/2326-6066.CIR-21-0062

Yao J, Caballero OL, Yung WK, Weinstein JN, Riggins GJ, Strausberg RL, Zhao Q. 2014. Tumor subtype-specific cancer-testis antigens as potential biomarkers and immunotherapeutic targets for cancers. *Cancer Immunol Res* **2**: 371–379. doi:10.1158/2326-6066.CIR-13-0088

Yazdanifar M, Barbarito G, Bertaina A, Airoldi I. 2020. Γδ T cells: the ideal tool for cancer immunotherapy. *Cells* **9**: 1305. doi:10.3390/cells9051305

Yin B. 2011. PRAME: from diagnostic marker and tumor antigen to promising target of RNAi therapy in leukemic

cells. *Leukemia Res* **35**: 1159–1160. doi:10.1016/j.leukres
.2011.04.018

Yu AL, Gilman AL, Ozkaynak MF, London WB, Kreissman
SG, Chen HX, Smith M, Anderson B, Villablanca JG,
Matthay KK, et al. 2010. Anti-GD2 antibody with
GM-CSF, interleukin-2, and isotretinoin for neuroblasto-
ma. *N Engl J Med* **363**: 1324–1334. doi:10.1056/NEJ
Moa0911123

Yvon ES, Burga R, Powell A, Cruz CR, Fernandes R, Barese
C, Nguyen T, Abdel-Baki MS, Bollard CM. 2017. Cord
blood natural killer cells expressing a dominant negative
TGF-β receptor: implications for adoptive immunotherapy for glioblastoma. *Cytotherapy* **19**: 408–418. doi:10
.1016/j.jcyt.2016.12.005

Zayed H, Petersen I. 2018. Stem cell transcription factor
SOX2 in synovial sarcoma and other soft tissue tumors.

Pathol Res Pract **214**: 1000–1007. doi:10.1016/j.prp.2018
.05.004

Zhang QW, Liu L, Gong CY, Shi HS, Zeng YH, Wang XZ,
Zhao YW, Wei YQ. 2012. Prognostic significance of tumor-associated macrophages in solid tumor: a meta-analysis of the literature. *PLoS ONE* **7**: e50946. doi:10.1371/
journal.pone.0050946

Zhang BL, Qin DY, Mo ZM, Li Y, Wei W, Wang YS,
Wang W, Wei YQ. 2016. Hurdles of CAR-T cell-based
cancer immunotherapy directed against solid tumors.
Sci China Life Sci **59**: 340–348. doi:10.1007/s11427-016-
5027-4

Zheng X, Liu X, Lei Y, Wang G, Liu M. 2022. Glypican-3: a
novel and promising target for the treatment of hepatocellular carcinoma. *Front Oncol* **12**: 824208. doi:10.3389/
fonc.2022.824208

Cite this article as *Cold Spring Harb Perspect Med* doi: 10.1101/cshperspect.a041636

Epigenetic Therapies

Wallace Bourgeois, Scott A. Armstrong, and Emily B. Heikamp

Division of Hematology/Oncology, Boston Children's Hospital, Department of Pediatric Oncology, Dana-Farber Cancer Institute, and Harvard Medical School, Boston, Massachusetts 02215, USA

Correspondence: emilyb_heikamp@dfci.harvard.edu

Epigenetic therapies are emerging for pediatric cancers. Due to the relatively low mutational burden in pediatric tumors, epigenetic dysregulation and differentiation blockade is a hallmark of oncogenesis in some childhood cancers. By targeting epigenetic regulators that maintain tumor cells in a primitive developmental state, epigenetic therapies may induce differentiation. The most well-studied and clinically advanced epigenetic-targeted therapies include azacitidine and decitabine, which inhibit DNA methylation through competitive inhibition of the enzymatic activity of the DNA methyltransferase family enzymes. These DNA hypomethylating agents are Food and Drug Administration (FDA) approved for hematologic malignancies. The discovery that DNA hypermethylation occurs in patients with isocitrate dehydrogenase (IDH) mutations has led to the development and FDA approval of IDH inhibitors for hematologic and solid tumors. Epigenetic dysregulation in pediatric tumors is also driven by changes in the "histone code" that either promote oncogene expression or repress tumor suppressors. Cancers whose chromatin landscape is characterized by such aberrant histone posttranslational modifications may be amenable to targeted therapies that inhibit the chromatin-modifying enzymes that read, write, and erase these histone modifications. Small molecules that inhibit the enzymatic activity of histone deacetylases, acetyltransferases, and methyltransferases have been approved for the treatment of some adult cancers, and these agents are currently under investigation in various pediatric tumors. Chromatin regulatory complexes can be hijacked by oncogenic fusion proteins that are produced by chromosomal translocations, which are common drivers in pediatric cancer. Small molecules that disrupt oncogenic fusion protein activity and their associated chromatin complexes have demonstrated remarkable promise, and this approach has become the standard treatment for a subset of leukemias driven by the PML-RARA oncogenic fusion protein. A deeper understanding of the mechanisms that drive epigenetic dysregulation in pediatric cancer may hold the key to future success in this field, as the landscape of druggable epigenetic targets is also expanding.

The "quiet" genome of pediatric tumors, which harbor relatively few mutations in known oncogenic drivers compared to adult cancers, underlies the importance of epigenetics in pediatric oncology. Chromosomal rearrangements that generate oncogenic fusion proteins are sometimes the only detectable genetic lesion in pediatric tumors. Such oncogenic fusion pro-

Cite this article as *Cold Spring Harb Perspect Med* doi: 10.1101/cshperspect.a041637

teins are often the result of translocations involving at least one partner gene with epigenetic functionality, and their role in oncogenesis is to promote genome-wide epigenetic dysregulation (Lawrence et al. 2013; Brien et al. 2019). This review will provide a framework for understanding how targeted epigenetic therapies can address epigenetic dysregulation in pediatric cancer.

Aberrant chromatin organization leading to oncogenic gene expression is a hallmark of cancer. Mutations in epigenetic regulators may result in changes in DNA methylation, histone modifications, and dysfunction of multi-subunit protein complexes that bind to and regulate chromatin architecture. The basic unit of chromatin is composed of 146 bp segments of DNA that are wrapped around histone octamers, forming a nucleosome. Repeated units of nucleosomes constitute chromatin (Bates 2020; Zhao et al. 2021). Histone proteins in nucleosome octamers include H2A, H2B, H3, and H4. Posttranslational modifications of histone tails and DNA methylation constitute an "epigenetic code" that is associated with either increased or decreased levels of gene expression, depending on the cellular and genomic context (Strahl and Allis 2000; Jenuwein and Allis 2001). Chromatin modifiers include readers, writers, and erasers of this epigenetic code. Chromatin can become dysregulated in various cancers, and a basic understanding of chromatin biology is helpful to understand epigenetic therapies that target chromatin. In discussing the targets of various epigenetic therapies, we will aim to cover normal chromatin biology, how it is dysregulated in cancer, and how epigenetic therapies can restore normal function.

DNA METHYLATION

Methylation of DNA is perhaps the most well-studied epigenetic modification, and it is the target of the first epigenetic therapies developed for clinical use. Both hypermethylation and hypomethylation of DNA have been implicated in oncogenesis (Ehrlich 2002). Methylation of DNA occurs at CpG islands, which refer to regions that are enriched for cytosine and guanine nucle-

otides, particularly with the dinucleotide CpG (5′-Cytosine-phosphate-Guanine-3′), that are found primarily at sites of transcription initiation (Saxonov et al. 2006). CpG islands tend to be unmethylated; however, in cancers that have aberrant DNA methylation, CpG island promoters may become hypermethylated (Fernandez et al. 2012). DNA methylation occurs on cytosine, and these marks are deposited and maintained by the DNA methyltransferase (DNMTs) family of enzymes, which includes *DNMT1*, *DNMT3A*, and *DNMT3B*, and results in repression of transcription and gene silencing (Baylin and Jones 2011; Yang et al. 2015). *DNMT3A* mutations are found in ~22% of acute myeloid leukemia (AML) patients, suggesting a key role in disease development (Ley et al. 2010). However, the relationship between *DNMT3A* and CpG island hypermethylation is complex: in patients with wild-type *DNMT3A*, *DNMT3A* is responsible for the observed CpG island hypermethylation, whereas *DNMT3A* mutations in AML are dominant negative and cause focal methylation loss (Spencer et al. 2017).

Azacitidine and decitabine are cytidine nucleoside analogs that inhibit DNA methylation (Fig. 1). They were developed long before their DNMTs activity was known. They were initially developed because at high doses they are cytotoxic, which also confers an unfavorable toxicity profile that led to rejection of the initial FDA application (Egger et al. 2004; Ganesan et al. 2019). It was subsequently discovered that at lower doses they inhibit DNMTs by becoming incorporated into DNA and forming covalent bonds with DNMTs, which causes depletion of DNMTs and subsequent demethylation of replicated DNA (Derissen et al. 2013).

Azacitidine has been used in various hematologic malignancies and has FDA approvals for several indications in AML (FDA 2022; Jen et al. 2022; Kayser and Levis 2022), myelodysplastic syndrome (MDS) (Kaminskas et al. 2005), and juvenile myelomonocytic leukemia (JMML) (Rubio-San-Simón et al. 2023). Decitabine is also FDA approved for use in MDS and AML (Kantarjian et al. 2012; Steensma et al. 2019). See Table 1 for a list of FDA-approved therapies in adult and childhood cancers.

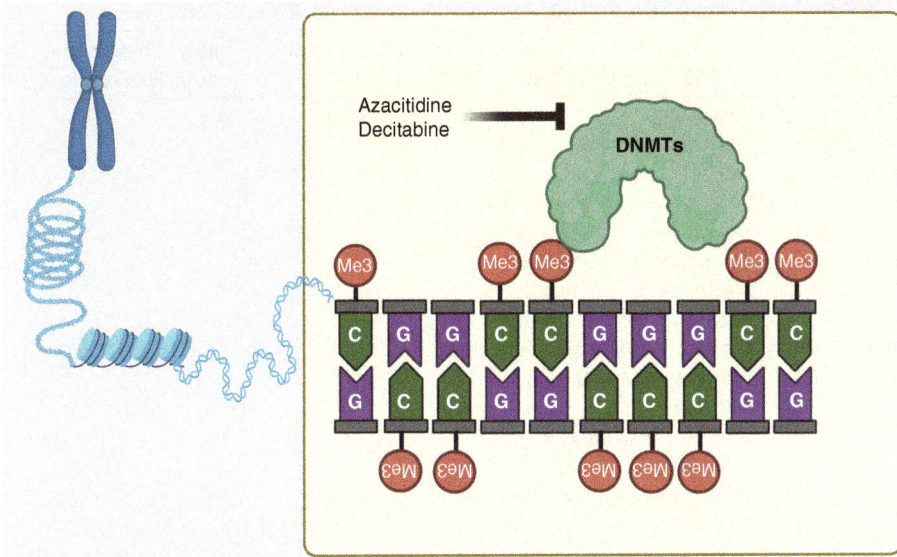

Figure 1. Mechanism of action of hypomethylating agents. Small-molecule inhibitors of DNA methyltransferase (DNMTs) family enzymes include azacitidine and decitabine, which act as cytidine nucleoside analogs that inhibit DNA methylation of CpG dinucleotides.

DNA hypermethylation is also found in patients with mutations in isocitrate dehydrogenase (*IDH1* and *IDH2*), and Tet methylcytosine dioxygenase 2 (*TET2*). *IDH1* and *IDH2* mutations have been found in more than 70% of WHO grade II and III anaplastic astrocytomas, oligodengrogliomas, and oligoastrocytomas (Hartmann et al. 2009; Yan et al. 2009), in addition to 15%–30% of cytogenetically normal AML (Mardis et al. 2009; Marcucci et al. 2010). *IDH1* and *IDH2* mutations disrupt Kreb's cycle such that instead of isocitrate conversion to α-ketoglutarate, α-ketoglutarate is reduced to 2-hydroxyglutarate (Ward et al. 2010). In AML, *IDH1* and *IDH2* mutations are associated with aberrant increase in promoter methylation and hypermethylation of genes (Figueroa et al. 2010).

The metabolite 2-hydroxyglutarate also competitively inhibits TET family enzymes (Xu et al. 2011). *IDH1* and *IDH2* mutations are mutually exclusive with *TET2* mutations, which catalyzes cytosine 5-hydroxymethylation and induces DNA demethylation; observed *TET2* loss of function thus results in DNA hy-

permethylation (Tahiliani et al. 2009; Ito et al. 2010). This mutual exclusivity implies that *IDH1/2* mutations and *TET2* mutations are redundant, as suggested by their common feature of hypermethylation and the fact that *IDH1/2* mutants inhibit TET2-induced cytosine 5-hydroxymethylation. Despite their shared hypermethylated profile, responses in patients with *IDH1/IDH2* and *TET2* mutations to the hypomethylating agents azacytidine and decitabine have shown mixed results (Bejar et al. 2014; Dinardo et al. 2014). Inhibitors of *IDH1* and *IDH2* —ivosidenib and enasidenib, respectively—have shown promise in AML patients harboring *IDH* mutations with relapsed/refractory disease, demonstrating a 30.4% complete remission or complete remission with partial hematologic recovery rate among those with *IDH1*-mutant disease who received the recommended phase 2 dose (RP2D) of ivosidenib (Dinardo et al. 2018). A similar group of patients with *IDH2* mutations had a 19.8% CR/CRi rate when receiving enasidenib (Stein et al. 2017). Based on these data, ivosidenib and enasidenib both have FDA approvals in relapsed/refractory AML pa-

Table 1. Food and Drug Administration (FDA)-approved therapies in adult and childhood cancers

Drug(s)/drug class	FDA approval adults	FDA approval children	Key clinical trial results for non-FDA-approved drugs
Hypomethylating agents: azacitidine and decitabine	Azacitidine 2004: myelodysplastic syndrome (MDS)[a] 2018/2020: Newly diagnosed acute myeloid leukemia (AML) patients age ≥75 or unfit to receive intensive induction therapy[b] 2020: AML patients who achieve CR but are unable to complete intensive, curative therapy[c] Decitabine 2006/(2020): MDS[d] (oral formulation[e]) 2018/2020: Newly diagnosed AML patients age ≥75 or unfit to receive intensive induction therapy[f]	Azacitidine 2022: newly diagnosed JMML[g]	
Isocitrate dehydrogenase (*IDH1* or *2*) inhibitors	Ivosidenib: *IDH1* inhibitor 2018: Relapsed/refractory AML with *IDH1* mutation[h] 2021: Refractory/metastatic/ unresectable cholangiocarcinoma with an *IDH1* mutation[i] 2023: Relapsed/refractory MDS with *IDH1* mutation[j] Olutasidenib: *IDH1* allosteric inhibitor 2022: Relapsed/refractory AML with *IDH1* mutation[k] Enasidenib: *IDH2* inhibitor 2017: Relapsed/refractory AML with *IDH1* mutation[l]		Vorasidenib: dual *IDH1* + *IDH2* inhibitor, CNS penetrant 2023: Phase 3 trial (NCT04164901) in residual/ recurrent grade 2 glioma demonstrating superior progression-free survival with vorasidenib (27.7 mo) vs. placebo (11.1 mo)[m]
Histone deacetylase inhibitors (HDACs)	Vorinostat 2006: Cutaneous T-cell lymphoma (CTCL) in patients who have received at least two prior therapies[n] Romidepsin 2009: CTCL patients who have received at least one prior therapy[o] Belinostat 2014: Relapsed or refractory PTCL[p] Panobinostat 2015: MM in combination with dexamethasone and bortezomib for patients who have received at least two prior therapies[q]		

Continued

Cite this article as *Cold Spring Harb Perspect Med* doi: 10.1101/cshperspect.a041637

Table 1. *Continued*

Drug(s)/drug class	FDA approval adults	FDA approval children	Key clinical trial results for non-FDA-approved drugs
Bromodomain inhibitors			NCT02259114 Biradbresib (BRD2, BRD3, BRD4 inhibitor), phase 1 study enrolling solid tumor patients, nine patients with NUT midline carcinoma, three partial responses, three patients with stable disease[r] NCT01587703 Molibresib (BRD2, BRD3, BRD4 inhibitor), phase 1 + 2 study, 19 patients with NUT-carcinoma, 11% overall response rate, inferior response in non-NUT-carcinoma patients[s] NCT03936465 Study of the bromodomain (BRD) and extraterminal domain (BET) inhibitors BMS-986158 and BMS-986378 in pediatric cancer, phase 1 trial enrolling patients with NUT midline carcinoma, *MYCN* or *MYC* translocation or high copy number gain, translocation involving *BRD3* or *BRD4*
Tazemetostat	2020: Relapsed/refractory follicular lymphoma whose tumors are positive for an EZH2 mutation[t] 2020: Patients 16 years and older with metastatic or ineligible-for-resection epithelioid sarcoma[u]		
DOT1L inhibitors			NCT01684150 Phase 1 study of EPZ-5676/ pinometostat in *KMT2A*-r leukemias, 2 CRs among 51 patients, MTD not reached[v]
KAT6A inhibitors			NCT04606446 Phase 1 study in ER$^+$HER2$^-$ metastatic breast cancer with PF-07248144 demonstrated a 30% overall response rate and 10-month progression-free survival

Continued

Table 1. *Continued*

Drug(s)/drug class	FDA approval adults	FDA approval children	Key clinical trial results for non-FDA-approved drugs
Arsenic trioxide and all-*trans* retinoic acid	<u>All-*trans* retinoic acid (ATRA)</u> 1995: Single agent in relapsed/refractory acute promyelocytic leukemia (APL) with *t*(15;17) translocation[w] <u>Arsenic trioxide (ATO)</u> 2000: Single agent in relapsed/refractory APL with *t*(15;17) translocation[x] <u>ATRA + ATO</u> 2018: Combination therapy for patients with non-high-risk, de novo APL		
Menin inhibitors			<u>NCT04065399</u> Phase 1 study of SNDX-5613/revumenib in *KMT2A*-r and *NPM1c* acute leukemias, 30% CR/CRh rate[y] <u>NCT04067336</u> Phase 1/2 study of KO-539/ziftomenib in *KMT2A*-r and *NPM1c* acute leukemias, 25% CR/CRh rate[z]

[a]www.pubmed.ncbi.nlm.nih.gov/15793220.

[b]www.fda.gov/drugs/resources-information-approved-drugs/fda-grants-regular-approval-venetoclax-combination-untreated-acute-myeloid-leukemia.

[c]Jen et al. 2022.

[d]www.accessdata.fda.gov/drugsatfda_docs/label/2010/021790s006lbl.pdf.

[e]Kim et al. 2022.

[f]www.fda.gov/drugs/resources-information-approved-drugs/fda-grants-regular-approval-venetoclax-combination-untreated-acute-myeloid-leukemia.

[g]www.fda.gov/drugs/resources-information-approved-drugs/fda-approves-azacitidine-newly-diagnosed-juvenile-myelomonocytic-leukemia#:~:text=On%20May%2020%2C%202022%2C%20the,juvenile%20myelomonocytic%20leukemia%20(JMML).

[h]Norsworthy et al. 2019.

[i]Casak et al. 2022.

[j]www.fda.gov/drugs/resources-information-approved-drugs/fda-approves-ivosidenib-myelodysplastic-syndromes#:~:text=On%20October%2024%2C%202023%2C%20the,by%20an%20FDA%2Dapproved%20test.

[k]www.fda.gov/drugs/resources-information-approved-drugs/fda-approves-olutasidenib-relapsed-or-refractory-acute-myeloid-leukemia-susceptible-idh1-mutation.

[l]www.fda.gov/news-events/press-announcements/fda-approves-new-targeted-treatment-relapsed-or-refractory-acute-myeloid-leukemia.

[m]Mellinghoff et al. 2023.

[n]Mann et al. 2007.

[o]www.accessdata.fda.gov/drugsatfda_docs/label/2009/022393lbl.pdf.

[p]Lee et al. 2015.

[q]Raedler 2016.

[r]Lewin et al. 2018.

[s]Piha-Paul et al. 2019.

[t]www.fda.gov/drugs/fda-granted-accelerated-approval-tazemetostat-follicular-lymphoma#:~:text=On%20June%2018%2C%202020%2C%20the,approved%20test%20and%20who%20have.

[u]www.fda.gov/drugs/resources-information-approved-drugs/fda-approves-tazemetostat-advanced-epithelioid-sarcoma.

[v]Stein et al. 2018.

[w]www.accessdata.fda.gov/drugsatfda_docs/label/2023/020438s007s008lbl.pdf.

[x]www.accessdata.fda.gov/drugsatfda_docs/label/2015/021248s013lbl.pdf.

[y]Issa et al. 2023.

[z]Wang et al. 2024.

tients with *IDH1* and *IDH2* mutations, respectively, with ivosidenib having an additional approval in AML patients who are elderly or unfit for intensive chemotherapy. Another IDH1 inhibitor, olutasidenib, was well-tolerated and efficacious in trials with and without azacitidine, and also recently received FDA approval for *IDH1*-mutant MDS and AML (Watts et al. 2023). Importantly, with these monotherapy agents, resistance mutations in *IDH1* and *IDH2* that abrogate the activity of these compounds has been described in AML patients (Intlekofer et al. 2018).

IDH inhibitors have shown activity outside of AML. *IDH1* mutations are found in cholangiocarcinomas (13% of intrahepatic cholangiocarcinomas). Ivosidenib holds FDA approval for metastatic or unresectable, locally advanced cholangiocarcinoma based on the results of a phase 3 trial comparing ivosidenib to placebo control, wherein ivosidenib conferred a several month progression-free and overall survival advantage (Abou-Alfa et al. 2020; Zhu et al. 2021). A CNS-penetrant, dual-IDH1 and IDH2 inhibitor, vorasidenib, was recently developed and has progressed through clinical trials for grade 2 *IDH*-mutant glioma. Encouragingly, a phase 3, double-blind trial comparing oral vorasidenib to placebo demonstrated a 27.7-month progression-free survival for patients receiving vorasidenib as compared to 11.1 months for patients receiving placebo (Mellinghoff et al. 2023).

HISTONE MODIFICATIONS

Histone Acetylation and Deacetylation

Histone acetylation is generally associated with active transcription, while deacetylation of histones is associated with silencing of transcription. One example of an acetyl histone modification that is often found at active promoters and enhancers is acetylated histone H3, lysine 27 (H3K27Ac) (Creyghton et al. 2010; Raisner et al. 2018). Histone acetyltransferases (HATs) add acetyl moieties to lysine residues on histone tails, resulting in chromatin decompaction by decreasing the interaction between positively charged lysine and negatively charged

DNA (Roth et al. 2001). Removal of acetyl histone marks is performed by a family of enzymes known as histone deacetylases (HDACs).

Histone Deacetylases

HDACs are "erasers" that remove acetyl groups from histones, however, they also deacetylate nonhistone proteins (Fig. 2; West and Johnstone 2014). First insights into the role of HDACs came from the observation that there is global loss of monoacetylation on histone H4 in cancer cells (Fraga et al. 2005). HDAC inhibitors cause hyperacetylation of histone (and nonhistone) proteins, resulting in the expression of repressed genes that can induce apoptosis, cell-cycle arrest, and differentiation, among other causes (Bondarev et al. 2021). In some cancers, mutations in various HDACs are thought to promote tumorigenesis, a discovery that prompted the development of HDAC inhibitors (somatic *HDAC1* mutations are found in 8% of dedifferentiated liposarcoma) (Taylor et al. 2011) and work in a conditional mouse model has demonstrated that loss of *Hdac1* and *Hdac2* induces tumorigenesis (Heideman et al. 2013). Defining the precise mechanism by which HDAC inhibitors halt oncogenesis has remained elusive because both HDACs and their inhibitors are promiscuous. Many HDACs are recruited to different multiprotein complexes, they have multiple substrates, and HDAC inhibitors often affect several HDACs simultaneously (Bradner et al. 2010; West and Johnstone 2014). There are currently four HDACs approved by the FDA: vorinostat and romidepsin for cutaneous T-cell lymphoma, belinostat and romidepsin for cutaneous and peripheral T-cell lymphomas, and panobinostat for multiple myeloma (Bondarev et al. 2021).

Histone Acetyltransferase Inhibitors

Lysine acetyltransferase 6A (*KAT6A*) is recurrently fused to *CREBBP* in <1% of AML cases (Döhner et al. 2017). In non-KAT6A-rearranged AML, *KAT6A* has a functional interaction with ENL and has been proposed to be a therapeutic target in AML (Yan et al. 2022). It is

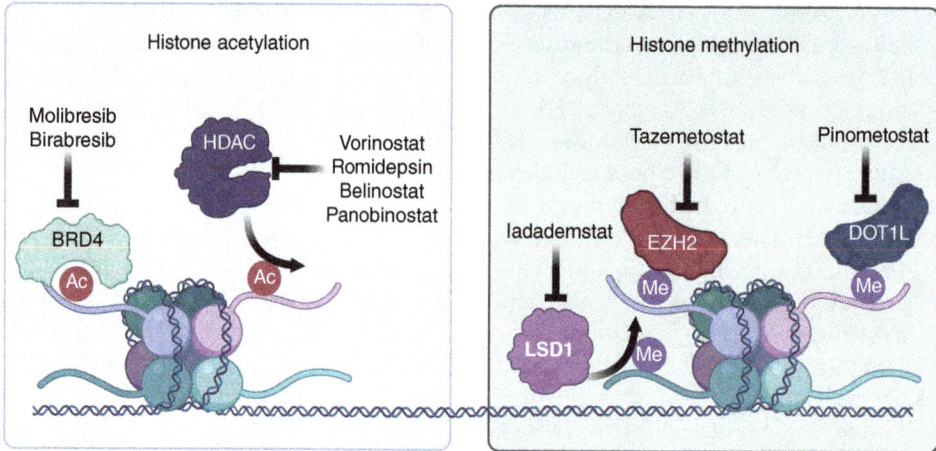

Figure 2. Histone acetylation and methylation targets of epigenetic cancer therapies. Chromatin regulators of histone acetylation include histone deacetylases (HDACs), which are inhibited by vorinostat, romidepsin, belinostat, and panobinostat. Inhibitors of histone methyltransferases, such as LSD1, EZH2, and DOT1L, include small-molecule iadademstat, tazemetostat, and pinometostat.

also amplified in 12%–15% of breast cancer (Turner-Ivey et al. 2014). HAT inhibitors have been developed, shown strong preclinical activity in breast cancer models (Sharma et al. 2023), and have entered clinical trials (PF-07248144) in breast cancer. A phase 1 clinical trial in ER⁺HER2⁻ metastatic breast cancer with PF-07248144 recently demonstrated a 30% overall response rate and 10-month progression-free survival (Mukohara et al. 2024).

Bromodomain Inhibitors

Histone acetylation serves as a docking site for proteins with "reader" domains (Xu and Vakoc 2017). Perhaps the most well-characterized of these "readers" are bromodomain-containing proteins, which are essential for guiding protein complexes to chromatin to mediate the regulation of gene expression (Sanchez and Zhou 2009; Filippakopoulos et al. 2012; Filippakopoulos and Knapp 2014). There are four BET (bromodomain and extraterminal domain-containing) family members, including BRD2, BRD3, BRD4, and BRDT. *BRD4* is fused to *NUTM* in the aggressive NUT carcinoma (in addition to rarer fusion partners, including BRD3 [Dickson

et al. 2018], which has no effective therapy and short survival [French et al. 2003; Eagen and French 2021]). In this context, BRD4 recruits p300 to chromatin, resulting in large "megadomains" of chromatin enriched with BRD4-NUT, p300, and acetylated histones. These domains activate transcription, particularly *MYC*, *SOX2*, and *TP63*. There has been extensive drug development targeting bromodomains. The first molecules targeting these were pan-BET inhibitors that target BRD2, BRD3, and BRD4: JQ-1 and iBET (Filippakopoulos et al. 2010; Nicodeme et al. 2010). Many BET inhibitors have been evaluated in clinical trials, but efficacy has been limited and thrombocytopenia has been a prominent toxicity; more than a dozen phase 1 trials have been undertaken (Doroshow et al. 2017; Pearson et al. 2021). Recently, the BET inhibitors molibresib (GSK525762) and birabresib (MK-8628/OTX015) have demonstrated activity in NUT carcinoma (Fig. 2; Lewin J et al. 2018; Piha-Paul et al. 2020).

Histone Methylation and Demethylation

The outcome of histone methylation is context-dependent, as methyl marks can either activate or repress transcription depending on which amino acid is modified. For example, H3K27Me3 is a repressive histone mark that is associated with decreased transcriptional activity, chromatin compaction, and epigenetic silencing (Cao et al. 2002; Jiao and Xin 2015). In contrast, H3K4Me3 is associated with open chromatin and active transcription (Santos-Rosa et al. 2002). Methyl marks are deposited by a family of histone methyltransferase "writers," several of which have been implicated in oncogenesis, including EZH2, NSD1, SETD2, and DOT1L. Mutations in lysine demethylase "erasers" such as LSD1 can also promote tumorigenesis by failing to remove methyl marks from tumor suppressors, for example (Karakaidos et al. 2019).

EZH2 Inhibitors

Enhancer of zeste homolog 2 (EZH2) is the catalytic subunit of the polycomb repressive complex 2 (PRC2). EZH2 is a histone "writer" that trimethylates histone 3 lysine 27 (H3K27) at gene promoters, resulting in transcriptional silencing (Fig. 3; Kim and Roberts 2016). Gain-of-function mutations, loss-of-function mutations, and overexpression of EZH2 have been described, in addition to mutations in factors that antagonize EZH2, such as inactivating mutations in the H3K27 demethylase UTX, H3K27 mutations, as well as mutations in SWItch/sucrose nonfermentable (SWI-SNF) complex members (van Haaften et al. 2009; Kim and Roberts 2016; Duan et al. 2020). EZH2 overexpression was first described in prostate cancer and breast cancer and several others thereafter (Varambally et al. 2002; Kleer et al. 2003). Gain-of-function mutations in EZH2 have been most prominently reported in non-Hodgkin's lymphomas, where initial studies found Y641 mutations in more than 20% of germinal center diffuse large B-cell lymphoma and 7% of follicular lymphomas. These EZH2 mutations occur in the SET domain of the EZH2 protein and are activating (Morin et al. 2010); however, loss-of-function mutations in myeloid neoplasms have also been reported (Ernst et al. 2010). Given the overexpression and gain-of-function mutations,

enzymatic inhibitors of EZH2 have entered clinical trials in various cancer types, with phase 2 trials initiated in non-Hodgkin's lymphoma, bladder cancer, endometrial cancer, ovarian cancer, mesothelioma, malignant rhabdoid tumors/INI1-negative tumors, and epithelioid sarcoma (Duan et al. 2020). While tazemetostat is currently the most advanced EZH2 inhibitor in development, valemetostat, CPI-1205, and CPI-0209 are also in early-phase clinical trials. Tazemetostat is FDA approved for relapsed/refractory follicular lymphoma with EZH2 mutations following the results of a phase 2 trial wherein the objective response rate for patients with EZH2 mutations was nearly 70% compared to 45% in the EZH2 wild-type cohort (Morschhauser et al. 2020). It also has an FDA approval in relapsed refractory epithelioid sarcoma after a phase 2 basket study showed that patients with advanced epithelioid sarcoma (and loss of SMARCB2/INI1 by immunohistochemistry or biallelic SMARCB2/INI1 loss) tolerated tazemetostat well with 15% of patients having an objective response and 26% of patients having disease control (Gounder et al. 2020). SMARCB2/INI1 is one of the core subunits of the SWI/SNF ATP-dependent chromatin remodeling complex, which is a chromatin "mover" that slides nucleosomes to open chromatin and promote transcription (Centore et al. 2020). Another subunit of the SWI/SNF complex, SS18, is recurrently translocated with the SSX gene in synovial sarcoma, and EZH2 inhibitors are also being investigated in this aggressive but rare neoplasm (NCT02601950) (McBride et al. 2018).

DOT1L Inhibitors

DOT1L is the only known histone methyltransferase that catalyzes H3K79 methylation, a mark associated with increased transcriptional activity (Steger et al. 2008). In leukemias driven by chromosomal rearrangements of the lysine methyltransferase 2A gene (KMT2A, also known as mixed lineage leukemia 1 [MLL1]), the amino terminus of KMT2A is fused to nearly 100 different carboxy-terminal fusion partners (Meyer et al. 2023), and is part of a large, chromatin-bound protein complex essential for leukemic transfor-

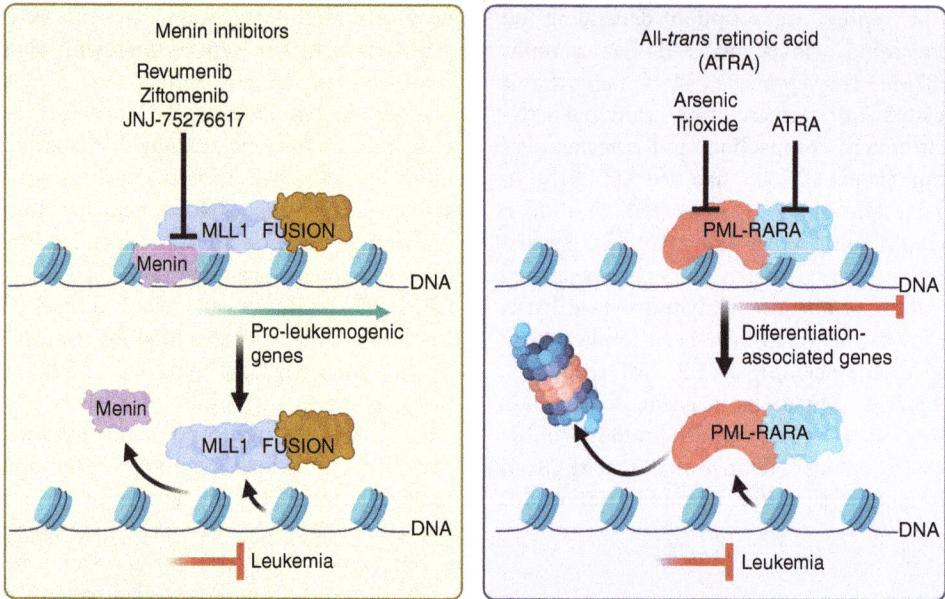

Figure 3. Oncogenic fusion proteins as targets for epigenetic therapies. Oncogenic fusion proteins such as KMT2A/mixed lineage leukemia 1 (MLL1)-fusion proteins or PML-RARA can be targeted using small molecules that disrupt chromatin binding. Revumenib, ziftomenib, and JNJ-75276617 inhibit the protein–protein interaction between the MLL-fusion protein and menin. In acute promyelocytic leukemia (APL), the PML-RARA fusion protein is degraded upon combination treatment with all-*trans* retinoic acid (ATRA) and arsenic.

mation and maintenance (Brien et al. 2019). Several KMT2A-fusion partners recruit DOT1L (Okada et al. 2005; Mueller et al. 2007). Studies have demonstrated that *KMT2A*-rearranged leukemias (*KMT2A*-r) are dependent on DOT1L-mediated H3K79 methylation (Bernt et al. 2011), and this discovery has led to the development of small molecules targeting the enzymatic activity of DOT1L. A phase 1 clinical trial of the DOT1L inhibitor pinometostat demonstrated tolerability and minimal activity (two CRs among 54 patients) although a maximally tolerated drug dose was not attained (Stein et al. 2018). Nonetheless, this study demonstrated the therapeutic potential of targeting DOT1L, and agents with better pharmacokinetic properties have shown encouraging preclinical activity (Perner et al. 2020).

LSD1 Inhibitors

Prior to the discovery of the demethylase function of LSD1 on K4-methylated histone H3, his-

tone methylation was considered to be a permanent modification (Shi et al. 2004). LSD1 was later found to demethylate K4-methylated histone H3, which marks enhancers and is regulated during cell fate decisions (Metzger et al. 2005; Kouzarides 2007). LSD1 thus suppresses gene expression. Initial interest in targeting demethylases came from studies suggesting a role in drug resistance and self-renewal (Sharma et al. 2010; Mohammad et al. 2019). LSD1 has been shown to be overexpressed in bladder cancer, associated with migration and invasion in non-small-cell lung cancer, and overexpression is associated with a poor prognosis in breast cancer (Hayami et al. 2011; Lv et al. 2012; Nagasawa et al. 2015). LSD1 inhibitors have subsequently been developed, and preclinical work showed the ability of iadademstat/ORY-1001 to differentiate blasts and efficacy in a PDX model of T-cell acute leukemia (Maes et al. 2018). Iadademstat then entered a phase 1 clinical trial in relapsed/refractory AML and demonstrated tolerability and a complete response with incomplete count recovery in one patient (Salamero et al. 2020). Combination studies are currently underway, and early efficacy in combination with azacitidine has been demonstrated (Salamero et al. 2022).

TARGETING CHROMATIN COMPLEXES

The targets and therapies discussed thus far have focused on either direct or indirect inhibition of enzymatic activity of DNA or histone-modifying enzymes. Another approach is to use epigenetic therapies to restore normal function by targeting chromatin complexes. To this end, small molecules have been developed that disrupt essential protein–protein interactions, which is thought to lead to disassembly of chromatin complexes and their eviction from chromatin.

All-*trans* Retinoic Acid and Arsenic Trioxide

Acute promyelocytic leukemia (APL) is an example of a malignancy driven by an oncogenic fusion protein that can be dismantled using targeted agents. APL constitutes 5%–15% of all leukemia cases in children and adults and is most often characterized by translocations of chromosome 15 and 17, resulting in fusion of the *PML* and *RARA* genes (Larson et al. 1984; de Thé et al. 1990; Creutzig et al. 2016; Rérolle et al. 2024). With cytotoxic chemotherapy regimens, 30%–40% of patients could achieve cure in early trials; however, early mortality from bleeding complications and eventual relapse were common (Kantarjian et al. 2021). Pioneering work in the 1980s showed that retinoic acid could differentiate APL in vitro (Breitman et al. 1981). This observation led to clinical trials of retinoic acid, which demonstrated remarkable results: 24 patients (eight relapsed/refractory, 16 de novo) were able to attain remissions and morphologic evidence of differentiation, though with eventual relapses in many patients (Huang et al. 1988). Concurrently, it was discovered that arsenic was the active ingredient in an anticancer remedy used by practitioners of traditional Chinese medicine, and two-thirds of patients

with APL receiving arsenic trioxide could attain complete response, and nearly 30% of patients had durable cures lasting over 10 years (Chen et al. 1996; Shen et al. 1997). Subsequent studies uncovered that the *PML::RARA* fusion acts as a transcriptional repressor, recruiting corepressors and ultimately blocking myeloid differentiation and leading to defects in apoptosis (de Thé et al. 2015). Mechanistically, all-*trans* retinoic acid (ATRA) leads to protein degradation of RARA and PML-RARA, but not PML, while arsenic trioxide led to protein degradation of PML and PML-RARA, but not RARA (Yoshida et al. 1996; Shao et al. 1998; Zhang et al. 2010). However, neither ATRA nor arsenic can be used as single agents, since resistance mutations in both PML and RARA may arise with monotherapy (Tomita et al. 2013).

Combination therapy with arsenic trioxide and ATRA has been successful. The first major study demonstrated that ATRA in addition to cytarabine and daunorubicin profoundly increased overall survival rates (Powell et al. 2010). Thereafter, it was demonstrated that patients with low-to-intermediate risk APL could not only be cured with ATRA and arsenic trioxide without chemotherapy (100% CR rate, 97% 2-yr-survival rate), but that this therapy outperformed ATRA + chemotherapy (95% CR rate, 86% 2-yr-survival rate) (Lo-Coco et al. 2013).

Menin Inhibitors

In *KMT2A*-r leukemias, the carboxy-terminal methyltransferase domain of KMT2A is lost, and the amino terminus is fused to nearly 100 different partner genes. The amino-terminal portion of KMT2A interacts with menin, an interaction that is essential for transformation in *KMT2A*-r leukemia (Yokoyama et al. 2005). Early work demonstrated that genetic loss of *Men1* (encoding menin) decreases *Homeobox* (*HOX*) genes in mouse embryos (Hughes et al. 2004). Blocking the menin-KMT2A interaction with a dominant-negative inhibitor led to decreased expression of *MEIS1*, a *HOX* gene cofactor essential for KMT2A-fusion protein-mediated leukemogenesis (Caslini et al. 2007). Several small molecules have been developed that target the menin-KMT2A protein–protein interaction, and the first inhibitors demonstrated antileukemic effects with decreased expression of *HOXA9* and *MEIS1* (Grembecka et al. 2012).

Preclinical work with menin-KMT2A inhibitors has demonstrated that blocking this interaction can halt oncogenesis in leukemias harboring *KMT2A*-r (Krivtsov et al. 2019; Klossowski et al. 2020), *Nucleophosmin 1* mutations (Kühn et al. 2016; Uckelmann et al. 2020), *Nucleoporin 98* rearrangements (Heikamp et al. 2022), and *UBTF* tandem duplications (Barajas et al. 2024). Mechanistically, menin-KMT2A inhibition results in the eviction of menin globally from chromatin, KMT2A loss from select loci, and *MEIS1* down-regulation. These leukemias share a *HOX-MEIS* gene expression program, and loss of the menin–KMT2A interaction results in down-regulation of these stem cell–associated genes and up-regulation of differentiation markers (Fig. 3). Menin-KMT2A inhibitors have subsequently entered clinical trials and have demonstrated promising results thus far, with the phase 1 AUGMENT-101 trial showing a 30% complete remission/complete remission with incomplete recovery rate (Issa et al. 2023). However, up to 40% of patients developed *MEN1* mutations after at least two cycles of therapy, which both validates the menin-KMT2A interaction as a therapeutic target but also highlights the need for combination therapies (Perner et al. 2023).

DISCUSSION

Epigenetic therapies play an important role in anticancer therapy, both through the rational design of therapies that target epigenetic drivers of disease and through the discovery that existing therapies mediate their effects through epigenetic functions. For example, in one of the more remarkable successes in cancer therapy, many patients with APL can be cured without chemotherapy, and instead receive arsenic trioxide and ATRA, which specifically and cooperatively degrades the oncogenic fusion protein (*PML::RARA*) that drives the disease. The combination of arsenic trioxide and ATRA in APL

also serves as an important example of the need for combination therapy, as many patients treated with either of these targeted agents alone will eventually become resistant and mutations that specifically abrogate each of these therapies have been defined. In fact, several targeted, rationally designed epigenetic therapies, such as IDH1/IDH2 inhibitors and menin inhibitors, have demonstrated promising responses, but they also are hampered by resistance mutations. Discerning which combination therapies can enhance the efficacy of these compounds while also preventing resistance mechanisms will be critical. Such work will require fundamental understanding of biologic mechanisms that lead to epigenetic dysregulation.

REFERENCES

*Reference is also in this subject collection.

Abou-Alfa GK, Macarulla T, Javle MM, Kelley RK, Lubner SJ, Adeva J, Cleary JM, Catenacci DV, Borad MJ, Bridgewater J, et al. 2020. Ivosidenib in IDH1-mutant, chemotherapy-refractory cholangiocarcinoma (ClarIDHy): a multicentre, randomised, double-blind, placebo-controlled, phase 3 study. *Lancet Oncol* **21:** 796–807. doi:10.1016/S1470-2045(20)30157-1

Barajas JM, Rasouli M, Umeda M, Hiltenbrand R, Abdelhamed S, Mohnani R, Arthur B, Westover T, Thomas III ME, Ashtiani M, et al. 2024. Acute myeloid leukemias with *UBTF* tandem duplications are sensitive to menin inhibitors. *Blood* **143:** 619–630. doi:10.1182/blood.2023021359

Bates SE. 2020. Epigenetic therapies for cancer. *N Eng J Med* **383:** 650–663. doi:10.1056/NEJMra1805035

Baylin SB, Jones PA. 2011. A decade of exploring the cancer epigenome—biological and translational implications. *Nat Rev Cancer* **11:** 726–734. doi:10.1038/nrc3130

Bejar R, Lord A, Stevenson K, Bar-Natan M, Pérez-Ladaga A, Zaneveld J, Wang H, Caughey B, Stojanov P, Getz G, et al. 2014. TET2 mutations predict response to hypomethylating agents in myelodysplastic syndrome patients. *Blood* **124:** 2705–2712. doi:10.1182/blood-2014-06-582809

Bernt KM, Zhu N, Sinha AU, Vempati S, Faber J, Krivtsov AV, Feng Z, Punt N, Daigle A, Bullinger L, et al. 2011. MLL-rearranged leukemia is dependent on aberrant H3K79 methylation by DOT1L. *Cancer Cell* **20:** 66–78. doi:10.1016/j.ccr.2011.06.010

Bondarev AD, Attwood MM, Jonsson J, Chubarev VN, Tarasov VV, Schiöth HB. 2021. Recent developments of HDAC inhibitors: emerging indications and novel molecules. *Br J Clin Pharmacol* **87:** 4577–4597. doi:10.1111/bcp.14889

Bradner JE, West N, Grachan ML, Greenberg EF, Haggarty SJ, Warnow T, Mazitschek R. 2010. Chemical phyloge-

Cite this article as *Cold Spring Harb Perspect Med* doi: 10.1101/cshperspect.a041637

netics of histone deacetylases. *Nat Chem Biol* **6**: 238–243. doi:10.1038/nchembio.313

Breitman TR, Collins SJ, Keene BR. 1981. Terminal differentiation of human promyelocytic leukemic cells in primary culture in response to retinoic acid. *Blood* **57**: 1000–1004. doi:10.1182/blood.V57.6.1000.1000

Brien GL, Stegmaier K, Armstrong SA. 2019. Targeting chromatin complexes in fusion protein-driven malignancies. *Nat Rev Cancer* **19**: 255–269. doi:10.1038/s41568-019-0132-x

Cao R, Wang L, Wang H, Xia L, Erdjument-Bromage H, Tempst P, Jones RS, Zhang Y. 2002. Role of histone H3 lysine 27 methylation in Polycomb-group silencing. *Science* **298**: 1039–1043. doi:10.1126/science.1076997

Casak SJ, Pradhan S, Fashoyin-Aje LA, Ren Y, Shen YL, Xu Y, Chow ECY, Xiong Y, Zirklelbach JF, Liu J, et al. 2022. FDA approval summary: ivosidenib for the treatment of patients with advanced unresectable or metastatic, chemotherapy refractory cholangiocarcinoma with an IDH1 mutation. *Clin Cancer Res* **28**: 2733–2737. doi:10.1158/1078-0432.CCR-21-4462

Caslini C, Yang Z, El-Osta M, Milne TA, Slany RK, Hess JL. 2007. Interaction of MLL amino terminal sequences with menin is required for transformation. *Cancer Res* **67**: 7275–7283. doi:10.1158/0008-5472.CAN-06-2369

Centore RC, Sandoval GJ, Soares LMM, Kadoch C, Chan HM. 2020. Mammalian SWI/SNF chromatin remodeling complexes: emerging mechanisms and therapeutic strategies. *Trends Genet* **36**: 936–950. doi:10.1016/j.tig.2020.07.011

Chen GQ, Zhu J, Shi XG, Ni JH, Zhong HJ, Si GY, Jin XL, Tang W, Li XS, Xong SM, et al. 1996. In vitro studies on cellular and molecular mechanisms of arsenic trioxide (As2O3) in the treatment of acute promyelocytic leukemia: As2O3 induces NB4 cell apoptosis with downregulation of Bcl-2 expression and modulation of PML-RARα/PML proteins. *Blood* **88**: 1052–1061. doi:10.1182/blood.V88.3.1052.1052

Creutzig U, Zimmermann M, Reinhardt D, Rasche M, von Neuhoff C, Alpermann T, Dworzak M, Perglerová K, Zemanova Z, Tchinda J, et al. 2016. Changes in cytogenetics and molecular genetics in acute myeloid leukemia from childhood to adult age groups. *Cancer* **122**: 3821–3830. doi:10.1002/cncr.30220

Creyghton MP, Cheng AW, Welstead GG, Kooistra T, Carey BW, Steine EJ, Hanna J, Lodato MA, Frampton GM, Sharp PA, et al. 2010. Histone H3K27ac separates active from poised enhancers and predicts developmental state. *Proc Natl Acad Sci* **107**: 21931–21936. doi:10.1073/pnas.1016071107

Derissen EJ, Beijnen JH, Schellens JH. 2013. Concise drug review: azacitidine and decitabine. *Oncologist* **18**: 619–624. doi:10.1634/theoncologist.2012-0465

de Thé H, Chomienne C, Lanotte M, Degos L, Dejean A. 1990. The t(15;17) translocation of acute promyelocytic leukaemia fuses the retinoic acid receptor α gene to a novel transcribed locus. *Nature* **347**: 558–561. doi:10.1038/347558a0

de Thé H, Zhu J, Nasr R, Ablain J, Lallemand-Breittenbach V. 2015. PML/RARA as the master driver of APL pathogenesis and therapy response. In *Targeted therapy of acute myeloid leukemia. Current cancer research* (ed. Andreeff M). Springer, New York.

Dickson BC, Sung YS, Rosenblum MK, Reuter VE, Harb M, Wunder JS, Swanson D, Antonescu CR. 2018. NUTM1 gene fusions characterize a subset of undifferentiated soft tissue and visceral tumors. *Am J Surg Pathol* **42**: 636–645. doi: 10.1097/PAS.0000000000001021

DiNardo CD, Patel KP, Garcia-Manero G, Luthra R, Pierce S, Borthakur G, Jabbour E, Kadia T, Pemmaraju N, Konopleva M, et al. 2014. Lack of association of *IDH1*, *IDH2* and *DNMT3A* mutations with outcome in older patients with acute myeloid leukemia treated with hypomethylating agents. *Leuk Lymphoma* **55**: 1925–1929. doi:10.3109/10428194.2013.855309

DiNardo CD, Stein EM, de Botton S, Roboz GJ, Altman JK, Mims AS, Swords R, Collins RH, Mannis GN, Pollyea DA, et al. 2018. Durable remissions with ivosidenib in *IDH1*-mutated relapsed or refractory AML. *N Eng J Med* **378**: 2386–2398. doi:10.1056/NEJMoa1716984

Döhner H, Estey E, Grimwade D, Amadori S, Appelbaum FR, Büchner T, Dombret H, Ebert BL, Fenaux P, Larson RA, et al. 2017. Diagnosis and management of AML in adults: 2017 ELN recommendations from an international expert panel. *Blood* **129**: 424–447. doi:10.1182/blood-2016-08-733196

Doroshow DB, Eder JP, LoRusso PM. 2017. BET inhibitors: a novel epigenetic approach. *Ann Oncol* **28**: 1776–1787. doi:10.1093/annonc/mdx157

Duan R, Du W, Guo W. 2020. EZH2: a novel target for cancer treatment. *J Hematol Oncol* **13**: 1–12. doi:10.1186/s13045-020-00937-8

Eagen KP, French CA. 2021. Supercharging BRD4 with NUT in carcinoma. *Oncogene* **40**: 1396–1408. doi:10.1038/s41388-020-01625-0

Egger G, Liang G, Aparicio A, Jones PA. 2004. Epigenetics in human disease and prospects for epigenetic therapy. *Nature* **429**: 457–463. doi:10.1038/nature02625

Ehrlich M. 2002. DNA methylation in cancer: too much, but also too little. *Oncogene* **21**: 5400–5413. doi:10.1038/sj.onc.1205651

Ernst T, Chase AJ, Score J, Hidalgo-Curtis CE, Bryant C, Jones AV, Waghorn K, Zoi K, Ross FM, Reiter A, et al. 2010. Inactivating mutations of the histone methyltransferase gene EZH2 in myeloid disorders. *Nat Genet* **42**: 722–726. doi:10.1038/ng.621

FDA. 2022. FDA approves ivosidenib in combination with azacitidine for newly diagnosed acute myeloid leukemia. U.S. Food & Drug Administration, Silver Spring, MD. https://www.fda.gov/drugs/resources-information-approved-drugs/fda-approves-ivosidenib-combination-azacitidine-newly-diagnosed-acute-myeloid-leukemia [accessed Nov. 20, 2024].

Fernandez AF, Assenov Y, Martin-Subero JI, Balint B, Siebert R, Taniguchi H, Yamamoto H, Hidalgo M, Tan AC, Galm O, et al. 2012. A DNA methylation fingerprint of 1628 human samples. *Genome Res* **22**: 407–419. doi:10.1101/gr.119867.110

Figueroa ME, Abdel-Wahab O, Lu C, Ward PS, Patel J, Shih A, Li Y, Bhagwat N, Vasanthakumar A, Fernandez HF, et al. 2010. Leukemic IDH1 and IDH2 mutations result in a hypermethylation phenotype, disrupt TET2 function, and impair hematopoietic differentiation. *Cancer Cell* **18:** 553–567. doi:10.1016/j.ccr.2010.11.015

Filippakopoulos P, Knapp S. 2014. Targeting bromodomains: epigenetic readers of lysine acetylation. *Nat Rev Drug Discov* **13:** 337–356. doi:10.1038/nrd4286

Filippakopoulos P, Qi J, Picaud S, Shen Y, Smith WB, Fedorov O, Morse EM, Keates T, Hickman TT, Felletar I, et al. 2010. Selective inhibition of BET bromodomains. *Nature* **468:** 1067–1073. doi:10.1038/nature09504

Filippakopoulos P, Picaud S, Mangos M, Keates T, Lambert JP, Barsyte-Lovejoy D, Felletar I, Volkmer R, Müller S, Pawson T, et al. 2012. Histone recognition and large-scale structural analysis of the human bromodomain family. *Cell* **149:** 214–231. doi:10.1016/j.cell.2012.02.013

Fraga MF, Ballestar E, Villar-Garea A, Boix-Chornet M, Espada J, Schotta G, Bonaldi T, Haydon C, Ropero S, Petrie K, et al. 2005. Loss of acetylation at Lys16 and trimethylation at Lys20 of histone H4 is a common hallmark of human cancer. *Nat Genet* **37:** 391–400. doi:10.1038/ng1531

French CA, Miyoshi I, Kubonishi I, Grier HE, Perez-Atayde AR, Fletcher JA. 2003. BRD4-NUT fusion oncogene: a novel mechanism in aggressive carcinoma. *Cancer Res* **63:** 304–307.

Ganesan A, Arimondo PB, Rots MG, Jeronimo C, Berdasco M. 2019. The timeline of epigenetic drug discovery: from reality to dreams. *Clin Epigenetics* **11:** 1–17. doi:10.1186/s13148-019-0776-0

Gounder M, Schöffski P, Jones RL, Agulnik M, Cote GM, Villalobos VM, Attia S, Chugh R, Chen TWW, Jahan T, et al. 2020. Tazemetostat in advanced epithelioid sarcoma with loss of INI1/SMARCB1: an international, open-label, phase 2 basket study. *Lancet Oncol* **21:** 1423–1432. doi:10.1016/S1470-2045(20)30451-4

Grembecka J, He S, Shi A, Purohit T, Muntean AG, Sorenson RJ, Showalter HD, Murai MJ, Belcher AM, Hartley T, et al. 2012. Menin-MLL inhibitors reverse oncogenic activity of MLL fusion proteins in leukemia. *Nat Chem Biol* **8:** 277–284. doi:10.1038/nchembio.773

Hartmann C, Meyer J, Balss J, Capper D, Mueller W, Christians A, Felsberg J, Wolter M, Mawrin C, Wick W, et al. 2009. Type and frequency of IDH1 and IDH2 mutations are related to astrocytic and oligodendroglial differentiation and age: a study of 1,010 diffuse gliomas. *Acta Neuropathol* **118:** 469–474. doi:10.1007/s00401-009-0561-9

Hayami S, Kelly JD, Cho HS, Yoshimatsu M, Unoki M, Tsunoda T, Field HI, Neal DE, Yamaue H, Ponder BA, et al. 2011. Overexpression of LSD1 contributes to human carcinogenesis through chromatin regulation in various cancers. *Int J Cancer* **128:** 574–586. doi:10.1002/ijc.25349

Heideman MR, Wilting RH, Yanover E, Velds A, de Jong J, Kerkhoven RM, Jacobs H, Wessels LF, Dannenberg JH. 2013. Dosage-dependent tumor suppression by histone deacetylases 1 and 2 through regulation of c-Myc collaborating genes and p53 function. *Blood* **121:** 2038–2050. doi:10.1182/blood-2012-08-450916

Heikamp EB, Henrich JA, Perner F, Wong EM, Hatton C, Wen Y, Barwe SP, Gopalakrishnapillai A, Xu H, Uckelmann HJ, et al. 2022. The menin-MLL1 interaction is a molecular dependency in NUP98-rearranged AML. *Blood* **139:** 894–906. doi:10.1182/blood.2021012806

Huang ME, Ye YC, Chen SR, Chai JR, Lu JX, Zhoa L, Gu LJ, Wang ZY. 1988. Use of all-*trans* retinoic acid in the treatment of acute promyelocytic leukemia. *Blood* **72:** 567–572. doi:10.1182/blood.V72.2.567.567

Hughes CM, Rozenblatt-Rosen O, Milne TA, Copeland TD, Levine SS, Lee JC, Hayes DN, Shanmugam KS, Bhattacharjee A, Biondi CA, et al. 2004. Menin associates with a trithorax family histone methyltransferase complex and with the hoxc8 locus. *Mol Cell* **13:** 587–597. doi:10.1016/S1097-2765(04)00081-4

Intlekofer AM, Shih AH, Wang B, Nazir A, Rustenburg AS, Albanese SK, Patel M, Famulare C, Correa FM, Takemoto N, et al. 2018. Acquired resistance to IDH inhibition through *trans* or *cis* dimer-interface mutations. *Nature* **559:** 125–129. doi:10.1038/s41586-018-0251-7

Issa GC, Aldoss I, DiPersio J, Cuglievan B, Stone R, Arellano M, Thirman MJ, Patel MR, Dickens DS, Shenoy S, et al. 2023. The menin inhibitor revumenib in KMT2A-rearranged or NPM1-mutant leukaemia. *Nature* **615:** 920–924. doi:10.1038/s41586-023-05812-3

Ito S, D'Alessio AC, Taranova OV, Hong K, Sowers LC, Zhang Y. 2010. Role of Tet proteins in 5mC to 5hmC conversion, ES-cell self-renewal and inner cell mass specification. *Nature* **466:** 1129–1133. doi:10.1038/nature09303

Jen EY, Wang X, Li M, Li H, Lee SL, Ni N, Przepiorka D, Vallejo J, Leong R, Ma L, et al. 2022. FDA approval summary: oral azacitidine for continued treatment of adults with acute myeloid leukemia unable to complete intensive curative therapy. *Clin Cancer Res* **28:** 2989–2993. doi:10.1158/1078-0432.CCR-21-4525

Jenuwein T, Allis CD. 2001. Translating the histone code. *Science* **293:** 1074–1080. doi:10.1126/science.1063127

Jiao L, Xin L. 2015. Structural basis of histone H3K27 trimethylation by an active polycomb repressive complex 2. *Science* **350:** aac4383. doi:10.1126/science.aac4383

Kaminskas E, Farrell A, Abraham S, Baird A, Hsieh LS, Lee SL, Leighton JK, Patel H, Rahman A, Sridhara R, et al. 2005. Approval summary: azacitidine for treatment of myelodysplastic syndrome subtypes. *Clin Cancer Res* **11:** 3604–3608. doi:10.1158/1078-0432.CCR-04-2135

Kantarjian HM, Thomas XG, Dmoszynska A, Wierzbowska A, Mazur G, Mayer J, Gau JP, Chou WC, Buckstein R, Cermak J, et al. 2012. Multicenter, randomized, open-label, phase III trial of decitabine versus patient choice, with physician advice, of either supportive care or low-dose cytarabine for the treatment of older patients with newly diagnosed acute myeloid leukemia. *J Clin Oncol* **30:** 2670–2677. doi:10.1200/JCO.2011.38.9429

Kantarjian H, Kadia T, DiNardo C, Daver N, Borthakur G, Jabbour E, Garcia-Manero G, Konopleva M, Ravandi F. 2021. Acute myeloid leukemia: current progress and future directions. *Blood Cancer J* **11:** 41. doi:10.1038/s41408-021-00425-3

Karakaidos P, Verigos J, Magklara A. 2019. LSD1/KDM1A, a gate-keeper of cancer stemness and a promising therapeutic target. *Cancers (Basel)* **11:** 1821. doi:10.3390/cancers11121821

Kayser S, Levis MJ. 2022. Updates on targeted therapies for acute myeloid leukaemia. *Br J Haematol* **196:** 316–328. doi:10.1111/bjh.17746

Kim KH, Roberts CW. 2016. Targeting EZH2 in cancer. *Nat Med* **22:** 128–134. doi:10.1038/nm.4036

Kim N, Norsworthy KJ, Subramaniam S, Chen H, Manning ML, Kitabi E, Earp J, Ehrlich LA, Okusanya OO, Vallejo J, et al. 2022. FDA approval summary: decitabine and cedazuridine tablets for myelodysplastic syndromes. *Clin Cancer Res* **28:** 3411–3416. doi:10.1158/1078-0432.CCR-21-4498

Kleer CG, Cao Q, Varambally S, Shen R, Ota I, Tomlins SA, Ghosh D, Sewalt RG, Otte AP, Hayes DF, et al. 2003. EZH2 is a marker of aggressive breast cancer and promotes neoplastic transformation of breast epithelial cells. *Proc Natl Acad Sci* **100:** 11606–11611. doi:10.1073/pnas.1933744100

Klossowski S, Miao H, Kempinska K, Wu T, Purohit T, Kim EG, Linhares BM, Chen D, Jih G, Perkey E, et al. 2020. Menin inhibitor MI-3454 induces remission in MLL1-rearranged and NPM1-mutated models of leukemia. *J Clin Invest* **130:** 981–997. doi:10.1172/JCI129126

Kouzarides T. 2007. Chromatin modifications and their function. *Cell* **128:** 693–705. doi:10.1016/j.cell.2007.02.005

Krivtsov AV, Evans K, Gadrey JY, Eschle BK, Hatton C, Uckelmann HJ, Ross KN, Perner F, Olsen SN, Pritchard T, et al. 2019. A menin-MLL inhibitor induces specific chromatin changes and eradicates disease in models of MLL rearranged leukemia. *Cancer Cell* **36:** 660–673.e11. doi:10.1016/j.ccell.2019.11.001

Kühn MW, Song E, Feng Z, Sinha A, Chen CW, Deshpande AJ, Cusan M, Farnoud N, Mupo A, Grove C, et al. 2016. Targeting chromatin regulators inhibits leukemogenic gene expression in *NPM1* mutant leukemia. *Cancer Discov* **6:** 1166–1181. doi:10.1158/2159-8290.CD-16-0237

Larson RA, Kondo K, Vardiman JW, Butler AE, Golomb HM, Rowley JD. 1984. Evidence for a 15;17 translocation in every patient with acute promyelocytic leukemia. *Am J Med* **76:** 827–841. doi:10.1016/0002-9343(84)90994-x

Lawrence MS, Stojanov P, Polak P, Kryukov GV, Cibulskis K, Sivachenko A, Carter SL, Stewart C, Mermel CH, Roberts SA, et al. 2013. Mutational heterogeneity in cancer and the search for new cancer-associated genes. *Nature* **499:** 214–218. doi:10.1038/nature12213

Lee HZ, Kwitkowski VE, Del Valle PL, Ricci MS, Saber H, Habtemariam BA, Bullock J, Bloomquist E, Li Shen Y, Chen XH, et al. 2015. FDA approval: belinostat for the treatment of patients with relapsed or refractory peripheral T-cell lymphoma. *Clin Cancer Res* **21:** 2666–2670. doi:10.1158/1078-0432.CCR-14-3119

Lewin J, Soria JC, Stathis A, Delord JP, Peters S, Awada A, Aftimos PG, Bekradda M, Rezai K, Zeng Z, et al. 2018. Phase Ib trial with birabresib, a small-molecule inhibitor of bromodomain and extraterminal proteins, in patients with selected advanced solid tumors. *J Clin Oncol* **36:** 3007–3014. doi:10.1200/JCO.2018.78.2292

Ley TJ, Ding L, Walter MJ, McLellan MD, Lamprecht T, Larson DE, Kandoth C, Payton JE, Baty J, Welch J, et al. 2010. DNMT3A mutations in acute myeloid leukemia. *N Eng J Med* **363:** 2424–2433. doi:10.1056/NEJMoa1005143

Lo-Coco F, Avvisati G, Vignetti M, Thiede C, Orlando SM, Iacobelli S, Ferrara F, Fazi P, Cicconi L, Di Bona E, et al. 2013. Retinoic acid and arsenic trioxide for acute promye-

locytic leukemia. *N Eng J Med* **369:** 111–121. doi:10.1056/NEJMoa1300874

Lv T, Yuan D, Miao X, Lv Y, Zhan P, Shen X, Song Y. 2012. Over-expression of LSD1 promotes proliferation, migration and invasion in non-small-cell lung cancer. *PLoS ONE* **7:** e35065. doi:10.1371/journal.pone.0035065

Maes T, Mascaró C, Tirapu I, Estiarte A, Ciceri F, Lunardi S, Guibourt N, Perdones A, Lufino MM, Somervaille TC, et al. 2018. ORY-1001, a potent and selective covalent KDM1A inhibitor, for the treatment of acute leukemia. *Cancer Cell* **33:** 495–511.e12. doi:10.1016/j.ccell.2018.02.002

Mann BS, Johnson JR, Cohen MH, Justice R, Pazdur R. 2007. FDA approval summary: vorinostat for treatment of advanced primary cutaneous T-cell lymphoma. *The Oncologist* **12:** 1247–1252.

Marcucci G, Maharry K, Wu YZ, Radmacher MD, Mrózek K, Margeson D, Holland KB, Whitman SP, Becker H, Schwind S, et al. 2010. *IDH1* and *IDH2* gene mutations identify novel molecular subsets within de novo cytogenetically normal acute myeloid leukemia: a Cancer and Leukemia Group B study. *J Clin Oncol* **28:** 2348–2355. doi:10.1200/JCO.2009.27.3730

Mardis ER, Ding L, Dooling DJ, Larson DE, McLellan MD, Chen K, Koboldt DC, Fulton RS, Delehaunty KD, McGrath SD, et al. 2009. Recurring mutations found by sequencing an acute myeloid leukemia genome. *N Eng J Med* **361:** 1058–1066. doi:10.1056/NEJMoa0903840

McBride MJ, Pulice JL, Beird HC, Ingram DR, D'Avino AR, Shern JF, Charville GW, Hornick JL, Nakayama RT, Garcia-Rivera EM, et al. 2018. The SS18-SSX fusion oncoprotein hijacks BAF complex targeting and function to drive synovial sarcoma. *Cancer Cell* **33:** 1128–1141.e7. doi:10.1016/j.ccell.2018.05.002

Mellinghoff IK, van den Bent MJ, Blumenthal DT, Touat M, Peters KB, Clarke J, Mendez J, Yust-Katz S, Welsh L, Mason WP, et al. 2023. Vorasidenib in IDH1- or IDH2-mutant low-grade glioma. *N Engl J Med* **389:** 589–601. doi:10.1056/NEJMoa2304194

Metzger E, Wissmann M, Yin N, Müller JM, Schneider R, Peters AH, Günther T, Buettner R, Schüle R. 2005. LSD1 demethylates repressive histone marks to promote androgen-receptor-dependent transcription. *Nature* **437:** 436–439. doi:10.1038/nature04020

Meyer C, Larghero P, Almeida Lopes B, Burmeister T, Gröger D, Sutton R, Venn NC, Cazzaniga G, Corral Abascal L, Tsaur G, et al. 2023. The KMT2A recombinome of acute leukemias in 2023. *Leukemia* **37:** 988–1005. doi:10.1038/s41375-023-01877-1

Mohammad HP, Barbash O, Creasy CL. 2019. Targeting epigenetic modifications in cancer therapy: erasing the roadmap to cancer. *Nat Med* **25:** 403–418. doi:10.1038/s41591-019-0376-8

Morin RD, Johnson NA, Severson TM, Mungall AJ, An J, Goya R, Paul JE, Boyle M, Woolcock BW, Kuchenbauer F, et al. 2010. Somatic mutations altering EZH2 (Tyr641) in follicular and diffuse large B-cell lymphomas of germinal-center origin. *Nat Genet* **42:** 181–185. doi:10.1038/ng.518

Morschhauser F, Tilly H, Chaidos A, McKay P, Phillips T, Assouline S, Batlevi CL, Campbell P, Ribrag V, Damaj GL, et al. 2020. Tazemetostat for patients with relapsed or refractory follicular lymphoma: an open-label, single-arm, multicentre, phase 2 trial. *Lancet Oncol* **21:** 1433–1442. doi:10.1016/S1470-2045(20)30441-1

Mueller D, Bach C, Zeisig D, Garcia-Cuellar MP, Monroe S, Sreekumar A, Zhou R, Nesvizhskii A, Chinnaiyan A, Hess JL, et al. 2007. A role for the MLL fusion partner ENL in transcriptional elongation and chromatin modification. *Blood* **110:** 4445–4454. doi:10.1182/blood-2007-05-090514

Mukohara T, Park YH, Sommerhalder D, Yonemori K, Hamilton E, Kim SB, Kim JH, Iwata H, Yamashita T, Layman RM, et al. 2024. Inhibition of lysine acetyltransferase KAT6 in ER⁺HER2⁻ metastatic breast cancer: a phase 1 trial. *Nat Med* **30:** 2242–2250. doi:10.1038/s41591-024-03060-0

Nagasawa S, Sedukhina AS, Nakagawa Y, Maeda I, Kubota M, Ohnuma S, Tsugawa K, Ohta T, Roche-Molina M, Bernal JA, et al. 2015. LSD1 overexpression is associated with poor prognosis in basal-like breast cancer, and sensitivity to PARP inhibition. *PLoS ONE* **10:** e0118002. doi:10.1371/journal.pone.0118002

Nicodeme E, Jeffrey KL, Schaefer U, Beinke S, Dewell S, Chung CW, Chandwani R, Marazzi I, Wilson P, Coste H, et al. 2010. Suppression of inflammation by a synthetic histone mimic. *Nature* **468:** 1119–1123. doi:10.1038/nature09589

Norsworthy KJ, Luo L, Hsu V, Gudi R, Dorff SE, Przepiorka D, Deisseroth A, Shen YL, Sheth CM, Charlab R, et al. 2019. FDA approval summary: ivosidenib for relapsed or refractory acute myeloid leukemia with an isocitrate dehydrogenase-1 mutation. *Clin Cancer Res* **25:** 3205–3209. doi:10.1158/1078-0432.CCR-18-3749

Okada Y, Feng Q, Lin Y, Jiang Q, Li Y, Coffield VM, Su L, Xu G, Zhang Y. 2005. hDOT1L links histone methylation to leukemogenesis. *Cell* **121:** 167–178. doi:10.1016/j.cell.2005.02.020

Pearson AD, DuBois SG, Buenger V, Kieran M, Stegmaier K, Bandopadhayay P, Bennett K, Bourdeaut F, Brown PA, Chesler L, et al. 2021. Bromodomain and extra-terminal inhibitors—A consensus prioritisation after the Paediatric Strategy Forum for medicinal product development of epigenetic modifiers in children—ACCELERATE. *Eur J Cancer* **146:** 115–124. doi:10.1016/j.ejca.2021.01.018

Perner F, Gadrey JY, Xiong Y, Hatton C, Eschle BK, Weiss A, Stauffer F, Gaul C, Tiedt R, Perry JA, et al. 2020. Novel inhibitors of the histone methyltransferase DOT1L show potent antileukemic activity in patient-derived xenografts. *Blood* **136:** 1983–1988. doi:10.1182/blood.2020006113

Perner F, Stein EM, Wenge DV, Singh S, Kim J, Apazidis A, Rahnamoun H, Anand D, Marinaccio C, Hatton C, et al. 2023. MEN1 mutations mediate clinical resistance to menin inhibition. *Nature* **615:** 913–919. doi:10.1038/s41586-023-05755-9

Piha-Paul SA, Hann CL, French CA, Cousin S, Braña I, Cassier PA, Moreno V, de Bono JS, Harward SD, Ferron-Brady G, et al. 2019. Phase 1 study of molibresib (GSK525762), a bromodomain and extra-terminal domain protein inhibitor, in NUT carcinoma and other solid tumors. *JNCI Cancer Spectrum* **4:** pkz093. doi:10.1093/jncics/pkz093

Powell BL, Moser B, Stock W, Gallagher RE, Willman CL, Stone RM, Rowe JM, Coutre S, Feusner JH, Gregory J,

et al. 2010. Arsenic trioxide improves event-free and overall survival for adults with acute promyelocytic leukemia: North American Leukemia Intergroup Study C9710. *Blood* **116:** 3751–3757. doi:10.1182/blood-2010-02-269621

Raedler LA. 2016. Farydak (Panobinostat): first HDAC inhibitor approved for patients with relapsed multiple myeloma. *Am Health Drug Benefits* **9**(Spec Feature): 84–87.

Raisner R, Kharbanda S, Jin L, Jeng E, Chan E, Merchant M, Haverty PM, Bainer R, Cheung T, Arnott D, et al. 2018. Enhancer activity requires CBP/P300 bromodomain-dependent histone H3K27 acetylation. *Cell Rep* **24:** 1722–1729. doi:10.1016/j.celrep.2018.07.041

* Rérolle D, Wu HC, de Thé H. 2024. Acute promyelocytic leukemia, retinoic acid, and arsenic: a tale of dualities. *Cold Spring Harb Perspect Med* **14:** a041582. doi:10.1101/cshperspect.a041582

Roth SY, Denu JM, Allis CD. 2001. Histone acetyltransferases. *Annual Rev Biochem* **70:** 81–120. doi:10.1146/annurev.biochem.70.1.81

Rubio-San-Simón A, van Eijkelenburg NK, Hoogendijk R, Hasle H, Niemeyer CM, Dworzak MN, Zecca M, Lopez-Yurda M, Janssen JM, Huitema ADR, et al. 2023. Azacitidine (Vidaza) in pediatric patients with relapsed advanced MDS and JMML: results of a phase I/II study by the ITCC Consortium and the EWOG-MDS Group (Study ITCC-015). *Pediatr Drugs* **25:** 719–728. doi:10.1007/s40272-023-00588-5

Salamero O, Montesinos P, Willekens C, Pérez-Simón JA, Pigneux A, Récher C, Popat R, Carpio C, Molinero C, Mascaró C, et al. 2020. First-in-human phase I study of iadademstat (ORY-1001): a first-in-class lysine-specific histone demethylase 1A inhibitor, in relapsed or refractory acute myeloid leukemia. *J Clin Oncol* **38:** 4260–4273. doi:10.1200/JCO.19.03250

Salamero O, Somervaille TC, Molero A, Acuña-Cruz E, Perez-Simon JA, Coll R, Arnan M, Merchan B, Perez A, Cano I, et al. 2022. Iadademstat combination with azacitidine is a safe and effective treatment in first line acute myeloid leukemia. Final results of the ALICE trial. *Blood* **140:** 1711–1713. doi:10.1182/blood-2022-168945

Sanchez R, Zhou MM. 2009. The role of human bromodomains in chromatin biology and gene transcription. *Curr Opin Drug Discov Dev* **12:** 659–665.

Santos-Rosa H, Schneider R, Bannister AJ, Sherriff J, Bernstein BE, Emre NT, Schreiber SL, Mellor J, Kouzarides T. 2002. Active genes are tri-methylated at K4 of histone H3. *Nature* **419:** 407–411. doi:10.1038/nature01080

Saxonov S, Berg P, Brutlag DL. 2006. A genome-wide analysis of CpG dinucleotides in the human genome distinguishes two distinct classes of promoters. *Proc Natl Acad Sci* **103:** 1412–1417. doi:10.1073/pnas.0510310103

Shao W, Fanelli M, Ferrara FF, Riccioni R, Rosenauer A, Davison K, Lamph WW, Waxman S, Pelicci PG, Lo Coco F, et al. 1998. Arsenic trioxide as an inducer of apoptosis and loss of PML/RARα protein in acute promyelocytic leukemia cells. *J Natl Cancer Institute* **90:** 124–133. doi:10.1093/jnci/90.2.124

Sharma SV, Lee DY, Li B, Quinlan MP, Takahashi F, Maheswaran S, McDermott U, Azizian N, Zou L, Fischbach MA, et al. 2010. A chromatin-mediated reversible drug-tolerant state in cancer cell subpopulations. *Cell* **141**: 69–80. doi:10.1016/j.cell.2010.02.027

Sharma S, Chung CY, Uryu S, Petrovic J, Cao J, Rickard A, Nady N, Greasley S, Johnson E, Brodsky O, et al. 2023. Discovery of a highly potent, selective, orally bioavailable inhibitor of KAT6A/B histone acetyltransferases with efficacy against KAT6A-high ER⁺ breast cancer. *Cell Chem Biol* **30**: 1191–1210.e20. doi:10.1016/j.chembiol.2023.07.005

Shen ZX, Chen GQ, Ni JH, Li XS, Xiong SM, Qiu QY, Zhu J, Tang W, Sun GL, Yang KQ, et al. 1997. Use of arsenic trioxide (As2O3) in the treatment of acute promyelocytic leukemia (APL). II: Clinical efficacy and pharmacokinetics in relapsed patients. *Blood* **89**: 3354–3360. doi:10.1182/blood.V89.9.3354

Shi Y, Lan F, Matson C, Mulligan P, Whetstine JR, Cole PA, Casero RA, Shi Y. 2004. Histone demethylation mediated by the nuclear amine oxidase homolog LSD1. *Cell* **119**: 941–953. doi:10.1016/j.cell.2004.12.012

Spencer DH, Russler-Germain DA, Ketkar S, Helton NM, Lamprecht TL, Fulton RS, Fronick CC, O'Laughlin M, Heath SE, Shinawi M, et al. 2017. Cpg island hypermethylation mediated by DNMT3A is a consequence of AML progression. *Cell* **168**: 801–816.e13. doi:10.1016/j.cell.2017.01.021

Steensma DP, Baer MR, Slack JL, Buckstein R, Godley LA, Garcia-Manero G, Albitar M, Larsen JS, Arora S, Cullen MT, et al. 2009. Multicenter study of decitabine administered daily for 5 days every 4 weeks to adults with myelodysplastic syndromes: the alternative dosing for outpatient treatment (ADOPT) trial. *J Clin Oncol* **27**: 3842–3848. doi:10.1200/JCO.2008.19.6550

Steger DJ, Lefterova MI, Ying L, Stonestrom AJ, Schupp M, Zhuo D, Vakoc AL, Kim JE, Chen J, Lazar MA, et al. 2008. DOT1L/KMT4 recruitment and H3K79 methylation are ubiquitously coupled with gene transcription in mammalian cells. *Mol Cell Biol* **28**: 2825–2839. doi:10.1128/MCB.02076-07

Stein EM, DiNardo CD, Pollyea DA, Fathi AT, Roboz GJ, Altman JK, Stone RM, DeAngelo DJ, Levine RL, Flinn IW, et al. 2017. Enasidenib in mutant IDH2 relapsed or refractory acute myeloid leukemia. *Blood* **130**: 722–731. doi:10.1182/blood-2017-04-779405

Stein EM, Garcia-Manero G, Rizzieri DA, Tibes R, Berdeja JG, Savona MR, Jongen-Lavrenic M, Altman JK, Thomson B, Blakemore SJ, et al. 2018. The DOT1L inhibitor pinometostat reduces H3K79 methylation and has modest clinical activity in adult acute leukemia. *Blood* **131**: 2661–2669. doi:10.1182/blood-2017-12-818948

Strahl BD, Allis CD. 2000. The language of covalent histone modifications. *Nature* **403**: 41–45. doi:10.1038/47412

Tahiliani M, Koh KP, Shen Y, Pastor WA, Bandukwala H, Brudno Y, Agarwal S, Iyer LM, Liu DR, Aravind L, et al. 2009. Conversion of 5-methylcytosine to 5-hydroxymethylcytosine in mammalian DNA by MLL partner TET1. *Science* **324**: 930–935. doi:10.1126/science.1170116

Taylor BS, DeCarolis PL, Angeles CV, Brenet F, Schultz N, Antonescu CR, Scandura JM, Sander C, Viale AJ, Socci ND, et al. 2011. Frequent alterations and epigenetic silencing of differentiation pathway genes in structurally rearranged liposarcomas. *Cancer Discov* **1**: 587–597. doi:10.1158/2159-8290.CD-11-0181

Tomita A, Kiyoi H, Naoe T. 2013. Mechanisms of action and resistance to all-trans retinoic acid (ATRA) and arsenic trioxide (As2O3) in acute promyelocytic leukemia. *Int J Hematol* **97**: 717–725. doi:10.1007/s12185-013-1354-4

Turner-Ivey B, Guest ST, Irish JC, Kappler CS, Garrett-Mayer E, Wilson RC, Ethier SP. 2014. KAT6A, a chromatin modifier from the 8p11-p12 amplicon is a candidate oncogene in luminal breast cancer. *Neoplasia* **16**: 644–655. doi:10.1016/j.neo.2014.07.007

Uckelmann HJ, Kim SM, Wong EM, Hatton C, Giovinazzo H, Gadrey JY, Krivtsov AV, Rücker FG, Döhner K, McGeehan GM, et al. 2020. Therapeutic targeting of preleukemia cells in a mouse model of *NPM1* mutant acute myeloid leukemia. *Science* **367**: 586–590. doi:10.1126/science.aax5863

van Haaften G, Dalgliesh GL, Davies H, Chen L, Bignell G, Greenman C, Edkins S, Hardy C, O'Meara S, Teague J, et al. 2009. Somatic mutations of the histone H3K27 demethylase gene *UTX* in human cancer. *Nat Genet* **41**: 521–523. doi:10.1038/ng.349

Varambally S, Dhanasekaran SM, Zhou M, Barrette TR, Kumar-Sinha C, Sanda MG, Ghosh D, Pienta KJ, Sewalt RG, Otte AP, et al. 2002. The polycomb group protein EZH2 is involved in progression of prostate cancer. *Nature* **419**: 624–629. doi:10.1038/nature01075

Wang ES, Issa GC, Erba HP, Altman JK, Montesinos P, DeBotton S, Walter RB, Pettit K, Savona MR, Shah MV, et al. 2024. Ziftomenib in relapsed or refractory acute myeloid leukaemia (KOMET-001): a multicentre, open-label, multi-cohort, phase 1 trial. *Lancet Oncol* **25**: 1310–1324. doi:10.1016/S1470-2045(24)00386-3 [published correction appears in *Lancet Oncol* 2024; **25**: e542. doi:10.1016/S1470-2045(24)00584-9].

Ward PS, Patel J, Wise DR, Abdel-Wahab O, Bennett BD, Coller HA, Cross JR, Fantin VR, Hedvat CV, Perl AE, et al. 2010. The common feature of leukemia-associated IDH1 and IDH2 mutations is a neomorphic enzyme activity converting α-ketoglutarate to 2-hydroxyglutarate. *Cancer Cell* **17**: 225–234. doi:10.1016/j.ccr.2010.01.020

Watts JM, Baer MR, Yang J, Prebet T, Lee S, Schiller GJ, Dinner SN, Pigneux A, Montesinos P, Wang ES, et al. 2023. Olutasidenib alone or with azacitidine in IDH1-mutated acute myeloid leukaemia and myelodysplastic syndrome: phase 1 results of a phase 1/2 trial. *Lancet Haematol* **10**: e46–e58. doi:10.1016/S2352-3026(22)00292-7

West AC, Johnstone RW. 2014. New and emerging HDAC inhibitors for cancer treatment. *J Clin Invest* **124**: 30–39. doi:10.1172/JCI69738

Xu Y, Vakoc CR. 2017. Targeting cancer cells with BET bromodomain inhibitors. *Cold Spring Harb Perspect Med* **7**: a026674. doi:10.1101/cshperspect.a026674

Xu W, Yang H, Liu Y, Yang Y, Wang P, Kim SH, Ito S, Yang C, Wang P, Xiao MT, et al. 2011. Oncometabolite 2-hydroxyglutarate is a competitive inhibitor of α-ketoglutarate-dependent dioxygenases. *Cancer Cell* **19**: 17–30. doi:10.1016/j.ccr.2010.12.014

Yan H, Parsons DW, Jin G, McLendon R, Rasheed BA, Yuan W, Kos I, Batinic-Haberle I, Jones S, Riggins GJ, et al. 2009. IDH1 and IDH2 mutations in gliomas. *N Eng J Med* **360**: 765–773. doi:10.1056/NEJMoa0808710

Yan F, Li J, Milosevic J, Petroni R, Liu S, Shi Z, Yuan S, Reynaga JM, Qi Y, Rico J, et al. 2022. KAT6A and ENL form an epigenetic transcriptional control module to drive critical leukemogenic gene-expression programs. *Cancer Discov* **12**: 792–811. doi:10.1158/2159-8290.CD-20-1459

Yang L, Rau R, Goodell MA. 2015. DNMT3A in haematological malignancies. *Nat Rev Cancer* **15**: 152–165. doi:10.1038/nrc3895

Yokoyama A, Somervaille TC, Smith KS, Rozenblatt-Rosen O, Meyerson M, Cleary ML. 2005. The menin tumor suppressor protein is an essential oncogenic cofactor for MLL-associated leukemogenesis. *Cell* **123**: 207–218. doi:10.1016/j.cell.2005.09.025

Yoshida H, Kitamura K, Tanaka K, Omura S, Miyazaki T, Hachiya T, Ohno R, Naoe T. 1996. Accelerated degradation of PML-retinoic acid receptor α (PML-RARA) oncoprotein by all-*trans*-retinoic acid in acute promyelo-cytic leukemia: possible role of the proteasome pathway. *Cancer Res* **56**: 2945–2948.

Zhang XW, Yan XJ, Zhou ZR, Yang FF, Wu ZY, Sun HB, Liang WX, Song AX, Lallemand-Breitenbach V, Jeanne M, et al. 2010. Arsenic trioxide controls the fate of the PML-RARα oncoprotein by directly binding PML. *Science* **328**: 240–243. doi:10.1126/science.1183424

Zhao S, Allis CD, Wang GG. 2021. The language of chromatin modification in human cancers. *Nat Rev Cancer* **21**: 413–430. doi:10.1038/s41568-021-00357-x

Zhu AX, Macarulla T, Javle MM, Kelley RK, Lubner SJ, Adeva J, Cleary JM, Catenacci DV, Borad MJ, Bridgewater JA, et al. 2021. Final overall survival efficacy results of ivosidenib for patients with advanced cholangiocarcinoma with *IDH1* mutation: the phase 3 randomized clinical ClarIDHy trial. *JAMA Oncol* **7**: 1669–1677. doi:10.1001/jamaoncol.2021.3836

Pediatric Cancer Drug Development: Leveraging Insights in Cancer Biology and the Evolving Regulatory Landscape to Address Challenges and Guide Further Progress

Rosane Charlab,[1] Ruby Leong,[1] Stacy S. Shord,[1] and Gregory H. Reaman[2]

[1]Office of Clinical Pharmacology, Office of Translational Sciences, U.S. Food and Drug Administration, Silver Spring, Maryland 20993, USA

[2]Division of Cancer Treatment and Diagnosis, National Cancer Institute, Bethesda, Maryland 20892, USA

Correspondence: Rosane.CharlabOrbach@fda.hhs.gov

The discovery and development of anticancer drugs for pediatric patients have historically languished when compared to both past and recent activity in drug development for adult patients, notably the dramatic spike of targeted and immune-oncology therapies. The reasons for this difference are multifactorial. Recent changes in the regulatory landscape surrounding pediatric cancer drug development and the understanding that some pediatric cancers are driven by genetic perturbations that also drive disparate adult cancers afford new opportunities. The unique cancer-initiating events and dependencies of many pediatric cancers, however, require additional pediatric-specific strategies. Research efforts to unravel the underlying biology of pediatric cancers, innovative clinical trial designs, model-informed drug development, extrapolation from adult data, addressing the unique considerations in pediatric patients, and use of pediatric appropriate formulations, should all be considered for efficient development and dosage optimization of anticancer drugs for pediatric patients.

THE HISTORICAL EXPERIENCE WITH CANCER DRUG DEVELOPMENT HAS NOT BENEFITED CHILDREN

Even though the earliest studies of cancer drugs in children served to initiate the systematic discovery and development of anticancer therapies, there is a striking inequity between children and adults in the number of oncology drugs approved by the U.S. Food and Drug Administration (FDA) for the treatment of cancer. Less than 10% of all oncology products are approved for use in the pediatric population, which provides testimony to the long historical disparity between children and adults in both cancer drug discovery and development (Smith and Reaman 2022). The notable investigations of aminopterin and later methotrexate in children with acute leukemia by Farber established the role of antifolates in the treatment of cancer. Subsequent studies in children with leukemia by Frei, Holland, and Freireich at the Cancer Chemotherapy National Service Center at the National Institutes of Health (NIH) established the principle of combination chemotherapy and led to the first of the national clinical trial cooperative groups, the Acute Leukemia Chemotherapy Cooperative Study Group A (ALCCSGA), which later became the Children's Cancer Study Group (Smith and Reaman 2022). It could be stated that the first "targeted" anti-cancer therapy, L-asparaginase, was developed in children.

Nonetheless, the pediatric patient population has been historically excluded from the clinical cancer drug development enterprise for several reasons, including clinical trial challenges due to the fortunate rarity of cancer in the pediatric age group and the unfavorable market forces due to lack of a positive return on investment for both large pharmaceutical companies, as well as emerging small biotechnology companies, to provide sufficient financial incentive from the efforts and resources required for drug discovery and development for children with cancer (Adamson et al. 2014; Ho 2022). As a result, most of the labeling of approved oncology drugs developed for the treatment of malignancies in adults, although commonly used off label for the treatment of children with cancer, lack specific information to adequately guide their appropriate use in the pediatric population (Benning et al. 2021).

Not only are there insufficient data or evidence to support efficacy or safety of the drugs in pediatric patients, but also in those situations where one might assume efficacy, there are limited to absent data on the age, body surface area, or weight-based dosing strategies, short- and long-term safety, or information on age-appropriate formulations for pediatric patients. The availability of formulations appropriate for children, particularly very young children, is becoming increasingly important due to the current emphasis on orally administered small-molecule inhibitors of kinases and other molecular targets involved in cell signaling pathways.

LEGISLATIVE PROGRAMS DIRECTED AT FACILITATING PEDIATRIC DRUG DEVELOPMENT HAVE HAD LIMITED IMPACT

Cancer drug development occurs within the highly regulated environment in which drug development in general operates. Legislation dictating the licensing and use of safe and effective drug products in the United States originated in large part due to catastrophic events occurring predominantly in children. These included the deaths in children due to tetanus from contaminated typhoid vaccines, deaths due to unknown toxic substances in patent drugs, and deaths due to diethylene glycol poisoning from elixir of sulfanilamide, which ultimately resulted in the U.S. Food, Drug, and Cosmetic (FD&C) Act of 1938 (Hirschfeld and Ward 2013). More than two decades later, yet another tragic occurrence ensued involving children, the development of limb appendage abnormalities in newborns exposed prenatally to thalidomide (Vargesson 2015).

Although the policies resulting in legislation did not specifically address the participation of children in clinical research, the nature of the tragedies predominantly observed in children were such that they essentially led to their exclusion from clinical trials of new drug efficacy and safety giving rise to the characterization of children as therapeutic orphans (Shirkey 1968). The continued exposure of children to new drugs without adequate safety and dosing information for their use and the continued occurrence of unexpected adverse events led to the passage of a Final Rule on Pediatric Labeling in 1994 (www.govinfo.gov/content/pkg/FR-1994-12-13/html/94-30238.htm). This final rule required manufacturers of approved products to review existing data on the experience in children to potentially expand labeling information to include use in children, which provided little new pediatric labeling information.

Key pieces of legislation directed at facilitating pediatric drug development were subsequently enacted (Fig. 1). Efforts by advocates, clinical investigators, and pharmaceutical company representatives resulted in a program to incentivize companies to conduct studies in

Cite this article as *Cold Spring Harb Perspect Med* doi: 10.1101/cshperspect.a041656

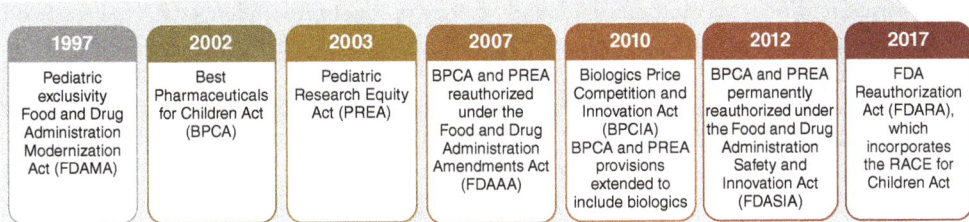

Figure 1. Timeline of key legislative programs directed at facilitating pediatric drug development. The RACE for Children Act: The Research to Accelerate Cures and Equity (RACE) for Children Act.

children, and which led to the 1997 passage of the Food and Drug Administration Modernization Act (FDAMA) that added Sec. 505A to the FD&C Act. This legislation includes a provision for granting a 6-month extension of exclusivity to manufacturers whose product was voluntarily studied under an FDA-issued written request (FDAMA 1997). A complimentary piece of legislation was passed, the Pediatric Rule, which required manufacturers to conduct studies of new drugs in children to support pediatric use. This and the exclusivity provision (Sec. 505A) were intended to drive pediatric development of appropriate new drugs (Pediatric Rule 1998).

The Pediatric Rule was later struck down by the Federal Court of the District of Columbia on the grounds that it exceeded the authority of the FDA to require the potential expansion of an indication, in this case to children, of an approved drug product. Later in 2002, the Best Pharmaceuticals for Children Act (BPCA) was passed, reauthorizing the exclusivity provision in Sec. 505A, which also included a process for investigation of off-patent drugs by the NIH (BPCA 2002). In 2003, Congress passed the Pediatric Research Equity Act (PREA), which incorporated most of the provisions of the Pediatric Rule; however, requirements for pediatric studies were limited to the

same indication for which the drug was approved in adults, and drugs for orphan indications were exempted (PREA 2003). There was no requirement for a proposed timeline for the planning and initiation of pediatric studies. In 2007, under the Food and Drug Administration Amendments Act (FDAAA), BPCA was amended to allow the NIH to recommend to the FDA those pediatric studies to be performed under a written request issued by the FDA to a manufacturer of a product of interest. In 2010, the pediatric exclusivity provision of BPCA was extended to include biologics as well as drugs by the Biologics Price Competition and Innovation Act (BPCIA) (Title VII of the Patient Protection and Affordable Care Act 2010; www.congress.gov/111/plaws/publ148/ PLAW-111publ148.pdf), and in 2012, statutory language was added under the Food and Drug Administration Safety and Innovation Act (FDASIA) to require sponsors to submit an Initial Pediatric Study Plan (iPSP) by 60 days after the end-of-phase 2 meeting, and to reach an agreement with the FDA on the plan prior to submission of a New Drug Application (NDA) or Biologics Licensing Application (BLA) in an attempt to expedite pediatric investigations (FDASIA 2012).

These key pieces of legislation demonstrated the Federal Government's shift in emphasis from protecting children from clinical research to protecting children through clinical investigation, thereby facilitating approval of and access to safe and effective new medicines. Both the BPCA and PREA had sunset provisions requiring reauthorization, which was accomplished in 2007 under the FDAAA. Both were permanently reauthorized under the FDASIA in 2012.

The combination of mandates and incentives has had a dramatic impact on pediatric labeling information for new drugs approved for a variety of adult indications with more than 650 (Avant et al. 2018) approved drugs as of September 2016 having product labeling demonstrating the safe and effective use of the drugs in children. However, these legislative programs have had far less of an impact on extending the use to the pediatric population of oncology drugs. Most cancers that occur in adults rarely, if ever, are observed in children. Therefore, the requirements for pediatric assessments of new cancer drugs are waived because studies would be impossible or highly impracticable, since the specific clinical indication for which the drug is approved in adults occurs so infrequently in children. Furthermore, those cancers seen in adults that do also occur in the pediatric population have an incidence (<200,000 patients/ year) that warrants orphan designation, thereby exempting sponsors from the PREA requirement for pediatric assessments. Thus, only the BPCA has provided a stimulus, albeit voluntary, for the development of oncology products for children. Since the initiation of the BPCA through the end of 2021, 40 written requests for pediatric studies of oncology drugs and biologics (herein referred to collectively as drugs) were issued by the FDA resulting in the approval of nine oncology drugs for use in a pediatric cancer indication (Akalu et al. 2021). Prior to 2020, however, there had been no PREA-required assessment of a new oncology drug in children.

THE EVOLVING REGULATORY FRAMEWORK FOR PEDIATRIC CANCER DRUG DEVELOPMENT

Concerted, focused, and persistent advocacy efforts have recently reversed the unintended exclusion of children with cancer from the facilitated drug development benefits afforded by the

PREA. Under Sec. 504 of the FDA Reauthorization Act (FDARA) of 2017, which incorporates the RACE (Research to Accelerate Cures and Equity) for Children Act, pediatric trials for oncology products are no longer waived solely on infeasibility of studies in children due to the rarity of the adult indication for which the drug is being developed or exempted by orphan designation. Early pediatric assessment of dose, tolerability, and signal of activity is required for drugs directed at molecular targets substantially relevant to the growth or progression of a pediatric cancer. Original marketing applications, both NDAs and BLAs submitted to the FDA as of August 18, 2020, for new active ingredients intended for the treatment of cancer(s) in adults that are directed at a molecular target that the FDA has determined to be relevant to the growth or progression of a cancer that occurs in children, must be preceded by plans for which FDA agreement is required for pediatric cancer investigations. The results of these pediatric studies are expected to be submitted with the new product's application for approval (FDARA 2017).

Thus, the pediatric drug development timeline for oncology drugs is expected to be coordinated with the development plan for adults as part of an overall drug development plan. This amendment to the PREA also makes the timelines for submission of plans for pediatric oncology drug development a requirement in the United States, the iPSPs, more in line with pediatric investigation plans (PIPs) submitted to the European Medicines Agency (EMA), as required by the European regulations, although the contents and specific requirements differ (Reaman et al. 2020). The global nature of cancer drug development in general and the increasing requirement for international collaboration for pediatric clinical trials due to rare patient populations, now amplified by the identification of genomic subtypes, requires focused efforts to align, to the extent possible, international regulatory requirements for the assessment of new cancer drugs in children. In fact, simultaneous submission of iPSPs and PIPs to their respective regulatory agencies has been recommended to facilitate coordinated and collaborative review of potential pediatric development plans by regulatory agencies early to accelerate timelines (Reaman et al. 2020).

The impact of the amendments to the PREA on pediatric cancer drug development cannot yet be fully assessed given the fact that although the legislation was passed in 2017, its full implementation only began in late 2020, as stipulated in the statute. The time required for adequate study planning, initiation, analysis of results, and their incorporation in an application package for submission to the FDA is insufficient to see a causal relationship in the number of new cancer drugs whose use has been extended to the pediatric population. Nonetheless, the increased awareness by industry of and their positive response to new requirements for early pediatric investigation have resulted in earlier communications between industry, clinical investigators, patient advocates, and regulatory agencies. The report of the U.S. Government Accounting Office (GAO) on the impact of Sec. 504 of FDARA, which details the FDA's concurrence with sponsors' proposed pediatric cancer investigations outlined in submitted iPSPs and the significant increase in the number of approved cancer drugs with a post-marketing requirement for a pediatric assessment during the year following the legislation's full implementation of FDARA as compared to years prior (GAO-23-105947 2023), provides early indication that the legislation has had a positive impact on the acceleration of pediatric investigations of new cancer drugs and expectations that children with cancer will also benefit from advances in precision oncology.

ADVENT, EXPECTATIONS, AND CHALLENGES FOR PRECISION ONCOLOGY IN PEDIATRIC CANCER

Pediatric Cancers Are Not the Same as Adult Cancers

Children are not "little adults" from a physiological and pharmacological standpoint. Also, differences between adult and pediatric cancers, including differences in cancer genetics, transcriptional profiles, and tumor microenvironment, support that pediatric cancers are not adult cancers, but rather represent a unique and diverse subset with specific therapeutic challenges (Kattner et al. 2019).

The array of pediatric cancers is essentially distinct from that found in adults (Jones et al. 2019; Sweet-Cordero and Biegel 2019). Adult cancers are mostly of epithelial origin, while many pediatric cancers originate from mesoderm or neuroectoderm (Pfister et al. 2022). Pediatric cancers and adult cancers differ in etiology and mutational processes. Compared to adult cancers, pediatric cancers have a higher frequency of germline alterations (Brodeur et al. 2017; Sweet-Cordero and Biegel 2019; Alonso-Luna et al. 2023) and higher transcriptional diversity (Comitani et al. 2023). Even more than for adult cancers, and because most studies of pediatric cancer genomes used diagnostic samples, processes through which pediatric cancers progress, metastasize, and resist therapeutic approaches are largely uncharacterized (Sweet-Cordero and Biegel 2019).

Moreover, pediatric cancers are, by definition, rare diseases. The rarity and limited understanding of disease biology make pediatric cancer drug development extremely challenging, including from operational, statistical, and ethical standpoints (Shebley et al. 2019; Subbiah 2023). Approximately 80%–85% of children with cancer who live in high-income countries tend to survive 5 years or more after diagnosis. However, these statistics drop to 40% or less in most low- and middle-income countries. Survivors of pediatric cancers often have long-term effects of treatment, indicating the need for more effective and less toxic targeted products to treat pediatric patients with cancer (Adamson 2015; Kattner et al. 2019; Langenberg et al. 2021).

Particularly complex are pediatric central nervous system (CNS) tumors, which constitute a heterogeneous and diverse group at the molecular level and reiterate the inadequacy of one-drug-fits-all models. Disease heterogeneity also creates many hurdles to identify the right drugs and dosages, and in the case of CNS tumors, integration of CNS pharmacokinetics (PK) into treatment strategies is a major stumbling block (Findlay et al. 2022).

Molecular Landscape: Pediatric versus Adult Cancers and Precision Oncology

The advancement of sequencing technologies along with the increase in publicly available tumor genomic databases created a unique field of opportunities to probe the underlying molecular mechanisms of pediatric cancers (He et al. 2021). As a result of pancancer analyses (Gröbner et al. 2018; Ma et al. 2018), the genetic profile of pediatric cancers has been characterized as "quiet" because, compared to adult cancers, pediatric cancers typically have a lower tumor mutational burden (TMB). The pancancer analyses also found differences in the spectrum of mutated genes with cancer driver potential; still, ~30%–45% of somatically mutated genes matched those identified in adult pancancer analyses. These results, and the ever-expanding number of targeted and immunotherapies available or in development for adult cancers, permeated the field with hope that similar precision-based approaches to adult cancers could be used toward a spectrum of molecular targets also associated with pediatric cancer. Yet, clearly defined genomic targets have not been identified in many cases, and other vulnerabilities have been explored such as the expression of specific cell surface antigens (Laetsch et al. 2021). Candidate targets are therefore diverse and include targets with specific molecular alterations, associated with cell lineage determinants, present on normal immune cells and cellular components of the tumor microenvironment, and/or associated with specific oncogenic pathways or functional mechanisms (Charlab et al. 2023).

As discussed in the previous section, FDARA provisions further engendered a conducive environment to molecularly targeted pediatric cancer investigations. Rapidly evolving precision-based approaches have demonstrated activity in the refractory/relapsed setting (Butler et al. 2021) and, in some instances, also in the previously untreated setting for metastatic disease. Approved targeted agents now include a growing number of small-molecule kinase inhibitors, monoclonal antibodies, bispecific T-cell engagers, antibody-drug conjugates (ADCs), and genetically engineered cell therapies (CAR T cells) in consort with the older cytotoxic chemotherapy drugs, a testimony of the paradigm shift in pediatric cancer treatment (Butler et al. 2021; Laetsch et al. 2021; U.S. FDA Oncology Center of Excellence Pediatric Oncology Drug Approvals; www.fda.gov/about-fda/oncology-center-ex cellence/pediatric-oncology-drug-approvals). Notable examples of drugs approved as single agents or in combination for the treatment of solid tumors, hematological malignancies, and for tumor agnostic indications are listed in Table 1. Figure 2 highlights examples of specific oncogenic drivers (i.e., ALK and NTRK fusions and BRAF V600E mutations) identified in different adult and pediatric cancers, and for which the same targeted drugs were approved (Fig. 2). Aspects contributing to successful developments include readiness of age-appropriate formulations and availability of appropriate biomarkers to allow for enriched enrollment of specific pediatric populations (Laetsch et al. 2021).

However, while there are reasons for optimism on some fronts, there are also more than a few aspects remaining to be addressed. From a molecular target standpoint, several tumor dependencies are still unique to pediatric cancer (Dharia et al. 2021). In contrast to adult cancers, apart from a few outlier instances, high-frequency molecular alterations are not found within a given tumor type, creating precision medicine implementation challenges (Carroll et al. 2015; Bandopadhayay and Meyerson 2018). Moreover, pediatric cancers include difficult targets to counteract, such as transcription factor fusions and the epigenome (Carroll et al. 2015; Panditharatna and Filbin 2020; Laetsch et al. 2021). Therefore, just repurposing adult oncology drugs will likely be insufficient to significantly improve pediatric cancer outcomes. These differences emphasize that in certain settings, pediatric cancers need to be considered independently to inform therapeutic approaches.

Of note, different molecular alterations (e.g., different point mutations, fusion variants, amplification, overexpression) affecting a specific target gene or different nodes of a biological pathway are not necessarily equivalent in terms of oncogenic potential, prognosis, and/or drug sensitivity (Ou and Nagasaka 2020; Kennedy and Lowe 2022; Robinson et al. 2023). Due to rarity, feasibility challenges, and gaps in biological understanding, many pediatric cancer studies either do not select patients based on defined molecular alterations or group several alteration types when defining the eligibility criteria (e.g., fibroblast growth factor receptor [FGFR] alterations), potentially hampering the efficacy assessment in molecular subgroups of patients most likely to respond (Laetsch et al. 2021). Likewise, for potential candidate gene targets shared between adult and pediatric cancers, the spectrum of alterations affecting a gene and/or the pattern of expression of a target protein (e.g., cytoplasmatic vs. membranous) in pediatric versus adult cancers may differ. Therefore, the use of targeted gene panels to identify preselected alterations common to adult cancers may not be immediately translated to pediatric cancers, and, in such cases, assay customization to pediatric cancers needs to be considered (Carroll et al. 2015). In addition, the identification of potential targets with adequate levels of surface expression in pediatric tumors is key to tailor antibody-based therapies to pediatric cancers (Hingorani et al. 2022). It is also important to highlight that precision-based approaches do not replace routine histopathology, but rather complement it with the joint purpose of further uncovering tumoral processes that can be exploited therapeutically (Cifci et al. 2023).

Immunotherapy Approaches; Hindrance of Low Tumor Mutational Burden

Hot or Cold: Understanding the Immune Context

Pediatric and adult cancers are also different in what relates to immune context (Simon et al. 2015; Hutzen et al. 2019; Terry et al. 2020; Xiao et al. 2021; Long et al. 2022; Thakur et al. 2022). Tumors exploit the immune system to their advantage, and several characteristics are recognized as relevant to cancer-immune response interactions. These include, but are not limited

to, tumor infiltration and activation status of immune cells, expression of immune checkpoints on tumor and immune cells, expression of tumor neoantigens, metabolic pathway dysregulation, dysregulation of DNA repair mechanisms, and, in children, immune system ontogeny. Tumors have been classified in immune hot or cold based on some of these characteristics such as T-cell infiltration in the tumor microenvironment that ranges from high in hot tumors to absent or nearly absent in cold tumors. High T-cell infiltration found in hot tumors, and high TMB have been associated with sensitivity to immune checkpoint inhibitor (ICI) therapies such as anti-PD-1/PD-L1 antibodies, especially as single agents. TMB is often considered a surrogate marker for tumor neoantigen load, and high TMB would represent enough "foreignness" by the tumor to elicit an immune response once the cancer-immunosuppressive mechanisms that disrupt T-cell infiltration or activity are counteracted with immunotherapies (Blank et al. 2016). Similarly, PD-L1 expression, microsatellite instability (MSI) status, and DNA mismatch repair deficiency (dMMR) have been extensively evaluated as biomarkers of response for anti-PD-1/PD-L1 treatment (Terry et al. 2020). dMMR/MSI-H is a tumor phenotype found in a subset of colorectal and other cancers, and which is associated with high generation of neoantigens and high TMB, likely contributing to shape an immune-active tumor microenvironment.

While several adult tumors, especially of epithelial origin, are hot tumors typically characterized by higher TMB, the more common pediatric cancers are typically cold tumors, with lower TMB and fewer infiltrating T cells (Casey and Cheung 2020; Butler et al. 2021; Xiao et al. 2021), reflecting differences in underlying oncogenic and immune escape mechanisms between adult and pediatric cancers. In most cases, ICI activity against unselected pediatric cancers has been discouraging. Notable exceptions include pediatric classic Hodgkin lymphoma (Table 1), where overexpression of PD-1 ligands by pathologic Reed–Sternberg cells provide biologic rationale for susceptibility to anti PD-1/PD-L1 therapies (Long et al. 2022). Instances of molecularly defined cancers where the underlying bi-

Table 1. Pediatric cancer drug approvals of targeted agents and immunotherapies

Drugs approved for pediatric cancers	Approved pediatric indication(s)	Drug target(s) primarily associated with indication	Year approved
Hematological malignancies			
Kinase inhibitors			
Bosutinib	Chronic phase Ph⁺ CML, newly diagnosed or resistant or intolerant to prior therapy	BCR-ABL	2023
Crizotinib	R/R, systemic ALK-positive ALCL	ALK	2021
Dasatinib	Ph⁺ CML in chronic phase; newly diagnosed Ph⁺ ALL in combination with chemotherapy	BCR-ABL	2017 (CML), 2018 (ALL)
Imatinib	Newly diagnosed Ph⁺ CML in chronic phase; newly diagnosed Ph⁺ ALL in combination with chemotherapy	BCR-ABL	2006 (CML), 2013 (ALL)
Nilotinib	Newly diagnosed Ph⁺ CML in chronic phase; prior TKI resistant or intolerant chronic phase and accelerated phase Ph⁺ CML	BCR-ABL	2018
Monoclonal antibodies			
Pembrolizumab	R/R cHL; R/R PMBCL	PDCD1 (PD-1)	2017 (cHL-relapsed after ≥3 prior lines), 2020 (cHL-relapsed after ≥2 prior lines), 2018 (PMBCL)
Rituximab	Previously untreated, advanced stage CD20-positive DLBCL, BL, BLL or mature B-AL in combination with chemotherapy	MS4A1 (CD20)	2021
Antibody-drug conjugates			
Brentuximab vedotin	Previously untreated high-risk cHL in combination with doxorubicin, vincristine, etoposide, prednisone, and cyclophosphamide	TNFRSF8 (CD30)	2022
Gemtuzumab ozogamicin	R/R CD33-positive AML; newly diagnosed CD33-positive AML	CD33	2017 (R/R), 2020 (newly diagnosed)
Bispecific T-cell engagers			
Blinatumomab	R/R CD19-positive B-cell precursor ALL; CD19-positive B-cell precursor ALL in first or second complete remission with MRD ≥ 0.1%	CD19/CD3	2017 (R/R), 2018 (in first or second complete remission with MRD ≥ 0.1%)

Continued

Table 1. *Continued*

Drugs approved for pediatric cancers	Approved pediatric indication(s)	Drug target(s) primarily associated with indication	Year approved
Chimeric antigen receptor (CAR) T-cell therapies			
Tisagenlecleucel	R/R B-cell precursor ALL	CD19	2017
Targeted cytotoxin			
Tagraxofusp-erzs	BPDCN	IL3RA (CD123)	2018
Depletion enzymes			
Asparaginase	ALL, as a component of a multi-agent chemotherapeutic regimen	L-asparagine	1978
Asparaginase *Erwinia chrysanthemi* (recombinant)-rywn	ALL and LBL, for patients who have developed hypersensitivity to *E. coli*–derived asparaginase, as a component of a multi-agent chemotherapeutic regimen	L-asparagine	2021
Calaspargase pegol-mknl	ALL, as a component of a multi-agent chemotherapeutic regimen	L-asparagine	2018
Pegaspargase	ALL, as a component of a multi-agent chemotherapeutic regimen, for patients with hypersensitivity to native forms of L-asparaginase; ALL, first line, as a component of a multi-agent chemotherapeutic regimen	L-asparagine	1994, 2006 (first line)
Solid tumors			
Kinase inhibitors			
Crizotinib	Unresectable, recurrent, or refractory ALK-positive IMT	ALK	2022
Cabozantinib	Locally advanced or metastatic DTC that has progressed following prior VEGFR-targeted therapy and who are radioactive iodine-refractory or ineligible	Multiple receptor tyrosine kinases including but not limited to VEGFR, MET, AXL, RET	2021
Dabrafenib + Trametinib	BRAF V600E mutation-positive LGG	BRAF + MAP2K1/2 (MEK1/2)	2023
Everolimus	Subependymal giant cell astrocytoma (SEGA)	MTOR	2010
Pralsetinib	Advanced or metastatic RET mutation-positive MTC; advanced or metastatic RET fusion-positive TC	RET	2020

Continued

Table 1. *Continued*

Drugs approved for pediatric cancers	Approved pediatric indication(s)	Drug target(s) primarily associated with indication	Year approved
Selpercatinib	Advanced or metastatic RET mutation-positive MTC; advanced or metastatic RET fusion-positive TC	RET	2020
Selumetinib	Neurofibromatosis type 1 (NF1) with symptomatic, inoperable plexiform neurofibromas (PNs)	MAP2K1/2 (MEK1/2)	2020
Monoclonal antibodies			
Atezolizumab	Unresectable or metastatic ASPS	CD274 (PD-L1)	2022
Avelumab	Metastatic MCC	CD274 (PD-L1)	2017
Denosumab	Giant cell tumor of bone that is unresectable or where surgical resection is likely to result in severe morbidity	TNFSF11 (RANKL)	2013
Dinutuximab	High-risk neuroblastoma in combination with granulocyte-macrophage colony-stimulating factor (GM-CSF), interleukin-2 (IL-2), and 13-*cis*-retinoic acid (RA)	GD2	2015
Ipilimumab ± Nivolumab	Unresectable or metastatic melanoma[a]	CTLA4 ± PDCD1 (PD-1)	2017 (single agent), 2023 (combination)
Ipilimumab + Nivolumab	MSI-H or dMMR metastatic CRC	CTLA4 + PDCD1 (PD-1)	2018
Naxitamab-gqgk	R/R high-risk neuroblastoma in the bone or bone marrow, in combination with GM-CSF	GD2	2020
Nivolumab and Relatlimab-rmbw	Unresectable or metastatic melanoma	PDCD1 (PD-1) + LAG3	2022
Nivolumab ± Ipilimumab	MSI-H or dMMR metastatic CRC; unresectable or metastatic melanoma[a]	PDCD1 (PD-1) ± CTLA4	2017 (CRC-single agent), 2018 (CRC-combination), 2023 (melanoma)
Nivolumab	Completely resected stage IIB, stage IIC, stage III, or stage IV melanoma (adjuvant setting)	PDCD1 (PD-1)	2023
Pembrolizumab	Recurrent locally advanced or metastatic MCC; stage IIB, IIC, or III melanoma following complete resection (adjuvant setting)	PDCD1 (PD-1)	2018 (MCC), 2021 (melanoma)

Continued

Table 1. *Continued*

Drugs approved for pediatric cancers	Approved pediatric indication(s)	Drug target(s) primarily associated with indication	Year approved
Epigenetic modifiers			
Tazemetostat	Metastatic or locally advanced epithelioid sarcoma	EZH2	2020
Tumor agnostic			
Kinase inhibitors			
Dabrafenib + Trametinib	BRAF V600E mutation-positive solid tumors	BRAF + MAP2K1/2 (MEK1/2)	2022 (patients ≥ 6 yr old), 2023 (expanded to patients ≥ 1 yr old)
Entrectinib	NTRK gene fusion-positive solid tumors	NTRK	2019 (patients ≥ 12 yrs old), 2023 (expanded to patients > 1 mo old)
Larotrectinib	NTRK gene fusion-positive solid tumors	NTRK	2018
Monoclonal antibodies			
Pembrolizumab	Unresectable or metastatic MSI-H or dMMR solid tumors; unresectable or metastatic TMB-H (≥10 mutations/megabase [mut/Mb]) solid tumors	PDCD1 (PD-1)	2017 (MSI-H or dMMR solid tumors), 2020 (TMB-H solid tumors)

Listing is not comprehensive. Indications are abbreviated, and adult indications for the listed drugs are not shown. For the most recent information and full indications, including pediatric age groups for which drugs were approved, refer to Drugs@FDA (www.fda.gov/drugsatfda) and U.S. FDA, Oncology Center of Excellence, Pediatric Oncology Drug Approvals (www.fda.gov/about-fda/oncology-center-excellence/pediatric-oncology-drug-approvals).

(BCR-ABL) BCR::ABL kinase fusion, (ALK) ALK receptor tyrosine kinase, (PDCD1) (PD-1): programmed cell death 1, (MS4A1) (CD20): membrane spanning 4-domains A1, (TNFRSF8) (CD30): TNF receptor superfamily member 8, (CD33) CD33 molecule, (CD19) CD19 molecule, (CD3) γ subunit of T-cell receptor complex, (IL3RA) (CD123): interleukin 3 receptor subunit α, (ALK) ALK receptor tyrosine kinase, (VEGFR) vascular endothelial growth factor, (MET) proto-oncogene, receptor tyrosine kinase, (AXL) AXL receptor tyrosine kinase, (RET) ret proto-oncogene, (BRAF) B-Raf proto-oncogene, serine/threonine kinase, (MAP2K1/2) (MEK1/2): mitogen-activated protein kinase kinase 1/2, (MTOR) mechanistic target of rapamycin kinase, (MAP2K1/2) (MEK1/2): mitogen-activated protein kinase kinase 1/2, (CD274) (PD-L1): CD274 molecule, (TNFSF11) (RANKL): TNF superfamily member 1, (GD2) disialoganglioside GD2, (LAG3) lymphocyte-activating 3, (CTLA4) cytotoxic T-lymphocyte-associated protein 4, (EZH2) enhancer of zeste 2 polycomb repressive complex 2 subunit, (NTRK) neurotrophic receptor tyrosine kinase 1/2/3, (ALCL) anaplastic large cell lymphoma, (Ph+ ALL) Philadelphia chromosome-positive acute lymphoblastic leukemia, (Ph+ CML) Philadelphia chromosome-positive chronic myeloid leukemia, (TKI) tyrosine kinase inhibitor, (R/R) relapsed or refractory, (cHL) classical Hodgkin lymphoma, (PMBCL) primary mediastinal large B-cell lymphoma, (DLBCL) diffuse large B-cell lymphoma, (BL) Burkitt lymphoma, (BLL) Burkitt-like lymphoma, (mature B-AL) mature B-cell acute leukemia, (B-cell precursor ALL) B-cell precursor acute lymphoblastic leukemia, (MRD) minimal residual disease, (BPDCN) blastic plasmacytoid dendritic cell neoplasm, (LBL) lymphoblastic lymphoma, (IMT) inflammatory myofibroblastic tumor, (DTC) differentiated thyroid cancer, (LGG) low-grade glioma, (MTC) medullary thyroid cancer, (TC) thyroid cancer, (ASPS) alveolar soft part sarcoma, (MCC) Merkel cell carcinoma, (MSI-H) microsatellite instability-high, (dMMR) mismatch repair deficient, (CRC) colorectal cancer, (TMB-H) tumor mutational burden-high.

[a]Combination of ipilimumab + nivolumab in melanoma: indication duplicated for clarity.

Pediatric indication	Targets/alterations	Adult indication
ALK-positive ALCL[*]	ALK fusions[1]	ALK-positive NSCLC
BRAF V600E mutation-positive LGG	BRAF V600E + MEK[2]	BRAF V600E mutation-positive NSCLC, ATC
NTRK gene fusion-positive solid tumors[#]	NTRK fusions[3]	NTRK gene fusion-positive solid tumors[#]

Figure 2. Selected examples of specific oncogenic drivers shared between different pediatric and adult cancers for which targeted therapies have been approved. [*]Include young adults, [#]Spectrum of NTRK fusion-positive pediatric tumors is different from that in adults (Hsiao et al. 2019). Approved drugs: 1: crizotinib; 2: dabrafenib + trametinib; 3: larotrectinib, entrectinib. Indications approved for the same cancers in pediatric and adults are not shown. For the most recent information and full indications refer to Drugs@FDA (www.fda.gov/drugsatfda) and U.S. FDA, Oncology Center of Excellence, Pediatric Oncology Drug Approvals (www.fda.gov/about-fda/oncology-center-excellence/pediatric-oncology-drug-approvals). (ALK) ALK receptor tyrosine kinase, (BRAF) B-Raf proto-oncogene, serine/threonine kinase, (MEK) (MAP2K1/2): mitogen-activated protein kinase kinase 1/2, (NTRK) neurotrophic receptor tyrosine kinase 1/2/3, (ALCL) anaplastic large cell lymphoma, (LGG) low-grade glioma, (NSCLC) non-small-cell lung cancer, (ATC) anaplastic thyroid cancer.

ology appears to support a role for ICI therapies include pediatric patients with hypermutated (Das et al. 2022) and with SWI/SNF-related, matrix-associated, actin-dependent regulator of chromatin, subfamily b, member 1 (SMARCB1)-deficient cancers, likely due to increased presence of neoantigens in these subsets, and ongoing trials are evaluating the efficacy of ICIs for patients with these pediatric cancers (Long et al. 2022). Moreover, potential combination therapy approaches to convert cold into hot tumors are being considered; most trials are yet for adult patients (Terry et al. 2020). In this respect, and as discussed above for targeted gene panels, immunoprofiling assays developed for adult cancers may need to be optimized to pediatric cancers.

Beyond TMB and PD-L1 Expression: In Search of Response Biomarkers

Clinical data in adult cancers, nevertheless, show that there are no known unambiguous, stand-alone ICI response biomarkers. Determinants of sensitivity to ICIs are complex and multifactorial (Blank et al. 2016; Zabransky et al. 2023) and may involve different characteristics within and across cancer types. Despite a high TMB, not all patients with non-small-cell lung cancer (NSCLC) or melanoma respond to ICI therapy. Compared to NSCLC, small-cell lung cancer exhibits higher TMB and other ICI sensitivity features, but clinical results with ICIs are inferior to

what is observed in NSCLC and in other cancer types known to respond to ICIs (Lum and Alamgeer 2019; Schwendenwein et al. 2021). In contrast, Merkel cell polyomavirus-positive Merkel cell carcinoma exhibits among the highest responses to ICIs despite a low overall TMB, a phenomenon attributable to the recognition of "high-quality" viral antigens by the immune system (Zabransky et al. 2023). This suggests that no one single scenario is decisive, rather different determinants in the tumor and tumor microenvironment may play a role in predicting response to ICIs in different cancer contexts. Likewise, the suitability of using PD-L1 expression levels on the tumor and/or tumor microenvironment as a predictive biomarker for anti-PD-1/PD-L1 antibodies is influenced by additional biological and methodological aspects (Fisher et al. 2022).

In this regard, a series of potential predictive biomarkers of response to an anti-PD-L1 antibody were recently identified across refractory pediatric cancers by using immunohistochemistry, RNA sequencing, comprehensive genomic profiling, and T-cell receptor sequencing, reiterating there is no one universal predictive biomarker. Based on the results of these exploratory analyses, a multimodal approach to identify pediatric patients most likely to benefit from ICI therapy in prospective trials has been suggested (Nabbi et al. 2023). Another aspect is to pinpoint mechanisms of primary and acquired resistance to ICIs (Long et al. 2022). Worth highlighting, the role of gut microbiome is emerging as an important biomarker of response to immunotherapies. This may have implications to pediatric patients, as the gut microbiome changes from birth to adults (Stewart et al. 2018; Villemin et al. 2023).

Therefore, ascertaining the immune context of tumors—and the underlying reasons to why they are "hot" or "cold"—is important to inform immunotherapy choices in both adult and pediatric cancers.

Everything Must Be Considered: Unraveling Tumor Strategic Interactions

Of note, for each pediatric cancer type/therapeutic approach that is being considered, the available evidence supporting the known or potential cancer driver (or drivers), relevant pathways, co-alterations that may impact the phenotype, immune context, and cooperating molecular mechanisms should be carefully evaluated. As these aspects likely contribute to tumor pathogenesis, prognosis, and response to therapy together rather than individually, it is important to consider the totality of these aspects as much as the available information allows. This may be especially important when there is no one reliable predictive biomarker of efficacy of the potential treatment. Also, tumors typically have multiple alterations of unknown significance that can be passengers and ascertaining which are oncogenic drivers is challenging. In view of the complexity and plasticity of cancers, this "holistic" assessment can help inform the design of treatment strategies. These aspects could be relevant into determining which population responds best to a specific treatment or whether different doses and/or schedules might be more adequate for different molecular subsets of disease (Peck 2018; Mercher and Schwaller 2019). Moreover, efforts should be made to unravel mechanisms of resistance and toxicity in response to therapy (Sweet-Cordero and Biegel 2019).

CONTINUING CHALLENGES AND THE REQUIREMENT FOR INNOVATION

The delay in initial pediatric investigation of new cancer drugs, recently highlighted by the observation of delays in the first-in-human studies to first-in-children studies of >6 yr (Neel et al. 2019), has had a profoundly negative impact on the pediatric cancer drug development paradigm. As discussed, recent changes in the regulatory framework internationally related to cancer drug development now require sponsors to propose plans for early clinical investigation and possible development in the pediatric space. As a result, there may be multiple targeted agents in the same class or multiple agents directed at different targets in the same pediatric cancer for which parallel development plans are proposed competing for constrained study populations (DiMartino 2022). This anticipated yet unintended consequence is addressed by the FDA in its FDARA Implementation Guidance (FDA 2021). Specific advice is provided to sponsors to request waivers of pediatric studies for investigation of a new same in class agent unless there is evidence of improved activity or effectiveness in a specific cancer, a more favorable toxicity profile, preferential PK parameters including CNS penetration, preferred route of administration, or superior pediatric appropriate formulations. (Also refer to Pediatric Oncology Subcommittee of the Oncologic Drugs Advisory Committee Meeting materials [FDA 2022d].)

Another challenge given the emphasis on precision medicine and rare patient populations is the need for innovation in the consideration of study designs. Key considerations in the selection of specific study designs in pediatric cancer include extent of unmet clinical need, existence of standard of care, complexity of disease biology, rarity of populations for study, and extent of knowledge of disease natural history. The recognized frequentist approach for regulatory approval has been the randomized control trial, albeit either rarely blinded given the complexity of cancer therapies, or placebo controlled. Trial efficiency while maintaining evidentiary standards for drug approval requires thoughtful engagement of investigators, sponsors, regulators, and advocates. Reductions in sample sizes may be accomplished in the setting of Bayesian adaptive designs that borrow from adult or previous pediatric data, although feasibility mandates both data availability and quality and requires modeling and simulation to mitigate bias (Sridhara et al. 2023). Recently, the interest in the use of external controls using real-world evidence (RWE) provided by real-world data (RWD) has increased given the timelines associated with randomized controlled studies, and the ever-shrinking numbers of patients for study due to the increasing subclassification of already rare diseases based on clinical prognostic factors, biology, and genomic characterization (Cooner et al. 2023; Sridhara et al. 2023). Ensuring comparability of a control population requires the availability of high-quality, curated, patient-level data that is as contemporaneous and propensity score-matched as possible. Early discussion between study sponsors and regulators and agreement with a prospective statistical analysis plan is critical to fulfill the objective of accelerating pediatric development using designs that are based partially (hybrid designs) or fully on an external control (FDA 2023a).

DOSAGE SELECTION IN PEDIATRIC PATIENTS

Several approaches may be taken to identify a recommended dosage for pediatric patients and provide substantial evidence to support safe and effective use of a drug or biological product in a pediatric population(s). The recommended dosage refers to both the dose (defined as the specific amount of a drug taken at one time) and schedule (defined as a specific amount of drug administered at a specific frequency over a certain duration, if applicable) (FDA 2023b).

Leveraging Data from Adults and Pediatric Populations

One possible approach involves leveraging data observed in adults (and sometimes older pediatric patients); this approach is often referred to as pediatric extrapolation, which assumes that the disease and the expected response to the drug or biological product would be sufficiently similar in the pediatric population to that of the adults (or other pediatric populations) (FDA 2022b). The data necessary to support efficacy when pediatric extrapolation is considered will depend

upon the continuum of similarity or differences of available data/knowledge gaps and the response to treatment (Fig. 3; FDA 2022c). Clinical studies and model-informed approaches may be used to address these knowledge gaps and characterize the PK and safety of the drug or biological product in the relevant pediatric populations. This approach requires that a dosage be identified that results in an exposure range or distribution comparable to what has been observed in the reference population (such as adults). To that end, inherent physiological differences between the relevant pediatric population and the reference population play an important role in identifying the dosage. Maturation and size-related differences can cause differences in drug absorption, distribution, metabolism, and excretion between pediatric and adult patients and across pediatric age groups, such as infants, children, and adolescents (Kearns et al. 2003).

These physiological differences impact small molecule drugs and biological products differently. As an example, developmental differences in drug transporters, drug-metabolizing enzymes, and drug elimination by the liver or kidneys can substantially alter the absorption, distribution, metabolism, and excretion of small molecule drugs; these differences appear more notable in pediatrics 2 yr and younger (e.g., several cytochrome P450 enzymes reach levels observed in adults by 2 yr of age) (Kearns et al. 2003). Comparatively, developmental differences in tissue water content, tissue perfusion rate, capillary permeability, and immunoglobulin subclasses may alter the distribution and elimination of biological products (Temrikar et al. 2020). Subsequently, the dosage selected for an adult population, or an older pediatric population, may be different than that for younger pediatric patients. Additional dose-finding studies or model-informed approaches that account for these inherent differences can be used to select an appropriate dosage for a pediatric population.

Dose-Finding Studies in Pediatric Patients

Pediatric extrapolation may not be appropriate if the factors that play a role in the development

Cite this article as *Cold Spring Harb Perspect Med* doi: 10.1101/cshperspect.a041656

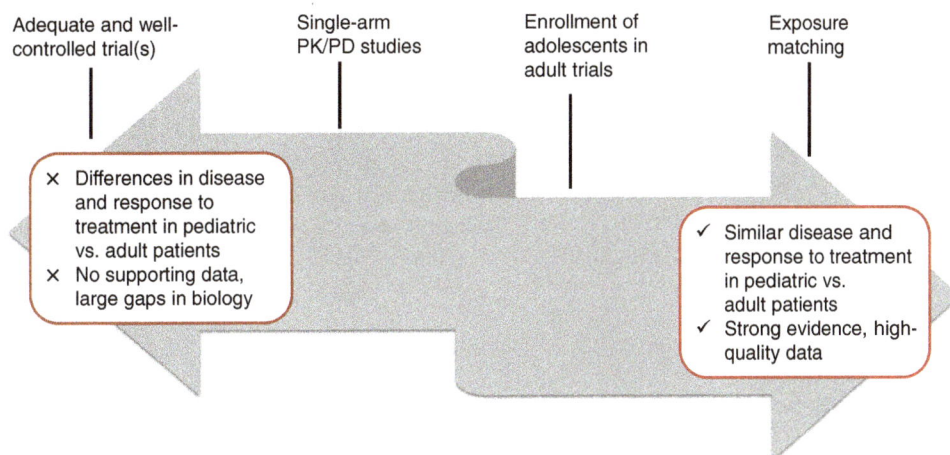

Figure 3. Strength of evidence supporting potential pediatric trial designs or extrapolation. (Figure adapted from E11A pediatric extrapolation, FDA 2022c.) (PK) Pharmacokinetics, (PD) pharmacodynamics.

and presentation of the cancer are different in the pediatric population compared to that being investigated or approved in adults, or if the progression or initiation of the cancer in pediatrics differs from that in adults (i.e., environmental or lifestyle factors) (Shebley et al. 2019). Extrapolation may also not be feasible if the safety profile of the drug is likely to be different in the relevant pediatric age groups compared to that of adults (such as impact on physical or sexual maturation). In these cases, a more extensive pediatric development program is warranted. This approach involves conducting clinical studies in the relevant pediatric populations to adequately evaluate the PK, pharmacodynamics (PD), safety, and efficacy and characterize the dose- and exposure-response relationships to identify a dosage and provide substantial evidence of effectiveness and safety to support a marketing application.

Designing Dose-Finding Trials in Pediatric Patients

History of Dosage Selection

Dose-finding trials for oncology products have historically been designed to identify the maximum tolerated dose (MTD) and select the recommended phase 2 dose (RP2D) (Le Tourneau et al.

2009; Nie et al. 2016). These trials typically evaluate stepwise, increasing doses in a small number of patients at each dose level for short periods of time until a prespecified rate of severe or life-threatening dose-limiting toxicities (DLTs). This approach was initially developed for cytotoxic chemotherapy; however, this approach does not adequately consider other information, such as low-grade symptomatic toxicities, dosage modifications, dose- and exposure-response relationships, and relevant specific populations (defined by age, organ impairment, concomitant medications, or concurrent illnesses) when selecting dosages to be evaluated in subsequent trials. Alternative approaches that consider all relevant nonclinical and clinical data, including long-term safety data, as well as the dose- and exposure-response relationships, are currently being implemented to select a dosage(s) for targeted therapies (such as kinase inhibitors, monoclonal antibodies, and ADCs) (Sachs et al. 2016; Shah et al. 2021; Fourie Zirkelbach et al. 2022). These alternative approaches are warranted given targeted therapies typically demonstrate a different dose- or exposure-response relationship compared to that of cytotoxic chemotherapy. Additionally, the treatment duration with targeted therapies is typically longer than that of cytotoxic chemotherapy increasing the relative importance of lower grade toxicities to patient quality of life.

Last, the MTD may never be reached with a targeted therapy and serious toxicities may occur only after several months of treatment. Innovative clinical trial designs and model-informed drug development, in conjunction with well-established strategies, can be used to optimize dosages for these targeted therapies in pediatric patients.

Selecting the Starting Dosage for Dose-Finding Trials

Given drug development in pediatric patients with cancer often occurs after there is considerable experience with the drug in adult patients, data from adults are routinely used to help identify the dosages to be evaluated in pediatric trials. The starting dose in pediatric dose-finding trials is often selected as 80% of the recommended dose in adults, adjusted for body weight or body surface area. This approach of adjusting for body size was historically implemented for cytotoxic chemotherapy due to narrow therapeutic indices (Shebley et al. 2019). Other scaling approaches (e.g., allometric scaling) that account for developmental differences, as well as modeling and simulation (i.e., physiologically based PK [PBPK] or quantitative systems pharmacology [QSP]) are now commonly used. The starting dose in pediatrics that is equal to that of the adult RP2D, adjusted for body weight or body surface area, may also be selected, especially when the adult MTD exceeds the RP2D.

Dosage escalation for dose-finding trials: In pediatric oncology trials, incremental dose increases are usually relatively small (with dose increases often in the 25%–30% range) and a limited number of dose levels are typically explored since the pediatric RP2D is generally similar to the adult RP2D (Smith et al. 1998). One published review of pediatric dose-finding trials of molecularly targeted therapies reported that the pediatric RP2D ranged between 90% and 130% of the adult RP2D for 13 (69%) of the 19 trials reviewed; the majority of differences between the adult and pediatric RP2Ds occurred in trials of targeted therapies in which DLTs were not observed and the MTD could not be determined (Doussau et al. 2016).

Unique Considerations Affecting Dosage Selection in Pediatric Patients

There are several unique considerations in pediatric oncology that can influence the approach to dosage optimization. As compared to adults, different dosages and strategies may be needed across the span of pediatric age groups due to potential developmental differences that may impact PK and PD. Appropriate PK and PD sampling schedules that account for blood sampling volume limits should also be considered. For oral drugs, a pediatric patient's ability to swallow, palatability, and food effect (e.g., formula or milk for infants as well as soft foods), should be considered. The potential for long-term effects on growth, cognition, sexual development, and other late effects for some molecularly targeted therapies should be evaluated. Due to the rarity and complexity of pediatric cancers, thoughtful integration of dosage optimization early into drug development is especially important.

Model-Informed Drug Development Supporting Dosage Optimization

Model-informed drug development (MIDD) approaches play an important role in pediatric drug development and can be used to support pediatric extrapolation and dosage(s) selection earlier in development of new drugs for pediatric cancers given the rarity and resource challenges. These approaches can also be used for pediatric dosage selection and optimization by incorporating information on pediatric ontogeny, identifying important covariates such as weight and body surface area, and predicting PK in various age groups. Model-informed approaches can also be used to fill in gaps in knowledge by leveraging prior knowledge from adults or other drugs from the same class and known exposure-response relationships for efficacy and safety. Another utility of modeling is to inform clinical trial design including appropriate dosages for evaluation across the pediatric age range, sample size, and optimal PK or PD sampling for pediatric patients. A variety of statistical and other quantitative tools, such as

PK/PD, PBPK modeling, and QSP models can be utilized.

Additional Considerations for Combination Therapy

Dosage optimization is an important consideration when investigating combination therapies. Important considerations include potential for overlapping toxicity (including long-term toxicities) and drug–drug interactions. For oral drugs, other considerations include different administration with regard to food.

AGE-APPROPRIATE FORMULATIONS

If the available drug product is not acceptable for the planned pediatric age groups, an age-appropriate formulation should be developed and tested, and approval should be sought for that formulation(s). Therefore, the availability of age-appropriate formulations is an important aspect of pediatric cancer drug development. For oral small-molecule drugs, early attention should be given to the development of an age-appropriate formulation and how it should be administered with regard to food. The parenteral formulation for small molecule drugs or biological products may be used in pediatric patients; however, the appropriateness of the final concentration and volume in pediatric patients should be carefully considered before using the available formulation.

For oral small molecule drugs, an age-appropriate formulation may be needed to evaluate a drug in pediatric populations. The dosage strengths, size, and number of tablets or capsules required for an individual dose should be considered. For example, the dosage strengths and number of tablets or capsules should be able to provide the dosages to be evaluated for the relevant pediatric population identified based on body size or age. Selumetinib and crizotinib are examples where the approved pediatric dosages are provided by body surface area; dabrafenib and trametinib are examples where the approved pediatric dosages are provided by body weight. Alternative formulations (such as liquid dosage forms) or administration means may be needed

if relevant pediatric population(s) cannot swallow intact solid dosage forms. For updated labeling and additional details on these drugs refer to Drugs@FDA. For a new pediatric formulation, a relative bioavailability study of the pediatric and adult formulations and a dedicated food effect study of the pediatric formulation should be conducted in adults, as early as possible during drug development (FDA 2022a). When a formulation that is approved for use in adults is approved for use in a pediatric population ≥12 yr of age, a separate food effect study is not necessary. Furthermore, a separate food effect study might not be necessary if a pediatric formulation is very similar to the adult formulation, and if the pediatric formulation is approved based on a biowaiver approach.

Small quantities of liquids or soft foods (e.g., 5–15 mL) can be used as vehicles for pediatric drug delivery. Soft foods for use as vehicles should be identified and qualified by in vitro assessments to demonstrate a lack of potential physicochemical interactions between the soft food and the drug product (intact or manipulated), including interactions of the drug substance or excipients with the vehicle (FDA 2018). In vitro assessments are used to identify and qualify potential vehicles but do not replace the need for an in vivo evaluation to permit instructions for drug administration with qualified soft food vehicles in drug labeling. For the drug labeling to indicate that the drug can be sprinkled on soft foods, additional in vivo, relative bioavailability studies should be performed using the soft foods listed as vehicles in the proposed drug labeling (FDA 2022a).

Manipulating solid dosage forms, such as crushing tablets or opening capsules and administering with a vehicle to make an extemporaneous preparation, is generally discouraged as it can impact drug exposure leading to safety or efficacy concerns. There are cases where extemporaneous preparation is reasonable to facilitate the conduct of studies with drugs that can offer an advantage to pediatric patients, while the age-appropriate formulation is being developed; however, the potential differences in exposure with the extemporaneous preparation should be evaluated to select an appropriate dosage(s) and minimize effects on efficacy or safety of the drug. When it is necessary to administer oral drug products via an enteral feeding tube, compatibility may be an important consideration.

TAILORING PEDIATRIC CANCER DRUG DEVELOPMENT TO PEDIATRIC CANCER

Pediatric cancer drug development faces many challenges, including disease rarity and the requirement for innovative study designs, ontogeny aspects, potential long-term effects of therapy, biology gaps, and need for adequate dosing strategies and appropriate formulations (Fig. 4). Despite the many differences between pediatric and adult cancers, pediatric cancer drug development is mostly reliant on drug development for adult cancers. Finding common ground between adult and pediatric cancers contributes to the implementation of available adult cancer therapies to pediatric cancers. Finding differences is also key, as it accentuates the areas where specific

strategies are warranted. Innovative study designs, exploration of RWD applicability, and new approaches focusing on unique drivers of pediatric cancers are crucial to address challenges faced by pediatric cancer drug development. As previously discussed, considerations on conducting studies of same-in-class molecularly targeted products should be carefully weighed. Partnerships and alliances should also be considered. Last, continuous attempts to understand the biology of pediatric cancer should be undertaken (Sweet-Cordero and Biegel 2019; Subbiah 2023).

DISCLAIMER

The opinions expressed in this article are those of the authors and should not be interpreted as the position of the U.S. Food and Drug Administration or the National Cancer Institute. For the most recent version of an FDA guidance cited in this review, see https://www.fda.gov/regulatory-information/search-fda-guidance-documents.

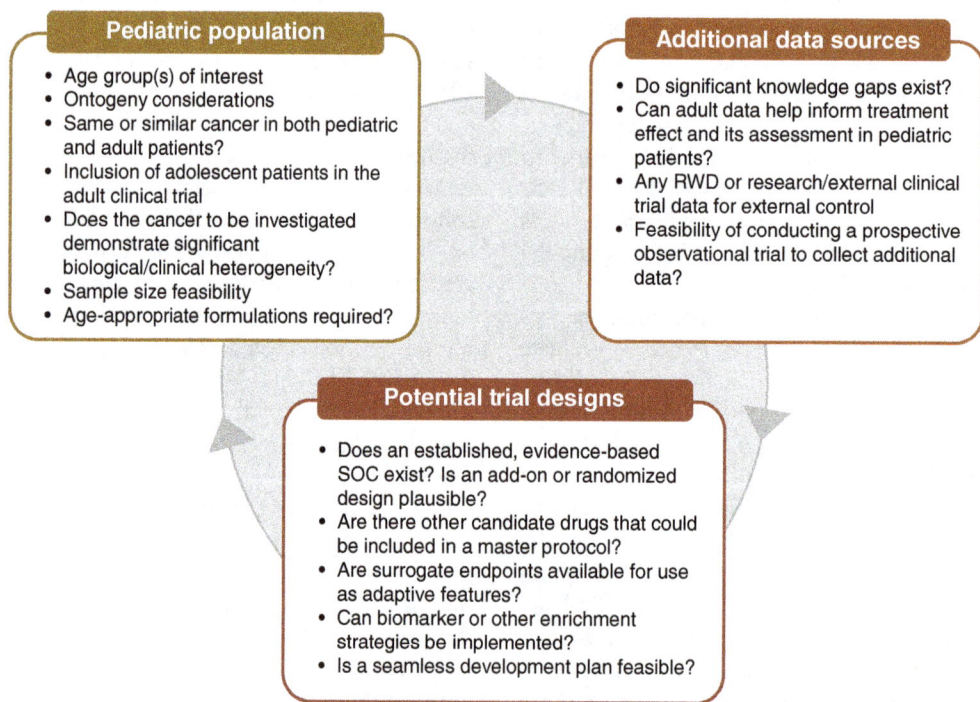

Pediatric population

- Age group(s) of interest
- Ontogeny considerations
- Same or similar cancer in both pediatric and adult patients?
- Inclusion of adolescent patients in the adult clinical trial
- Does the cancer to be investigated demonstrate significant biological/clinical heterogeneity?
- Sample size feasibility
- Age-appropriate formulations required?

Additional data sources

- Do significant knowledge gaps exist?
- Can adult data help inform treatment effect and its assessment in pediatric patients?
- Any RWD or research/external clinical trial data for external control
- Feasibility of conducting a prospective observational trial to collect additional data?

Potential trial designs

- Does an established, evidence-based SOC exist? Is an add-on or randomized design plausible?
- Are there other candidate drugs that could be included in a master protocol?
- Are surrogate endpoints available for use as adaptive features?
- Can biomarker or other enrichment strategies be implemented?
- Is a seamless development plan feasible?

Figure 4. Important factors when considering the design of pediatric cancer clinical trials. (RWD) Real-world data, (SOC) standard of care. (Figure modified and reprinted from Cooner et al. 2023 with permission from Taylor & Francis © 2023.)

ACKNOWLEDGMENTS

All authors contributed to writing the manuscript. The authors thank Giang N. Ho, PharmD, Office of Clinical Pharmacology, Center for Drug Evaluation and Research, U.S. Food and Drug Administration, for helping with the design of Figures 1, 2, and 4. The authors declare no competing interests for this work. No funding was received for this work.

REFERENCES

Adamson PC. 2015. Improving the outcome for children with cancer: development of targeted new agents. *CA Cancer J Clin* **65**: 212–220. doi:10.3322/caac.21273

Adamson PC, Houghton PJ, Perilongo G, Pritchard-Jones K. 2014. Drug discovery in paediatric oncology: roadblocks to progress. *Nat Rev Clin Oncol* **11**: 732–739. doi:10.1038/nrclinonc.2014.149

Akalu AY, Meng X, Reaman GH, Ma L, Yuan W, Ye J. 2021. A review of the experience with pediatric written requests issued for oncology drug products. *Pediatr Blood Cancer* **68**: e28828. doi:10.1002/pbc.28828

Alonso-Luna O, Mercado-Celis GE, Melendez-Zajgla J, Zapata-Tarres M, Mendoza-Caamal E. 2023. The genetic era of childhood cancer: identification of high-risk patients and germline sequencing approaches. *Ann Hum Genet* **87**: 81–90. doi:10.1111/ahg.12502

Avant D, Wharton GT, Murphy D. 2018. Characteristics and changes of pediatric therapeutic trials under the best pharmaceuticals for children act. *J Pediatr* **192**: 8–12. doi:10.1016/j.jpeds.2017.08.048

Bandopadhayay P, Meyerson M. 2018. Landscapes of childhood tumours. *Nature* **555**: 316–317. doi:10.1038/d41586-018-01648-4

BPCA. 2002. Best Pharmaceuticals for Children Act. https://www.congress.gov/107/plaws/publ109/PLAW-107publ109.pdf

Benning TJ, Shah ND, Inselman JW, Van Houten HK, Ross JS, Wyatt KD. 2021. Drug labeling changes and pediatric hematology/oncology prescribing: measuring the impact of U.S. legislation. *Clin Trials* **18**: 732–740. doi:10.1177/17407745211030683

Blank CU, Haanen JB, Ribas A, Schumacher TN. 2016. Cancer immunology. The "cancer immunogram." *Science* **352**: 658–660. doi:10.1126/science.aaf2834

Brodeur GM, Nichols KE, Plon SE, Schiffman JD, Malkin D. 2017. Pediatric cancer predisposition and surveillance: an overview, and a tribute to Alfred G. Knudson Jr. *Clin Cancer Res* **23**: e1–e5. doi:10.1158/1078-0432.CCR-17-0702

Butler E, Ludwig K, Pacenta HL, Klesse LJ, Watt TC, Laetsch TW. 2021. Recent progress in the treatment of cancer in children. *CA Cancer J Clin* **71**: 315–332. doi:10.3322/caac.21665

Carroll WL, Raetz E, Meyer J. 2015. State of the art discovery with tumor profiling in pediatric oncology. *Am Soc Clin Oncol Educ Book* **35**: e601–e607. doi:10.14694/EdBook_AM.2015.35.e601

Casey DL, Cheung NV. 2020. Immunotherapy of pediatric solid tumors: treatments at a crossroads, with an emphasis on antibodies. *Cancer Immunol Res* **8**: 161–166. doi:10.1158/2326-6066.CIR-19-0692

Charlab R, Burckart GJ, Reaman GH. 2023. Fine-tuning the relevance of molecular targets to pediatric cancer: addressing additional layers of complexity. *Clin Pharmacol Ther* **113**: 957–959. doi:10.1002/cpt.2759

Cifci D, Veldhuizen GP, Foersch S, Kather JN. 2023. AI in computational pathology of cancer: improving diagnostic workflows and clinical outcomes? *Annu Rev Cancer Biol* **7**: 57–71. doi:10.1146/annurev-cancerbio-061521-092038

Comitani F, Nash JO, Cohen-Gogo S, Chang AI, Wen TT, Maheshwari A, Goyal B, Tio ES, Tabatabaei K, Mayoh C, et al. 2023. Diagnostic classification of childhood cancer using multiscale transcriptomics. *Nat Med* **29**: 656–666. doi:10.1038/s41591-023-02221-x

Cooner F, Ye J, Reaman G. 2023. Clinical trial considerations for pediatric cancer drug development. *J Biopharm Stat* **33**: 859–874. doi:10.1080/10543406.2023.2172424

Das A, Sudhaman S, Morgenstern D, Coblentz A, Chung J, Stone SC, Alsafwani N, Liu ZA, Karsaneh OAA, Soleimani S, et al. 2022. Genomic predictors of response to PD-1 inhibition in children with germline DNA replication repair deficiency. *Nat Med* **28**: 125–135. doi:10.1038/s41591-021-01581-6

Dharia NV, Kugener G, Guenther LM, Malone CF, Durbin AD, Hong AL, Howard TP, Bandopadhayay P, Wechsler CS, Fung I, et al. 2021. A first-generation pediatric cancer dependency map. *Nat Genet* **53**: 529–538. doi:10.1038/s41588-021-00819-w

DiMartino J. 2022. Future directions. In *Pediatric cancer therapeutics development* (ed. DiMartino J, et al.), pp. 177–181. Pediatric Oncology Series, Springer, Switzerland

Doussau A, Geoerger B, Jiménez I, Paoletti X. 2016. Innovations for phase I dose-finding designs in pediatric oncology clinical trials. *Contemp Clin Trials* **47**: 217–227. doi:10.1016/j.cct.2016.01.009

FDA. 2018. U.S. FDA Draft Guidance for Industry: Use of liquids and/or soft foods as vehicles for drug administration: general considerations for selection and in vitro methods for product quality assessments 2018. https://www.fda.gov/media/114872/download

FDA. 2021. FDARA Implementation Guidance for Pediatric Studies of Molecularly Targeted Oncology Drugs: Amendments to Sec. 505B of the FD&C Act Guidance for Industry. https://www.fda.gov/media/133440/download

FDA. 2022a. U.S. FDA Guidance for Industry: Assessing the effects of food on drugs in INDs and NDAs—clinical pharmacology considerations 2022. https://www.fda.gov/media/121313/download

FDA. 2022b. U.S. FDA Draft Guidance for Industry: General clinical pharmacology considerations for pediatric studies of drugs, including biological products. https://www.fda.gov/media/90358/download

FDA. 2022c. E11A pediatric extrapolation. https://www
.fda.gov/regulatory-information/search-fda-guidance-
documents/e11a-pediatric-extrapolation

FDA. 2022d. Pediatric Oncology Subcommittee of the
Oncologic Drugs Advisory Committee Meeting materials, May 11–12, 2022. https://www.fda.gov/advisory-
committees/advisory-committee-calendar/may-11-12-
2022-pediatric-oncology-subcommittee-oncologic-drugs-
advisory-committee-meeting#event-materials

FDA. 2023a. U.S. FDA Draft Guidance for Industry: Considerations for the design and conduct of externally controlled trials for drug and biological products. https://www
.fda.gov/media/164960/download

FDA. 2023b. U.S. FDA Draft Guidance for Industry: Dosage
and administration section of labeling for human pre-
scription drug and biological products—content and format. https://www.fda.gov/media/72142/download

FDAMA. 1997. Public Law 105-115. https://www.govinfo
.gov/app/details/PLAW-105publ115

FDARA. 2017. Public Law 115-52. https://www.congress
.gov/115/plaws/publ52/PLAW-115publ52.pdf

FDASIA. 2012. Public Law 112-144. https://www.congress
.gov/112/plaws/publ144/PLAW-112publ144.pdf

Findlay IJ, De Iuliis GN, Duchatel RJ, Jackson ER, Vitanza
NA, Cain JE, Waszak SM, Dun MD. 2022. Pharmaco-
proteogenomic profiling of pediatric diffuse midline glioma to inform future treatment strategies. *Oncogene* 41:
461–475. doi:10.1038/s41388-021-02102-y

Fisher KE, Ferguson LS, Coffey AM, Merritt BY, Curry JL,
Marcogliese AN, Major AM, Kamdar KY, Lopez-Terrada
DH, Curry CV. 2022. Programmed cell death ligand 1
expression in aggressive pediatric non-Hodgkin lymphomas: frequency, genetic mechanisms, and clinical significance. *Haematologica* 107: 1880–1890. doi:10.3324/hae
matol.2021.280342

Fourie Zirkelbach J, Shah M, Vallejo J, Cheng J, Ayyoub A,
Liu J, Hudson R, Sridhara R, Ison G, Amiri-Kordestani L,
et al. 2022. Improving dose-optimization processes used
in oncology drug development to minimize toxicity and
maximize benefit to patients. *J Clin Oncol* 40: 3489–3500.
doi:10.1200/JCO.22.00371

GAO-23-105947. 2023. Pediatric cancer studies: early results
of the Research to Accelerate Cures and Equity for Children Act. https://www.gao.gov/products/gao-23-105947

Gröbner SN, Worst BC, Weischenfeldt J, Buchhalter I,
Kleinheinz K, Rudneva VA, Johann PD, Balasubramanian
GP, Segura-Wang M, Brabetz S, et al. 2018. The landscape
of genomic alterations across childhood cancers. *Nature*
555: 321–327. doi:10.1038/nature25480

He J, Zhang Y, Zhu J, Tan H, Rössler J. 2021. Editorial:
molecular diagnostics of pediatric cancer. *Front Oncol*
11: 777662. doi:10.3389/fonc.2021.777662

Hingorani P, Zhang W, Zhang Z, Xu Z, Wang WL, Roth ME,
Wang Y, Gill JB, Harrison DJ, Teicher BA, et al. 2022.
Trastuzumab deruxtecan, antibody-drug conjugate tar-
geting HER2, is effective in pediatric malignancies: a re-
port by the pediatric preclinical testing consortium. *Mol
Cancer Ther* 21: 1318–1325. doi:10.1158/1535-7163
.MCT-21-0758

Hirschfeld S, Ward RM. 2013. History of children and the
development of regulations at the FDA. In *Pediatric drug*

development, 2nd ed. (ed. Mullber AE, et al.), pp. 6–15.
Wiley, Hoboken, NJ.

Ho P. 2022. The pharma/biotech model for drug develop-
ment: implications for pediatric cancer therapeutics. In
Pediatric cancer therapeutics development (ed. DiMartino
J, et al.), pp. 89–107. Pediatric Oncology Series, Springer,
Cham, Switzerland.

Hsiao SJ, Zehir A, Sireci AN, Aisner DL. 2019. Detection of
tumor NTRK gene fusions to identify patients who may
benefit from tyrosine kinase (TRK) inhibitor therapy. *J
Mol Diagn* 21: 553–571. doi:10.1016/j.jmoldx.2019.03
.008

Hutzen B, Ghonime M, Lee J, Mardis ER, Wang R, Lee DA,
Cairo MS, Roberts RD, Cripe TP, Cassady KA. 2019. Im-
munotherapeutic challenges for pediatric cancers. *Mol
Ther Oncolytics* 15: 38–48. doi:10.1016/j.omto.2019.08.005

Jones DTW, Banito A, Grünewald TGP, Haber M, Jäger N,
Kool M, Milde T, Molenaar JJ, Nabbi A, Pugh TJ, et al.
2019. Molecular characteristics and therapeutic vulnerabilities across paediatric solid tumours. *Nat Rev Cancer*
19: 420–438. doi:10.1038/s41568-019-0169-x

Kattner P, Strobel H, Khoshnevis N, Grunert M, Bartholo-
mae S, Pruss M, Fitzel R, Halatsch ME, Schilberg K, Sie-
gelin MD, et al. 2019. Compare and contrast: pediatric
cancer versus adult malignancies. *Cancer Metastasis Rev*
38: 673–682. doi:10.1007/s10555-019-09836-y

Kearns GL, Abdel-Rahman SM, Alander SW, Blowey DL,
Leeder JS, Kauffman RE. 2003. Developmental pharmacology—drug disposition, action, and therapy in infants
and children. *N Engl J Med* 349: 1157–1167. doi:10.1056/
NEJMra035092

Kennedy MC, Lowe SW. 2022. Mutant p53: it's not all one
and the same. *Cell Death Differ* 29: 983–987. doi:10.1038/
s41418-022-00989-y

Laetsch TW, DuBois SG, Bender JG, Macy ME, Moreno L.
2021. Opportunities and challenges in drug development
for pediatric cancers. *Cancer Discov* 11: 545–559. doi:10
.1158/2159-8290.CD-20-0779

Langenberg KPS, Looze EJ, Molenaar JJ. 2021. The landscape of pediatric precision oncology: program design,
actionable alterations, and clinical trial development.
Cancers (Basel) 13: 4324. doi:10.3390/cancers13174324

Le Tourneau C, Lee JJ, Siu LL. 2009. Dose escalation methods in phase I cancer clinical trials. *J Natl Cancer Inst* 101:
708–720. doi:10.1093/jnci/djp079

Long AH, Morgenstern DA, Leruste A, Bourdeaut F, Davis KL.
2022. Checkpoint immunotherapy in pediatrics: here, gone,
and back again. *Am Soc Clin Oncol Educ Book* 42: 1–14.

Lum C, Alamgeer M. 2019. Technological and therapeutic
advances in advanced small cell lung cancer. *Cancers
(Basel)* 11: 1570. doi:10.3390/cancers11101570

Ma X, Liu Y, Liu Y, Alexandrov LB, Edmonson MN, Gawad
C, Zhou X, Li Y, Rusch MC, Easton J, et al. 2018. Pan-
cancer genome and transcriptome analyses of 1,699 paediatric leukaemias and solid tumours. *Nature* 555: 371–
376. doi:10.1038/nature25795

Mercher T, Schwaller J. 2019. Pediatric acute myeloid leukemia (AML): from genes to models toward targeted therapeutic intervention. *Front Pediatr* 7: 401. doi:10.3389/
fped.2019.00401

Cite this article as *Cold Spring Harb Perspect Med* doi: 10.1101/cshperspect.a041656

Nabbi A, Danesh A, Espin-Garcia O, Pedersen S, Wellum J, Fu LH, Paulson JN, Geoerger B, Marshall LV, Trippett T, et al. 2023. Multimodal immunogenomic biomarker analysis of tumors from pediatric patients enrolled to a phase 1-2 study of single-agent atezolizumab. Nat Cancer 4: 502–515. doi:10.1038/s43018-023-00534-x

Neel DV, Shulman DS, DuBois SG. 2019. Timing of first-in-child trials of FDA-approved oncology drugs. Eur J Cancer 112: 49–56. doi:10.1016/j.ejca.2019.02.011

Nie L, Rubin EH, Mehrotra N, Pinheiro J, Fernandes LL, Roy A, Bailey S, de Alwis DP. 2016. Rendering the 3 + 3 design to rest: more efficient approaches to oncology dose-finding trials in the era of targeted therapy. Clin Cancer Res 22: 2623–2629. doi:10.1158/1078-0432.CCR-15-2644

Ou SI, Nagasaka M. 2020. A catalog of 5′ fusion partners in ROS1-positive NSCLC circa 2020. JTO Clin Res Rep 1: 100048. doi:10.1016/j.jtocrr.2020.100048

Panditharatna E, Filbin MG. 2020. The growing role of epigenetics in childhood cancers. Curr Opin Pediatr 32: 67–75. doi:10.1097/MOP.0000000000000867

Peck RW. 2018. Precision medicine is not just genomics: the right dose for every patient. Annu Rev Pharmacol Toxicol 58: 105–122. doi:10.1146/annurev-pharmtox-010617-052446

Pediatric Rule. 1998. 63 FR 66632—regulations requiring manufacturers to assess the safety and effectiveness of new drugs and biological products in pediatric patients. https://www.govinfo.gov/app/details/FR-1998-12-02/98-31902

Pfister SM, Reyes-Múgica M, Chan JKC, Hasle H, Lazar AJ, Rossi S, Ferrari A, Jarzembowski JA, Pritchard-Jones K, Hill DA, et al. 2022. A summary of the inaugural WHO classification of pediatric tumors: transitioning from the optical into the molecular era. Cancer Discov 12: 331–355. doi:10.1158/2159-8290.CD-21-1094

PREA. 2003. Public Law 108–155. https://www.congress.gov/108/plaws/publ155/PLAW-108publ155.pdf

Reaman G, Karres D, Ligas F, Lesa G, Casey D, Ehrlich L, Norga K, Pazdur R. 2020. Accelerating the global development of pediatric cancer drugs: a call to coordinate the submissions of pediatric investigation plans and pediatric study plans to the European Medicines Agency and US Food and Drug Administration. J Clin Oncol 38: 4227–4230. doi:10.1200/JCO.20.02152

Robinson BW, Kairalla JA, Devidas M, Carroll AJ, Harvey RC, Heerema NA, Willman CL, Ball AR, Woods EC, Ballantyne NC, et al. 2023. KMT2A partner genes in infant acute lymphoblastic leukemia have prognostic significance and correlate with age, white blood cell count, sex, and central nervous system involvement: a children's oncology group P9407 trial study. Haematologica 108: 2865–2871. doi:10.3324/haematol.2022.281552

Sachs JR, Mayawala K, Gadamsetty S, Kang SP, de Alwis DP. 2016. Optimal dosing for targeted therapies in oncology: drug development cases leading by example. Clin Cancer Res 22: 1318–1324. doi:10.1158/1078-0432.CCR-15-1295

Schwendenwein A, Megyesfalvi Z, Barany N, Valko Z, Bugyik E, Lang C, Ferencz B, Paku S, Lantos A, Fillinger J, et al. 2021. Molecular profiles of small cell lung cancer subtypes: therapeutic implications. Mol Ther Oncolytics 20: 470–483. doi:10.1016/j.omto.2021.02.004

Shah M, Rahman A, Theoret MR, Pazdur R. 2021. The drug-dosing conundrum in oncology—when less is more. N Engl J Med 385: 1445–1447. doi:10.1056/NEJMp2109826

Shebley M, Menon RM, Gibbs JP, Dave N, Kim SY, Marroum PJ. 2019. Accelerating drug development in pediatric oncology with the clinical pharmacology storehouse. J Clin Pharmacol 59: 625–637. doi:10.1002/jcph.1359

Shirkey H. 1968. Therapeutic orphans. J Pediatr 72: 119–120. doi:10.1016/s0022-3476(68)80414-7

Simon AK, Hollander GA, McMichael A. 2015. Evolution of the immune system in humans from infancy to old age. Proc Biol Sci 282: 20143085. doi:10.1098/rspb.2014.3085

Smith FO, Reaman GH. 2022. History of drug development for children with cancer. In Pediatric cancer therapeutics development (ed. DiMartino J, et al.), pp. 1–8. Pediatric Oncology Series, Springer, Cham, Switzerland.

Smith M, Bernstein M, Bleyer WA, Borsi JD, Ho P, Lewis IJ, Pearson A, Pein F, Pratt C, Reaman G, et al. 1998. Conduct of phase I trials in children with cancer. J Clin Oncol 16: 966–978. doi:10.1200/JCO.1998.16.3.966

Sridhara R, Marchenko O, Jiang Q, Barksdale E, Alonzo TA, Amatya A. 2023. Statistical considerations in pediatric cancer trials: report of American statistical association biopharmaceutical section open forum discussions. Stat Biopharm Res 15: 845–851. doi:10.1080/19466315.2023.2238650

Stewart CJ, Ajami NJ, O'Brien JL, Hutchinson DS, Smith DP, Wong MC, Ross MC, Lloyd RE, Doddapaneni H, Metcalf GA, et al. 2018. Temporal development of the gut microbiome in early childhood from the TEDDY study. Nature 562: 583–588. doi:10.1038/s41586-018-0617-x

Subbiah V. 2023. The next generation of evidence-based medicine. Nat Med 29: 49–58. doi:10.1038/s41591-022-02160-z

Sweet-Cordero EA, Biegel JA. 2019. The genomic landscape of pediatric cancers: implications for diagnosis and treatment. Science 363: 1170–1175. doi:10.1126/science.aaw3535

Temrikar ZH, Suryawanshi S, Meibohm B. 2020. Pharmacokinetics and clinical pharmacology of monoclonal antibodies in pediatric patients. Paediatr Drugs 22: 199–216. doi:10.1007/s40272-020-00382-7

Terry RL, Meyran D, Ziegler DS, Haber M, Ekert PG, Trapani JA, Neeson PJ. 2020. Immune profiling of pediatric solid tumors. J Clin Invest 130: 3391–3402. doi:10.1172/JCI137181

Thakur MD, Franz CJ, Brennan L, Brouwer-Visser J, Tam R, Korski K, Koeppen H, Ziai J, Babitzki G, Ranchere-Vince D, et al. 2022. Immune contexture of paediatric cancers. Eur J Cancer 170: 179–193. doi:10.1016/j.ejca.2022.03.012

Vargesson N. 2015. Thalidomide-induced teratogenesis: history and mechanisms. Birth Defects Res C Embryo Today 105: 140–156. doi:10.1002/bdrc.21096

Villemin C, Six A, Neville BA, Lawley TD, Robinson MJ, Bakdash G. 2023. The heightened importance of the microbiome in cancer immunotherapy. Trends Immunol 44: 44–59. doi:10.1016/j.it.2022.11.002

Xiao L, Yeung H, Haber M, Norris MD, Somers K. 2021. Immunometabolism: a "hot" switch for "cold" pediatric solid tumors. Trends Cancer 7: 751–777. doi:10.1016/j.trecan.2021.05.002

Zabransky DJ, Yarchoan M, Jaffee EM. 2023. Strategies for heating up cold tumors to boost immunotherapies. Annu Rev Cancer Biol 7: 149–170. doi:10.1146/annurev-cancerbio-061421-040258

Index